WITHDRAWN

This book is ~~~ ~~~~~ or before   6204

08 MAR 1990   24 JAN 1991   8 NOV 1992

28 MAR 1990   21 FEB 1991   0 9 DEC 1992

                 21 MAR 1991   1 9 APR 1993

17 MAY 1990   9 MAY 1991

12 JUN 1990   06 JUN 1991   4 JUN 1993

                 27 JUN 1991   8 DEC 1993

03 JUL 1990   1 OCT 1991

                 21 NOV 1991

03 OCT 1990

1 OCT 1990   14 JAN 1992

22 NOV 1990   10 MAR 1992   2 8 OCT 1995

631.5 BAR

PLANT SCIENCE

PLANT SCIENCE

JOHN A. BARDEN
Department of Horticulture
Virginia Polytechnic Institute
Blacksburg, Va.

R. GORDON HALFACRE
Department of Horticulture
Clemson University
Clemson, S.C.

DAVID J. PARRISH
Department of Agronomy
Virginia Polytechnic Institute
Blacksburg, Va.

McGRAW-HILL BOOK COMPANY

New York   St. Louis   San Francisco   Auckland   Bogotá   Hamburg   Johannesburg
London   Madrid   Mexico   Milan   Montreal   New Delhi   Panama   Paris
São Paulo   Singapore   Sydney   Tokyo   Toronto

Library of Congress Cataloging-in-Publication Data

Barden, John A.
  Plant science.

  1. Crops. 2. Plants. 3. Botany. I. Halfacre,
R. Gordon. II. Parrish, David J. III. Title.
SB91.B37 1987   631   86-7224
ISBN 0-07-003669-1

PLANT SCIENCE

1 2 3 4 5 6 7 8 9 0   HALHAL   8 9 4 3 2 1 0 9 8 7 6

ISBN 0-07-003669-1

ISBN 0-07-003669-1

This book was set in Melior by Monotype Composition Company, Inc. (ECU).
The editors were Mary Jane Martin and Barry Benjamin;
the designer was Rafael Hernandez; the cover art was done by Molly Conner Ogorzaly;
the production supervisor was Leroy Young.
The drawings were done by Wellington Studios Ltd.
Arcata Graphics/Halliday was printer and binder.

John Barden earned a B.S. at the University of Rhode Island. After receiving his M.S. and Ph.D. at the University of Maryland, John joined the Horticulture Department at Virginia Polytechnic Institute (VPI). John has taught several horticulture courses, guided graduate students, and conducted research in the area of fruit tree physiology. He has received the L. M. Ware Distinguished Teaching Award from the Southern Region of the American Society for Horticultural Science (ASHS) and the Teaching Award of Merit from the Virginia Tech Chapter of Gamma Sigma Delta. John served as Associate Editor for both ASHS journals from 1983 to 1986, as Vice President for Education in 1984–1985, and was named a Fellow of ASHS in 1984.

John is the son of John and Hazel Barden, and grew up on the family fruit farm in Rhode Island. John and his wife Irma have two children, one each in college and graduate school. Irma Barden works in the guidance department at Blacksburg High School.

Gordon Halfacre is a graduate of Clemson College and received a Ph.D. in Horticulture at Virginia Polytechnic Institute (VPI) and an MLA from North Carolina State University.

Throughout his professional career, Gordon has emphasized undergraduate teaching. He was named Outstanding Teacher at North Carolina State, received the L.M. Ware Distinguished Teaching Award from the Southern Region of the American Society for Horticultural Science (ASHS), and was given the Outstanding Teacher Award by the Clemson Chapter of Gamma Sigma Delta. Gordon has been active in the Education Division of ASHS.

His previous books include *Fundamentals of Horticulture, Keep 'Em Growing, Landscape Plants of the Southeast,* and *Horticulture.*

Gordon worked in a landscape nursery throughout his childhood in South Carolina. His parents are Lela and Harvey Halfacre. Gordon's wife Adrienne is Grounds Superintendent at Greenville Technical College, and his children Angela and Robert are high school students.

Dave Parrish was trained as a biologist and plant physiologist, receiving degrees (one each) at East Tennessee State, Wake Forest, and Cornell Universities. He put his training to work in the field of agronomy, specializing in crop physiology. He teaches courses in plant science and crop physiology and does research that deals with some basic and applied aspects of crop ecology, energy cropping, and seed physiology.

Dave lives with his family (one wife, two children, and two cats) on a one-sixth hectare, nonintensively-managed, suburban land holding in Blacksburg Virginia. There they practice principles of plant science on their 20 × 5 m garden, 15 fruit trees, and several flower beds.

Dave is one and a half generations removed from the farm; he largely owes to his parents, Leon and Evelyn Parrish, the respect he has for agriculture and for those who engage in this crucial profession.

TO OUR CHILDREN

Cindy and Jay Barden
Angela and Robert Halfacre
Joe and Nathan Parrish

And To All Children

CONTENTS

PREFACE

This book was written to offer an introduction to plant science at the collegiate freshman-sophomore level. Our goal was to offer the basics of plant science in a readable, readily understood presentation. We wrote it from an agricultural standpoint to emphasize that plant science is not an abstract subject but one that has a daily impact on us all.

Instructors may wish to use chapters and sections selectively to fit their particular needs and preferences. To cover the entire book would take two terms. Since most universities devote only one term to a plant science course, the instructor can choose those chapters which fit the particular course. For example, one teacher might emphasize the chapters which deal with the environmental factors; another might choose to spend most of the term on the basics of anatomy, morphology, chemistry, and genetics. The remainder can serve as a reference or as refresher reading.

Part One covers salient aspects of botany, plant physiology, biochemistry, and genetics. Although many books have been written on each of these topics, we have attempted to present an introduction to each sufficient for the student to grasp key principles. It is only on such a foundation that students can build their knowledge of the science and technology of modern plant agriculture.

Part Two encompasses the major environmental factors that affect plants. Light, temperature, water, soils, nutrients, and plant pests are discussed in separate chapters. All are considered from an ecological standpoint, but equally important are the agricultural implications and applications. Supplemental lighting, freeze protection, irrigation, soil management, fertilization, pest control, and a myriad of other environmental manipulations are used in crop production. For simplicity, the environmental factors are treated as discreet entities and then the concluding chapter in Part Two, Crop Ecology, ties them all together as a highly interactive set. One of the major goals of plant scientists today is to optimize environmental factors for maximum crop yield and quality.

Part Three explores the ways by which we manipulate plants to our advantage. The introduction of new and better cultivars is one approach; such work is aimed at increased productivity and quality, adaptation to mechanical harvest, greater pest resistance, adaptation to different soils and climates, and even tolerance of air pollutants. Although propagation of many crops today does not differ vastly from that practiced by the earliest farmers, new methods of propagation open up exciting possibilities. For example, tissue culture involves the production of hundreds of new plants from a small piece of tissue or even a few cells. A chapter on directing plant growth describes many of the techniques used to regulate plant size and structure, and the final chapter in Part Three, Cropping Systems, covers overall management schemes for crops both the traditional cropping systems used in developing countries in the tropics and the highly mechanized systems characterizing United States agriculture today.

Part Four introduces the major world crops—not only their basic botany and general production systems but also their contributions to our needs for food and fiber. The coverage of individual crops is necessarily brief but provides an idea of the tremendous diversity and importance of the plants on which we depend for food, fiber, ornamentals, and turf.

We offer thanks to our many colleagues who have contributed to this book in so many ways. At the same time, however, we assume full responsibility for its content, including any errors and shortcomings. In particular, we wish to thank for their contributions D. A. Wescott, R. L. Andersen, B. M. Cool, S. R. Chapman, E. W. Carson, M. L. Brinkley, and C. E. Zipper. To Jean Morris, Marion Halfacre, and Nancy Turner, our proficient and faithful typists, we are grateful. We also extend our thanks to the many individuals, companies, and organizations who provided photographs. Most especially, we thank our families whose love, patience, sacrifice, and understanding allowed us to complete this project.

PLANT SCIENCE

CHAPTER ONE

INTRODUCTION

Life on earth is sustained by energy derived from the sun, and photosynthetic plants are the connecting link between the sun and all forms of life. Because of this ultimate dependence of all living things, photosynthesis can realistically be called the most important chemical process on earth. Chlorophyll, the plant pigment which captures radiant energy, is sometimes referred to as the "green blood" of the earth. Without chlorophyll and the plants that produce it, life would cease. This fact alone makes plant science enormously important.

Plants have fed our world, replenished our air, cured our ills, provided materials for our industries, and even enriched our spiritual lives for many thousands of years. Human dependence on plants has naturally led to attempts to manipulate members of the plant kingdom. We increase their numbers and cultivate them for maximum production; we select and breed strains with superior qualities; we control their growth habit as well as their flowering and fruiting characteristics.

Plant science is such a broad subject that some initial orientation is appropriate. This book is written from a practical perspective. *Plant science* is first and foremost the science and technology of the production of crops. (A *crop* is any plant used by human beings.) Although this general definition is narrower than the definition of plant science as the study of plants, it stresses the major daily impact plant science has on every man, woman, and child.

Many disciplines are included under the umbrella of plant science. Agronomy, horticulture, forestry, and weed science are applied sciences dealing with particular crops or groups of crops. Anatomy, morphology, genetics, physiology, pathology, and biochemistry, which are primarily biologically oriented subdisciplines, may not be especially concerned with crops per se. Agronomists, horticulturists, and foresters are practitioners who may be scientific specialists within any of the subdiscipline areas.

Agronomists (from the Latin word for field) generally deal with field crops such as wheat, corn, soybean, cotton, and forages, all of which are grown on a large scale with relatively low-intensity management. Horticulturists (from the Latin word for garden) are concerned with fruits, vegetables, woody ornamentals, and floricultural crops. In horticultural spheres, plant production generally occurs on a smaller scale than in agronomy and with higher-intensity management. Foresters are plant scientists primarily interested in wood and pulp production along with recreation, wildlife, and watershed management in forested areas.

FOOD CHAINS

A food chain can be defined as the pattern of flow of energy-rich consumables through a biological community or ecosystem. Photosynthesizing plants, the only organisms on earth which can fix energy received from the sun, constitute the first vital link in food chains—whether on land or in water.

Consider a short food chain starting with a field of corn which converts radiant energy (sunlight) into chemical energy by the process of photosynthesis. The energy is stored and interconverted in the corn as carbohydrates (sugars, starch, cellulose, etc.), lipids (fats, oils), and proteins. When the corn is eaten by a herbivore, such as a steer, some of the energy stored in the corn is released by digestion and used to form muscle and fat, some is used by the animal for work and maintenance of body temperature, and some passes through as waste. Of the total energy in the corn eaten, less than 10 percent will be retained in a form available to the next member of the food chain, a carnivore. The same low efficiency in using food energy applies at each level in the chain.

It is partly because of the low efficiency of energy conversion at each step in a food chain that many people are largely vegetarian. For example, 100 kilograms (kg) of corn fed to a steer might produce about 10 kg of meat; and 10 kg of beef eaten by a consumer might produce only 1 kg of human body weight.† The same energy-conversion rate of 10 percent means that 100 kg of corn consumed directly could provide about 10 kg of human body weight.

A vegetarian diet is much more food-energy efficient, but there are obviously other considerations in choice of diet and food-chain patterns. Particularly in highly developed parts of the world, meat is prized not only for its high protein content but also for its flavor, texture, and the variety it adds to the diet. In addition, since cattle, sheep, deer, and many other animals are able to digest plant matter not nutritionally available to human beings, they can be raised largely on feed which people find unpalatable, indigestible, or both. Thus herbivores form a vital link in our food supply and permit the utilization of land and plants that would not otherwise enter human food chains.

Food chains operating in oceans, lakes, rivers, streams, and ponds, like those on land, start with photosynthesizing plants. In a water environment, the green plants are free-floating, often single-celled,

algae. The energy contained in algae is passed on to microscopic algae-eating animals and from them to a wide variety of more complex water organisms, including fish. Ultimately, human beings may obtain energy formerly fixed by the algae in the form of freshwater fish or seafood, ranging from clams and crabs to flounder and tuna.

To grasp the overall significance of a food chain, consider the energies involved, which start with the fixation of solar energy by plants. Despite its pivotal role, photosynthetic efficiency is amazingly low. Only 1 to 2 percent of the solar energy received at a particular place on the earth's surface is fixed by the plants growing there; and in many places, there are essentially no plants. The remainder of the sun's energy is reflected, transmitted, converted into heat and dissipated, or simply absorbed by the soil. When we further consider the small fractions of the energy passed on through each step of a food chain, we begin to realize how ominous the burgeoning of the world's population is.

One of the major aims of modern agricultural research is to find ways of improving the energetic efficiency of human food chains. This book should give you a better understanding of the basic processes involved in providing food for the world and how plants are grown and manipulated to maximize crop productivity.

ORIGINS OF AGRICULTURE

Although the human race has been linked to creatures who lived some 2 million years ago, *Homo sapiens*, the human species, has been on earth perhaps only 250,000 years. We are the "new kid" on an earth that is at least 5 billion years old.

Until about 10,000 years ago, people probably existed on food obtained by hunting, fishing, or gathering wild edible plants. It is likely that early human beings experienced periods both of abundance and of great scarcity of food due to their total dependence upon nature. Their hunting-gathering lifestyle probably constrained early people to exist in rather widely dispersed small groups, and periods of food shortage, disease, and overall poor nutrition tended to keep populations rather low.

The earliest agriculture, a systematic tending and harvesting of the earth's bounties, is thought to have originated about 10,000 years ago, the transition from

†With a few exceptions, units of measure used in this text are SI units. Readers unfamiliar with them should refer to the appendix.

a hunter-gatherer way of life to that of a food producer probably evolving over an extended period. Agriculture led to the domestication of selected species, both plant and animal. Where agriculture began is debated, but one likely region is in the Near East. Because of its varying environments, the area once known as Mesopotamia is the home of a wide diversity of plant and animal species. The use of modern techniques such as radiocarbon dating make increasingly accurate assessment of prehistoric events possible. From deposits in the Near East, it appears that by about 7000 B.C. people were not only cultivating grain but had started to domesticate animals. Some of the earliest crops to be cultivated were wheat and barley. Since wild fruits and nuts were abundant, cultivation of perennial crops, e.g., olive and fig, probably started later.

Agriculture in the new world had its origins in Mexico and Peru, perhaps 1000 years later than in the Near East, around 6000 B.C. Some of the earliest cultivated plants were squash, chili pepper, maize (corn), and avocado. As agriculture developed, additional plants, including bean, cotton, and various fruits were brought under cultivation. Agriculture had long been a part of Native American culture when the earliest European settlers arrived in North America.

ORIGINS OF CROP PLANTS

It is difficult, if not impossible, to pinpoint the exact origin of modern crop plants, in part because many of them were under cultivation long before the start of recorded history. It is also reasonable to assume that as the early agriculturists moved about, they carried plants with them far from their point of origin. Thanks to the selective processes of evolution, many of our crop plants are markedly different from their early ancestors. In fact, in the case of corn (maize), the earliest progenitors are thought to be extinct.

Despite the lack of solid evidence, the origin of many of our crop species is believed to be as shown in Fig. 1-1, the major centers of origin being southwestern and central Asia, the Mediterranean region, southeastern Asia, and the highlands of tropical America. Note that no major crops are shown as originating in the United States. Although not of major importance in world trade, the cranberry,

blueberry, and pecan are native to the United States, as are certain species of plum, grape, and strawberry.

Although plant origins may seem to be only of academic interest, such information can be of real value. Plant breeders strive to improve crop plants by incorporating a wide range of genetic traits, e.g., disease or insect resistance, tolerance to adverse environmental factors, or modified growth habit. Often the best place to find genetic material with the desired characteristic is in the region to which the crop is native. There one can often find the broadest possible base of available genetic diversity in the relatives and progenitors of the crop species.

PLANTS AS FOOD

Grains
Although the species vary with location and climate, the grain-yielding grasses are a major human food source in much of the world. In many temperate regions, for example, wheat is the primary grain staple (Fig. 1-2); in warmer, more humid areas, rice may fill this role. Other major grain crops in the grass family include corn, oats, rye, barley, millet, and grain sorghum. We discuss later how grains serve as a major feed source for animals destined for human consumption.

Starchy Foods
In addition to the cereal grains, starchy roots and other plant parts such as potato, sweet potato, cassava, yam, and taro (Fig. 1-3) have served as staple foods for centuries, especially in regions where grains are not well adapted. The potato has played a major role in world commerce and history. Cassava, also known as manioc, from which we get tapioca, has sustained people in the tropics for centuries. Although usually considered a fruit in the United States, the banana is a starchy staple in many tropical regions of the world.

Vegetables
Besides starchy products like the potato and sweet potato, many other vegetables contribute to the human diet. Favorite species vary with location in the world and even within the United States. Important vegetable crops include sweet corn, snap bean, pea, bean, lettuce, and tomato. These foods offer not only pleasing texture, flavor, color, and appearance but

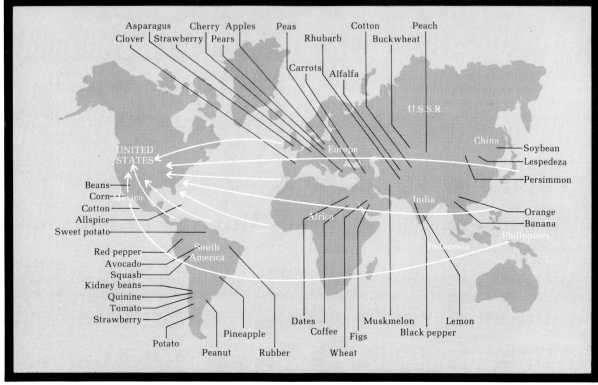

FIGURE 1-1
Some world contributions to the crop production of the United States. Many agronomic and horticultural crops originated in other countries but have been developed extensively through research in the United States. (*U.S. Department of Agriculture. Adapted from M. S. Kipps, Production of Field Crops, 6th ed., McGraw-Hill, New York, 1970.*)

vitamins and minerals as well. Although most are not staple foods, they constitute vital portions of our diet.

Fruits
The fruit crops of the world include a wide variety of species. The banana is a tremendously important fruit crop in the tropics, as are pineapple, mango, papaya, and citrus fruits such as orange, lemon, and lime (Fig. 1-4). They are important from the standpoint of enjoyment but also supply important dietary needs, particularly vitamin C. Popular fruits in the temperate zones are apple, pear, peach, cherry, plum, and apricot as well as the small fruits (grape, raspberry, strawberry, blackberry, blueberry, and currant).

Sugar Crops
In spite of its lack of protein, vitamins, and minerals, sugar is a major dietary component, especially in developed countries. The prime sources of table sugar in world trade are sugarcane (Fig. 1-5), grown in tropical and subtropical climates, and sugarbeet, grown in temperate climates. Many other plant products contain starch that can be converted into sugar, e.g., corn syrup.

Oil Crops
Plant oils are the basis for a rapidly expanding segment of plant agriculture. Recent concern with the relationship between human heart disease and animal fats has given tremendous impetus to the use of vegetable oils. Partly because of the demand for

FIGURE 1-2
Wheat harvest in eastern Washington. Vast areas of the United States are devoted to the production of this vitally important small grain crop. (*U. S. Department of Agriculture.*)

FIGURE 1-4
Citrus grove in Florida. Although confined to subtropical regions, citrus fruits are major crops in the United States.

oil, soybean production in the United States sky-rocketed in the second half of this century. Other important sources of oils are sunflower, peanut, corn, cottonseed, olive, coconut, safflower, and certain palms. Some of these crops are grown primarily for oil; with others, e.g., cotton and corn, the oil is an increasingly important secondary product.

Nut, Spice, and Beverage Crops

Other components of our everyday diets have their origin in a broad variety of plants—nuts, e.g., walnut and pecan; herbs and spices, e.g., vanilla, pepper, allspice, and oregano; and beverages, e.g., coffee, chocolate, tea, and cola.

FIGURE 1-3
Taro roots have long been used as a staple food in tropical areas of the Pacific. They are cooked and made into a gray paste called poi. The young tender leaves are also eaten.

FIGURE 1-5
Sugarcane is one of the most productive of the many crop plant species. Shown here is a field of sugarcane in Hawaii.

PLANTS AS FEED

Many of the world's inhabitants, particularly those in the developed countries, consume large amounts of meat and animal products. But remember the eggs at breakfast, the hamburger for lunch, and the glass of milk at dinner all had their energetic origin in plants. The entire animal industry depends on plants as the source of energy. The energy losses in each step of a food chain mean that tremendous quantities of plant material must be fed to animals, but animal scientists have made great strides in increasing the amount of output (meat, milk, eggs, etc.) per unit of input (feed).

The production of animal feed is a major part of plant agriculture. Consider for a moment the tremendous diversity of ways in which this feed is grown and fed to animals. Since the earliest settlers moved west, vast areas of rangeland have been used for cattle and sheep production. Long before the earliest settlers arrived, many Native Americans built their lives—diet and culture—around the buffalo. Although we have changed the species of herbivore, the food chain has remained almost unchanged for centuries. To satisfy today's consumer, many beef cattle are moved to special feed lots (Fig. 1-6) for a few weeks before slaughter; there they are fed grain rather than grass, both to accelerate their weight gain and to increase the amount of fat in the meat.

Dairy cattle consume vast quantities of plant material in a variety of forms; because they continually produce large amounts of energy-rich milk, dairy cows require more high-energy feed than beef cattle.

FIGURE 1-6
Modern feed lot in Colorado. Before slaughter, cattle are moved to feed lots where they are fed grain to increase fat content of the meat, improving both texture and flavor.

Hogs are fed plant products which range from grain to pasteurized municipal garbage. Farmers often let hogs salvage the portion of crops left in the field. For example, after a crop of peanuts has been harvested, frequently enough is left to produce a lot of ham, bacon, and sausage. Among the meat animals, the modern chicken and turkey are probably the most efficient at converting plant products into meat. This feed efficiency, together with major strides in other aspects of poultry production, makes eggs, chicken, and turkey some of the least expensive sources of animal protein.

NONFOOD USES OF PLANTS

Fiber Crops

Plant fibers have been used since prehistoric times to make cloth. The predominant sources are cotton, in which the fibers are attached to the seed, and flax, in which the usable fibers are in the stem. Hemp and other plants yield stem fibers used to manufacture rope. Natural fibers have been replaced to a large degree by synthetics such as nylon, rayon, and polyester. For many uses, a blend of a synthetic and a natural fiber is preferred, a cotton-polyester blend being a common fabric for shirts or blouses. People in the tropics and subtropics can testify, however, that 100 percent cotton is still the coolest, most comfortable material to wear because it allows perspiration to evaporate.

Timber, Fuel, and Pulp

Forests are a tremendously important renewable resource. They are economically important primarily because they provide wood products for so many uses. For centuries, forests have been exploited as if there were no end to the available timber. In recent decades, however, we have made major strides in forest management to improve the renewal of timberlands. With proper management, a forest can be harvested every 15 to 75 years, depending on climate, species, and intended use.

Primary forest products include lumber for construction, furniture, and pallets. The manufacture of cardboard and paper consumes vast quantities of pulpwood. The use of wood stoves, familiar to many older Americans but almost extinct during the 1950s and 1960s, were revived as an economical alternative during the energy crisis of the 1970s.

The impact of forest lands on our lives far exceeds that of the wood products they provide. The recreational opportunities offered by private and public forests (Fig. 1-7)—hunting, hiking, and camping—are only part of the story. Since forests stabilize watersheds, such activities as boating and fishing are tied closely to forested areas.

Aesthetic Uses of Plants

Using plants for aesthetic purposes dates back to the dynastic Egyptians, whose gardens and fields were sharply different from those of earlier times, when plants were simply sources of food and fuel. Since the early beginnings, many societies have placed emphasis on ornamental plants. Although garden designs vary with the region of the world and the time in history, they are of major significance, particularly in the highly developed parts of the world.

In addition to the use of plants outdoors, the production and marketing of flowers is a major industry in the United States, where cut flowers are used for weddings, funerals, birthdays, and other special occasions. Over the past decade, the use of foliage plants in houses and offices has become widespread (Fig. 1-8).

FIGURE 1-8
The increased use of plants indoors add greatly to the enjoyment of interior environments in homes, offices, and stores. (*Courtesy of Robert McDuffie.*)

FIGURE 1-7
Picnic area in a national forest. With modern management techniques, forests serve many purposes in addition to timber production. (*U.S. Forest Service.*)

An aspect of plant science directly affecting our daily lives is the use of various grass species as turf. The vast amounts of land devoted to lawns around houses, schools, offices, and factories make the environment more attractive and generate millions of business dollars in seed, fertilizer, equipment, and custom lawn care. The construction and maintenance of golf courses and athletic fields are highly specialized branches of plant science.

Other Plant Products

Plants are valued sources of many products other than food, fuel, and fiber—including medicines, drugs, perfumes, cosmetics, insecticides, and industrial chemicals. A few of the many plant-derived medicines are digitalis, a heart stimulant; morphine, a pain reliever; and quinine, an antimalarial. Hundreds of other herbs have been used to treat human ills. Although many of them are now considered to be folk remedies, more than half of all prescriptions in this country use plant-derived products. Drugs of plant origin include marijuana, cocaine, and nicotine. Expensive perfumes are usually based on essential oils extracted from flowers; many cosmetics include products from plants such as aloe. Plant-derived insecticides include rotenone, pyrethrins, and nicotine. Turpentine, drying oils, many plastics, and other industrial solvents and chemicals are derived from plants. The list of plant uses—historical and current—is almost endless.

PLANT SCIENCE IN WORLD AFFAIRS

As we approach the twenty-first century, many issues must be addressed by society in general and agriculture in particular. Consider the many "crises" we read or hear about almost daily: human malnutrition and starvation, declining energy supplies, as well as pollution of air, water, and the environment in general. Although less frequently cited, soil erosion and the loss of highly productive land to urbanization pose serious problems as well. In one way or another, each of these matters relates directly to agriculture.

Human Malnutrition and Starvation
The problem of increasing world population is a complex issue with political, economic, social, religious, and agricultural implications. Although the subject is far too vast and complicated for treatment here, we emphasize the pivotal role of plants in sustaining the world's inhabitants. Modern agricultural science and technology have largely prevented, or at least delayed, the mass starvations predicted in the early nineteenth century.

The United Nations estimated the world's population in 1980 at 4.4 billion. The annual rate of increase, which hovers around 2 percent, translates into 170 new people per minute or about 90 million per year. At that rate, the world's population will double in about 35 years. This means that food production must increase in the next 35 years by the same amount as it has since people first walked on earth. Whether or not food production can keep up with the population explosion remains to be seen, but the challenge is there. People in the United States are sometimes lulled into a false sense of security by national food surpluses, but in the worldwide context, a major shortage of food may well be imminent. A world food shortage would be of concern not only for humanitarian reasons but also because of its implications for international peace.

Energy Supplies
Declining energy reserves are of concern in much of the world, and not just in the highly developed countries, which depend heavily upon fossil fuels for agriculture, transportation, industry, heating, and generation of electricity. Like the world population–food balance issue, the energy problem has many sides. We briefly explore a few which relate to plant agriculture.

Modern agricultural production systems in developed countries are energy-intensive; on the other hand, production systems in developing countries continue to be labor-intensive (both human and animal). In developed countries, the energy requirements for manufacturing equipment, fertilizers, and pesticides are tremendous. In addition, sizeable quantities of fuel are needed to operate equipment and transport products to and from the farm. At the other end of the energy-use spectrum is subsistence agriculture in underdeveloped countries, where small farms are run with human and perhaps animal energy but little or no fossil-fuel input. The farm output is consumed at home or sold locally. In the early nineteenth century, when subsistence agriculture was the norm in the United States, each farm worker grew enough food for about 4 people; in the 1980s, this figure rose to about 75. Although we obviously cannot return to subsistence agriculture, we must strive to conserve energy while maintaining or increasing food production.

The potential use of plant products and residues as energy sources is another aspect of the energy situation. The use of wood as fuel is receiving renewed attention. A relatively new concept is the use of animal wastes and crop residues to generate "biogas," and crop residues are also finding increasing use as solid fuel. Another idea being explored is crop plants which produce large amounts of high-energy biomass for fermentation to ethyl alcohol (ethanol). Gasohol, now available in many areas, may reduce our dependence on nonrenewable petroleum supplies.

Adopting new production systems may enable us to save significant amounts of energy. Various reduced-tillage practices lower fuel use. Implementation of integrated pest-management schemes should reduce use of both pesticides and application equipment. These and other means of saving energy will be described in later chapters.

Environmental Pollution

We have become increasingly aware of the deterioration of our physical environment—air, water, and soil. Plant agriculture has been a contributor to pollution, but in some cases it can serve to relieve the problem. On the negative side, agricultural production has increased soil erosion by clearing trees, exposing the soil to both wind and water. Air-borne soil is an air pollutant; water-borne soil lowers water quality and can even clog streams, rivers, and bays. Runoff of water from heavily fertilized fields has added phosphates and other nutrients to lakes and rivers, leading to excessive growth of algae and other water plants, and ultimately to stagnation of the bodies of water.

Proper application of sound agricultural practices can and must minimize these farm-related problems. In a similar way, science and technology must be applied to nonagricultural sectors, e.g., construction sites and other areas where the soil is exposed.

Air pollution has a direct impact on plants, but we have only recently begun to assess the damage systematically. In addition, efforts are now under way to lessen the detrimental effects of aerial pollutants by a range of techniques, such as breeding of resistant cultivars.

Soil and Water Conservation

Not only does soil erosion contribute to environmental pollution, it also destroys prime natural resources for present and future generations. When land is denuded of plants, the soil is exposed to great danger. As topsoil and the layer of plant residue

is lost, more water tends to run off rather than soak in. Thus, instead of storing water, which would ultimately become available in streams and wells, the soil sheds it.

Indiscriminant cropping of land in the relatively dry, windy areas of the Great Plains and the southwestern United States helped create the dust bowl, which had devastating effects on both agriculture and people. In *The Grapes of Wrath*, John Steinbeck described the human consequences of ignoring the importance of plants in stabilizing the soil. Some scientists fear that lack of proper soil management on vast acreages of United States farmland has created a situation where an extended drought could lead to another dust bowl.

Different, but even more disastrous situations can be seen today in countries such as Haiti, where mountains have been stripped of almost all vegetation as a source of fuel. The topsoil has rapidly eroded away, leaving land that is useless for growing food. These and similar instances make it clear that water and soil management are vital parts of modern agricultural technology.

Loss of Agricultural Land

Unfortunately, many of us in the United States have not yet come to grips with our slowly declining supply of land—particularly prime agricultural land. The migration of people out of cities into suburban and rural areas requires land, not only for homes but also for schools, recreational areas, highways, and commercial enterprises. In many cases, the ideal land for such uses is also the best for crop production because of the terrain, drainage, and other soil characteristics. As an area changes from an agricultural to a residential community, pressures on farmers develop. Among them are increased property taxes, pressure from nonfarm families who may want to restrict normal farming practices, and diluted political power. The temptation to sell often overcomes the will to hold out and continue to farm. Effective solutions to the loss of farm land are not simple, but ignoring the problem indeed can only be a disservice to future generations.

PART ONE

FORM
AND
FUNCTION

CHAPTER TWO

PLANT MORPHOLOGY

Agriculturally important plants are tremendously diverse, not only in how they are grown and used, but also in form and structure. This chapter covers the basics of *plant morphology*, the scientific study and description of the form and structure of plants at all stages of their life cycle. Morphology deals with plant structure at the level of organs and organ systems. Chapter 3 will deal with plant anatomy, which describes plants at the cell and tissue level. This information and terminology will enable you to appreciate more fully the many differences and similarities between crop plants.

VEGETATIVE ORGANS

We have chosen to start our discussion of plant morphology with the vegetative plant parts, which include roots, stems, and leaves (Fig. 2-1). Described later in this chapter are the reproductive structures—flowers, fruits, and seeds.

Roots

Roots vary widely in form with different species but serve the same basic functions in all plants. Besides anchoring the plant in the soil, roots absorb water and nutrients and often store reserves for future use.

Roots are characterized by certain morphological features found in no other part of the plant. The *root cap* is a thimble-shaped mass of cells covering the root apex; it serves to protect the apex from mechanical injury as the root forces its way through the soil. The *root hairs*,† described in greater detail later, are extensions of surface cells and greatly increase the absorptive capacity of roots.

Roots may differ in origin. A *primary root* develops from the radicle (embryonic root) of a germinating seed (Fig. 2-2). It grows down into the soil and may branch repeatedly to produce *secondary*, or lateral, roots. The primary root often grows vertically whereas secondary roots may grow rather horizontally, away from the primary root. Some secondary roots turn downward. *Adventitious roots* arise from stems or even leaves rather than from another root, and are often beneficial; the prop roots of corn, which sprout from the lower nodes of the stem, serve as braces for the tall plant. Several ornamental species produce adventitious roots from their leaf tissues, which are used as a source of propagation (Chap. 15).

The characteristics of a plant's root system help determine how suitable the species is for a particular environment. If the primary root persists and maintains its dominance, it is referred to as a *taproot*. Plants with taproots that penetrate deeply, e.g., alfalfa and some trees, can obtain moisture from lower soil layers and therefore are less sensitive to brief droughts.

---

†Most technical terms are defined in the Glossary, which the student should consult when necessary, along with the index.

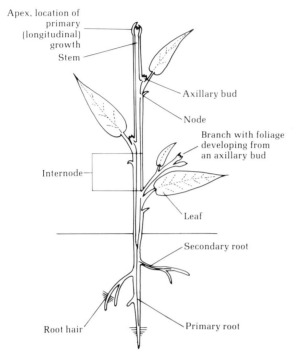

Apex, location of primary (longitudinal) growth

Stem

Axillary bud

Node

Branch with foliage developing from an axillary bud

Internode

Leaf

Secondary root

Root hair

Primary root

FIGURE 2-1
Longitudinal section of primary plant body of dicotyledon with basic lateral appendages. (*Adapted from A. J. Eames and L. H. MacDaniels, An Introduction to Plant Anatomy, 2d ed., McGraw-Hill, New York, 1947.*)

Pecan, parsley, cotton, and crownvetch are other examples of crops with dominant taproots. Plants with *fibrous roots*, which are shallower, tend to respond more rapidly to fertilizer and irrigation; they are better species for soil stabilization, but they tend to be more drought-sensitive.

Some biennial and perennial plants develop *fleshy storage* roots, in which a supply of reserve food is stored. Turnip, sugarbeet, and carrot are examples of food plants in which the taproot develops into a storage structure (Fig. 2-2). In dahlia, cassava, and sweet potato, branch roots develop into swollen storage structures. Food storage in roots of perennial plants such as alfalfa, asparagus, and many trees and shrubs is important for the plant's survival. These plants are normally able to overwinter, and the stored food in the roots provides energy for renewed growth in the spring.

Roots: Structure

Roots are cylindrical and usually whitish or tan. Root tips can be separated longitudinally into four functional zones: the root cap, the meristematic region, the region of cell elongation, and the region of maturation or differentiation (Fig. 2-3).

Cells of the *meristematic region* are arranged in rows parallel with the axis of the root; they carry on numerous cell divisions. Some cells formed at the apex of this region differentiate into layers of the root cap. As the root grows downward, the outer cells of the root cap are worn away and are replaced by underlying cells formed more recently by the rootcap meristem region. The rest of the cells formed by the meristem will migrate (or be dislocated by additional cell division) back into the root axis.

Directly behind the zone of cell division is the *region of cell elongation*. Here, the longitudinal axis of the cells increases more than the transverse axis. The combined effects of cell division in the meristem region and cell elongation in the zone of elongation cause the root to grow longer, an increase known as *primary growth*. The permanent tissues formed are

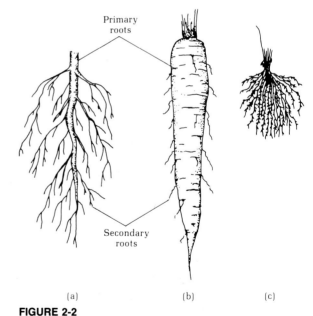

Primary roots

Secondary roots

(a)   (b)   (c)

FIGURE 2-2
Root systems. (*a*) taproot (pecan); (*b*) storage taproot (carrot); (*c*) fibrous roots (strawberry). (*From R. G. Halfacre and J. A. Barden, Horticulture, McGraw-Hill, New York, 1979.*)

Region of maturation	Region of elongation	Meristematic region	Root cap

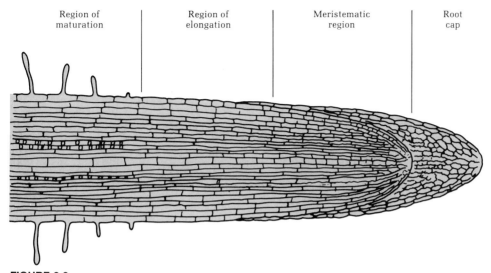

FIGURE 2-3
Root, showing rootcap, meristematic region, and regions of elongation and maturation.

known as *primary tissues* because they develop directly from the apical meristem (cells at the tip of the root).

After elongating, cells mature by taking on their final function and characteristics in the *region of differentiation* or maturity. The region of mature primary tissues is a zone of active water and mineral absorption. Water and minerals are also absorbed to some degree in the meristem and the region of elongation.

The outer walls of epidermal cells (the outer layer of cells) in the maturation region of the root form *root hairs*. These extensions of epidermal cells are responsible for much of the absorption of water and nutrients by the plant (Fig. 2-4). It is essential to the life of all typical land plants that they secure water from the soil at least equal to the quantity they lose to the air. If such plants are to grow and develop, rather than be merely sustained, additional water is necessary for the formation and growth of new cells. Root hairs provide much more surface area and potential absorptive area than the exterior of the root cylinder alone.

New root hairs are formed continuously as the roots extend into the soil, and older root hairs are sloughed off. How fast the plant develops new root hairs is important in determining success in transplanting bare-root plants (without soil attached to the roots). When plants are removed from the ground, the root hairs are usually broken or damaged, reducing the plant's ability to absorb water and minerals. Since tomato plants develop new root hairs from mature root tissues, they are easy to transplant. Squash and cucumber, however, seldom develop root hairs from mature roots, forming them only on newly developed roots; this makes them difficult to transplant bare-root. As root hairs die and decay, the epidermal cells around them develop thicker walls and may absorb less water and nutrients. Eventually, if the root exhibits secondary growth, i.e., growth in diameter, the epidermis cracks and disintegrates and cork or bark will take its place as the root covering.

Many plants which produce root hairs growing in moist air or soil will develop none when grown in water. This is one reason why it is not advantageous to root a cutting in a flask of water for later transplanting into soil. Many of the gymnosperms, e.g., the firs and pines, produce no root hairs at all. In many woody species, the function of root hairs appears to be replaced by fungal growths where fungi enter into a symbiotic relationship with the tree. The mycorrhizae, as they are called, appear to be able to

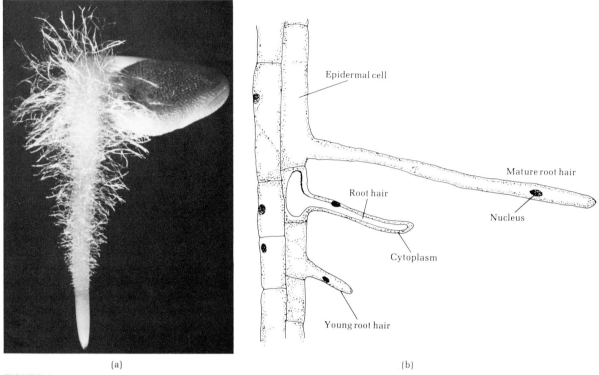

(a)   (b)

FIGURE 2-4
Root hairs. (a) Radish (*Raphanus*) seedling with root hairs. When very young, the root may produce these hairs along its entire length rather than only at its tip. (*Carolina Biological Supply Company*.) (b) Developing and mature root hair. Cytoplasm streams into hair and lies against the inner surface of its wall. The nucleus is carried in cytoplasm from cell into hair; ultimately it may become located nearer the tip of the hair. The central areas of both epidermal cell and hair are filled with vacuolar fluid. (*From R. G. Halfacre and J. A. Barden, Horticulture, McGraw-Hill, New York, 1979.*)

extract water and nutrients from the soil and provide them to the tree.

Stems

The *stem* is that part of the axis of a plant which develops from the *epicotyl* (embryonic shoot) of a seed or from a *bud* (shoot apical meristem) of an existing stem (Fig. 2-5). A stem is composed of *nodes* (the region of the stem where one or more leaves and buds are attached) and *internodes* (region of stem between nodes) (Fig. 2-6). The stem bears leaves and buds. The stems, leaves, and buds compose the plant *shoot*. Stems may have adventitious buds arising from their internodes.

Stems exhibit wide variation in form, size, and structure. They range from less than 1 mm to more than 120 meters (m) long. In thickness they vary from hairlike structures to tree trunks 18 m or more in circumference. Morphologically, tree stems can be divided into several parts. The *trunk* is the main axis of the stem system; a *branch* is a lateral portion of the tree that originates from the trunk or from another branch and gives rise to smaller branches and to leaves. The term *shoot* is sometimes used to describe current season's growth with leaves attached; and especially after leaf abscission, the first-year growth may be called a *twig*. The *terminal bud* of a stem is an apical meristem that has become inactive. Bud scales protect the dormant bud; they encircle the stem and leave scars when they are shed as growth starts in the spring. The bud-scale scars on a stem occur between one season's growth and the next. *Spurs*, which develop on some fruit trees, are short stems that bear leaves or fruit and leaves.

Plant stems exhibit a diversity in form primarily due to genetic differences; in addition, growing conditions alter stem growth, as do various pruning systems (Chap. 16). The stems of strawberry plants and many ferns are so short that all leaves seem to occur in a rosette; these short stems are sometimes collectively called a *crown*. Stems of common asparagus bear flattened, green branches called *cladophylls* which resemble leaves and perform their functions. The true leaves of asparagus are reduced to inconspicuous *bracts*, which are small, modified leaves. The stems of white clover and many other plants are prostrate. Grapes and bean "vines" are stems that require supporting structures to be erect. Modifications of stems or leaves called *tendrils* permit pea, grape, cucumber, English ivy, and other plants to attach to a support. In some xerophytes (plants adapted to surviving in areas with limited water), the stem tissue is both a storage area for water and the site of photosynthesis; in cacti, the *spines* are actually modified leaves. The stem of kohlrabi is enlarged and bulbous with stored food in its cells. These are just a sample of the tremendous variation in stem morphology.

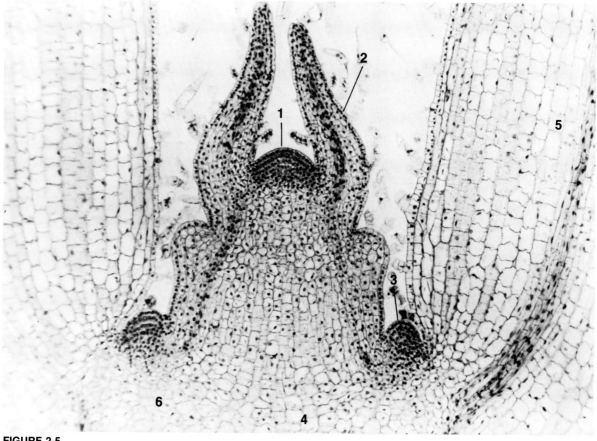

FIGURE 2-5
Longitudinal section of shoot apex of *Coleus blumei.* Apex with older leaf primordia (embryonic leaves) and older axillary bud primordia. 1, Apical meristem; 2, leaf primordia; 3, axillary bud primordia; 4, region of cell elongation in stem; 5, older embryonic leaves; 6, provascular strand. Leaves of coleus are opposite; hence the uniformity in origin of pairs of leaf primordia from the shoot apex. (*Carolina Biological Supply Company.*)

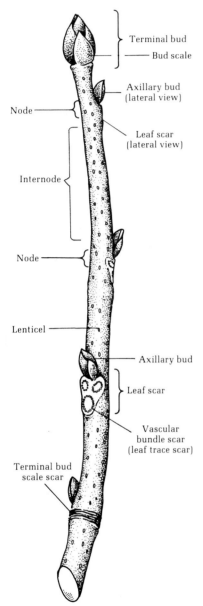

FIGURE 2-6
Diagram of stem showing node, internode, and bud arrangement. (*Drawn by John Norton.*)

Terminal bud
Bud scale
Axillary bud (lateral view)
Node
Leaf scar (lateral view)
Internode
Node
Lenticel
Axillary bud
Leaf scar
Vascular bundle scar (leaf trace scar)
Terminal bud scale scar

Modified Stems

Many plants have highly modified stems that function in storage and regeneration. *Offshoots* are short lateral stems which occur in whorls at the crown of stems and bear fleshy buds or leafy rosettes, as in the artichoke and pineapple. A *stolon,* or *runner,* is a lateral aboveground stem which lies on the soil or other surface and may form adventitious roots at the nodes. Plants with stolons include black raspberry, white clover, bermudagrass, and strawberry. *Rhizomes* are horizontal stems that grow partly or entirely underground; they are often thickened and serve as storage organs. Rhizomes may produce roots and shoots at their nodes and sometimes on their internodes as well. Plants with rhizomes include iris, many ferns, orchid, banana, bluegrass, quackgrass, asparagus, tall fescue, and kohleria (Fig. 2-7a). *Corms,* produced for example by timothy, gladiolus, and crocus, are short, fleshy underground stems with few nodes and very short thickened internodes (Fig. 2-7b). A *tuber* is a greatly enlarged tip of a fleshy underground stem, e.g., potato (Fig. 2-7c). The *eyes* of a potato tuber are its nodes and consist of a leaf scar and axillary bud or buds. A *bulb* is a budlike structure consisting of a small stem with closely crowded fleshy or papery leaves or leaf bases. Axillary bud primordia may occur in the axils of the leaves of some bulbs, e.g., lily, onion, and tulip (Fig. 2-7d).

Buds

According to their position on the stem and origin, buds are classified as terminal, axillary, accessory, or adventitious. *Terminal* buds are at the apex of stems and are responsible for primary growth; they are often larger and give rise to more vigorous shoots than other buds (Fig. 2-8). *Axillary buds* are borne laterally on the stem in the axils of leaves. *Accessory* or *supernumerary* buds occur laterally at the base of terminal buds and beside axillary buds. In the silver maple and peach, accessory buds can be produced on both sides of the axillary bud. *Adventitious buds* originate at other places on the plant than nodes or stem apices, e.g., on the internodes of stems or on roots. Adventitious buds may form shoots, called *watersprouts* when they develop on stems and *suckers* when they develop on roots.

Buds are also classified by the types of tissues in the bud. A *mixed bud* contains *primordia* (embryonic states) of both leaves and flowers. A *simple bud*

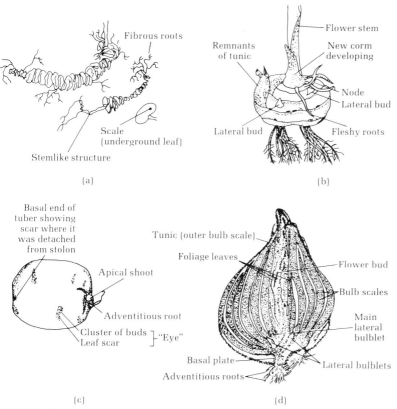

Fibrous roots

Scale
(underground leaf)

Stemlike structure

(a)

Remnants
of tunic

Flower stem

New corm
developing

Node
Lateral bud

Lateral bud

Fleshy roots

(b)

Basal end of
tuber showing
scar where it
was detached
from stolon

Apical shoot

Adventitious root

Cluster of buds ⎤ "Eye"
Leaf scar ⎦

Tunic (outer bulb scale)

Foliage leaves

Flower bud

Bulb scales

Main
lateral
bulblet

Basal plate

Adventitious roots

Lateral bulblets

(c)

(d)

FIGURE 2-7
Modified underground stem. (*a*) rhizome (Kohleria); (*b*) corm (gladiolus); (*c*) tuber (potato); (*d*)
bulb (tulip). (*b, c,* and *d* redrawn by permission from *H. T. Hartmann and D. E. Kester, Plant
Propagation: Principles and Practices, 4th ed., Prentice-Hall, Englewood Cliffs, N.J., 1983.*)

contains either leaf or flower primordia but not both. Apple and pear trees have simple vegetative (leaf) buds but mixed flower buds. Peach, plum, and cherry trees bear only simple buds, either vegetative or reproductive. Leaf buds are typically thinner and more pointed than either mixed or flower buds, which are rounded and plump.

Buds from certain plants are economically important horticultural products. Cabbage and head lettuce are examples of very large, terminal vegetative buds, and the vegetative axillary buds of brussels sprouts are the edible part of that plant. In broccoli (Fig. 2-9) and cauliflower not only the flower buds but portions of the stem and small leaves associated with the flower buds are edible. In the globe artichoke it is the fleshy basal portion of the bracts of the flower

and the end of the flower stalk (*receptacle*) that are edible.

Leaves

Leaves are the plant organ usually responsible for carrying on photosynthesis. Although the differences in form and shape between leaves and stems are striking, another important difference between them is their potential for growth. Most stems continue growing in length and girth because of generally *indeterminate growth*; i.e., the meristems continue to be active indefinitely. Leaves, however, are characterized by *determinate growth*. After periods of cell multiplication and enlargement, leaf cells stop dividing and the leaf ceases to grow. The mature leaf functions for a few weeks to several seasons, de-

(a)

(c)

FIGURE 2-8
Bud structure: stages in the opening of terminal foliage buds of hickory (*Carya* sp.): (*a*) closed bud with large pubescent (hairy) bud scales; (*b*) bud scales opening; (*c*) bud open, with young compound leaves expanding; bud scales have either dropped from bud or curled backward at base of expanded bud. (*Carolina Biological Supply Company.*)

(b)

pending on species and environment. The leaf then senesces, dies, and falls from the stem. Photosynthesis takes place primarily in the leaf, although it also occurs in other plant organs (grain heads of grasses and in herbaceous and young woody stems). Because of its large surface area, a leaf is ideally suited for photosynthesis, which requires the interception of light. Subsequent chapters will describe how the large surface area and minimal volume of a leaf make it efficient in exchanging carbon dioxide and oxygen with the atmosphere.

The typical dicot leaf consists of the *blade*, the flattened, expanded portion, and the *petiole*, or leaf stalk (Fig. 2-10). *Lamina* is another, more technical term for the leaf blade. *Stipules* are flattened, leaflike appendages often found at the petiole base. In some plants, the stipules remain attached to the stem as long as the petiole and blade remain; in others they abscise soon after the leaf unfolds. The petiole extends the blade outward, giving greater exposure

FIGURE 2-9
The flower buds of broccoli along with portions of the stem and small leaves are eaten. (Harris Seed Company.)

The parts of a grass plant, including its leaves, are shown in Fig. 2-11. The grass stem, called a *culm*, is like other stems in having conspicuous nodes and internodes with leaves developing from the nodes; however, the leaves do not have petioles, nor are they truly sessile. The leaf consists of a blade and a sheath. The *sheath* wraps around the culm, extending upward from its node until it expands into the blade. Where the blade and the sheath come together, there are frequently two distinctive structures. The *auricles* are appendages, often clawlike, extending around the culm. The *ligule* is a membranous or fringelike structure between the blade and the culm that arises from the *collar*, the junction of the blade and the sheath. The auricle and ligule are useful identifying characteristics when grasses are in the vegetative stage.

The conducting tissues in a leaf blade are referred to as *veins*; their pattern of arrangement in the blade is called *venation*. Leaves of plants that have *netted venation* are of two basic types. *Pinnately veined leaves* have secondary veins extending laterally from a single midrib, as in sunflower, apple, beech, and rose. *Palmately veined leaves* have several large veins radiating into various parts of the blade from

to light. The petiole also provides conducting tissues for transport of water, nutrients, and photosynthates between the stem and leaf blade. In some plants, such as celery and rhubarb, the petiole has a large quantity of storage tissue. In species where a petiole is lacking the leaf is said to be *sessile*.

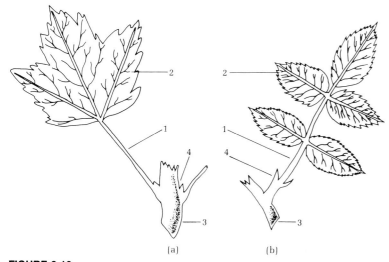

FIGURE 2-10
Parts of dicotyledon leaf: (*a*) simple palmate leaf: 1, petiole; 2, blade; 3, stem; 4, axillary bud; (*b*) compound pinnate rose leaf: 1, petiole; 2, leaflet (all leaflets make up the blade); 3, stem; 4, photosynthetic stipules. (*From R. G. Halfacre and J. A. Barden, Horticulture, McGraw-Hill, New York, 1979.*)

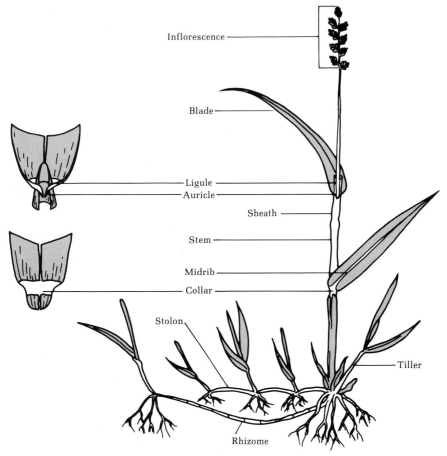

FIGURE 2-11
Parts of monocotyledon plant (grass).

the petiole at the level where petiole and blade join, as in cotton, sycamore, maple, and buckeye. In leaves with *parallel venation*, several rather large veins are essentially parallel to each other and connected by lateral veins. This type of venation is a characteristic of corn and other grasses; in banana and calla lily, the parallel veins run laterally from the midrib.

Simple, or *unifoliolate*, leaves are those whose blades form a single unit, sometimes with various types of marginal indentations. *Compound leaves* have blades divided into several individual leaflets growing from the *rachis* (leaf axis). Both simple and compound leaves may be either palmately or pinnately veined. Compound leaves may have a palmate or pinnate leaflet arrangement. Maple leaves are

simple with palmate venation (Fig. 2-10a); oak leaves are simple with pinnate venation; crownvetch and rose leaves are compound with pinnately arranged leaflets (Fig. 2-10b), and buckeye and red clover are palmately compound. Soybean and true clovers are trifoliolate, i.e., have three leaflets.

The arrangement of the leaves on the stem is related to the species. When two leaves arise from opposite sides of the same node, as in dogwood, sunflower, and potato, they are said to be *opposite*. *Alternate* leaf arrangement, as in rose, soybean, and walnut, is characterized by one leaf per node with each leaf located on the stem opposite the leaves directly above or below. In the *whorled* arrangement, as in the barberry, three or more leaves develop at a node. Other leaf characteristics useful in plant identification are leaf shapes, leaf apices, leaf bases, and leaf margins.

Leaf Modification and Adaptation

There are many modifications of the usual green photosynthetic leaves. Rhizomes and tubers may have *scales*, which are actually modified leaves (see Fig. 2-7). *Storage leaves* are found in bulbous plants such as lily and onion. *Succulent leaves*, which are thickened and fleshy, accumulate water and occur especially in plants of arid and semiarid regions. Many *tendrils* are slender modifications of twining leaf or stem parts used for support, as in pea, cucumber, and grape (Fig. 2-12); some tendrils are modified stipules. *Bracts*, which are also modified leaves, subtend many flowers or inflorescences and may appear to be a part of the flower. These structures are usually inconspicuous, but the bracts of the poinsettia and flowering dogwood are a most striking feature of these plants.

REPRODUCTIVE STRUCTURES

All seed-producing plants have a life cycle comprising both a vegetative and a reproductive phase. Gymnosperms (naked-seeded plants) produce seeds; angiosperms (flowering plants) produce fruits and

FIGURE 2-12
Leaves of grape (*Vitis rotundifolia*) modified into tendrils. (*Carolina Biological Supply Company.*)

FIGURE 2-13
Flower of the strawberry, bisected to show complete dicotyledon flower (as explained below, this is a perigynous flower). Sepals, petals, and stamens are attached to the receptacle around the ovary. 1, receptacle; 2, pistil; 3, filament; 4, petal; 5, sepal; 6, flower stalk; 7, anther; 8, stamen. (*Courtesy of Fred D. Cochran.*)

seeds during their sexually reproductive phase. This is of interest not only as a biological reproduction process but also because the flower, fruit, or seed is usually the reason—aesthetic or economic—for growing the plant.

Flowers

Flowers are important not only because so many of them are beautiful and grown for their own sakes but also because they are necessary for the development of such economically important products as apples, cucumbers, ears of corn, and grains of wheat.

The flower is a highly modified shoot of determinate growth and supported by a short stem, the *pedicel*. The *receptacle* is the enlarged apex of the pedicel (Fig. 2-13), from which the floral parts arise. The receptacle is sometimes called the *torus*, especially when it becomes conspicuously enlarged, as in a strawberry or raspberry. Bracts may occur at the base of the receptacle.

Flower Structure

A typical flower has four major types of parts: *sepals* (collectively the *calyx*), *petals* (collectively, the *corolla*), *stamens* (the *androecium*), and *pistil(s)* (the *gynoecium*) (Fig. 2-14). Sepals, petals, stamens, and pistils are leaflike units. Sepals and petals form the two outermost whorls of the flower and are collec-

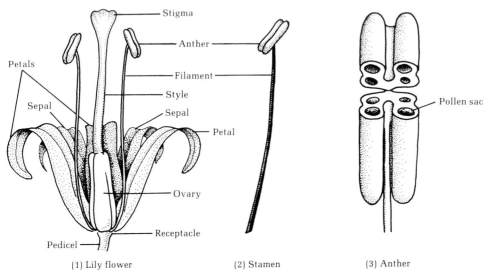

FIGURE 2-14
Parts of the flower: I, flower of lily with one sepal and four stamens removed; 2, view of stamen showing attachment of filament to anther; 3, an immature anther cut in half, showing the four pollen sacs and the connective. (*From J. B. Hill, H. W. Popp, and A. R. Grove, Botany, 4th ed., McGraw-Hill, New York, 1967.*)

tively called the *perianth*. Although not involved directly in sexual reproduction, they often attract insects and other pollinators either by color or nectar and may protect the inner gamete-producing structures. The sepals usually form the outermost whorl of the flower (Fig. 2-14) and are often green.

The stamen, the male reproductive organ, consists of an *anther*, or pollen sac, supported by a long, slender stalk, the *filament*. Inside the anther are produced single-celled pollen grains which contain male gametes (sperm) (Fig. 2-15).

The pistil is the female reproductive organ. The *stigma* of the pistil is the pollen-receptive structure (Fig. 2-16). The *style* connects the stigma to the *ovary*, an expanded structure containing the *ovules*. Inside each ovule, an *embryo sac*, the female gametophyte contains an *egg*, the female gamete. A *simple pistil* consists of a single *carpel*, a specialized leaf which has become modified for seed bearing. The ovary of a simple pistil, e.g., a pea pod, has only one cavity, or *locule*. *Compound pistils* have two or more carpels and therefore usually have two or more locules in the ovary. The compound pistil of the lily has 3 locules, the apple has 5, and flax has 4 to 10.

A *complete flower* is, by definition, one in which all four floral parts are present: sepals, petals, stamens, and pistils (Fig. 2-14). Some examples of plants with complete flowers are bean, cotton, peanut, tomato, pea, apple, raspberry, strawberry, and citrus fruits. An *incomplete flower* lacks one or more of the four kinds of floral parts. A *perfect flower* has both pistil(s) and stamens but may lack sepals or petals or both. An *imperfect flower* lacks either stamens or pistils. *Pistillate flowers* have only the female reproductive structures, and conversely, *staminate flowers* have stamens but no pistils. When neither stamens nor pistils are present, the flower, which is said to be *sterile*, does not enter directly into sexual reproduction. Corn has both pistillate (ear) and staminate (tassels) imperfect flowers. Corn "silks" are actually elongated styles whose tips have a sticky stigmatic surface for catching air-borne pollen.

Plants which produce separate staminate and pistillate flowers may be either *monoecious*, i.e., both staminate and pistillate flowers are borne on the same plant, as in corn (Fig. 2-17), oak, squash, pecan, pumpkin, and watermelon, or *dioecious*, i.e., stam-

FIGURE 2-15
Transverse view of a maize flower anther, a structure with four locules connected through a single vascular element to the rest of the plant. The photograph was taken late in the development of the anther. (*From J. Troughton and L. A. Donaldson, Probing Plant Structure,* McGraw-Hill, New York, 1972.)

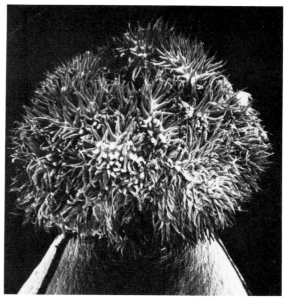

FIGURE 2-16
Stigma from a daphne flower (*Daphne* sp.). The stigma is responsible for trapping pollen as the first stage in fertilization. (*From J. Troughton and L. A. Donaldson, Probing Plant Structure, McGraw-Hill, New York, 1972.*)

(a)   (b)

FIGURE 2-17
A monoecious plant, sweet corn, showing (*a*) staminate and (*b*) pistillate flowers. (*Carolina Biological Supply Company.*)

inate and pistillate flowers are borne on separate plants, as in asparagus, buffalograss, hemp, cottonwood, ginkgo, date palm, papaya (Fig. 2-18), American holly, spinach, and some cultivars of grapes. *Andromonoecious* species, e.g., muskmelon, contain both perfect and imperfect staminate flowers on the same plant; *gynomonoecious* species, e.g., cucumber, contain both perfect and imperfect pistillate flowers on the same plant.

Flower Types

Flowers are classified as hypogynous, perigynous, or epigynous depending on where the different floral parts are attached to the receptacle, ovary, or both. If the sepals, petals, and stamens are attached to the receptacle below the ovary, the ovary is said to be *superior* and the flower is *hypogynous* (Fig. 2-19). If the receptacle is extended to form a cuplike structure around a portion of ovary, the flower is *perigynous* and again the ovary is superior (Fig. 2-13). If the perianth and stamens are attached above the ovary, the ovary is *inferior* and the flower is *epigynous* (Fig. 2-20).

Inflorescences

Flowers may occur singly, e.g., tulip, or in branched multiflowered associations known as *inflorescences*, e.g., oats. The main branch of an inflorescence system is called a *rachis*. The *peduncle* is a terminal branch

FIGURE 2-19
Hypogynous flower of tulip, bisected, showing superior ovary. Sepals, petals, and stamens are attached to the receptacle below the ovary. (*Carolina Biological Supply Company.*)

bearing the flowers. Single flowers of the inflorescence may branch off from a rachis. They may be connected to the rachis by a short *pedicel*, or they may be connected directly to the rachis (sessile). A bract may occur at the axil of each flower or flowering branch.

Inflorescences are classified according to the distribution and arrangement of individual flowers on the rachis (Fig. 2-21):

FIGURE 2-18
Papaya fruit produced on pistillate plant with staminate flowers on adjacent plant. (*From R. G. Halfacre and J. A. Barden, Horticulture, McGraw-Hill, New York, 1979.*)

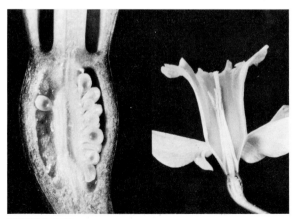

FIGURE 2-20
Epigynous flower of narcissus, showing inferior ovary. Sepals, petals, and stamens are attached to the receptacle above the ovary. (*Carolina Biological Supply Company.*)

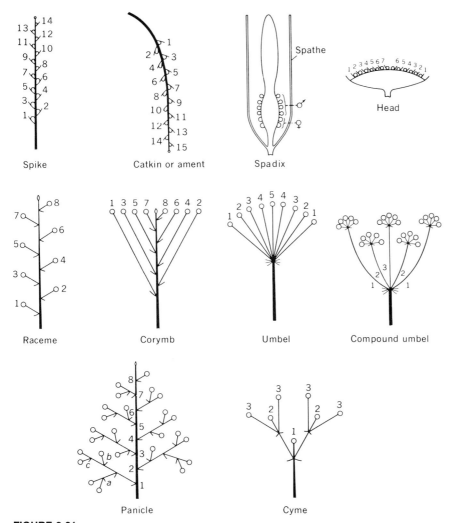

FIGURE 2-21
Types of inflorescences (*diagramatic*). Flowers are shown by small circles and bracts by short, slightly curved lines. Figures indicate the usual sequence of flower opening, number I opening first. (*From J. B. Hill, H. W. Popp, and A. R. Grove, Botany, 4th ed., McGraw-Hill, New York, 1967.*)

Raceme bears flowers on short pedicels on a single floral axis, e.g., hyacinth, snapdragon, alfalfa, and radish.

Spike has sessile flowers attached to the floral axis, e.g., wheat, rye, and plantain.

Panicle is a more highly branched inflorescence in which the flowers are more loosely arranged, e.g., oats.

The above types are indeterminate inflorescences,

since the flowers open from the bottom to the top or from the outer margins to the center. This blooming pattern permits further elongation of the inflorescences after flowering has started. In a determinate inflorescence the terminal flower is the first to form, preventing further elongation of the inflorescence. There are three basic determinate types:

Cyme—broad, more or less flat-topped inflorescence in which the central flowers bloom first—and the pedicels may be either alternate or opposite around the central axis, e.g., dogwood.

Glomerule—a cymelike inflorescence with a dense head, e.g., mint

Thyrse—a compact, condensed cyme or panicle, e.g., blackberry

Fruits

A fruit is the fertilized, developed ovary and the parts directly associated with it. The wall of a fertilized ovary commonly enlarges to become the fleshy portion of the fruit, and the ovules become the seeds. The ovary is usually considered to be a fruit when a rapid increase in growth occurs within

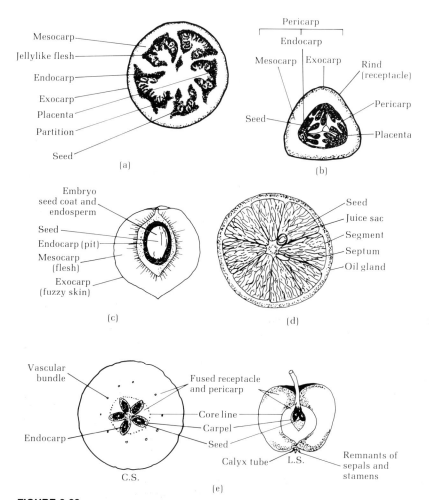

FIGURE 2-22
Simple fleshy fruits. (*a*) berry (tomato); (*b*) pepo (cucumber); (*c*) drupe (peach); (*d*) hesperidium (orange); (*e*) pome (apple). (*From R. G. Halfacre and J. A. Barden, Horticulture, McGraw-Hill, New York, 1979.*)

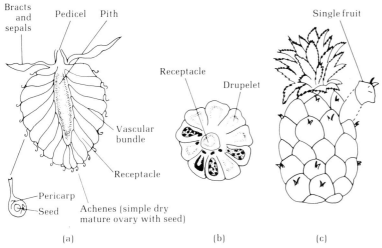

FIGURE 2-23
Aggregate fruits: (a) strawberry and (b) raspberry; (c) pineapple, a multiple fruit. (*From R. G. Halfacre and J. A. Barden, Horticulture, McGraw-Hill, New York, 1979.*)

it, whether or not seeds have also developed. A fruit may consist of not only the ovary but also the receptacle and/or other accessory parts of the flower.

The fruit wall, or *pericarp*, may have three distinct layers: typically, the outer skinlike region of the fruit is the *exocarp*, the center portion the *mesocarp*, and the inner area the *endocarp*. In a peach, for example, the skin is exocarp, the edible flesh is mesocarp, and the hard pit is endocarp.

Fruits can be classified into three main groups:

Simple fruits (Fig. 2-22), exemplified by the peach, develop from a single pistil and thus consist of a single matured ovary together with accessory structures.

Aggregate fruits (Fig. 2-23a and b) consist of a number of individual matured ovaries clustered as a unit on a common receptacle together with accessory structures. The entire fruit develops from one flower having many separate pistils, e.g., raspberry and blackberry.

Multiple fruits (Fig. 2-23c) consist of all the matured ovaries of an entire inflorescence grouped into a single mass, e.g., pineapple and mulberry.

Simple fruits are either fleshy or dry. *Fleshy fruits*, which include the berry, pepo, hesperidium, drupe,

and pome, have a pericarp that is soft and fleshy at maturity (Fig. 2-22). The entire pericarp of the *berry* is fleshy and usually edible, as in the tomato or blueberry. *Pepos* (pumpkin, winter squash, and watermelon) are fruits that have a hard rind, while *hesperidiums* (citrus fruits) have a leathery rind. The *drupe* (peach, cherry, and olive) has a thin exocarp, a thick and fleshy mesocarp, and a hard and stony endocarp. In the *pome* (apple and pear) the pericarp, which develops from the ovary, is enclosed by fleshy parts derived from other parts of the flower.

In *dry fruits*, the pericarp is often hard and brittle at maturity. *Dehiscence* is the opening of dry fruits at maturity which allows the seeds to be disseminated. Dry fruits are either *dehiscent*, in which case the carpel splits along definite seams or sutures at maturity, or *indehiscent*, in which case the fruit wall does not split at any definite seam at maturity. Many pulpy or nutlike fruits fall into this latter category. Their seeds are released only when the fruit decays or when the ovary wall is broken. A dehiscent fruit may contain several to many seeds, whereas an indehiscent fruit usually has only one or two. The legume pod (pea), follicle (macadamia), capsule (okra and poppy), and silique (mustard) are dehiscent fruits (Fig. 2-24). The achene (sunflower), caryopsis (wheat), samara (maple), and nut (chestnut) are indehiscent fruits (Fig. 2-25).

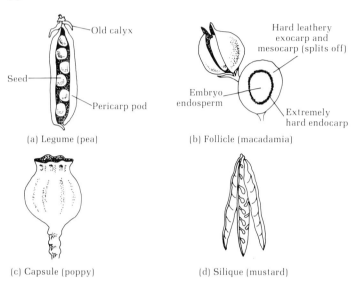

FIGURE 2-24
Dry dehiscent fruits. (*From R. G. Halfacre and J. A. Barden, Horticulture, McGraw-Hill, New York, 1979.*)

In the *legume*, the one locule splits along two sutures (soybean). The *follicle* has a locule which splits along one suture (macadamia and milkweed). The *capsule* has two or more locules which split in various ways (lily, poppy, flax, and cotton). There are two fused locules in a *silique* which separate at maturity, leaving a persistent partition between them (cabbage and mustard).

The *achene*, an indehiscent dry fruit, has only one seed, which is separable from the walls of the ovary except where it is attached to the inside of the pericarp (sunflower). The *caryopsis* or grain, has one

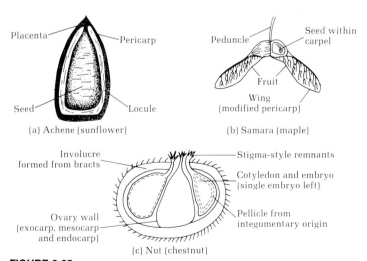

FIGURE 2-25
Dry indehiscent fruits. (*From R. G. Halfacre and J. A. Barden, Horticulture, McGraw-Hill, New York, 1979.*)

seed, which is completely fused to the inner surface of the pericarp (corn and wheat). A *nut* is similar to the achene except that the pericarp is hard throughout (pecan and walnut). The *samara* has either one or two seeds, and the pericarp bears a flattened winglike outgrowth (maple and elm).

Seeds

Seeds of gymnosperms are borne on the surface of *scales* of ovulate (female) cones, while seeds of angiosperms are ripened ovules borne inside ovaries that have ripened into fruit. Even though seeds of different species vary greatly in size and structure

(Fig. 2-26), they all consist of an embryo with associated stored reserves encased in a protective seed coat. All seeds have some stored foods, even though they may be quite limited in some cases.

The *seed coat*, or *testa*, which covers the seed may be thin and papery (peanut "skins"), or thick and tough (coconut "shells"). Some are readily permeated by water; others are quite resistant to water uptake. Some small-seeded legumes are noted for their ability to stay dry internally, even completely submerged. Such *hard-seeded* species remain dormant until the seed coat is *scarified*, or disrupted sufficiently to allow water to enter; the means of scarification may

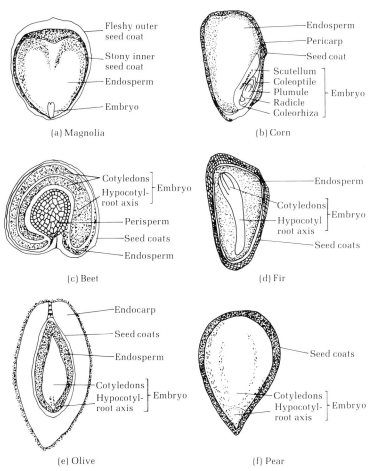

(a) Magnolia

(b) Corn

(c) Beet

(d) Fir

(e) Olive

(f) Pear

FIGURE 2-26
Seed structure of representative species. (*By permission from H. T. Hartmann and D. E. Kester, Plant Propagation: Principles and Practices, 4th ed., Prentice-Hall, Englewood Cliffs, N.J., 1983.*)

be biological (fungi or bacteria) or physical (mechanical or chemical).

The seed coats of most seeds, especially legumes, are generally uniform in appearance and texture over most of the seed surface, but certain distinctive regions or points help in identifying the species (Fig. 2-27). These morphological features may also have developmental or physiological significance. The *hilum* is essentially the navel of the seed, the scar left from its funiculus ("umbilical cord"), or attachment to the parent (fruit). The hilum is usually round or oval and often different in color or texture from the rest of the testa (seed coat). Its size may vary from 2 or 3 millimeters (mm) or more (beans) to 0.5 mm or less (some clovers). On one side of the hilum is a microscopic hole through the testa, called the *micropyle*; it is the site of the hole through which pollen tubes entered the ovule in fertilization. As the fertilized ovule grows into the seed, the micropyle remains evident. Another morphological feature of many seeds is the *raphe*, a ridge in the seed coat (usually on the opposite side of the hilum from the

micropyle); it marks the former location of the funiculus, which pressed against the developing ovule. Whether the hilum, micropyle, and raphe have important roles in germination is not always clear, but they are often helpful (along with size, color, and shape) in identifying seeds.

Inside the testa is the embryo, and there may also be *endosperm*, a nutritive substance which provides the embryo with energy and raw materials for its development. Most legumes use up all their endosperm by the time the seed reaches full maturity; only the embryo is found inside the seed coat of soybean and pea seeds. Examples of seeds with conspicuous endosperm are corn, wheat, and other cereal grains.

The embryo is an entire plant, composed of root, stem, and leaves (all in an embryonic form, of course). The embryonic root, or radicle, always lies adjacent to the micropyle.† When the seed begins to germinate, the radicle is usually the first organ to break through the seed coat. It grows downward and begins to produce root hairs and branches. At some point (more or less arbitrarily determined), it is no longer called the radicle but the primary root.

The embryonic leaves of most seeds are of two types, cotyledons and plumules. In legumes, the *cotyledons* are highly modified, usually fleshy structures which take up most of the space inside the seed and act as storage tissues from which the germinating seedling can draw its energy and nutrients. Cotyledons in other dicot families are often more "leafy" and green, with less storage function. Cotyledons of many legumes turn green also and can provide some new photosynthate for the growing seedling. The other type of leaves on an embryo are the *plumules*, small, unexpanded true leaves found at the stem apex. Plumule means "little feather," and these very young leaves often have a feathery appearance. Small and fragile as they are, they usually contain all the cells of the mature leaf. Leaf development of the first leaves in a seedling simply involves an expansion of the cells already present. Several leaves (with leaflets), completely formed but unexpanded, may be present as plumular leaves in the seed.

The embryonic leaves arise at the nodes of the embryonic stem. Cotyledonary and plumular nodes

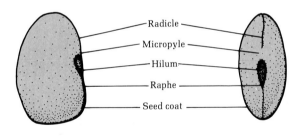

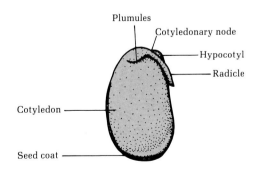

Embryo of the lima bean

FIGURE 2-27
Morphological features of a lima bean: (*above*) external, (*below*) dissected.

---

†It is usually easier to find the radicle and then the micropyle rather than vice versa.

are distinguishable. Internodes (extremely short ones) are also present in the seed. Perhaps because they are short and not easily distinguished, the embryonic stems are usually simply divided into a portion below the cotyledonary node, the *hypocotyl*, and one above, the *epicotyl*. The hypocotyl extends to the point (very indistinguishable in a seed) where the radicle begins. Because they are not readily distinguished, the stem and radicle are sometimes simply called the *embryonic axis*.

Seeds of many plants, e.g., corn, bean, and pea, serve as an important source of food. Oils extracted from some seeds, e.g., soybean, corn, and cotton, are used as food and in a variety of other products, including soaps, varnishes, and paints. Seeds have many uses as beverages, medicines, spices, and condiments. The economic importance of some seeds is such that the seed, rather than the whole plant, comes to mind when the crop is mentioned, e.g., soybean or pea.

CHAPTER THREE

PLANT ANATOMY

In spite of their tremendous diversity, all plants have the same basic structural and functional unit, the cell. Whether it is a towering redwood, a sugarbeet, or a marigold, a plant is a complex organism made up of billions or trillions of highly specialized and coordinated cells. All plant cells are similar in physical and biochemical properties in the early stages of their development, but as they mature, each develops into a specialized unit with distinctive form and function. The vitally important functions served by the wide array of cells in a plant include manufacture, transport, and storage of food; uptake and transport of water and nutrients; provision of structural strength, suppression of water loss; and, of course, division to form new cells.

Lower plants, such as algae, may be composed of only a few cells—even one. In these plants, cells generally exhibit little diversity of size, shape, or function. In more complex organisms, however, such as flowering plants, cells show great diversity. Through the process of *cell differentiation*, cells with specialized forms and functions develop from a pool of fairly simple, embryonic cells. The various differentiated cells can constitute tissues, which in turn make up organs and the entire plant. In this chapter we shall consider the major cell components, the types of cells, and how they are organized into plant tissues. This is the study of *plant anatomy*.

THE CELL

We have long known that the cell is the basic unit of living matter. As early as 1665, Robert Hooke observed the cellular structure of cork, using a primitive microscope. In 1838, Theodor Schwann and Matthias Schleiden independently proposed the theory that all life is based on the cell. Not until the advent of the electron microscope and other sophisticated methods, were we able to study in detail the intricacy, variety, and complexity of this basic unit. Much remains to be discovered because even with the advances of modern science, the cell continues to hold many mysteries.

Size and Shape of Cells

Typical vascular plants consist of hundreds of billions of cells. An average apple leaf, for example, is made up of approximately 50 million cells. Since most cells range between 10 and 100 micrometers (μm) in diameter, they can be examined only with a microscope. Plant cells vary greatly in size: the embryonic cells in root or stem tips are generally cuboidal and a few μm thick; the food- or water-conducting cells in stems, which are elongated cylinders, are among the cellular "giants," ranging in length from 2 to 7 mm and up to 1 mm in diameter.

Structure and Function of Cells

Plant cells (Fig. 3-1) are bounded by a membrane but, unlike animal cells, they are also surrounded by a more or less rigid wall. The cell is made up of the *cell wall* and the *protoplast,* the living portion of the cell. The protoplast is composed of *cytoplasm* and minute subcellular bodies, called *organelles,* that carry on specific functions. Each type of cell and each component has a special structure, chemical composition, and function. We shall consider the cell components in turn, paying particular attention to the relationship between form and function.

Cell Wall Cell walls give protection, support, and shape to each cell and collectively form the structural framework of the plant as a whole. Though it is produced by the living protoplast, the cell wall itself is nonliving. Thickness and other physical properties of cell walls vary greatly with age, type, and function of the cell. The cell wall is usually composed of three distinct layers: the primary layer, the secondary layer, and the middle lamella.

The *primary cell wall,* which is formed during early stages of growth, is composed of layers of cellulose, hemicellulose, and pectin—all discussed in Chap. 5. This layer is thin and water-permeable and, thanks to its high elasticity, stretches as the cell grows (if it did not, newly formed cells would be locked into a little box and growth would be impossible).

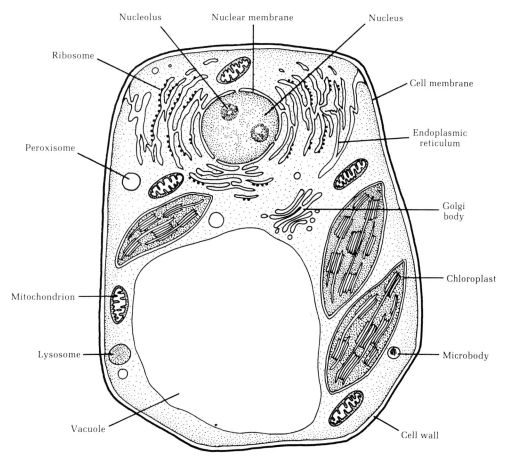

FIGURE 3-1
Cell and organelles. (*Drawn by John Norton.*)

The *secondary cell wall*, usually present in mature cells, is normally laid down against the inner surface of the primary wall after the cell has achieved its maximum expansion. With the synthesis of the secondary cell wall, *lignification* typically occurs, i.e., the deposition of a group of complex chemicals collectively known as lignin. The process makes cells with secondary walls rigid and unstretchable, but because of their strength, these cells can provide mechanical support for the plant.

Plant cells with secondary cell walls originally contain extremely small pores, or pits, through which cytoplasmic strands, the *plasmodesmata*, extend. These strands provide living connections between cells and make it easier for water, minerals, photosynthates, and hormones to move from one cell to another. During lignification, these pits are often plugged and the plasmodesmata severed; when this occurs, the cells are more or less waterproof.

The *middle lamella*, the outermost layer of a mature cell wall, consists of pectin that cements together primary cell walls of adjacent cells. Thus the middle lamella is the glue holding cells together. As fleshy fruits ripen, the pectin of the middle lamella is broken down enzymatically, the cells begin to separate, and the fruit softens. In many cases, cooking fruits and vegetables softens them by dissolving the middle lamella.

Protoplast The protoplast is the living portion of the cell inside the cell wall. It is surrounded by the *plasma membrane*, or *plasmalemma*, a selectively permeable, flexible membrane composed of lipids and proteins. This membrane regulates the movement of water as well as organic and inorganic materials into and out of the protoplast.

Cytoplasm The somewhat viscous matrix enclosed by the plasma membrane, the *cytoplasm*, contains water, proteins, carbohydrates, lipids, and inorganic salts. It usually has a consistency similar to raw egg white and may contain various pigments. Suspended in it are the organelles, which carry on all functions necessary to sustain life at the cellular level.

Endoplasmic Reticulum The endoplasmic reticulum is a long, membranous network which extends throughout large portions of the cytoplasm (Fig. 3-1). It apparently functions in the inter- and intracellular transport of cellular products, as a surface for protein synthesis, in separating enzymes and enzyme reactants, in moving cell components during cell division, and in physical support. The endoplasmic reticulum often appears to be connected to the nuclear membrane (Fig. 3-1) and the plasma membrane.

Ribosomes In the cytoplasm and often associated with the endoplasmic reticulum are extremely small, dense globular particles called *ribosomes*. The ribosomes contain ribonucleic acid (RNA) and proteins (see Chaps. 5 and 6) and are involved in the synthesis of proteins, including enzymes. *Polyribosomes* are aggregations of many ribosomes, either on the endoplasmic reticulum or floating free in the cytoplasm, where they also function in protein synthesis.

Plastids Among the organelles within plant cells are the plastids, whose functions range from photosynthesis to the storage of starches and fats. They are frequently classified by color. *Chloroplasts* (green), *chromoplasts* (yellow, red, or orange), and *leucoplasts* (colorless) appear to originate from common precursors called *proplastids*.

The chloroplasts are the dominant plastid in the leaves and other green parts of a plant. In a typical green cell of a leaf, there are 20 to 100 chloroplasts (Fig. 3-2). Chloroplasts contain chlorophyll and carotenoid pigments, but the yellow and orange carotenoids are masked by chlorophyll and do not become evident until the chlorophyll deteriorates. This effect is responsible for much of the color of autumn leaves.

Chloroplasts also contain the enzymes and intermediate substances required for conversion of radiant energy into chemical energy by photosynthesis (see Chap. 5). Each chloroplast may have a large number of disk-shaped *grana lamellae*, highly folded or convoluted layers of the *inner membrane* of the chloroplast (Fig. 3-2). Stacks of grana lamellae constitute a *granum*. These grana membranes contain chlorophyll molecules, which capture light energy from the sun. The actual transformation of carbon dioxide into more complex carbon-containing compounds occurs in the *stroma*. This liquid (nonmembrane) portion of the chloroplast is a viscous "soup" which contains enzymes and intermediates of the photosynthetic process.

Yellow and orange flowers, fruits, and seeds lack chlorophyll when mature, owing their color instead to chromoplasts. These variously shaped plastids

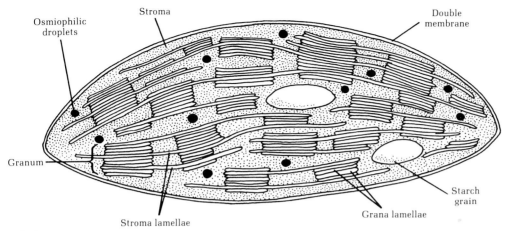

FIGURE 3-2
Chloroplast, showing a large number of disk-shaped grana lamella. Stacks of grana lamellae make up a granum. (*Drawn by John Norton.*)

synthesize and retain yellow (carotene), red (xantho-phyll), and orange (carotenoid) pigments rather than chlorophyll. Electron micrographs show chromo-plasts are less highly organized than chloroplasts.

Leucoplasts are nonpigmented organelles of dif-ferent shapes which are usually elastic and serve in the synthesis and storage of starch, oils, and proteins from smaller molecules. They are located in the colorless leaf cells of variegated leaves, stems, roots, and other storage organs, as well as in the underlying tissues of the plant not exposed to light.

Mitochondria The mitochondria, which are much smaller than most plastids, are pickle-shaped (Fig. 3-1). Each mitochondrion contains a semigel stroma of proteins (enzymes) within a two-layered mem-brane. The mitochondria serve as the cell's power-house and are the major sites for respiration. In mitochondrial respiration, various substances are broken down into carbon dioxide and water with the release of energy (see Chap. 5).

Dictyosomes The dictyosomes, or *Golgi bodies*, serve as collection centers for complex carbohy-drates; they are believed to be involved in cell-wall synthesis and the packaging of cellular products for excretion from the cell. They also develop numerous vesicles thought to serve as a source of plasma membrane components.

Microtubules Microtubules are proteinaceous structures located in the cytoplasmic matrix of non-dividing cells and in the spindle fibers of cells undergoing cell division. They are believed to be involved in the growth of cellulose fibers in the cell wall. During cell division, microtubules attach to the chromosomes and appear to draw them to the op-posite poles.

Glyoxysomes Glyoxysomes, or *peroxysomes*, spherical enzyme-filled organelles surrounded by a single membrane, serve several important metabolic roles including glycolic acid metabolism associated with photorespiration (see Chap. 5). These micro-bodies also have the enzymes necessary for the conversion of fats into carbohydrates during germi-nation of many seeds.

Lysosomes These organelles are bounded by a single membrane and contain enzymes capable of digesting various cellular components such as pro-teins and lipids. Their exact functions are unclear, but they are believed to be important in the devel-opment of those cells which are hollow at maturity. They may also be a defensive mechanism against invasion by pathogens.

Vacuoles Young meristematic cells have many minute vacuoles within the cytoplasm. As the cells grow and mature, these vacuoles fuse into a large central vacuole, which increases in size until it may occupy 80 or 90 percent of the intracellular volume. In most mature plant cells, the nucleus, plastids, and other subcellular inclusions are confined to the peripheral layer of cytoplasm; and the vacuole, which is larger, remains in the center. The vacuole is filled with *cell sap*, an aqueous solution of inorganic salts and organic solutes. The vacuole is often considered to be a storage compartment for materials which are no longer needed or which might be detrimental to metabolic processes in the cytoplasm. These are often stored as insoluble crystals in the cell sap. The vacuole is surrounded by a semipermeable membrane known as the *tonoplast*, which apparently regulates the pH of the cytoplasm and controls its chemical composition. The vacuole may also contain pigments such as the anthocyanins, which are responsible for some colors of flowers, fruits, and autumn leaves.

Nucleus The nucleus of mature plant cells is the largest organelle in the cell except for the vacuole. It is usually spherical or somewhat disk-shaped, 10 to 20 μm in diameter, and surrounded by a double membrane called the *nuclear envelope* (Fig. 3-1). In young cells, the nucleus is centrally located, but as the vacuole expands to fill most of the cell, the nucleus is usually pushed into one corner.

The nonparticulate portion of the nucleus is a gel-like nucleoplasm, comparable to the cytoplasm of the cell. It contains one or more spherical bodies, called *nucleoli*, and is composed of DNA, RNA, and proteins, collectively referred to as *chromatin*. When cell division begins, the chromatin condenses to form microscopically visible bodies called *chromosomes*. The DNA of the chromosomes contains genetic information and controls the synthesis of specific enzymes on the ribosomes; these enzymes in turn control the metabolic activities of the cell and thus determine the overall characteristics of the organism. These processes and relationships are discussed in more detail in Chaps. 5 and 6.

CELL PROCESSES

Cell Division and Growth

In plants, specific areas are active in cell division: the root tips, the stem tips, the cambium in dicoty-

ledons, and the intercalary meristem in monocotyledons.

In a multicellular plant, cells undergo many changes as they develop from a newly formed cell in a meristem to a functional mature cell. This *differentiation* during or after enlargement forms cells highly specialized in size, shape, and function. Some cells complete all structural changes in a few days; others take much longer. Some live for years; others die in a few hours. Many factors are involved in the differentiation and survival time of an individual cell.

Although the differentiation of cells is usually irreversible, living cells somtimes change their function. For example, the apical cells of a stem may initiate leaf primordia for a time, after which they may assume the function of floral precursors that give rise to the various parts of a flower. The cells of cortex or phloem (food-conducting tissue) may become cork cambium, which, by division, produces the cork of woody plants. Furthermore, the parenchyma cells (thin-walled cells with a protoplast present at maturity) around a wound often become meristematic and divide to form wound tissue.

Cell Enlargement and Differentiation

Both cell maturation and differentiation involve development and specialization of cells. The ultimate appearance and function of each cell depend on the changes it undergoes during differentiation. As a result, certain cells of a plant may have thin walls and function in food production or storage. Other cells may become specialized as conducting components; still others may develop thick cell walls that collectively furnish strength and rigidity to the plant. Many of the cells in a mature plant are dead but still are functional, e.g., the hollow water-conducting components of leaves, stems, and roots.

Some cells do not achieve full size immediately following cell division. In the young apple fruit, for example, cell division takes place for about 6 weeks after fertilization, producing most of the cells that will be found in the mature fruit. For many weeks thereafter, the fruit continues to grow, not by cell division but by expansion of the small cells already formed (Fig. 3-3). The same situation occurs in many leaves; a very young leaf at the stem tip contains most of the cells it will have at maturity.

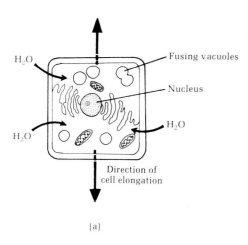

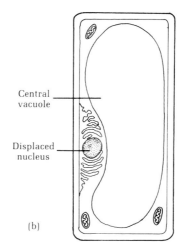

(a)

Increase in hormones causes
greater permeability, allowing
greater H₂O intake; vacuoles
fuse to form central vacuole.

(b)

Central vacuole now fills
center of cell, displacing
nucleus and cytoplasm so
they lie against inner surface
of cell wall; turgidity increases.

FIGURE 3-3
Early events in the elongation of a cell. (*Drawn by John Norton.*)

TISSUES

From a morphological standpoint, a *tissue* is an organized group of cells with common origin and function. Diversity may be found in cellular form and function within the tissue, but the cells that form the tissue are continuous and furnish some portion of the structural or functional basis of the plant. Included in various plant tissues are collections of cells specialized for growth, storage of food, protection from dehydration, support, conduction of water, synthesis of food, and absorption of water and mineral elements. Tissues composed entirely of one cell type are said to be *simple*; tissues made up of two or more cell types are *compound*.

Tissues can be classified as *meristematic* (tissues composed of cells whose main function is dividing) or *permanent* (tissues composed of cells in which growth and differentiation have been completed). The major cell types making up the permanent tissues of higher plants are parenchyma, collenchyma, and sclerenchyma (Fig. 3-4). These basic types constitute the conducting tissues (with components known as tracheids, vessels, and sieve and companion cells), the ground tissues (pith and cortex), and the dermal tissues (epidermis and bark).

Meristems

Meristems are tissues in which protoplasmic synthesis and tissue initiation occur, as these cells have retained their ability to divide. In fact, division is the primary function of meristematic cells. These cell factories occur in developing embryos, in the tips of roots and stems, and in leaf and floral primordia. *Apical*, or *primary*, *meristems*, at the tips of roots and stems, produce cells which result in *longitudinal*, or *primary*, *growth* of the plant. The embryonic cells are generally small and isodiametric, contain dense protoplasm with small vacuoles, and have thin walls consisting of only the primary cell-wall layer (Fig. 3-4a). Photomicrographs usually show a spherical nucleus at the center of each meristematic cell. After formation in the apical meristem, cells are displaced from their origin site and undergo elongation and differentiation. At maturity, they constitute the *primary tissues* of the plant.

At the base of grass leaves are the *intercalary meristems*, which produce new cells for the leaves, a growth pattern that allows repeated mowing or grazing of grasses without killing the plant. The *vascular cambium*, or *lateral meristem*, is a cylinder of fusiform (long and tapering) meristematic cells in the stems and roots of many plants. These are the

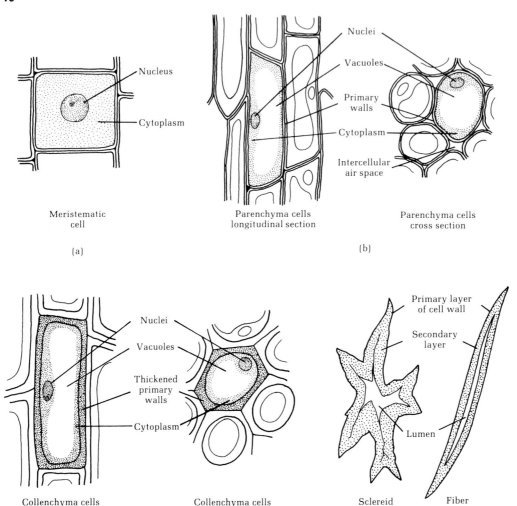

FIGURE 3-4
Basic plant cell types: (*a*) meristematic cell: has primary cell wall, carries on mitosis, and occurs in root and stem tips, leaf, and flower primordia; (*b*) parenchyma cell: is most abundant type in plants, has many modifications, and lacks all but primary layer of cell wall; (*c*) collenchyma: has some thickening on inner surface of angles of primary cell wall without lignification; (*d*) sclerenchyma (sclereid and fibers): secondary and primary layers of cell wall become lignified, resulting in death of protoplast. (*Drawn by John Norton.*)

dividing cells responsible for the increase in diameter, or *secondary growth*, of the plant.

Parenchyma

The term *parenchyma* refers to a general cell type rather than a specific kind of cell. Parenchyma cells are usually thin-walled and have protoplasts with large central vacuoles (Fig. 3-4*b*). They may be primary or secondary in origin. Meristematic cells are, by definition, parenchymatous. The plasticity of their walls allows for a rapid increase in size in developing leaves or fruits. The primary cell walls of parenchyma cells are occasionally strengthened by lignified secondary walls. Parenchyma cells are commonly many-sided, or polyhedral.

Parenchyma cells occur in all areas of the plant, including the pith (a tissue occupying the center of many stems), the cortex (the tissue lying between the epidermis and the vascular tissue of roots and stems), and the phloem of stems and roots as well as in the leaves, floral parts, seeds, and fruits. The epidermis is composed of modified parenchyma cells.

Parenchyma cells perform a variety of metabolic functions. *Nectaries* at the base of flowers consist of parenchyma cells secreting substances that attract insects. *Laticifer cells,* found in some plants, are parenchyma cells specialized to synthesize and store latex. Some parenchyma tissues function in food storage, e.g., the cortex of carrot and pith of sugarcane; others are involved in photosynthesis (mesophyll of leaves). When parenchyma cells contain chloroplasts, they are known as *chlorenchyma cells.* These are the principal sites for photosynthesis. Since parenchyma cells retain the latent ability to divide, they may be important in wound healing, regeneration, and initiation of adventitious roots on stem cuttings. They can also play a critical role in moving water and food within the plant.

Collenchyma
Collenchyma cells are a special type of parenchyma cell that functions as supportive tissue during the primary growth of some plants (Fig. 3-4c). Collenchyma cells are typically living and contain a large central vacuole. These cells lend mechanical support to the young plant while allowing it to bend slightly under stress and strain. This type of cell occurs in the midrib and veins of leaves and in the petioles of herbaceous vegetables such as celery and rhubarb. Although collenchyma cells provide the necessary support for such plants, they are also responsible for their stringiness.

Sclerenchyma
The term *sclerenchyma,* like parenchyma, describes a general cell type rather than a specific kind of cell. Sclerenchyma cells usually have no protoplast at maturity. They may develop in any or all parts of the primary and secondary plant body, and they may be formed more or less directly from meristematic cells or by later modification of parenchyma or

collenchyma cells. The principal characteristic of sclerenchyma cells is their thick, usually lignified walls; their function is usually conduction, mechanical support, or protection.

Supportive or protective sclerenchyma cells are diverse in shape and size and of two general types: fibers and sclereids. *Fiber cells* (Fig. 3-4d) are slender and elongated with tapering ends, which overlap and are often fused with each other. Their secondary walls are usually lignified, and the *lumen,* or cavity in the fiber, often lacks a protoplast at maturity. By virtue of their number and overlapping nature, fiber cells help hold stems erect and enable them to withstand the stress of bearing large loads of flowers and fruits. *Sclereids* (Fig. 3-4d) are variable in shape. The term stone cell is often used for certain almost isodiametric sclereids which are unbranched and are without uniform shape. The hardest part of seeds, nuts, and hard fruits is usually composed of stone cells of various types. In mass they afford hardness and mechanical protection, e.g., walnut shells and peach pits. The scattered occurrence of groups of stone cells lowers the quality of the fleshy part of pears in certain cultivars. Stone-cell formation is largely prevented by picking the pear before it is fully ripe.

TISSUE SYSTEMS

Cells that form the various parts of the plant can be grouped in terms of function and structure into three so-called *tissue systems:* the vascular, the dermal, and the fundamental.

Vascular System
The tissues in the conductive, or vascular, system are composed of parenchymatous and sclerenchymatous cells specialized for support and for the translocation of substances throughout all parts of the plant. In most mature perennial plants, the vascular system constitutes the bulk of the root and stem. There are two types of tissues in the vascular system, the *xylem* and the *phloem,* each of which may be considered to be a compound tissue because it is composed of several types of cells. Both xylem and phloem tissues are continuous throughout the plant. Xylem conducts water and minerals in solution from the roots upward, while phloem conducts complex organic materials in solution from the point

of their synthesis or storage to various areas of the plant. The organization and arrangement of the vascular system varies with the kind of plant as well as with individual parts. In a young dicotyledonous stem, the vascular system consists of a number of *vascular bundles* (small groups of xylem and phloem cells), which form a cylindrical pattern. Typically, the inner part of each bundle is composed of xylem; the outer portion is phloem. In monocotyledonous plants, the vascular bundles occur more or less randomly throughout the fundamental system of the plant (the pith and cortex). In young roots, the vascular system is the central portion of the root and is surrounded by the pericycle and endodermis. The *pericycle* is one or more layers of cells from which the lateral roots arise. Collectively, the vascular tissues and associated fundamental tissues of the root and stem are called the *stele*.

Xylem The xylem is composed of sclerenchymatous cells (tracheids, vessels, and fibers) and parenchymatous cells. Tracheids and vessels are non-living and thick-walled when mature. Xylem cells that become adapted for support or transport of water and minerals in solution often die, and their protoplasts disappear, but they remain as a permanent and functional part of the plant. The protoplast naturally dies only after the cell attains full size and the secondary cell wall has been formed. These specialized cells become tubes in an arrangement aligned parallel to the plant axis.

Tracheids, the fundamental water-conducting cell type in the xylem of conifers and ferns, occur in longitudinal tiers and are elongated, with tapered overlapping end walls (Fig. 3-5). In cross section, the tracheid is usually angular, although some rounded forms occur. Their secondary cell wall layers have various arrangements of thickened, lignified materials and contain pits and/or perforations which sometimes permit the ready movement of water between adjacent cells.

Vessel elements are similar to tracheids except that their end walls are broken down, giving one long passageway through a series of vertically adjacent cells. This dissolution of end walls as the protoplast dies forms a column through which water and dissolved minerals can move with even less resistance than through vertical rows of tracheids (Fig. 3-6). When young, before their end walls have

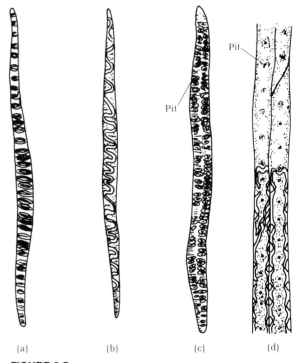

FIGURE 3-5
Longitudinal views of tracheids of the xylem tissue: (*a*) annular (ringed) tracheid; (*b*) helical (spiraled) tracheid; (*c*) scalariform tracheid; (*d*) pitted tracheid. (*From R. G. Halfacre and J. A. Barden, Horticulture, McGraw-Hill, New York, 1979.*)

disappeared, the cells that form vessels contain protoplasts which lay down their lignified walls. Vessels are formed only in angiosperms and provide them with some support but function primarily in conduction (Fig. 3-7). Vessels generally have a greater diameter than that of tracheids, allowing more efficient water movement.

Xylem fibers, typical of the sclerenchymatous fibers mentioned earlier, are elongated, strengthening cells with thickened walls. Their ends are pointed; at maturity they lack protoplasts. They differ from tracheids chiefly in their thicker walls and reduced number of pits. Fibers are common in the xylem tissue of angiosperms. The scattered *xylem parenchyma* cells are similar to parenchymatous cells in other areas of the plant in that they function in food storage and lateral transport. Sometimes their walls thicken and become lignified.

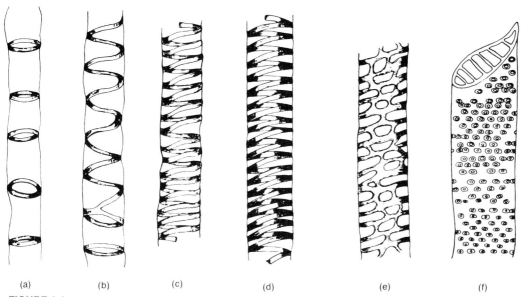

FIGURE 3-6
Longitudinal views of vessel segments of xylem tissue. (*a*) ringed vessel, (*b*) to (*d*) spiral vessel; (*e*) scalariform vessel; (*f*) pitted vessel with scalariform end walls. (*a*) to (*d*), from the stem of the balsam plant (*Impatiens sultani*); (*e*), from the stem of the sunflower (*Helianthus annuus*); (*f*), from the stem of the alder (*Alnus*). (*From J. B.Hill, H. W. Popp, and A. R. Grove, Botany, 4th ed., McGraw-Hill, New York, 1967.*)

Phloem Phloem is a compound tissue consisting of sieve cells, companion cells, fibers, and parenchyma cells (Fig. 3-8). *Sieve cells*, the main functional component of the phloem, are elongated, slender cells with thin, nonlignified walls; i.e., they are parenchymatous. In dicotyledons they are arranged in a vertical series to form sieve tubes. Each sieve cell in the series is a *sieve element*. A sieve element has a large central vacuole surrounded by a thin layer of cytoplasm pressed against the inner surface of the cell wall. Although living, mature sieve cells do not contain a nucleus. Perforations in the cell wall allow strands of cytoplasm to extend from one sieve element to another, apparently facilitating the transport of materials through the tissue. Callose, an accumulation of sugar polymers, may form over these perforations, interrupting the transport of food and rendering the cell nonfunctional.

One or more small, slender *companion cells* are associated with the sieve cells in the phloem of angiosperms. The companion cells, which are living and contain nuclei, may regulate the metabolic activities of the enucleate sieve cells. *Phloem fibers* are elongated sclerenchymatous cells occurring in groups in the phloem. *Phloem parenchyma* cells are living cells scattered among other cells of the phloem that enable substances to move laterally and may provide a means of storing limited quantities of food.

Dermal System

In young stems, roots, and leaves the outermost layer of cells, called the *epidermis*, forms the dermal system. The epidermis consists of specialized parenchyma cells, which are irregularly shaped and usually one cell layer thick. We discussed in Chap. 2 the peculiar nature of some root epidermal cells that produce root hairs. Sometimes the leaf's epidermis consists of several layers of cells, a *multiple epidermis*, e.g., leaves of fig, rubber plant, and peperomia. The outer walls of leaf and stem epidermal cells are often impregnated with *cutin*, a complex of waxy substances deposited inside the cell walls and

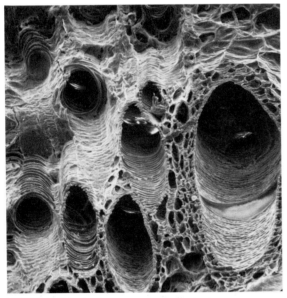

FIGURE 3-7
Xylem vessels in a cucumber stem. The internal cell wall surface may be smooth or pitted, but several differences in the appearance of the wall are related to the age of this tissue. The cell initially consists of primary cell walls, but as increasing amounts of secondary tissue are laid down, it progressively obscures the primary wall. In the earlier xylem elements the secondary wall may occur as rings; in the later stages the surface has a pitted appearance. This photograph shows annular and spiral types on the left and reticulate types on the right. (*From J. Troughton and L. A. Donaldson, Probing Plant Structure., McGraw-Hill, New York, 1972.*)

on their outer surface. Cutin is an effective but not absolute barrier to water loss.

The epidermis is replaced in the older portions of roots and stems (in which secondary growth occurs) by the bark, or *periderm*, which consists of cork (phellem), cork cambium (phellogen), and sometimes underlying layers of cells (phelloderm). The cork is composed of nonliving cells with walls formed by the cork cambium that are said to be *suberized*, i.e., filled with the waxy substance *suberin*, which water-proofs them. To some degree, the cork also serves to protect the underlying tissues of the plant. Cork oaks produce massive amounts of periderm which is harvested and cut into shapes for making gaskets, bottle corks, and other products.

Fundamental Tissue System

All tissues of the plant encased by the epidermis other than those of the vascular system constitute the *fundamental* or *ground*, tissue system. In the young roots, this system includes the cortex; in the young stem, it consists of the cortex and pith. Cortex and pith cells are parenchymatous and function from time to time in storage or wound healing. As a result of extensive secondary growth, the cortex in older woody stems and roots becomes stretched, crushed, and gradually destroyed. Generally, the cortex is composed of parenchyma cells, but collenchyma and sclerenchyma cells may also be present.

SOME ROOT-SPECIFIC ANATOMY

The general organization of roots was discussed in Chap. 2, but some anatomical distinctions remain to be made (Figs. 3-9 and 3-10). In the zone of cell maturation, epidermal cells expand outward to form *root hairs*. These distinctive protrusions represent a major site for the absorption of water and nutrients. As root growth continues, root hairs are sloughed off, being continually replaced on the younger portion of the root. Naturally the epidermis of subterranean roots is not cutinized, since a waxy layer would interfere with absorption.

Lateral (secondary) roots are generally initiated at a point proximal to the root hair zone, where the epidermis no longer functions so actively in absorption. The lateral root appears externally only after its growth is well begun. In angiosperms and gymnosperms, the new apical meristems arise in the pericycle just beneath the *endodermis*, a single layer of cells between the cortex and pericycle in roots. In the formation of lateral roots, several adjacent cells of the pericycle become meristematic, dividing to produce a new root apex. As the apex grows, it pushes out through the surrounding tissues (endodermis, cortex, and epidermis), finally emerging from the root.

The roots of most gymnosperms and dicotyledons have a vascular cambium that produces secondary growth in the mature region of the root. *Secondary xylem* is formed to the inside and *secondary phloem* to the outside of the vascular cambium. The secondary phloem pushes the primary phloem and cortex farther out. Gradually the pressure from underlying

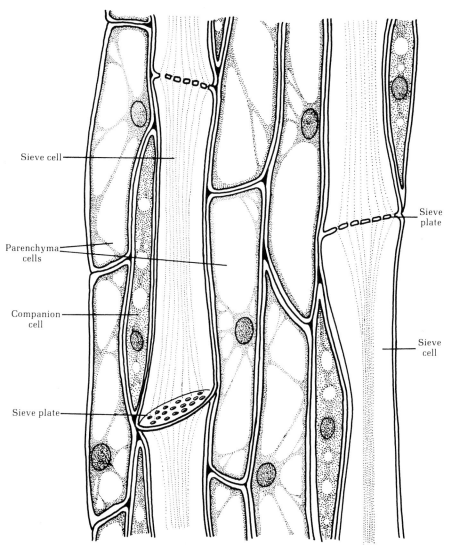

FIGURE 3-8
Longitudinal section through the phloem tissue of a stem, showing the cellular structure. (*Drawn by John Norton.*)

expanding tissues stretches and crushes the primary phloem, endodermis, cortex, and epidermis. The *cork cambium* is a cylindrical layer of cortex or phloem cells that becomes meristematic. It develops in many roots as well as stems, forming a layer of cork (bark) which performs the protective function of the epidermis.

SOME STEM-SPECIFIC ANATOMY

Surrounding the vascular bundles in young stems is the cortex, which is in turn surrounded by the epidermis (Fig. 3-11). Extending tangentially through and between each vascular bundle is a layer (or thin region) of cells which retain their ability to divide.

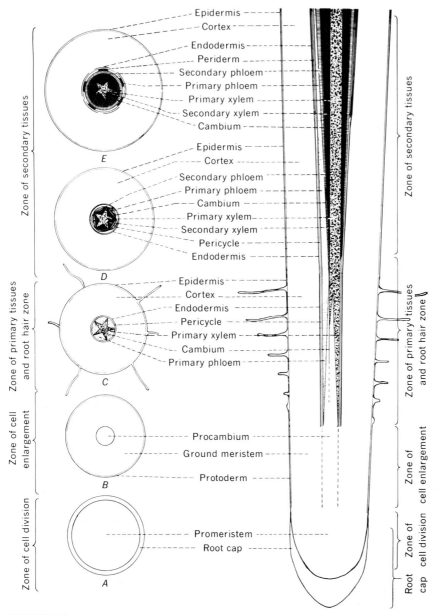

FIGURE 3-9

Diagrammatic representation of the origin and arrangement of primary and secondary tissues in a young root of a dicotyledonous plant; (*right*) longitudinal section of root; (*left*) transverse sections at different distances from the tip (a) to (e.) (*From J. B. Hill, H. W. Popp, and A. R. Grove, Botany, 4th ed., McGraw-Hill, New York, 1967.*)

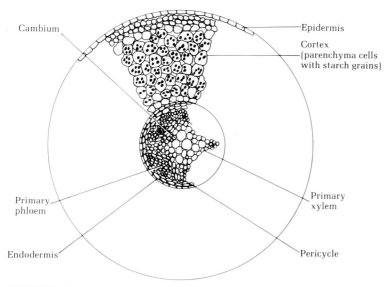

Cambium

Epidermis

Cortex
(parenchyma cells
with starch grains)

Primary
xylem

Primary
phloem

Endodermis

Pericycle

FIGURE 3-10
Cross section of Ranunculus root showing primary structures before secondary growth has
begun. (*From R. G. Halfacre and J. A. Barden, Horticulture, McGraw-Hill, New York, 1979.*)

Until they become active in cell division, these cells form a cylindrical layer known as the *procambium*, which later becomes the vascular cambium. After primary growth has ceased at a given level in the dicotyledon stem, this cambium begins to produce the secondary growth of the stem.

Dicots show obvious anatomical differences between herbaceous (Fig. 3-11) and woody plants (Figs. 3-12 and 3-13), especially in activity of the cambium and the relative amounts of parenchymatous and sclerenchymatous cells produced. Tree species produce large amounts of secondary xylem, the tissue of which is predominately fibers and tracheids or vessels. A potato, which is a stem, has secondary growth also (the faint outline of the phloem and cambium is often evident in a potato chip); but potatoes do not become woody because their secondary tissues are essentially all parenchymatous.

Characteristically, primary growth is all that occurs in monocot stems or culms (grass stems), since they are without a vascular cambium. Their vascular bundles occur more or less at random throughout the soft, parenchymatous tissue (Fig. 3-14). Any increase in circumference in such stems is accomplished by an enlargement of parenchymatous pith

cells originally formed by the apical meristem. Examples of monocotyledonous plants that have very diverse stems are banana, lily, corn, asparagus, rice, wheat, and date palm. The absence of secondary growth explains why the stem of a palm tree is essentially the same diameter for its entire length.

Gymnosperm and dicotyledonous trees and shrubs have woody stems (Fig. 3-13). The parts of a 1-year-old woody stem from center to periphery are the pith, which is composed of parenchyma cells; a cylindrical pattern of vascular bundles in which there is a cylindrical vascular cambium; narrow *pith rays*, which radiate from the pith and extend between adjacent vascular bundles; a cortex; and an epidermis. The vascular bundles are much closer together than in herbaceous dicot stems and constitute a much greater percentage of the total volume in a woody stem than in a herbaceous one. In a mature woody stem, all the tissues from the vascular cambium to the outer circumference form the bark. In a young woody stem, the bark includes the phloem of the vascular bundles, the cortex, and the epidermis. When the stem becomes older and increases in circumference as a result of secondary growth, cells of the cortex and epidermis are stretched, gradually

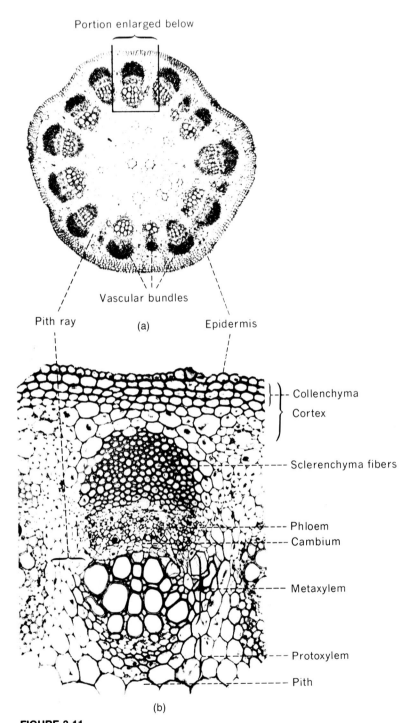

Portion enlarged below

Vascular bundles

Pith ray   (a)   Epidermis

Collenchyma
Cortex

Sclerenchyma fibers

Phloem
Cambium

Metaxylem

Protoxylem

Pith

(b)

FIGURE 3-11
The primary tissues of the stem of herbaceous dicotyledonous plant, sunflower (*Helianthus annuus*): (*a*) transverse section of entire stem; (*b*) enlarged portion of (*a*), showing structure of the individual tissues, including one vascular bundle. (*Photomicrograph by D. A. Kribs; from J. B. Hill, H. W. Popp, and A. R. Grove, Botany, 4th ed., McGraw-Hill, New York, 1967.*)

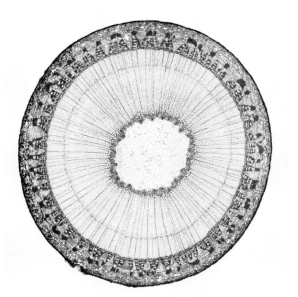

FIGURE 3-12
Cross section of 2-year-old woody dicotyledonous plant, tulip tree (*Liriodendron tulipifera*). (*Carolina Biological Supply Company.*)

FIGURE 3-13
Semidiagrammatic representation of a 1-year-old stem of Liriodendron, in transverse, radial, and tangential views. (*From J. B. Hill, H. W. Popp, and A. R. Grove, Botany, 4th ed., McGraw-Hill, New York, 1967.*)

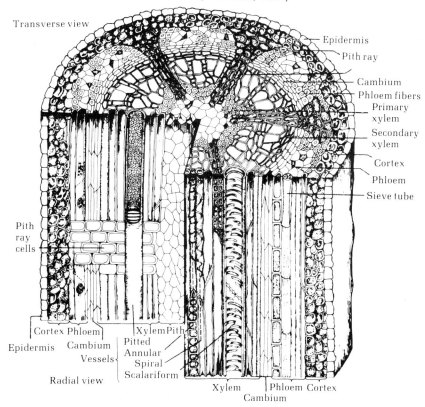

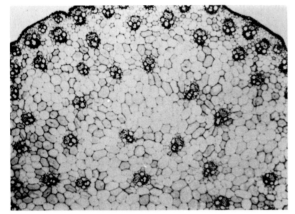

FIGURE 3-14
Cross section of monocotyledonous stem, showing random arrangement of vascular bundles. (*Carolina Biological Supply Company.*)

crushed, and eventually sloughed off. As this occurs, however, a cylindrical cell layer of the cortex or phloem develops (or reinstates) the ability to divide. This layer becomes the cork cambium, which frequently produces cells capable of depositing suberin on their outer walls to form the cork. Hence in these older stems the bark consists of phloem, cork cambium, and cork.

Increases in stem diameter are due to the activity of the vascular cambium. This activity varies seasonally, being greatest in spring and early summer. During this period of active cell division, the bark slips easily, i.e., is readily separated from the xylem. This is when propagation by grafting and budding are most successful (see Chap. 15). Because the vascular cambium generally produces larger cells in the spring and smaller, more tightly packed cells in the summer, species growing in temperate regions

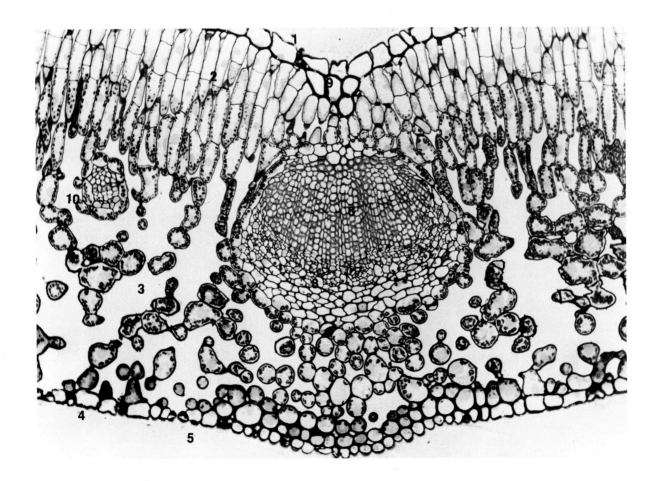

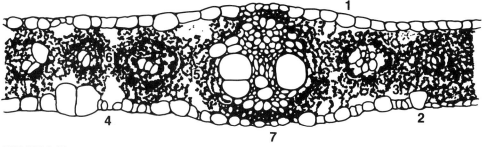

FIGURE 3-16

Leaf structure. Cross section of a monocotyledon (corn, *Zea mays*) leaf with large central midvein and smaller parallel veins. Cells of spongy mesophyll contain numerous chloroplasts. Pairs of guard cells with their stomata are in both upper and lower epidermal layers. Internal to each stoma is an enlarged area of intercellular space. Between the upper and lower epidermises and the midvein is a band of sclerenchyma cells. 1, upper epidermis; 2, lower epidermis; 3, spongy mesophyll; 4, guard cells and stoma; 5, midvein; 6, small vein paralleling midvein; 7, sclerenchyma cells. (*Carolina Biological Supply Company.*)

form annual layers (growth rings) of xylem. As new secondary phloem is formed on the outer face of the vascular cambium, however, the outer, older phloem is stretched and obliterated. Since the newly formed phloem cells replace the older degenerated phloem, distinctive annual layers of phloem are seldom produced.

The xylem of a tree can be divided into sapwood and heartwood. *Sapwood* is usually confined to the outer few annual layers of the woody stem and is the part of the xylem which functions in water and mineral transport. Year by year, the inner rings of sapwood slowly change into heartwood. *Heartwood* consists of older annual layers; its cells contain deposits of resins and gums which strengthen and preserve the tissue. Conversion of sapwood into heartwood generally occurs faster in slower-growing stems than in faster-growing ones. The ratio of heartwood to sapwood varies considerably with the species.

LEAF ANATOMY

Although leaf morphology varies greatly with species, leaf anatomy is surprisingly similar. The environment in which plants grow has some effect on leaf anatomy. A typical dicotyledonous leaf is presented in cross section in Fig. 3-15 and a typical monocotyledonous leaf in Fig. 3-16. The epidermis, usually one cell layer thick, is a protective layer covering the upper and lower surface of the leaf. It may be covered with a waxy layer called the *cuticle*, which, because it resembles paraffin in texture, reduces the loss of water. Cells of the epidermis may be thicker on the side exposed to the sun, but they

FIGURE 3-15

Leaf structure. Cross section of dicotyledon leaf (*privet, Ligustrum vulgare*) showing details of midvein and blade tissue on either side of midvein: 1, upper epidermis; 2, closely packed, multiple-layered palisade cells with numerous chloroplasts; 3, loosely packed spongy cells with chloroplasts (notice the large amount of intercellular space); 4, lower epidermis; 5, pair of guard cells and nearly closed stoma in lower epidermis; 6, radiating rows of xylem components of midvein; 7, phloem of midvein; 8, parenchyma cells associated with midvein (in some leaves, cells of this area develop into fibers with strikingly thick walls); 9, sclerenchyma cells just below upper epidermis and just above lower epidermis in midvein region; 10, branched vein enclosed in bundle sheath. Section cut 1.5 μm thick. (*Carolina Biological Supply Company.*)

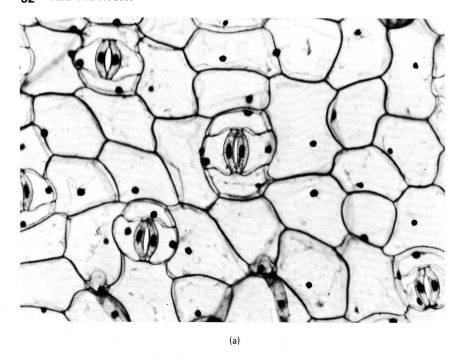

(a)

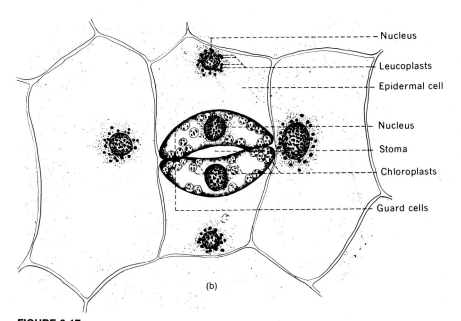

- - Nucleus

- - Leucoplasts

- Epidermal cell

- Nucleus

- Stoma

- Chloroplasts

- Guard cells

(b)

FIGURE 3-17
(a) Surface view of epidermis from a leaf of *Tradescantia*. Guard cells are bordered by subsidiary cells. (*Carolina Biological Supply Company.*) (b) Surface view of a stoma from a leaf of *Tradescantia*. (*From J. B. Hill, H. W. Popp, and A. R. Grove, Botany, 4th ed., McGraw-Hill, New York, 1967.*)

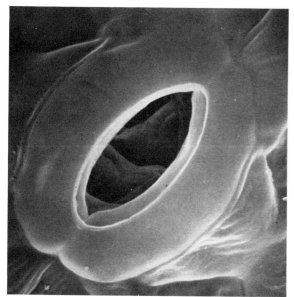

FIGURE 3-18

A stoma in the epidermis of a cucumber leaf, characteristic of stomata in dicotyledons because there are two clearly defined guard cells surrounding the pore. Note the mesophyll cells as seen through the stoma. The mesophyll cells in the leaf are important as the site from which water vapor evaporates and carbon dioxide goes into solution. These are spongy mesophyll cells because the stoma is on the lower side of the leaf. (*From J. Troughton and L. A. Donaldson, Probing Plant Structure, McGraw-Hill, New York, 1972.*)

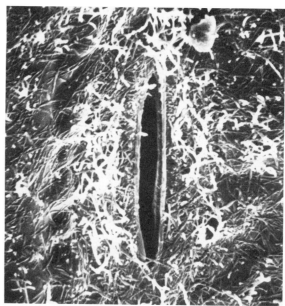

FIGURE 3-19

A stoma on the lower surface of a wheat leaf (cultivar Raven), a monocot. Wax on this leaf surface appears as long rodlets lying on the cuticle. The open stoma in wheat appears as a long slit, in contrast to the stomatal pore in cucumber. (*From J. Troughton and L. A. Donaldson, Probing Plant Structure, McGraw-Hill, New York, 1972.*)

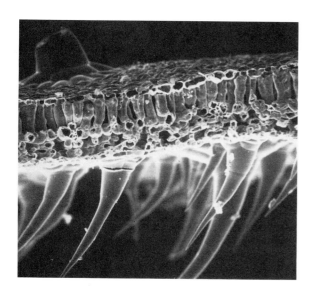

FIGURE 3-20

Transverse view of a leaf of cucumber, showing the large hairs (trichomes) that arise from both the upper and lower epidermal cells. A stoma (*s*) with substomatal cavity can be seen in the upper epidermis. The transport of water into and through the leaf occurs in the xylem vessels and subsequently via the cell walls. The xylem vessels are located beside the phloem tissue and collectively these two types of transport systems are termed the vascular system (*v*). (*From J. Troughton and L. A. Donaldson, Probing Plant Structure, McGraw-Hill, New York, 1972.*)

do not have chloroplasts (except for certain cells to be described next).

Small openings, *stomata* occur in the epidermis; they may be situated in both the upper and lower epidermis or only one surface. Each stoma is bounded by two chloroplast-containing *guard cells* (Fig. 3-17). The presence or absence of *turgor*, or water pressure, controls the opening and closing of stomata. When a pair of guard cells is swollen with water, turgor pressure causes the stoma to open. A decrease in turgor causes a reduction in size of the stomatal opening. The structure of the stomata in dicots (Fig. 3-18) is different from that in monocots (Fig. 3-19). Through the stomata, carbon dioxide, oxygen, and water vapor are exchanged with the atmosphere. The rate of gas exchange and water loss is affected by the number and area of stomata per unit of surface area as well as by such morphological features as the presence of *trichomes* (leaf hairs) (Fig. 3-20), the thickness of the cuticle, and the position of the stomata. Numerous environmental factors also play an important role in the activity of stomata, which are sensitive to light, carbon dioxide level, and water availability both in the leaf and in the surrounding atmosphere.

Between the upper and lower epidermal layers of the leaf lies the *mesophyll*, a tissue consisting of palisade cells and spongy parenchyma cells (Figs. 3-15, 3-16). *Palisade* mesophyll cells are elongated parenchymatous cells oriented perpendicular to the upper surface of the leaf. They contain many chloroplasts, and much of the photosynthetic activity of the leaf occurs in them. *Spongy* mesophyll cells, present between the palisade cells and lower epidermis, are loosely arranged, thin-walled, irregular cells of various sizes. Intercellular air spaces occur among all mesophyll cells but are especially prominent in the spongy parenchyma. Normally these spaces are filled with carbon dioxide, oxygen, water vapor, and other atmospheric gases. Chloroplasts are generally sparser in spongy cells, but photosynthesis occurs in them. Small vascular bundles (veins) occur through the mesophyll of the leaf. They are composed of xylem and phloem and extend from the midrib or main veins in the patterns of venation discussed in Chap. 2. In some plants they are encased in a *bundle sheath* composed of chloroplast-containing parenchyma cells. Water and nutrients in solution diffuse from the small veins to the mesophyll cells. Photosynthates are also transported in the veins.

CHAPTER FOUR

PLANT GROWTH AND DEVELOPMENT

COORDINATION AND SEQUENCE

Imagine what life must be like for a *unicellular* organism, its whole being wrapped up in a single cell (Fig. 4-1). Life is simple; it has to be for such a general-purpose cell. As a *generalized* cell, it is limited in its ability to respond to a changing environment. Its survival as an individual depends on being able to regulate *intracellular* processes and perhaps develop a relatively lifeless "resting" form which can tolerate unfavorable conditions temporarily.

On the other hand, higher plants are *multicellular* organisms (Fig. 4-2). Their different parts are *specialized* to provide different essential functions. This specialization makes the higher plants effective in acquiring energy and surviving environmental extremes. Although they have many advantages over single-celled organisms, they must solve a problem unicellular organisms do not face—*intercellular* coordination. Since the trillions of cells in a multicellular organism must all act in concert, a high degree of intercellular cooperation is required, not just between adjacent cells but between cells sometimes as far apart as the goals on a football field. The cells in the leaves of the tallest giant redwood are exposed to desiccation every day and must be continually resupplied with water or they will dry out and die. Water can flow through xylem cells in the stem and trunk to the leaves; but it must first be obtained by xylem cells in the roots as much as 125 m below the thirsty leaf cells. Such coordination works in both directions; the root cells benefit when they receive the energy-rich products of photosynthesis from the leaves (via phloem cells).

Cooperation between specialized cells within the plant results in the survival of all. Increased specialization of cells provides an opportunity for wider adaptation to various environmental situations. Unicellular organisms and other simple plants are generally aquatic or highly dependent upon readily available water, but vascular plants, with trillions of cells, can grow on land and some even thrive in arid deserts. A cactus has a distinct advantage over a unicellular alga in the Mohave Desert as long as its cell function is coordinated. This chapter discusses how that coordination is achieved.

Intercellular coordination in any mature plant is crucial, but the control of a plant's development from its earliest single-celled stage is equally important. The growth and development of zygotes into trillion-celled individuals capable of producing other zygotes is a straightforward if complex process. Plants develop in a highly predictable and tightly regulated sequence. For example, flowers normally do not form until a plant has gone through a well-defined period of preliminary development and has attained sufficient size to produce fruit and seed successfully.

The sequence of events in a plant's development (Fig. 4-3) is marked by a logical progression of stages, each arising from the previous one and giving rise

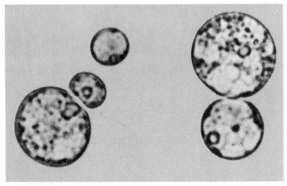

FIGURE 4-1

A unicellular organism. This single-celled alga is able to carry out photosynthesis, obtain nutrients from its environment, temporarily store materials, enter a drying-resistant stage, and act as a gamete for sexual reproduction. Intracellular coordination is a key to the organism's survival. (*From P. B. Weisz and M. S. Fuller. The Science of Botany. McGraw-Hill, New York, 1962.*)

FIGURE 4-2

A multicellular organism. Specialized groups of cells in the leaves carry out photosynthesis. Certain cells in the roots are specialized for absorption of water and nutrients. Others provide for movement of photosynthate, water, and nutrients throughout the plant body. Still other cells in the flowers are specialized for producing gametes, attracting pollinators, or protecting young embryos. All these trillions of cells function as a coordinated whole. Intercellular coordination is a key to the organism's survival.

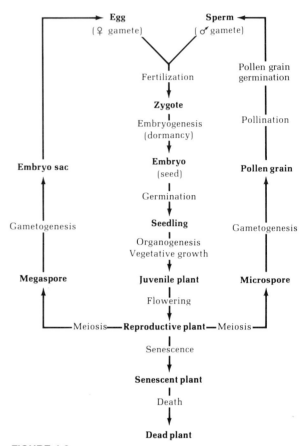

FIGURE 4-3

A generalized life cycle of flowering plants. Each morphological stage (boldface words) proceeds to the next via key physiological processes. Some but by no means all plants undergo senescence, an orderly process resulting in plant death.

to the succeeding stage in a steady march toward reproductive maturity and repetition of the *life cycle.* The eventual development of flowers or other reproductive structures assures the production and fusion of compatible *gametes* (egg and sperm). The resulting *embryo,* contained in a seed in higher plants, may be dormant for a time, but eventually it resumes activity to produce a young *seedling.* The seedling produces more roots, stems, and leaves and may develop into a rather large plant. Regardless of its size, however, the plant is considered a *juvenile* until it reaches sexual maturity. Juvenility ends with the appearance of flowers in *flowering plants* or other sexual structures in the nonangiosperms. The *ma-*

ture, or reproductive, plant produces flowers from which the next generation of the cycle begins.

In order, gametes, zygotes, embryos, seedlings, juvenile plants, and reproductive plants are stages in plant life cycles. Between all successive stages occurs an important transition. For example, the transition from the dormant embryo of a seed to the seedling is called *germination*. It is marked by many important changes in the anatomy, morphology, physiology, and biochemistry of the plant. Exactly how this transitional process is controlled is not fully known, but it is known to be highly regulated, as are the other developmental processes that cause a plant to pass from one developmental stage to the next. *Vegetative growth* produces new roots, stems, and leaves in a predictable way. *Flowering*, which marks the transition from juvenility to maturity, is a well-coordinated and regulated event. In this chapter on how developmental coordination is achieved in plants we focus especially on that category of plants we call crops, concentrating on the practical significance of their innate growth and development processes.

PHYTOHORMONES

Essentially every aspect of plant growth is under the direct or indirect control of plant hormones. These *phytohormones* are naturally occurring compounds that act in minute amounts, regulating the growth and development of plants. Hormones may regulate in either a positive (promotive) or negative (inhibitory) way. For example, certain hormones cause dormancy in seeds or buds while others bring on germination or bud growth.

Several characteristics make the study of phytohormones particularly intriguing. For example, natural plant growth regulators have not one but several functions in a plant. Auxin, a hormone that helps control stem growth, also helps regulate fruit development and activate the vascular cambium in the spring. The same hormone also can help coordinate processes as diverse as root initiation, sex determination in flowers, and phototropism (bending of stems and leaves toward the light). Gibberellin, another hormone found in higher plants, can also be involved in controlling stem elongation, fruit development, root initiation, cambium activation, and sex determination. Some species require both auxin and

gibberellin for a particular response, which is often determined by the ratio of one hormone to the other, e.g., male flowers when auxin predominates and female flowers when gibberellin does.

The multiple, overlapping, and interacting functions of hormones provide a challenging area of study, in which the multiplicity of responses and roles has created diversity of opinion among investigators. Much of what we now know about plant hormones was unknown 5 years ago, and some of what we now know contradicts what we believed 5 years ago. This is not a criticism; it is the nature of science continually to uncover new information and reevaluate old theories. The life of ideas in plant science is rather short but no shorter than in any other area of active research. This section will focus on those ideas which have withstood the ravages of time or scientific scrutiny and which are of the greatest practical importance in agriculture.

Phytohormones are commonly grouped into five categories. In two of them are put regulators with similar chemical compositions but sometimes different biological properties. Although each of the other three categories is represented essentially by only one naturally occurring substance, for convenience and clarity we shall treat the hormones in the traditional categories—auxin, gibberellins, cytokinins, ethylene, and abscisic acid (Fig. 4-4).

Auxin

This category is represented in most plants by a single naturally occurring substance, *indoleacetic acid* (IAA). In the early years of plant science, auxin was thought to be the master hormone and simply called the *growth substance*. Auxin is certainly involved in a number of dramatic and vital growth responses, but it does not act alone. Even in growth processes for which it is most responsible, other hormones appear to play a supporting role.

The natural auxin IAA was identified in the 1920s, but evidence for its existence came from work dating back to Darwin. A series of discoveries led (sometimes obliquely) from Darwin's initial observations in 1880 to the most recent findings on the molecular basis for the auxin response. The historical record is interesting and, at points, humorous. For example, one of the first dependable sources of auxin was human urine. (Your body probably produces more IAA than the average plant, but its role in human beings is unclear.)

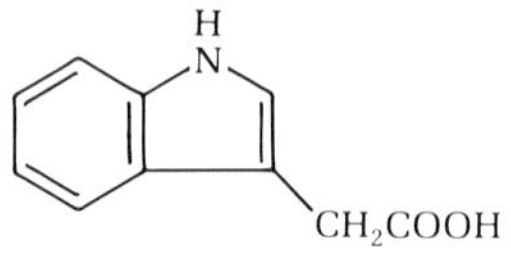

Indoleacetic acid (auxin)

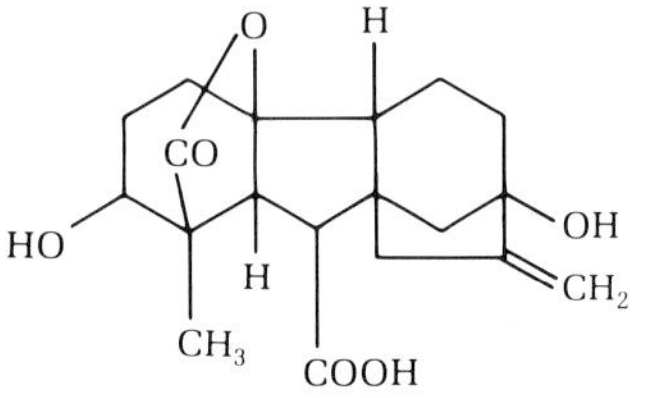

Gibberellic acid 3, a gibberellin

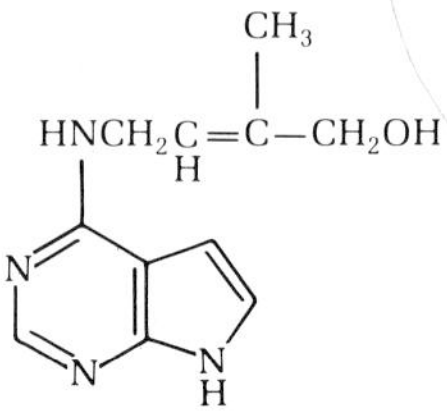

Zeatin, a cytokinin

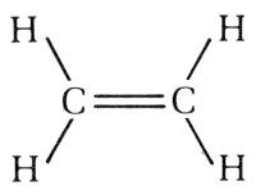

Ethylene

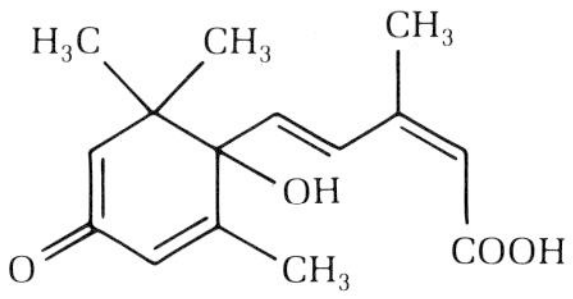

Abscisic acid

FIGURE 4-4
Structural formulas of representatives from each of the five different phytohormone categories.

Auxin as IAA is now believed to occur in all higher plants and is known to be present in many algae and fungi as well. In most if not all of these species, growth is triggered within a few minutes of the introduction of IAA. How a chemical as simple as IAA can almost instantly stimulate cell growth is not known with certainty, but evidence is accumulating. A currently favored idea is that auxin triggers the secretion of protons (H^+) from the protoplast, lowering pH in the cell wall. Acidification of the cell wall softens it, allowing it to stretch under the tension exerted by turgor (Fig. 4-5). Although the theory is scientifically attractive and supported by considerable evidence, not all plant scientists are convinced. Even researchers who favor the theory suggest that auxin must be acting in other ways as well.

Functions Auxin promotes the *elongation growth* of stems and roots and is involved in the *enlargement* of many fruits and tubers. In such cases, the cell walls in the tissues stretch in one or more directions, and auxin appears to trigger the expansion. In the process of cell growth and maturation, new materials must be added to the cell walls; auxin appears to be involved here as well.

Auxin promotes *cell division* in some tissues, especially in the vascular cambium. The *reactivation of the cambium* in spring is apparently triggered by an auxin signal that moves down the stem from the swelling buds. Auxin is believed to stimulate the cambium's production of secondary xylem cells, while another hormone (gibberellin) may cause secondary phloem cells to be produced. In some situations, auxin causes previously formed parenchymatous pith and cortex cells to differentiate into sclerenchymatous xylem cells. This process can be especially important in *wound healing* and in forming a vascular union between grafted plant parts.

Root formation on plant cuttings and on intact plant parts is at least partially under the control of auxin. IAA or IAA-like compounds can be applied to cuttings to stimulate root initiation. Paradoxically, the concentrations of auxin that are applied to promote root formation are often sufficient to retard subsequent root elongation. Such concentrations do not appear to develop under normal conditions, however.

Auxin acts as a natural and normal *inhibitor* of growth and development in areas other than the root. Some stems are kept from growing at all, and others are caused to develop quite differently from most branches. The formation of modified stems, e.g., short fruiting spurs in apples and tillers in small grains is partially under the control of auxin.

In addition to enforcing the growth of some stems, auxin also inhibits the *abscission* of leaves and fruits; strange as it may seem, abscission is a process that must constantly be overcome or prevented. The leaf or fruit must send a continuous auxin signal to the base of its petiole or pedicel to prevent formation of an abscission layer, a zone of easily broken cells. If the auxin supply to the petiole or pedicel is interrupted even briefly, the abscission layer forms. Auxin production or its movement to the zone where the abscission layer would form may be halted by disease, injury, or natural developmental processes.

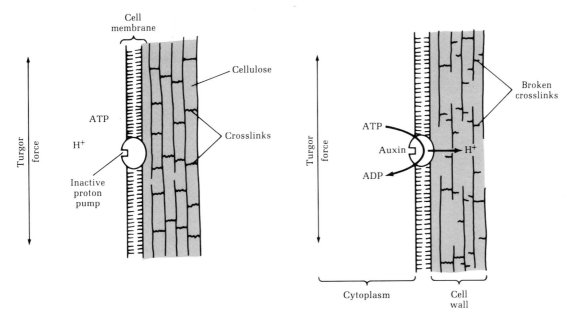

Without auxin, the proton pump is inactive and the cell wall is kept from stretching because of interlocking crosslinks between cellulose molecules.

Auxin activates the proton pump, making the cell wall more acidic. This breaks crosslinks and allows the wall to stretch under the influence of turgor.

FIGURE 4-5
The acid-secretion theory for auxin-promoted cell enlargement. Auxin (IAA) presumably associates with a receptor on the plasma membrane and activates a "proton pump," acidifying the cell wall. Auxin must be present to set the whole sequence in motion.

Leaves drop off in autumn partially because they cease to make auxin. Apples and some other fruits may drop from the plant as a result of auxin activity (actually auxin inactivity). When auxin production in the fruits declines and an abscission layer is formed, the pedicels of the fruits detach. Knowledge of this relationship has been put to practical use in the fruit industry.

Use in Agriculture Artificial applications of auxin are useful in several agricultural situations. Rooting of cuttings, prevention of premature fruit drop, and even the promotion of fruit drop, or thinning, can be achieved with auxin treatments. The chemical used in such cases is seldom IAA, however. IAA is more difficult to produce and frequently less effective than other chemicals resembling the natural auxin. The synthetic auxins include some familiar names: 2,-4-dichlorophenoxyacetic acid (2,4-D) and 2,4,5-

trichlorophenoxyacetic acid (2,4,5-T). These and other chemical *analogs*† of auxin mimic the action of IAA (Fig. 4-6). They are often more effective than IAA because unlike the natural auxin, which is quickly degraded both inside and outside the plant, the synthetics may remain active for some time. Synthetic auxins are included in a larger group of chemicals known as *plant growth regulators* (PGRs), i.e., phytohormones and synthetic substances that behave like phytohormones in that very small amounts are active in regulating or modifying plant growth and development.

When the PGR auxins 2,4-D and 2,4,5-T are applied in small doses, they can act to stimulate growth or to delay fruit drop; in larger doses, they are herbicidal. Many herbicides in use today have synthetic

---

†Chemical analogs are compounds with similar chemical structures and often similar activity.

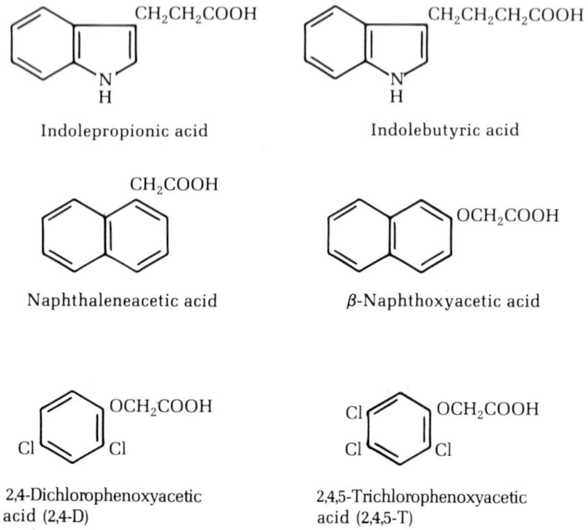

CH$_2$CH$_2$COOH

Indolepropionic acid

CH$_2$CH$_2$CH$_2$COOH

Indolebutyric acid

CH$_2$COOH

Naphthaleneacetic acid

OCH$_2$COOH

β-Naphthoxyacetic acid

OCH$_2$COOH

Cl

Cl

2,4-Dichlorophenoxyacetic
acid (2,4-D)

Cl

Cl

OCH$_2$COOH

Cl

2,4,5-Trichlorophenoxyacetic
acid (2,4,5-T)

FIGURE 4-6
Some auxin analogs with plant growth regulator activity. The
most common use of these compounds is as herbicides.

auxins as the major active ingredient. Broadleaf weed killers are commonly auxin-based; since young broadleaf (dicot) plants have a greater leaf area than monocots, they receive more of the herbicide, but the *selectivity* of the auxinical herbicides also stems from metabolic differences in plants. For example, grass species can metabolize 2,4-D and block its herbicidal activity.

That synthetic auxins would have a herbicidal effect could perhaps have been predicted from a knowledge of the activity of native IAA. Early workers noted that when it was concentrated and reapplied to roots and other growing tissues, it sometimes inhibited growth almost completely. There is as yet no entirely satisfactory explanation for the inhibitory action of IAA or the synthetic auxins. Undesirable tumorlike growths caused by excess auxin sometimes disrupt functions and lead to a plant's demise, but the nature of other lethal effects is unknown.

Gibberellins

The gibberellins have been investigated intensively for the same length of time as auxin. In fact, gibberellin's ability to promote growth was demonstrated in 1926, a year before auxin was discovered. However, all the early work on the gibberellins, which was carried out in Japan and reported only in Japa-

nese, did not come to the attention of western scientists until after World War II.

Gibberellins were discovered by a young Japanese researcher concerned with a disease of rice called *bakanae*, which means "foolish seedling." Plants affected with the disease grew excessively tall and usually fell over and died. The fungus that caused the disease, *Gibberella fujikuroi*, was shown to produce a substance that triggered the unwanted growth. When the substance was identified several years later, it was named *gibberellic acid* (GA). At least 50 naturally produced chemicals, all slightly different from the originally described GA, are now known to be formed by various plants including fungi, algae, ferns, conifers, and angiosperms. Together they compose the category of phytohormones called the gibberellins.

Functions The list of growth processes in which gibberellins are involved is at least as long as that for auxin; in fact, almost the same list could be used. Gibberellins affect stem and leaf growth, fruit development, flowering, cell division, dormancy, senescence, germination, and more. Almost every process in plant growth and development seems to be directly or indirectly modified by gibberellins (and auxins) in one species or another. Despite the functional similarity between auxin and gibberellins, the two are very different in chemical structure (Fig. 4-4), point of production, method of movement, and mode of action. The next few paragraphs review a few of the gibberellin responses of agricultural significance.

Organ enlargement is one of the major effects of gibberellins. Stems, fruits, and leaves of many species expand under the influence of these hormones. The gibberellins cause growth by triggering *cell expansion* of previously produced cells, stimulating *cell division*, or both. Applications of gibberellin can produce the same, or even greater, responses as those naturally present in a plant. (On the other hand, auxin applications to whole plants often cause inhibitory responses—perhaps because of excessive amounts when combined with the naturally produced portion.) For example, GA is used commercially to promote enlargement of Thompson Seedless grapes; at the same time, it causes the stem portions of the grape clusters to enlarge so that the larger grapes do not crush each other as they grow. Applied to sugarcane, GA promotes the growth of the stems and increases yields.

Gibberellins can stimulate *flowering* in some species. Many carrot cultivars flower naturally only after an extended cold period, but if GA is administered without the cold, the plants will flower. After a cold period, GA is produced naturally in such species; artificial application of the hormone apparently does away with the low-temperature requirement.

The *germination* of many seeds is partially or even largely controlled by gibberellin. The gibberellins produced by the embryo move into the storage area of the seed and "turn on" the enzymes that make the stored starches, proteins, and oils available to the embryonic plant. Gibberellin's role in the germination of the cereal grains has been extensively studied. In this case the stored material is largely starch, and once the starch-digesting enzymes have been activated by GA, they solubilize the sugars needed by the embryo. The *malting* of barley, an early step in brewing, is a consequence of gibberellin-triggered enzyme activity. The grains are germinated for a day or so, causing the starches to be broken down into simple sugars. Germination is stopped by heating the malt, and the sugars can then be fermented by adding yeast.

Gibberellins have not been used on a large scale agronomically. In fact, aside from herbicides, few PGRs have ever found use on the large acreages usually associated with field crops, partly because such chemicals are so expensive and partly because of the unreliability of hormones or hormonelike PGRs, which often evoke many other responses besides the desired one. An "antigibberellin" PGR known as chlormequat is sometimes applied to wheat and other small grains to reduce stem length.

Cytokinins

The group of hormones called cytokinins were so named because of their ability to promote cell division (cytokinesis). In early stages of research, cytokinins were assumed to be *the* hormonal regulator of all cell division in the plant. We now know that auxin and gibberellins also affect cell multiplication and that cytokinins are involved in other processes as well.

Cytokinins were discovered in the mid-1950s while plant scientists were looking for compounds that would stimulate the growth of *tissue cultures*, cells or tissue fragments removed from various organs and maintained apart from the donor plant. By manipulating the medium in which such tissues live, plant scientists can cause them to divide and produce new plants. It was discovered that cytokinins added to tissue cultures help stimulate cell division and differentiation.

The original source of purified cytokinin was almost as strange as the first source of purified auxin; in 1954 a substance that sharply promoted cell division in tissue cultures was isolated from autoclaved† herring sperm DNA. Other early sources of cytokinins were tomato juice, coconut milk, and malt extract. We now know that some of the early cytokinins were accidentally created in the isolation procedures: herring sperm DNA that has not been autoclaved contains no native cytokinins. Many naturally occurring, chemically similar cytokinins have been isolated and identified over the past few years, and new ones still occasionally come to light (Fig. 4-7).

Functions Besides promoting mitosis and cell division in tissue cultures or plant meristems, cytokinins are involved in a number of other crucial growth and developmental processes. Cell expansion in leaves, stems, and roots of some species is controlled partially by cytokinins. They are also involved in the formation of buds on stems, on tissue cultures, and on leaf cuttings. African violets, for example, form buds more quickly when cytokinins are applied to the leaves being propagated.

In the control of many growth processes the interaction between cytokinins and auxin is subtle but very important. The formation of roots and shoots on tissue cultures depends upon the *balance* between cytokinin and auxin in the medium. When their ratio shifts in favor of the cytokinins, shoots are formed; in the reverse situation, roots are formed. In many species the relationship of auxin to cytokinin is also critical in the control of branching. Cytokinins promote growth of axillary buds, while auxin suppresses it.

Cytokinins promote germination of some seeds. These hormones also can act as a "fountain of youth" in preventing senescence of leaves, keeping them

---

†An autoclave is an airtight sterilizer used in laboratories and industrial processes.

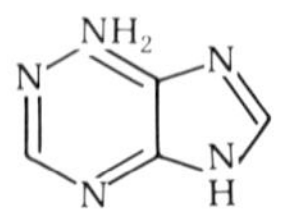

Adenine

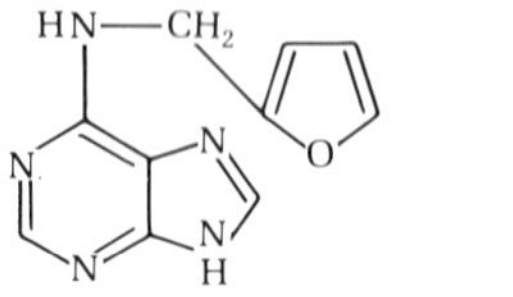

Kinetin

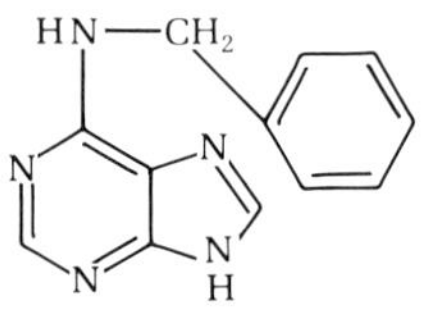

Benzyladenine

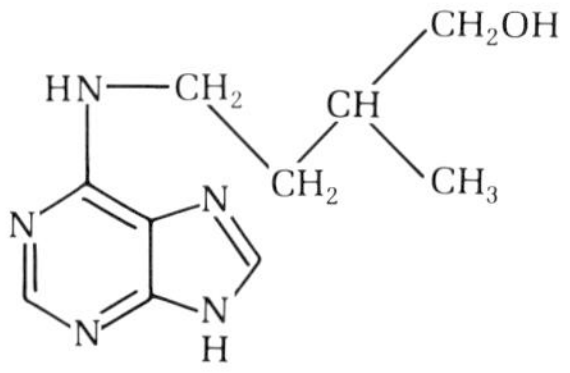

Dihydrozeatin

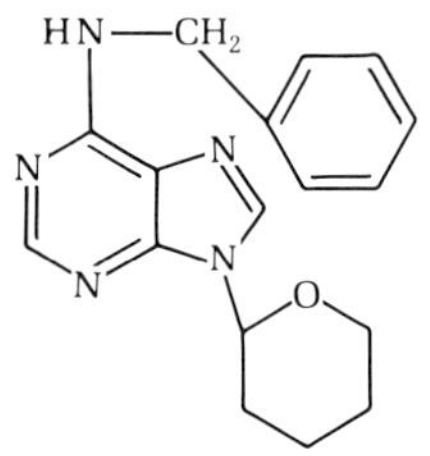

Tetrahydropyranylbenzyladenine

FIGURE 4-7
Some cytokinins. Adenine and dihydrozeatin are naturally occurring; the others are synthetic analogs.

green and active after other leaves on the plant have yellowed or died. Several other effects of cytokinin may be related to those already mentioned. For example, the increased respiration associated with cytokinin applications may be the result of increased premitotic activity. A tissue-protective effect provided by the hormone to plants infected with certain diseases may be related to its effects on cell division and senescence.

Although there are many synthetic cytokinins, few of them are in use on a commercial basis in agriculture. Cytokinin PGRs are used regularly in tissue culture to promote cell division and shoot differentiation, and modified plant-tissue-culture techniques are used commercially to produce large numbers of certain plants. Aside from this, cytokinin-like PGRs have occasionally been used as sprays to improve fruit shape and size of apples and pears.

Ethylene

Ethylene is a peculiar hormone; its most unusual property is that it is a gas. Because of its peculiarities, it did not gain acceptance as a phytohormone until long after many of its growth-regulatory properties had been demonstrated. When the original studies on ethylene were made by adding it to plants, researchers assumed that the effects were entirely unnatural. Experiments showed that ethylene in very low concentrations can make potato tubers sprout, fruits ripen, leaves abscise, and pineapples flower. In the mid-1930s it was demonstrated that ethylene is produced naturally by many plants, but the time was not ripe for its acceptance as a phytohormone, since western plant physiologists then believed that auxin was the chief (perhaps only) plant hormone. Not until the late 1960s did most plant scientists list ethylene among the phytohormones.

Ethylene is a gas and a very simple molecule—2 carbons and 4 hydrogens. How could something so simple regulate growth, especially when it is a gas? Many animal hormones are relatively complex chemicals that are synthesized in the endocrine glands and circulated to other parts of the body, where they exert their major influence. Auxins, gibberellins, and cytokinins resemble animal hormones, even though the exact sites of their synthesis and the methods of their movement are not fully known. On the other hand, a simple gaseous substance does not seem to be in the same category.

In spite of its unlikely nature, ethylene is a potent and important regulator of many developmental processes. Fruit ripening, seed germination, petal and leaf abscission, wound healing, inhibition of stem and bud growth, promotion of stem and fruit growth, radial enlargement of stems and roots, flowering, and rooting—in at least some species all these have been shown to be under the natural control of ethylene.

Ethylene could formerly be applied artificially only in enclosures of some type, but development of ethylene-generating chemicals that can be applied to plants in spray form has initiated some large-scale agricultural uses. Although commercial applications at present are limited to controlling pineapple flowering, the height of small grains, and fruit ripening, other uses are likely in the near future, especially as we learn more about how ethylene acts in the plant.

Abscisic Acid

It is generally agreed that abscisic acid is the only substance in this final category of phytohormones, although some similar, naturally occurring compounds may be involved in controlling growth. Abscisic acid (ABA) is primarily thought of as an *inhibitor* of growth processes; nevertheless it has positive, promotive functions in some cases. In some species, dormancy of buds and seeds and perhaps the cambium is under at least partial control of ABA. It can act as an antagonist to many of the promotive functions of auxin, gibberellins, and cytokinins.

Abscisic acid got its name from its association with the abscission of leaves and fruits, although the response may not be widespread. The hormone is also involved in the development of *hardiness* (resistance to cold) in some plants. It is perhaps a universal regulator of stomatal closure in drought-stressed plants.

Generally less is known about ABA than about the other hormones, and its role in plant development is not yet clearly understood. It was identified in the late 1960s, and we are still finding new, natural functions for it in the plant.

VEGETATIVE GROWTH AND DEVELOPMENT

Beginning with germination, the seedling embarks on a venture that may lead to maturity—whether as an apple-bearing tree, a soybean-producing plant, or a sweet potato-yielding vine. This journey can be slowed by any number of causes, but as long as a plant is alive, it characteristically continues to increase in size, complexity, and mass—a phenomenon called *growth*. The growth between germination and the development of sexual structures (flowers, for most crops) is often called *vegetative growth* as distinguished from *reproductive growth*, in which flowers and fruit are produced.

Several crops are important only for their vegetative parts (roots, stems, or leaves) and are harvested before they reach reproductive maturity. Cabbage, sugarbeet, lettuce, and carrot normally grow only vegetatively until harvested. Some other crops have

weak flowering capacities under typical growing conditions. Potato and sweet potato may or may not produce flowers, but only their vegetative portions are harvested. With some crops, the grower blocks or removes reproductive growth to accentuate vegetative growth. Tobacco producers often go to considerable trouble to remove flowers to produce more leaves of better quality. By stressing vegetative growth and avoiding reproductive growth in sugarcane, yields of sugar have been dramatically increased. Again, for hay and pasture plants the primary concern is with the vegetative parts, especially leaves.

For most crops, however, the seeds or fruits are the harvested portion, and reproductive growth is essential. Production of coconut, cacao, peanut, and soybean relies on the plant's ability to flower and produce a fruit. Even such "vegetables" as tomato, cucumber, and squash are botanically fruits and thus dependent on flowering for their production. In these plants and many more, sexual reproduction of the species results in the formation of a useful commodity. Although we usually grow and tend these plants to accentuate their reproductive potential, the production of flowers and fruits cannot be promoted to the exclusion of vegetative growth. Reproductive growth clearly depends on the vegetatively produced roots, stems, and leaves to support the later development of flowers and fruits. Much of what the grower does is intended to favor vegetative growth so that subsequent reproductive growth will be improved.

The following discussion of various stages of plant growth and development will stress critical processes and transitions that occur along the way (Fig. 4-3). When the hormones controlling the processes and transitions are known, we shall comment on them. Emphasis throughout will be on coordination and regulation of developmental processes of special importance in the life cycles of the angiosperms, or flowering plants, economically the most important plant group. These processes are: (1) development and expansion of vegetative organs, (2) production of flowers, (3) development of fruits and seeds, and (4) senescence, a crucial final phase in the development of many plants or plant parts.

Increase in Size

Any increase in size or weight of a plant must ultimately be due to changes in the cells. Cell division, cell expansion, or both must occur for the plant body to become larger and heavier. Development in a root (Chap. 2) demonstrates how both cellular division and expansion contribute to growth. At the growing point, or apical meristem, of a root, the region just behind and surrounded by the root cap (Fig. 2-3), is a group of cells whose function is to divide and form new cells. As new cells are formed, the cells formed earlier lie farther and farther from the meristem and the root becomes longer.

A second and more dramatic cause of root growth occurs behind the root meristem, in the region of elongation, where the small cells produced by meristematic divisions begin to expand. Most of the cells expand primarily in the dimension parallel to the long root axis. Since they elongate to many times their original length while expanding only slightly in diameter, the result is a tremendous extension that drives the root tip through the soil.

Apical meristem cell division and subsequent cell expansion both contribute to the primary growth of the root, which makes the root longer. Later, secondary growth increases the diameter of the root as new cells are formed in concentric layers within the tissue. The characteristic growth rings produced by this process are most obvious in woody plants that live for several years in climates with definite growing seasons. Secondary growth is the result of cell divisions of the cambium and expansion of the new cells.

Stems grow in length and diameter by a method similar to that in roots: cells are continually formed in an apical meristem and then elongate below to produce primary growth. Subsequent cell division by the cambium and expansion of the new cells result in thicker stems. Leaves grow in a more complex fashion: meristems produce new cells that expand in two dimensions to create the flat leaf blade.

Hormones regulate cell division during growth of roots, stems, and leaves. The cytokinins are believed to be involved in promoting the division of cells in apical meristems at root and stem tips and in leaf meristems. Some plant scientists think that cytokinins are produced in the roots and then moved to meristems throughout the plant, but what controls the production and movement of these crucial chemical messengers is largely unknown. A crucial role in the regulation of cambium activity is played by auxin, which promotes cell division to the inside of

the cambium, producing secondary xylem. Cambium activity is also partly under the control of gibberellin, which causes production of secondary phloem to the outside of the cambium.

Auxin triggers cell lengthening in stems and roots of many species. (The cells stretch in only one dimension because of the unique arrangement of cellulosic fibers in their cell walls.) Auxins also appear to promote the synthesis of new cell wall materials in conjunction with the stretching. The cell wall thickens as the secondary cell wall layer is laid down after the cell has reached its full size. Gibberellins often act in concert with auxin to promote elongation of roots and stems, but the nature of the interaction is not as clear as auxin's solitary role.

Tropisms

Imagine the result if auxin, which promotes elongation of cells in roots and stems, were not distributed equally across the region of elongation. What would happen if there were more auxin on one side than on the other? If you predict that cells on the side with more auxin would grow faster and that the tip of the stem or root would bend away from the faster-growing side, you are at least half right. The situation is not hypothetical; it is the basis for some crucial growth responses.

Tropisms are the directional, or oriented, growth responses of plant organs resulting from a directional stimulus or signal. When a plant is illuminated from one side, the stem tends to grow toward the light, a *positive phototropism* (Fig. 4-8). Auxin is involved in phototropistic growth responses of many species. For reasons that are unclear, the auxin concentration increases on the shaded side of stems illuminated from one side. The shaded side of the stem then elongates faster than the lighted side, thus bending the stem tip toward the light.

Geotropisms are growth responses to the pull of gravity. Stems are generally *negatively* geotropic and grow *away* from the gravitational force. A plant placed on its side will begin to turn up at the stem tip by bending upward in the region of elongation. This occurs because, as you can guess, auxin becomes more concentrated on the lower side of the stem and causes it to grow faster than the upper side.

Roots generally exhibit a *positive geotropism*, growing downward, *toward* the gravitational pull of the earth. Geotropic responses in roots result from a

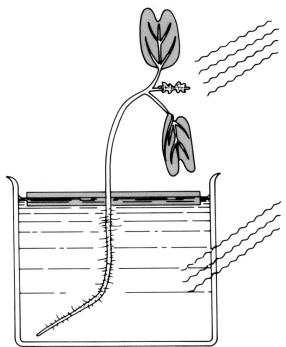

FIGURE 4-8
Phototropism. This plant is orienting its leaves toward a light coming from the right (positive phototropism) and its root away from the light (negative phototropism). (*From E. W. Sinnott and K. S. Wilson. Botany. McGraw-Hill, New York, 1955.*)

different set of events from that in stems but one no more clearly understood. It is known, however, that the root cap is involved in sensing the direction of gravity's pull. If the root cap is removed, roots grow horizontally or even turn upward. How the root cap senses gravity is still not proved to everyone's satisfaction, but everyone agrees that it does. When the root cap detects a gravitational pull, it sends to the region of elongation a signal that is translated into an unequal growth of the cells, the lower side growing more slowly. Plant physiologists disagree about what the signal from the root cap is and how it is translated into growth. Is auxin directly involved? Is growth of the lower side really slowed down, or is growth of the upper side accelerated? The answers to these and similar questions will probably vary with the species being considered.

The adaptive value of phototropisms and geotropisms is clear when one considers the alternatives.

Without some method for bending and orienting toward the sun, the leaves borne on a stem would not receive as much direct light. Positive geotropism keeps roots growing generally downward, where they can function properly in absorbing water and nutrients. Negative geotropism helps the germinating seedling to find its way out of the soil into the light and the shoot to keep properly oriented aboveground.

Dwarfing and Bolting

External influences, e.g., light and gravity, can cause stems to grow unequally; internal factors can have just as dramatic effects. The phenomenon of genetic dwarfing shows how crucial the internal control can be.

Crops that are normally tall, such as corn (up to 3 or 4 m) or peas (with vines up to 2 m or more), occasionally produce a plant that is only a few tenths of a meter high. The plants are normal in all obvious respects except for their very short stems. They have normal reproductive structures and often the same number and size of leaves as normal plants. When these plants are crossed with another dwarf, their offspring are usually also dwarfs. Because the phenomenon is genetically controlled, it is called *genetic dwarfing*.

Since many genetically dwarfed plants produce less gibberellin than normal plants, the obvious suspicion is that not enough gibberellin is present to cause adequate stem elongation. The hypothesis seems borne out by the fact that many genetic dwarfs grow to normal size when extra gibberellin is supplied.

Dwarfing is not necessarily undesirable. Corn plants generally do better if they are tall and their leaves are distributed along a longer stem, but many plants are grown more advantageously if they are dwarfed. Dwarf squash vines take up less room in the garden. Dwarf peas tend to have a bushlike growth instead of growing as a vine (Fig. 4-9). Some bush beans (dwarfs) are different from some pole beans (vines commonly grown on poles) only in that one genetic trait. The wild Chinese progenitors of our modern soybean were viney, an unsatisfactory growth habit for row crops, especially when combines are used for harvesting, but early in their domestication, the genetic dwarfing character was introduced.

A number of other crops are normally low-growing because of short stems. During vegetative growth, cabbage, radish, spinach, and sugarbeet produce all

FIGURE 4-9
Pea vines such as these have elongated, climbing stems.
Other cultivars of pea are genetically dwarfed.

normally do not bolt until they have experienced cold weather; after overwintering, such plants will bolt and produce flowers. The bolting appears advantageous to the developing flowers; perhaps elevating them makes them more conspicuous for bees or exposes them to air currents for wind pollination.

In many species bolting is largely under the hormonal control of gibberellins. At the time of natural bolting, gibberellin content increases in the plants, and adding gibberellins to rosetted plants will cause them to bolt prematurely. The effect of adding gibberellin is similar to that observed in dwarf corn or pea. The stems of rosetted plants and some genetically dwarfed plants are short because they do not

FIGURE 4-10
Bolting has occurred in this lettuce plant, producing an erect leafy stem from what was an essentially stemless rosette of leaves.

their leaves very close to the ground in a *rosette*, somewhat like the crowded petals in a rose bud. In cabbage and other plants, each succeeding leaf arises slightly higher on the stem, but the internode, or stem segment between leaves, is very short, often less than 1 or 2 mm.

Under certain conditions, the short internodes in lettuce, radish, sugarbeet, and other rosette plants are triggered to elongate and the plant grows into a tall, stemmy plant with leaves distributed all along its length and frequently flowers at the top. This *bolting* occurs naturally in many rosette plants when they begin to flower. Spring-planted radish, spinach, and lettuce often bolt and flower after hot weather arrives (Fig. 4-10). Sugarbeet, carrot, and cabbage receive enough gibberellin to stimulate internode elongation. In the one case, the plant appears genetically to have lost the ability to make enough of the hormone; in the other, the plant has retained its ability to make gibberellin but does not do so until some environmental condition triggers production of the hormone.

Organogenesis

The development of a plant from germination to full size involves more than just the division and expansion of cells from previously formed root, stem, and leaf meristems. The embryonic plant in a seed contains the rudimentary forms of a single root (the radicle), one stem (the hypocotyl and epicotyl), and a few leaves (the cotyledons and plumules); but the mature plant has many branched roots, a much-divided shoot, many leaves, and some new organs— the flowers. It is obvious that in growth and development the plant must form new *growing points* for additional roots, stems, and leaves. Making new vegetative parts is called *organogenesis*.

The Origin of Roots

Roots are the product of a root apical meristem (Chap. 2). Branches of roots develop when pericycle cells organize themselves into new root apical meristems. Branches of a root normally form some distance away from its tip.

Under some circumstances, cells in leaves and stems may also develop into root apical meristems and produce *adventitious roots*. Such roots form naturally on the leaves or stems of many plants when the aerial organs are brought in contact with a moist medium. We have learned to take advantage of this natural tendency and propagate many plants, especially horticultural species, by taking cuttings and allowing them to root. The formation of new root growing points on the leaves or stems appears to be largely under the control of auxin. At least, auxin clearly promotes the formation of adventitious roots when it is artificially supplied to cuttings.

In corn and certain other plants, one or more whorls of adventitious roots may form on the lower portion of the stem (Fig. 4-11). The roots grow outward (horizontally) for a few centimeters, in seeming defiance of their normal geotropic response. They then turn down and grow into the soil. Once underground, they become typical, normally functioning roots. In corn such roots are called *prop*, or

FIGURE 4-11
Prop, or brace, roots of corn. These adventitious roots emerge from the lower internodes and then grow down into the soil. Once there, they branch repeatedly and become a major source of nutrients, water, and physical support for the plant.

brace, roots. They can prevent tall stalks from falling over, especially in moist, loose soils or under windy conditions, and may even become the most important part of the root system for water and nutrient absorption.

The Origin of Leaves

New leaves normally form only at the growing tip of a stem. A leaf primordium, the meristem of a leaf, develops as a small bump on the side of the apical meristem. The cells in the bump divide rapidly to form an embryonic leaf. Before its final expansion, the leaf is only a few millimeters long and frequently still inside the bud at the stem tip. The rapid growth of leaves, which can take them from this almost microscopic form to their mature size in a few days, is largely the result of cell expansion. Most or all of the cells in a full-sized leaf are already present in the embryonic leaf. Auxin and in some cases cytokinins appear to be involved in this rapid expansion.

Growth Habit and Branching

The *growth habit* of a plant refers to its overall form or the pattern of stem development. Some stems are climbing *vines*; others lie *prostrate*; others are self-supporting and *erect*. The last type may be highly branched, with a shrublike appearance, or unbranched, producing one major stem without side shoots, or intermediate in the degree of branching.

...wth habit depends on several genetic and ...mental factors. In addition to factors deter-...g the strength of the stem (woodiness and ...ness), the elongation of internodes and the ...vity of axillary buds also help determine the final ...rm. The tendency of a lateral shoot to orient vertically or at some angle also varies with species. We have already discussed the elongation of stems and its effect on plant height, but stem length is only part of the picture; there is a great difference between a pine tree, an apple tree, and a pineapple plant even when they are all the same height.

The Origin of Branches

Roots form their branches from new meristems at positions well back from the root tip, but stems form potential branches right at the stem tip. At the base of each leaf primordium, develops a second bump, or mound, of actively dividing cells called the *bud primordium*. These cells become organized into an apical meristem that looks very much like the apical meristem from which they arose, one which can grow to produce more leaf and bud primordia. In other words, at every node there is a potential branch, whether in a highly branched species like boxwood shrubs or a single-stemmed, unbranching species like sunflower and corn.

In nonbranching species, axillary buds remain *dormant*; the buds develop to an embryonic stage and become inactive. The inactivity of these potential branches is partially under the control of auxin coming from the main growing point, or stem apex, but cytokinins from the roots may also be involved. If cytokinin content of the axillary buds is raised (naturally or artificially), or if the auxin concentration at the dormant buds is lowered (by disease, apex removal, or increasing stem length), the axillary buds may break dormancy and develop into branches. The control of axillary bud development (branching) by the stem apex is called *apical dominance*. As expected, the presence or absence of apical dominance dramatically affects the growth habit of a plant. A peculiar disease of some plants known as *witches-broom* causes dormant buds to grow and produce a dense cluster, or broom, of stems. The disease organism, a bacterium, is known to produce cytokinin, and artificial application of cytokinins can produce identical brooms in noninfected plants.

Apical dominance is exhibited strongly in many grasses. Although the aboveground stems of such plants as corn, wheat, rice, and sugarcane are rarely branched, careful inspection of the node at the base of each leaf reveals dormant axillary buds, which under different circumstances can develop into branches. They remain dormant in part because of the influence of the main apical meristem.

The *tillers* of many grass species result from belowground branching (Fig. 4-12). The first several internodes of seedlings of grasses like corn, wheat, rice, rye, barley, sorghum, and tall fescue elongate only a few millimeters. Therefore, the first several nodes, with their associated axillary buds, remain below the soil surface. During early development, the apex of the main stem may not entirely suppress the axillary buds at these lower nodes, and the buds can grow out as branches (tillers) from underground. These branches orient horizontally for a time before turning and growing upward. Tillers are important in grain crops because each new tiller can potentially produce a head of grain. The shortened lower internodes of tillers also bear underground axillary buds which can give rise to even more tillers and so on. Although the control of tillering is not clearly understood, it is evident that the plant can modify tillering in response to its environment. Wheat or rice plants growing in a thinly planted, well-fertilized stand produce more tillers than crowded or nutrient-deficient plants. This compensation effect often results in equal numbers of seed heads and total grain being produced from widely different planting rates.

FIGURE 4-12
Tillers of wheat, a member of the grass family. This cluster of stems was produced from a single original stem that branched (tillered) repeatedly at the soil level.

Corn is unlike other grain crops in its tillering, and there are differences between the various types of corn. The most common type, dent, or field, corn, has been selected over the years for its low tillering capacity. Because modern dent corn seldom tillers, adjusting seeding rates or plant populations is much more critical for corn producers than for growers of wheat, rice, or other small grains. Each kernel of corn planted usually produces only one stalk. Fields with widely spaced plants and high fertility may produce additional stems by tillering, but these suckers, as they are commonly called, rarely produce a large ear and are considered undesirable.

Some important crops have a natural but undesirable tendency to branch. The branches, or suckers, of tobacco pose a major problem for growers because they reduce the quantity and quality of marketable leaf. Since tobacco breeders have not been able to develop successful nonbranching types, the grower must either remove the suckers physically or suppress them chemically. Removal by hand is costly; although good suppression of axillary bud development can be achieved with chemical sprays, some of the most effective chemicals for sucker control are residual contaminants in the leaf and have come into disfavor, especially in Europe, a major market.

Tomato is another crop where excessive branching is common. The home gardener who pinches out the suckers will usually be rewarded with larger tomatoes and a more manageable vine, but physical removal of unwanted branches on a large scale—even a single hectare—is impractical.

Branch Orientation

Different orientation of the stem can obviously change the growth habit of a plant, completely. Many stems grow erect, negative geotropism and positive phototropism keeping them more or less vertical. Many other plants produce stems that seem almost oblivious to gravity and light; the stems may grow horizontally or lie on or under the surface with no tendency to grow upward, although the leaves are commonly thrust upward. The stems of many such plants are strong enough to be erect; and in fact the branches may be woody. Some prostrate shrubs and weeping trees are mutants of more normal-looking plants. The mutations apparently affect the ability of the stem tip to respond to normal tropistic stimuli. Because of their unusual and often useful forms, many of these plants are valuable ornamentals.

Horizontal stems are the norm in many species, many or all of whose stems are prostrate or even subterranean. In the examples of stolons and rhizomes discussed in Chap. 2 and in many other cases, horizontal stems play an important part in determining the growth habit of the plant and contribute to the natural asexual reproduction of the species.

REPRODUCTIVE GROWTH AND DEVELOPMENT

Flowering

A plant that is growing only by the expansion or addition of roots, stems, and leaves is juvenile. Since *flowering* marks the transition from juvenility to reproductive maturity in angiosperms, it is a tremendously important process in the plant's development. From the standpoint of the plant, flowering offers the opportunity to make seeds, providing for dispersal of the species and survival during seasons unsuitable for growth. From an agricultural standpoint, flowering is the process leading to the production of fruits and seeds, the harvested material of most crops. Rice, soybean, wheat, apple, tomato, vanilla, and cotton are among the many crops whose harvested portion results from flowering and reproductive development. The floricultural industry, of course, is built around this unique feature of angiosperms.

Four basically different *patterns of floral production* are observed among our important agronomic, horticultural, and forest species:

Annuals—These plants bloom and die naturally by the end of the first season of their growth. Their life cycle is completed within the year. Annual crops include soybean, corn, and pea.

Winter Annuals—These plants live for part of two seasons and flower after they have overwintered. They are usually planted in late summer or early fall. Crops commonly grown as winter annuals incude wheat, rye, barley, and crimson clover.

Biennials—These plants usually live for most or all of two growing seasons. In the first, the plants remain vegetative but may become relatively large. After overwintering, the biennial produces flowers and dies during the second growing season. Economically important biennials include carrot, sugarbeet, cabbage, and some sweet clovers. Of course, carrot, cabbage, and sugarbeet are not left to overwinter and flower unless they are being grown for seed.

Perennials—These plants live and bloom year after year. Production of flowers and fruits does not lead to the death of the plant, as it does in the other three types. All forest crops are perennials, as are many forage, fruit, and foliage plants.

You know enough about biology by now to realize that such categories as the above are not perfectly inclusive or exclusive. Exceptions and misfits are common. Grain sorghum and tomato are technically perennials and do not die automatically after producing fruit, but since they are killed by even a mild freeze, they behave like annuals in temperate areas. Soybean and rice, on the other hand, are true annuals, dying naturally before growing conditions turn bad. Some perennial plants produce annual or biennial shoots. Grown in a moderate climate, potato, alfalfa, bermudagrass, and chrysanthemum will come back each year from perennial underground parts. The canes or shoots of raspberry and other brambles are biennial; their rootstocks are perennial. The individual canes of most cultivars remain vegetative until they overwinter, after which they produce fruit and die.

The Timing of Flowering

Regardless of the pattern of flowering (annual, biennial, or perennial), its timing is closely regulated. Not only do flowers of particular species appear at the same time each year, all flowers of that species bloom almost simultaneously over a rather extended area. Crocuses appear about the same time all over town, then come the daffodils, followed by the tulips. In the fields, rye normally produces a visible grain head all over the county within a few days; then comes barley, usually followed by wheat. Soybeans of the same cultivar bloom almost simultaneously from one field to the next, even when the fields were planted several days apart. Somehow, most species "know" when to flower, and every member of the species does it more or less together.

At least one more easily observed fact provides evidence that flowering, especially in its timing, is neither left to chance nor automatic. With most perennials, young plants are juvenile and do not produce flowers. Only after attaining a particular size do apple and avocado begin to bloom. Likewise, the winter annuals and biennials do not flower during the first growing season. Even among annuals, seedlings do not normally begin to develop reproductively until they attain a certain size. Corn and tomato plants do not start making flowers as soon as they germinate but normally go through a distinct phase of vegetative growth before flowers appear. Flowering is therefore something that can be "turned on" and the "turning on" is closely regulated. Some method for timing and synchronizing flower production is operating in many different species.

Before discussing how flowering is scheduled, consider from the plant's standpoint why such coordination would be desirable. Synchronization and seasonality of flowering have several positive effects:

1. Seasonal flowering can ensure the production of flowers when enough growing season remains to produce good seed. What good would it do an apple tree to flower in August if it takes 3 to 5 months to develop viable seeds after fertilization? That sort of flowering might be considered promiscuous or at least intemperate.

2. Synchronizing the flower production in all the members is critical in open-pollinating species. What would happen if corn, which depends on pollination by wind currents, produced its tassels and silks at widely different times or a few plants flowered much earlier or later than the rest? Obviously, there is great value in making and releasing the pollen en masse. Pines and oaks do much the same thing. Tremendous volumes of pollen are released throughout the forest in just a few days, and the female receptive surfaces are ready at the same time. Ragweed pollen, a source of misery to many hay fever sufferers, spreads over the countryside in late summer. The pollen grains are produced in such numbers that many unfortunately find their way to human nostrils as well as ragweed pistils.

3. Another benefit is that a delay permits greater vegetative development before reproductive growth begins. The more vegetative growth (roots, leaves, and stems) a plant develops the more flowers, fruits, and seeds it can support during reproductive growth. If flowers appear too soon after germination, the plant may not be able to provide enough photosynthates to sustain additional vegetative or reproductive growth.

With good reasons for regulating when flowers appear, it remains to ask how the plants can synchronize and schedule flowering so precisely. No single explicit answer serves for all plants, but a general answer applies to many: plants take their cue for flowering from an environmental stimulus—some common, widespread, specific signal that recurs at the same time each year. Few natural phenomena meet these criteria. Rainfall, wind, barometric pressure, and specific temperatures are not generally unique or precise enough to act as developmental triggers (although rains come infrequently enough in deserts to act as an excellent stimulus for germination and flowering in that environment). The two environmental characteristics widespread and specific enough to be useful in synchronizing and scheduling flowering are day length, or photoperiod, and seasonality itself, especially the extended cold of winter.

Plants which respond to changes in day length are said to be *photoperiodic* or to exhibit *photoperiodism*. Photoperiodism controls developmental responses besides flowering, but flowering has been studied most extensively. How different day lengths affect flowering is discussed in Chap. 7.

The regulation of flowering by temperatures comes not from a single exposure to a particular temperature but usually from an extended period of cold. *Vernalization* is the developmental process by which flowering is triggered or hastened as a result of exposure to low temperatures for a few to several weeks. (High temperatures cause some plants, including lettuce, to bolt and flower.) Chapter 8 discusses the environmental aspects of flowering in more detail; here we want simply to bring out the logical value of a system like vernalization for providing a cue for flowering. Many perennial, biennial, and winter annual species use overwintering as a signal for initiating or accelerating flowering. As the warm weather of spring returns, the plants can move rapidly into the reproductive phase because the internal machinery for flowering has been activated.

The word "machinery" is used both to convey the complexity of the flowering processes and to cover our ignorance about what is happening. When flowering is turned on by photoperiod or vernalization, some sort of internal signal reaches the stems' growing points; some or all of them stop producing leaves and start producing flower primordia instead. The leaves receive the photoperiodic stimulus and produce some substance or condition transmitted to the growing points. The growing points themselves may be the receptive site for the vernalizing stimulus.

Plant physiologists have long speculated that some flowering hormone, or *florigen*, causes flowers to be formed by the apical meristems. Although experiments have demonstrated that some sort of transmissible flowering signal moves from leaves to apex, no natural and universal chemical that causes flowering has been discovered. That does not disprove the presence of florigen, but if it is there, it has eluded a lot of serious and dedicated seekers.

In some species, a known hormone does stimulate the flowering response. Gibberellins, auxin, and ethylene act as florigens in some plants. Artificially applied, the hormones trigger the floral development that would normally occur only after exposure to the proper photoperiod or vernalization. Flowering and subsequent fruiting of pineapple plants can be synchronized and hastened by applications of auxin or ethylene, a fact exploited for many years in commercial pineapple production. By inducing simultaneous flowering in plantings of large acreages, many fruits can be brought to maturity at one time, with great economic advantage.

In the categories of flowering plants (annuals, winter annuals, biennials, and perennials), flowering and later reproductive events appear to be under the immediate control of hormones or similar substances; however, it is largely environmental events that trigger the hormonal changes. By following the proper environmental cues the plant flowers when its chances for successfully producing good seed are best. We have learned how to alter the flowering pattern of some plants to economic advantage either by manipulating the environment or by administering chemicals that act as florigens, e.g., the flowering of pineapple. In the floricultural industry, flowering is often regulated by controlling the plant's environment to ensure bloom at the desired time. Easter lily, poinsettia, chrysanthemum, and many other greenhouse plants are forced into bloom at the time of peak seasonal demand by regulating temperatures or photoperiods or both.

The Origin of Flowers

Once it has received the signal to produce flowers, an apical meristem in a growing point or axillary bud stops producing leaves and axillary buds and begins making flowers and inflorescences (Fig. 4-13). From a morphological standpoint, flowers are highly modified shoots; sepals, petals, stamens, and pistils are highly modified leaves on a much-shortened

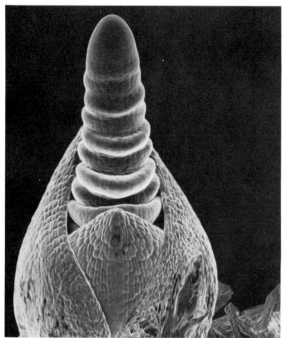
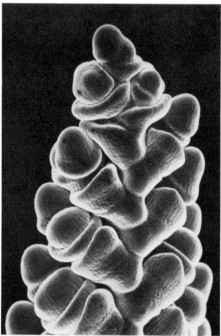

FIGURE 4-13
An apical meristem of wheat in (*left*) a vegetative and (*right*) reproductive stage. The rings and ridges on the vegetative meristem are leaf primordia. The grain head or spike with its florets is taking shape in the reproductive meristem. (*From J. Troughton and L. A. Donaldson. Probing Plant Structure. McGraw-Hill, New York, 1972.*)

stem. The apical meristems that become floral meristems continue to produce primordia, but they make sepal, petal, stamen, and pistil primordia, which in time develop into recognizable flowers.

Flowering Patterns

In some annuals and biennials, essentially all the potential flower-producing growing points convert to floral production at once; no more leaves and stems are produced once flowers start to form. These plants are said to be *determinate*, since flowering terminates all vegetative growth. Other annuals and all perennials retain some growing points for production of new leaves and shoots. Vegetative growth can proceed alongside reproductive. These are *indeterminate* plants. In indeterminate annuals and perennials, e.g., rose, the period of flowering may be spread out over the growing season, and fruits can mature on a continuous basis. (This is obviously not true of some perennials, e.g., peach, which restrict their flowering to a brief, specific period.) Plants with determinate flowering usually produce all their flowers more or less simultaneously, and the resulting fruits mature together. Plants that do not fit cleanly into either the determinate or indeterminate category are known as *semideterminate*. Cotton is only semideterminate at best, and that can pose a problem to the grower.

Flowering pattern can be very important in plants grown commercially especially if they are harvested mechanically. Most picking machines cannot distinguish between ripe and immature tomatoes and the machines destroy the vines as they harvest the fruit; furthermore, it is cheaper to harvest a field only once. Growers want to harvest all the fruit in a single pass, whether green beans, sweet corn, or canning tomatoes (Fig. 4-14). One time through means less wear on the machinery, less time, and less fuel. For

FIGURE 4-14
A commercial tomato harvester. The machine harvests the whole plant and then shakes off the fruit, which must then be sorted to remove immature tomatoes.

this reason plant breeders have worked to develop determinate cultivars of several horticultural crops.† The determinate cultivars are especially valuable for larger-scale commercial plantings, where harvesters destroy the plants; but nondestructive harvesters are available for some indeterminate crops that must be harvested repeatedly, e.g., pickling cucumbers.

Anthesis and Postanthesis

Anthesis, the opening of flowers or inflorescences, generally marks the stage when the flower is primed to carry out sexual reproduction. Once the flower has opened, pollination and fertilization become the crucial events. Unless the pollen is released when pistils are ready to receive it, all is lost. The lifespan of a pollen grain, once released, is usually short. Events occurring after a pollen grain successfully arrives on a receptive stigma are equally crucial. These processes are carefully regulated in the plant by hormonal and physical signals, which can be somewhat modified in response to a changing environment.

---

†Since most agronomic crops, on the other hand, are not harvested fresh, absolute determinacy is less important. All the plants will eventually die and dry down to a harvestable stage. Cotton is a sometimes difficult exception among agronomic crops.

The natural sequence of events in flower development includes the death and shedding of most floral parts; typically, only the ovary and the structures surrounding and supporting it during fruit growth are retained. After fertilization has occurred, the stamens and petals normally wither and abscise. Parts of the pistil that are no longer necessary, the stigma and the style, are often shed as well. This loss is in no sense accidental. The process is orderly and from the plant's standpoint desirable because once the petals and anthers have accomplished their purpose, they become a liability.

The natural *senescence* and *abscission* of floral parts after anthesis is under hormonal control. Naturally produced ethylene appears to play a major role in many plants, hastening the demise of the flowers. Whether it is produced in the flower or applied exogenously,† ethylene causes petals, stamens, and other floral parts to abscise within days or even hours. Ethylene produced by any biological tissue can cause senescence of flowers in its vicinity. For example, senescing flowers can hasten the senescence and abscission of younger flowers when the ethylene gas diffuses from one to the other, a phenomenon of great importance in the floricultural industry, where the florist or consumer wishes to keep flowers at peak condition for as long as possible.

Anyone who tries to keep flowers at full bloom is trying to make time stand still. It is natural and normal for petals to senesce, wilt, and then abscise. That is part of the *postanthesis* sequence of events in the overall process of reproduction, as compelling in plants as it is in animals. Although senescence cannot be stopped it can be slowed down. Lowering the temperature slows all biochemical processes, including those causing petals to senesce. Florists put cut flowers in large coolers and do their best to keep them away from ethylene, whether produced by older flowers or other biological and nonbiological sources. Good ventilation with free circulation of clean air is important in prolonging the life of flowers. (Corsages can usually be worn more than once if they are kept in the refrigerator before and between wearings.)

---

†*Exogenous* materials are those produced or supplied by an external source. This distinguishes them from *endogenous* or internally-produced substances.

FRUITING

Flowering is the economic endpoint of development for floricultural plants since the flowers are the marketable crop, but in many other crops flowering is just a means to the desired end of fruit and seed production. The harvested portions of tomato, soybean, corn, orange, rice, and grape, just to name a few, are either fruits or seeds or derived from them. Fruits and seeds develop after flowers have bloomed and their ovules have been fertilized. Ovules inside the ovary develop into seeds, and the ovary itself (or closely allied floral parts) develops into the fruit.

Fruit Set

Fruit set is the term used for what happens in a young ovary to prepare for its further development into a fruit with seeds. Fertilization does not automatically lead to fruit development since in many species fertilized pistils may spontaneously abort. The pistil on a young ear of corn or in a head of wheat, for example, may or may not develop into a kernel even if fertilization has occurred. Many apple flowers abort, ovary and all, after they have been pollinated, and other species may also thin themselves well into fruiting, especially during periods of drought.

It appears that the plant "decides" after fertilization whether it can support the development of the ovary into a fruit or not. The plant must expend a tremendous amount of its energy in growing fruits; if there are too many of them, there may not be enough photosynthate and other assimilates to permit any of the fruits to develop sufficiently to produce good seed. (Remember that from the plant's standpoint, the whole reason for flowering and fruiting is to make good seed, and thus there is a survival logic in retaining only as many ovaries as can be nurtured to maturity.)

Apparently, hormonal balances in the plant, as affected by various environmental conditions, help determine the level of fruit set. Auxin is active in causing ovaries of some species to set, but reduced levels of abscisic acid seem important in many cases as well. Auxin promotes fruit set by delaying abscission, while ABA may promote fruit abscission. Paradoxically, synthetic auxins have been used exogenously to reduce fruit set. These higher concentrations of the auxin-like PGRs cause young fruits of apples and some other fruit crops to abscise, probably

by stimulating ethylene production by the tree. This intentional *chemical thinning* of the tree produces fewer but larger apples.

Synthetic auxins are used to cause fruit thinning early in the season, but the same chemicals can be used later to delay undesirable preharvest fruit drop. Since the effect resembles the prevention of leaf abscission by auxin, it is likely that the mechanism is the same. An abscission layer can form in the pedicel or stemlike attachment of the fruit when the auxin supply is disrupted or ethylene or ABA reach critical levels. The dramatic increase in ABA levels during drought stress probably explains fruit abscission during such times.

Fruit Growth

The enlargement of the ovary into a fruit following pollination and fertilization is governed by hormones. In a number of species, auxins, gibberellins, and cytokinins—alone or in combination—may cause cells of the ovary to divide and expand. Often the hormones triggering ovary enlargement are produced by the developing seeds. If the ovules are not fertilized, the seeds do not develop and the fruit does not receive the hormonal stimulus for its growth. When only a few of the seeds develop, insufficient hormonal stimulation can cause apples and other pome fruits to be smaller or misshapen. The fleshy receptacles of strawberries are greatly reduced in size if fertilization of their pistils is limited; however, adding auxin to the receptacle will cause its expansion even if the pistils have been removed (Fig. 4-15). Evidence like this helped establish the important role auxin plays in some fruit development.

A number of plants have fruits that will develop even though their ovules are not fertilized, e.g., pineapples, some oranges, and bananas. Such seedless fruits are called *parthenocarpic* (from the Greek word for "virgin"). Parthenocarpic fruits expand under the influence of hormones—but obviously not those coming from developing seeds. If the hormone supplied endogenously is supplemented with an extra, exogenous dose, parthenocarpic development is often increased dramatically.

Although hormones are important triggers for fruit set and development, other factors are equally critical. The amount of photosynthesis being carried on in the plant determines how much photosynthate (sugar) will be available to fill the fruits. Hence, environmental conditions that reduce photosyn-

FIGURE 4-15
Auxin and expansion of the fruiting receptacle in strawberry. When the seedlike achenes on the strawberry are removed (center), the receptacle does not expand but when achenes are removed and auxin is supplied in their place (right), the receptacle expands into a fleshy fruit. (*After J. P. Nitsch. American Journal of Botany 37:212. 1950.*)

thetic output of the plant will limit fruit development. Cloudy days, drought, mineral deficiencies, defoliation by insects, and disease may all slow down photosynthesis and thereby slow down or limit fruit expansion. The result will be smaller fruits and grains (or fewer fruits and grains if the plant aborts some).

As a general rule, the sugars that fill fruits and seeds are produced during fruit development. The plant has only a limited capability to store photosynthates vegetatively for subsequent use in the reproductive period. As a result, a drought can be much more devastating to yield if it occurs during reproductive growth stages than if it occurs during vegetative growth. Sometimes the grower can intervene to ensure a moisture supply during the critical seed-fill period by carefully timing irrigation (if water for that purpose is limited) to keep photosynthesis going during the critical period.

Unlike photosynthate, the other major component of most fruits and seeds, the proteins, are largely synthesized before active fruit growth. The amino acids that are assembled into proteins in fruits and seeds are normally produced and stored in the plant during its vegetative period and moved into the fruit later. Some plant scientists speculate that annual plants, which die after fruiting, do so because their developing fruits drain such an enormous amount of amino acids from the rest of the plant.

SENESCENCE AND POSTHARVEST PHYSIOLOGY

Senescence

Many field crops turn yellow and die before growing conditions become unfavorable. Wheat, oats, barley, and rye reach peak development and then die in early summer. Soybean and corn plants usually mature and die before the first killing frost. The death of such crops is not attributable to disease, injury, drought, temperature extremes, or any other environmental stress: they die because they are programmed to do so. The maturation and death of annuals and biennials are predetermined developmental events. For such plants, death following fruit development is just as much a part of normal development as flowering (Fig. 4-3).

In senescence, the plant undergoes a series of *regulated* changes that ultimately lead to death. The changes are triggered by, or at least coincident with, fruit and seed production (Fig. 4-16). Among other things that happen during senescence, the metabolic machinery in the plant is gradually dismantled, and many of the vital components of the vegetative organs are transferred into the developing fruits. In fact the movement of nutrients and other vital components from leaf and stem to fruit has been suggested as the cause for senescence. According to this theory, maturing fruits and seeds sap the vegetative portion of

FIGURE 4-16
The effect of soybean fruits on soybean plant senescence. The plants on the right were allowed to develop normally, while those on the left had their flowers and pods removed continuously. The plants on the left will stay green until they are killed by frost; the others have already senesced and died.

the plant by diverting materials coming from the roots away from the leaves.

Senescence is not an accident; perennials make fruit and seed without senescing. There must be survival value for some species in programming themselves to senesce and die once they have produced fruit. We know that senescence is genetically programmed because some mutants of normally senescing species can make fruit and not senesce (Fig. 4-17). In sorghum, a species that does not typically senesce, plant breeders have been able to breed in a senescence-causing gene. Although we can only speculate on the adaptive advantages of a planned senescence that leads to the death of a seed-bearing plant, a number of different possibilities come to mind:

1. The next generation (the seed) will have space to grow.

2. The parental generation will return its nutrients to the soil for the next generation's growth.

3. The dispersion of seeds is sometimes accomplished if the parent plant senesces, e.g., tumbleweed.

4. A plant may be able to produce more seeds if the seeds could draw off nutrients and assimilates otherwise retained in vegetative organs.

5. Senescence may hasten and coordinate the development of fruits and maturation of seeds before bad weather arrives.

FIGURE 4-17
Nonsenescing peas. This indeterminate plant has continued to produce new vegetative parts and flowers and set fruit for several weeks. It is a mutant of the normal pea, which produces fruits for only a few weeks at most before it senesces and dies. (Courtesy of *W. Proebsting.*)

Whatever the potential or actual value of senescence to a plant in the wild, the phenomenon is of tremendous value to us. We might still be tribal societies of nut and berry gatherers and hunters if plants did not senesce. The rise of cities depended on the principle of a division of labor, with certain people providing certain goods and services to the rest of the society. Growers of such major food crops as the bread grains and the legumes could never have raised them on a large scale if the crops did not senesce. Imagine trying to harvest 300 hectares† (or even 1 ha) of wheat or rice if the grains of individual plants reached ripeness over a period of several months and the plants did not dry down for threshing. Such conditions would have made it impossible for 3 percent of the population to provide food and fiber for the remainder. Because wheat plants senesce, a farmer in South Dakota can produce almost single-handedly enough grain each year to make a million or more loaves of bread. Because soybean plants senesce, they can be grown in fields stretching from horizon to horizon and yet be harvested in a few days by a handful of men and women and their machines.

Thus for annual crops, especially those harvested on a large scale as dry seeds, senescence is absolutely essential. Modern combines are designed to harvest and thresh only dry materials. Green, leafy matter clogs up the machinery and prevents good separation of the grain, seed, or fiber (Fig. 4-18). But harvestability is not the only value of senescence from the grower's standpoint. Senescence also ensures that the seeds will reach maturity before frost kills the crop. Soybean or corn fields hit by an early freeze, before maturity is reached, often yield crops that are drastically lower both in quantity and quality. If the plant is killed before it can finish the orderly business of senescence, the seeds or grains will not fill out completely and will often be of poor nutritional and keeping quality.

Maturity and Growing Season

While rice, wheat, soybean, and corn senesce and dry to a suitable harvest moisture as long as no calamity interrupts the process, the length of time to reach maturity can vary tremendously with different cultivars or hybrid lines. Some types of corn

take as little as 80 or 90 days to reach full maturity; others reach a harvestable stage only after 200 days or more. Many cultivars of rice require 150 days or more to mature, but newer lines that mature in about 100 days have been developed. The grower can choose a type of corn or rice that fits the local growing season. In Canada, *short-season* (also called *early-maturing* or simply *early*) lines of corn must be planted. In Georgia, *longer-season* (also called *later-maturing* or simply *late*) lines are planted to take advantage of the longer growing season.

As expected, plants that grow for a longer season normally have a higher yield than short-season cultivars. (Plants that grow longer vegetatively before flowering are typically bigger and able to support more reproductive growth, i.e., grain or seed.) Thus there is a temptation for growers to plant a cultivar that will not senesce and mature until just before the date of average first frost. If the season is cooler than average and growth is slowed, or if the first frost comes much earlier than usual, the crop can be seriously damaged or lost entirely.

Development of early-maturing rice cultivars was not spurred by a need to avoid frosts, since rice is generally grown in frost-free areas or areas with long growing seasons. The faster-maturing types offer one great advantage: in certain places with sufficient water to support rice production, three crops can be produced in one year. When only the longer-season types were available, no more than two crops could be grown per year. Three harvests of the earlier-maturing types can produce more grain than two harvests of the later-maturing ones.

Senescence of Plant Parts

Perennial plants, by definition, do not undergo senescence following seed production, or at least the whole plant does not senesce and die. However, many perennial plants have parts that are annual. For example, the leaves of temperate *deciduous* plants are borne for only one season. Toward the end of the growing season, the leaves senesce and usually abscise. These events appear to be much like those occurring in soybean and other annuals: the metabolic machinery in the leaf is disassembled, and many of the vital nutrients and components are moved out of the leaf to storage before it detaches. The peculiar chemistry associated with senescence of leaves also makes for vivid coloration in many perennials. The red, yellow, and orange hues evident

---

†The hectare (ha) is the SI unit of land area, equivalent to 2.471 acres.

FIGURE 4-18
Cotton defoliation. The leaves on cotton often do not senesce and abscise in a timely fashion for mechanical harvest. Chemical defoliants or desiccants are often used to kill the leaves and hasten opening of all the fruits (bolls). Before defoliants became available, multiple hand-picked passes had to be made to get opened bolls before they were spoiled by rain. (Courtesy of U.S. Department of Agriculture.)

in the autumn appear to be accidental, without any survival value for the species. The senescence and abscission process itself is desirable from the plant's standpoint. Leaves that cannot tolerate the cold or that would drain water from the plant uselessly are shed before winter.

Fruit Ripening and Senescence

The fruits of most perennial plants undergo a process similar to senescence. The later stages of fruit ripening are marked by changes in the fruit and seed that favor survival or dispersion of the new generation. In many species, the fruits dry, split open, and release the seeds. Many nut trees produce fruits that dry out during the later stages of development. In this dried state, the fruits or seeds can remain dormant but are still very much alive. Accessory parts of some fruits, e.g., the husks of walnuts and pecans, may eventually slough off and leave the nut which can survive extended periods of cold or dryness. The harvestable portion of many perennial crops, including nut trees and several spices, is a *dry fruit* or seed.

Most common perennial crops produce fruits that still contain a high percentage of water in their tissue at harvest maturity. Apples, blueberries, and tomatoes are examples of such *fleshy fruits*. They are 80 percent or more water and still very much alive when they naturally abscise or are picked. Abscission however, in no way ends the development of the fruit, since critical metabolic events can occur after the fruit has been shed. These processes appear to enhance the chances of seed dispersal and subsequent germination.

Some chemical changes in mature fleshy fruits may make them more edible. We normally wait for the changes during ripening of fruits because unripened fruits are typically hard, sour, or astringent. During ripening, the tissues soften, acidic chemicals decline, and starch is converted into soluble sugars. Although these ripening events are biochemically complex and varied, the outcome is similar in many species; sugar content increases or acid content declines (or both) while the tissues of the fruit begin to soften. The softening involves a breakdown of the middle lamella between cells. The pectin cementing the cell walls of adjacent cells together is dissolved by enzymes released during ripening. As this occurs, the tissues become softer and more edible. Green apples and unripe bananas are firm-textured, but ripening begins a process that can continue until the tissues almost dissolve into liquid.

The Climacteric

In many species ripening is greatly accelerated by a senescence-hastening process called the *climacteric*. It is marked by a final burst of ripening activity just before or some time after abscission. A sharp upturn in respiration signals the beginning of the end for the fruit (Fig. 4-19). The ripening processes, which had been proceeding at an even pace, now move quickly toward the death and dissolution of the fruit. The seeds do not die, however. In fact, the climacteric is probably the natural way of liberating seeds from the self-destructing fruit.

Ethylene, the phytohormone most involved in ripening and the climacteric, is produced endogenously by fruits entering the climacteric. Exogenous ethylene can cause preclimacteric fruits to enter the climacteric and ripen more quickly, a fact of great significance in commercial fruit production.

Banana and tomato, among other fruits, are commonly harvested in an unripened, preclimacteric state. (Avocado is always harvested in a preclimacteric condition, since its fruits will not ripen on the tree.) In this stage, the fruits are firm and can be handled with less danger of bruising. They can be stored much longer than fruits already starting to ripen when they are harvested. As preripe fruits, such crops can be held in refrigerated storage to reduce microbial spoilage and retard the onset of the climacteric. Bananas and tomatoes are often shipped to the grocer under refrigeration and then placed in a warmer area to permit natural ripening. The process can be hastened by exposing the fruits briefly to ethylene. Once the climacteric has been triggered by ethylene, it proceeds without further exogenous applications.

Plant breeders and physiologists have found cultivars of tomato that will not ripen. The mutant trait apparently derives from an inability to produce ethylene endogenously. The fruits remain green and firm unless they are exposed to exogenous ethylene. This characteristic is of great commercial interest, since grocers could hold such fruits for long periods without expensive refrigeration and then ripen them as needed.

The activity of ethylene can be as undesirable in some situations as it is essential in others. Most people prefer to eat apples that are crisp or firm. Crunchy, slightly acidic but sweet apples demand a better price than mushy or mealy ones. Since soft-

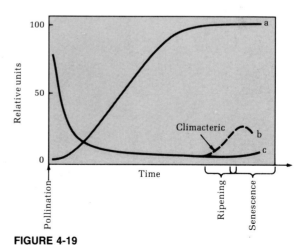

FIGURE 4-19
Growth (*a*) and respiration patterns typical of climacteric (*b*) and nonclimacteric (*c*) fruit. Ripening processes begin after the fruit reaches full size. Senescence follows quickly after the climacteric.

ening of the apple tissue is a natural consequence of the climacteric, apples are often picked as preclimacteric fruits, and great effort is expended to retard the climacteric. Storage under refrigeration and in *controlled atmospheres* helps. The cooler conditions slow down the metabolic processes leading to the climacteric, and the atmosphere is enriched with carbon dioxide and lowered in oxygen. With such treatment, apples can be kept crisp and essentially fresh-tasting for 12 months or more.

There is truth in the old saying that one rotten apple spoils the barrel. Bruises, cuts, or wounds can cause ethylene to be generated endogenously and start the climacteric rise in respiration. During the climacteric, ethylene is produced endogenously to escape and act exogenously. Thus one spoiled apple can cause a chain reaction, triggering the climacteric—and more ethylene production—in other fruits. Refrigerators and plastic bags, which tend to trap ethylene, are not good places to store apples if even only one of them is bad. Without resorting to controlled-atmosphere conditions, it is possible to improve apple storage life by circulating fresh air in the storage area as much as possible and by removing fruits as soon as they begin to soften. If plastic bags must be used for storage, they should be perforated. A corsage might also suffer from being kept in a refrigerator containing an ethylene-generating bad apple. On the other hand, storing a bromeliad in a plastic bag with a ripe apple will trigger its flowering.

The fleshy fruits of many crops are subject to climacteric (Table 4-1). Fruits of pear, tomato, and

TABLE 4-1

Some Common Climacteric and Nonclimacteric Fruits

Climacteric	Nonclimacteric
Apple	Bean
Avocado	Cherry
Banana	Grape
Cantaloupe	Grapefruit
Fig	Lemon
Peach	Lime
Pear	Orange
Tomato	Pineapple

banana are not "designed" to dry down at maturity the way corn, wheat, or rice do. Instead, the fleshy fruits ripen and self-destruct if left unharvested. The prime period for harvesting individual fruits of some of these crops may last only a day or two. The grower has to work quickly in such cases. Other fleshy fruits may be able to last without special care. Pumpkins and other winter squashes have no tendency to undergo climacteric. Stored in a cool environment dry enough to prevent fungal growth, the winter squashes can easily be kept over the winter. Nor do citrus fruits undergo a climacteric; they can be harvested over a 2-month period after reaching ripeness. Their deterioration in storage is usually associated with microbial spoilage, not senescence via the climacteric.

Chapter 5

THE CHEMISTRY OF PLANTS

Intelligent use of crops requires some knowledge of basic plant processes; the plant will grow only if light, nutrients, and water are supplied at appropriate times to the appropriate parts. In general, the more growers know about the internal processes of a plant, the better they can manage and use it. The result of such intelligent management is improved performance and greater yields.

CHARACTERISTICS OF BIOCHEMISTRY

The chemistry associated with life differs enough from test-tube chemistry to be given its own name, *biochemistry*. Nevertheless, biochemistry does not differ fundamentally from the chemistry of nonliving things. The same elements that occur in rocks and air are found in roses and aardvarks; and the same chemical principles that govern large-scale industrial production of plastics also govern the activity of chemicals in a corn leaf. Despite their basic similarities, however, some important differences distinguish typical test-tube chemical reactions from metabolic processes. Biochemical reactions are generally distinguished by:

1. The *speed* with which reactions occur at relatively low temperatures

2. The *precision* with which reactions occur

3. The *complexity* in size and number of reactants and products

4. The *regulation* apparent within and between many metabolic processes

5. The total dependence upon *organic* (carbon-containing) molecules

If one mixed together all the chemical components from a cell and heated the mixture gently to speed up the reaction (Fig. 5-1), the resulting combination would contain some products that occur in living cells but many that do not. Some essential molecules would no longer be present. Furthermore, every time the experiment is tried, heating the biochemical components from a cell will give a different mixture of products because the chemical events take place randomly *in vitro* (literally "in glass").

In vivo (living) chemistry, on the other hand, takes place with great speed, precision, complexity, and order at relatively low temperatures thanks to the participation of *enzymes* in biochemical events.

Enzymes

Enzymes are the facilitators, or *catalysts*, of metabolic reactions. They not only increase reaction rates but also determine the precision and specificity with which all the complex reactions necessary for life take place. An enzyme is a catalyst consisting of protein, a type of large molecule made up of different

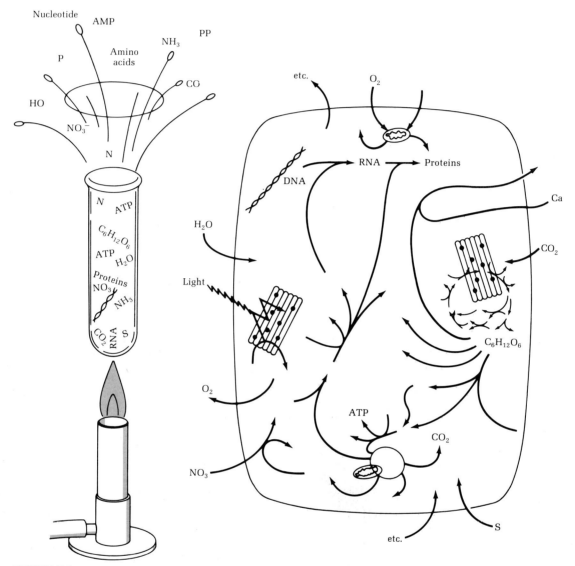

FIGURE 5-1

In vitro and in vivo chemistry: a simplified representation of the relative efficiency, effectiveness, and control inherent in the two types of systems. A test-tube mixture of biochemicals, when heated, produces rapid reactions; but the outcome is very unpredictable and destructive. In the cell, an orderly, interrelated progression of events (metabolism) takes place at rates sufficient to support life processes.

chemical units called *amino acids.* A few amino acids are shown in Fig. 5-2 and all 20 common ones in Fig. 5-27. The chemical and physical properties of amino acids allow them to (1) link together in a long chain, (2) form physical or chemical bonds between nonadjacent units in the chain, and (3)

interact in various ways with other chemicals in their vicinity. A particular group of active amino acids is brought together at some point in an enzyme molecule as a result of the way it folds (Fig. 5-3). Their peculiar physical and chemical properties enable this assemblage of amino acids to attract

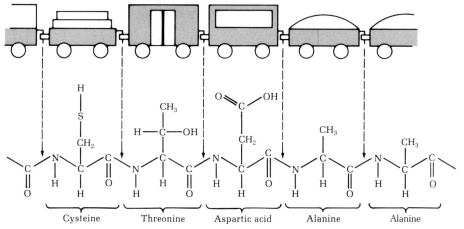

FIGURE 5-2
A portion of a protein chain showing a few of the amino acid subunits joined end to end. Each amino acid has different chemical and physical properties affecting the physical and chemical (catalytic) attributes of the resultant protein molecule. (*Adapted from A. Roller, Discovering the Basis of Life, McGraw-Hill, New York, 1974.*)

molecules of other chemicals. The *active site* attracts *substrates*, i.e., molecules with specific physical characteristics, which are then catalyzed in a specific fashion. The ability of enzymes to associate with specific substrates accounts for the specificity of the reactions they catalyze and in part for the complexity. Since usually only one chemical can enter the active site, only the reaction involving that chemical can

be catalyzed. Some enzymes bring two or more specific chemicals together and combine them into one product. More will be said about the specificity of enzymes in the next chapter.

The increased speed of reaction marking enzyme-catalyzed reactions is the consequence of bringing two or more potentially reactive molecules together. Many molecules will combine chemically (nonen-

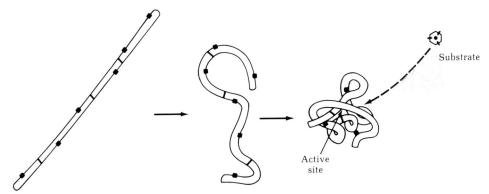

FIGURE 5-3
A protein folding into an active enzyme. Certain amino acids (*squares*) are critical in chain folding and others (*bars*) are involved in forming the active, or catalytic, site. In the active, properly folded form, the chemical to be catalyzed (the substrate) can fit into the active site just as a key fits into a lock. (*Adapted from A. Roller, Discovering the Basis of Life, McGraw-Hill, New York, 1974.*)

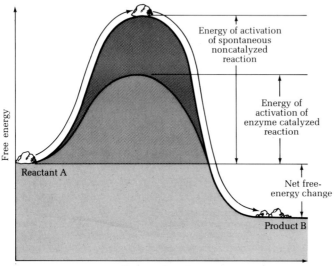

FIGURE 5-4

Enzyme catalysis and thermodynamics. Spontaneous reactions are those in which energy is released; the final products will contain less energy than the entering reactants, but before the reaction will proceed spontaneously, some energy must be put into the reactant. This energy of activation is analogous to the effort put into rolling a rock off of a cliff. Enzymes reduce the energy of activation, allowing more molecules to attain the necessary activation level more rapidly. (*Adapted from D.T. Plummer, An Introduction to Practical Biochemistry, McGraw-Hill, New York, 1978.*)

zymatically) whenever they happen to collide with sufficient energy and the proper molecular orientation. Enzymes facilitate reactions by increasing the chance that otherwise reactive molecules will come together in the critical way.

Enzymes may also hasten reaction rates by reducing the energy required to get molecules to react. All chemical reactions, whether biochemical or not, require bonds to be formed or broken between atoms. Even though energy may be given off as a result, some energy is usually necessary to initiate the bonding. This so-called *activation energy* often comes from the force of a collision between reacting substances.† In enzymatic reactions, the energy required to activate chemical bonding is lowered (Fig. 5-4), enabling it to take place at a temperature compatible with living matter.

---

†This is why heat is applied in many chemical reactions. Molecules move more rapidly at higher temperatures and therefore collide more frequently and more energetically.

Some Cellular Energy Crises

Many chemical reactions will not take place unless energy is added to the reactants. Heating a reaction mixture in a test tube can provide the energy for such *endergonic*, or energy-absorbing, reactions. Since enzyme molecules lack energy that can be transferred to substrates to cause reactions, they are able to catalyze only *exergonic*, or energy-releasing, reactions. By analogy, enzymes can push rocks off cliffs, but they cannot bring them from the bottom of the cliff to the top. All biochemical reactions are exergonic: the products contain less energy than the substrates. For this reason living systems could soon experience an energy crisis as each energy-releasing reaction brings the biochemicals to lower and lower energy states (the rocks will end up as rubble at the bottom of the cliff).

Living, growing systems, however, do not run down because, although enzymes cannot provide energy for endergonic reactions directly, they can transfer energy from other exergonic processes. Some of the energy released in an enzyme-catalyzed exergonic reaction may be *coupled* to an energy-absorbing enzyme-catalyzed reaction (Fig. 5-5). In this way enzymes transfer energy to reactants and catalyze energy-requiring processes, a crucial step in generating the molecules necessary for life.

The transfer of energy from exergonic to endergonic reactions, or coupling, involves an unavoidable loss of energy. The amount of energy released from substrate A (Fig. 5-5) is always greater than the amount of energy trapped or conserved in product Z, i.e., the energy transfer or coupling process is not 100 percent efficient; part of the energy released in energy-transferring biochemical reactions is lost, usually as heat, from the living system.

The inescapable loss of energy in the biochemistry of living systems raises the specter of another energy crisis. The organism must obtain enough energy to drive endergonic reactions, to provide the constant need for complex, energy-rich chemicals. In animals, this means finding and eating energy-rich materials (biochemicals) from plants or other animals (the familiar concept of a food chain). As the matter from living things is broken down and its energy used endergonically (coupled) to make new living matter, inevitably the new living matter will contain less energy than the food consumed. Unless energy is "pumped" into the biochemicals of living things, life is going to run down. (*All* the atomic and molecular "rocks" will be lying in rubble at the bottom of the energy cliff.) Fortunately, such an energy pump does exist. Green plants gather light energy from the sun and convert it into chemicals by *photosynthesis*. The plant then uses the chemical energy (exergonically) to generate (endergonically) new living matter. Using energy captured from sunlight, a green plant can push atoms and molecules up the energy hill again and combine them into complex biochemicals. All other life forms rely directly or indirectly on plants for the energy-rich matter required to generate their own particular type of molecular order. The bio-chemical activity of green plants winds the main-spring of life.

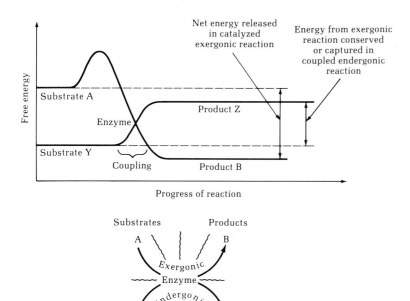

FIGURE 5-5
Coupling: an essential characteristic of some enzyme-catalyzed reactions. Energy released from an exergonic reaction may be partially absorbed by a simultaneous endergonic reaction. Each reaction takes place in the environment created by an enzyme that permits the transfer or capture of energy.

PHOTOSYNTHESIS

No biochemical process is more important to *all* living things than the reactions that occur in sunlit chloroplasts, where the energy of light is converted into chemical energy. In chemical form, the energy is available to other living things to power life itself by providing the energy needed to bring critical order to atoms and molecules.

Photosynthesis consists of two distinct processes. In one, light energy is captured, transformed into a chemical form, and used to create temporary energy-storing compounds. In the other, the energy from the temporary storage forms is used to make sugar ($C_6H_{12}O_6$) from CO_2. The overall equation for the two phases is

$$light + 6CO_2 + 6H_2O \longrightarrow C_6H_{12}O_6 + 6O_2 \quad (5\text{-}1)$$

Sugar, the photosynthetic end product, is a relatively long-lived energy-storing molecule, but it can be broken down, releasing energy exergonically, whenever needed, yielding energy for endergonic life processes.

The two phases of photosynthesis, energy gathering and sugar making, can be distinguished (1) by what is happening energetically and chemically, (2) by whether or not they depend on light, and (3) by their location in the chloroplast. The initial processes that capture and transform energy depend on illumination. These *light reactions* occur in chloroplast membranes. Since the second phase, the biochemical production of sugar from CO_2, does not require illumination, its stages are referred to as the *dark reactions*. They all take place in the stroma, the complex solution filling the chloroplast. The enzymes that catalyze the dark reactions exist in solution in the stroma, along with the chemical intermediates involved in sugar synthesis.

The Light Reactions

Chloroplast membranes contain several types of colored molecules called *pigments*. By far the most important and abundant are the chlorophylls, which make the chloroplasts and parts containing them look green. Chlorophylls *a* and *b* (Fig. 5-6) are the major light-capturing components in most photosynthetic organisms.

The energy in light occurs in packets called *photons*. Photons move at the speed of light and vibrate with a frequency that imparts a particular color or

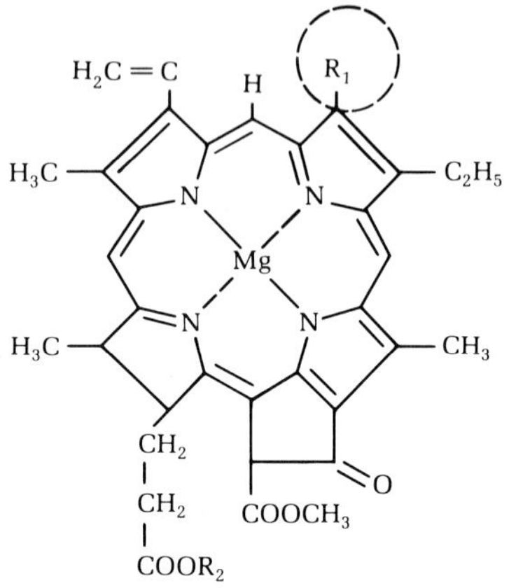

FIGURE 5-6
Chemical structure of the two major types of chlorophyll. In chlorophyll *a*, $R_1 = $ —CH_3; in chlorophyll *b*, $R_1 = $ —CHO. R_2 represents phytol ($C_{20}H_{39}$), a long-chain hydrocarbon.

wavelength. Photons of red or blue light are peculiar in being able to knock an electron away from the chlorophyll molecule when they strike it. The electron expelled by the force of the collision gains energy from the photon. The "excited" electron either returns to its original position in the chlorophyll molecule, releasing its extra energy as light and heat, or becomes involved in a reaction with other molecules (Fig. 5-7). It is this latter possibility, in which the chlorophyll's excited electron is passed to another molecule, that is productive photosynthetically. The first photosynthetic phase, the light reactions, begins with the excitation of a chlorophyll electron and includes the electron's progression through a whole series of reactions forming intermediate energy-rich compounds.

The path followed by electrons given up by chlorophyll molecules and boosted to a higher energy level is a tortuous route, both energetically and physically. The name *z scheme* has been given to the whole series of reactions taking place in the chloroplast membranes as the excited electrons move from one molecule to another. A 90° turn of Fig. 5-8 reveals the reason for the name. In this figure there are two separate excitation events, i.e., two points at which photons are involved in boosting energy levels of electrons. Those two sites, or *photosystems*, involve special chlorophyll molecules which are so oriented with respect to other molecules that transfer of the excited electrons can take place.

To follow the sequence of light-reaction events, it is best to start in the middle of the z scheme. Light striking a chlorophyll *a* molecule of photosystem I ejects an electron. This electron can be captured by a different molecule, which passes it on like a hot potato (which in an energetic sense it is) until it reaches nicotinamide adenine dinucleotide phosphate (NADP). This compound accepts pairs of high-energy electrons along with a proton (H^+), becoming NADPH, the chemically reduced form of NADP. Since NADPH can retain the high-energy electrons it receives from chlorophyll and give them up later, it is an *energy-rich intermediate*.

The chlorophyll *a* of photosystem I, in giving up an electron, becomes oxidized; but, because it has only one such electron to contribute to the z scheme, it must fill the vacancy, i.e., be chemically reduced by regaining an electron, before it can donate another excited electron. This electron comes from another photochemical event taking place in photosystem II (Fig. 5-8), where the slightly different molecule of chlorophyll *b* also loses an electron after being struck by a photon. The photon's energy is partially trans-

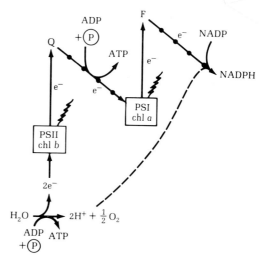

FIGURE 5-8

The z scheme of electron flow in the light reactions. Light strikes chlorophyll *a* (chl *a*) molecules in photosystem I (PS I), and the photoexcited e⁻ is passed to substance F, which in turn passes the e⁻ through additional intermediates to nicotinamide adenine dinucleotide phosphate (NADP). Reduction of NADP produces NADPH. Chlorophyll *b* in photosystem II (PS II) can also lose a photoexcited e⁻, which passes through various intermediates to the chl *a* of PS I. PS II regains electrons for subsequent photoexcitations from molecules of water (H_2O). The removal of e⁻ from H_2O produces two protons (H^+) and an atom of oxygen ($\frac{1}{2} O_2$).

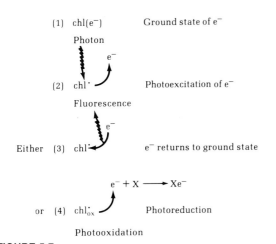

FIGURE 5-7

Photochemical events involving chlorophyll (chl). It has an electron (e⁻) which can be boosted to a higher energy state when struck by a photon of blue or red light (2). The photoexcited e⁻ can return to its original position in the chlorophyll molecule (3), releasing the energy as fluorescence (light). Alternatively, because of its high energy state, the e⁻ may react with some other molecule in its vicinity (4). The chlorophyll is *oxidized* when it loses the e⁻, and any molecule gaining the e⁻ is *reduced*.

ferred to the electron and can be passed along a series of reactions until it arrives at photosystem I. Along the way, part of the energy in the electrons is used to make another energy-rich intermediate, adenosine triphosphate (ATP). The electrons do not leave the z scheme, but part of their energy is passed to the ATP molecule. The net result of the photoexcitation of photosystem II by a photon is the production of ATP and the replacement of the electron lost earlier from photosystem I.

The electron replacements for chlorophyll *b* in photosystem II come from water. The chloroplast membrane at a site close to photosystem II contains an enzyme that can extract electrons from water molecules, pass them on to photosystem II, and use the light energy captured by photosystem II to generate more ATP. Removal of electrons from a water molecule splits it, producing two hydrogen ions (protons, H^+) and an atom of oxygen.

The net result of the light reactions is the production of ATP, NADPH (containing energy-rich electrons), and oxygen:

$$12H_2O + 12NADP + 18ADP + 18P + light$$
$$\longrightarrow 12NADPH + 6O_2 + 18ATP \qquad (5\text{-}2)$$

For the chloroplast, the oxygen is an unneeded byproduct of the splitting of water; it simply diffuses away and becomes the crucial part of the air we breathe. The major result of the z scheme is that light energy has been converted into chemical energy in the form of NADPH and ATP. These energy-rich intermediates produced in the chloroplast membranes are useful in the next phase of photosynthesis, the production of sugar in the stroma.

The Dark Reactions

The end product of photosynthesis is sometimes considered to be glucose (although in a system that is constantly turning over, the product of one reaction can immediately become the substrate for another and there really is no *end* product). Glucose contains 6 carbon atoms and much energy (686 kilocalories per gram molecular weight, or mole, written kcal mol^{-1}).† The energy, which is associated with electrons in the sugar molecules, derives ultimately from photons via the membrane-associated light reactions. In the dark reactions, the energy stored in NADPH and ATP is released exergonically and used endergonically to build carbon dioxide molecules into glucose.

The whole series constituting the dark reactions begins with adding CO_2 onto a 5-carbon compound. Besides making glucose, the reaction series regenerates more of the same 5-carbon substance. The resulting cycle of reactions is named for one of its discoverers, Nobel laureate Melvin Calvin. The *Calvin cycle* of dark reactions is shown in simplified form in Figure 5-9.

The 5-carbon compound to which CO_2 is added, or *fixed*, is a sugar called *ribulose*. Before CO_2 fixation, ribulose is combined with two phosphate groups. The resulting *ribulose bisphosphate* (RuBP)‡

---

†The calorie and kilocalorie are not SI units but are retained as being more familiar than their replacement, the joule (J).

‡The names *ribulose diphosphate* and *ribulose biphosphate* are also current in the literature.

can combine with CO_2 when catalyzed by the enzyme *RuBP-carboxylase*, one of the most important and most abundant enzymes in the world. The addition of CO_2 to RuBP results in an unstable compound that immediately splits up into two identical 3-carbon molecules of phosphoglyceric acid (PGA). The PGA can then be changed chemically into phosphoglyceraldehyde (PGAl) using the electrons of NADPH. PGAl is a simple sugar formed by the chemical reduction of PGA.

As equal numbers of RuBP and CO_2 molecules react with each other and the resultant PGA molecules are reduced, a pool of PGAl accumulates for further metabolism. Two PGAl molecules can be enzymatically combined to produce one 6-carbon sugar, the "end product" of photosynthesis. Other PGAl molecules enter a series of reactions leading to molecules with up to 7 carbons. As a result of various additions and divisions of these compounds, ribulose can be regenerated and more ATP is expended to create a pool of RuBP. The RuBP can combine with CO_2 in the presence of RuBP carboxylase and start through the Calvin cycle again. Since glucose has 6 carbon atoms, six CO_2 molecules must be fixed to six RuBP molecules before the net production of one molecule of glucose occurs. Most of the PGAl pool must be used to recycle carbons into RuBP. The net equation for the production of a glucose molecule in the dark reactions is

$$6CO_2 + 12NADPH + 18ATP \longrightarrow$$
$$C_6H_{12}O_6 + 6H_2O + 18ADP + 18P \qquad (5\text{-}3)$$

Adding Eqs. (5-2) and (5-3) algebraically gives Eq. (5-1). If you noticed that the hydrogens do not balance in Eqs. (5-2) and (5-3), the explanation is that since hydrogen ions (protons, H^+) are readily available in the cellular solution, we are ignoring them.

Photosynthetic Pathways

For most plants, photosynthesis occurs in the sequence we have just described: light reactions occur in the membranes while the Calvin cycle operates simultaneously in the stroma. The overall process (Fig. 5-10) is called the C_3 pathway, for reasons to be explained shortly. While the C_3 path is common and important agriculturally, there are other pathways and there is another biochemical process that interferes with the C_3 pathway significantly.

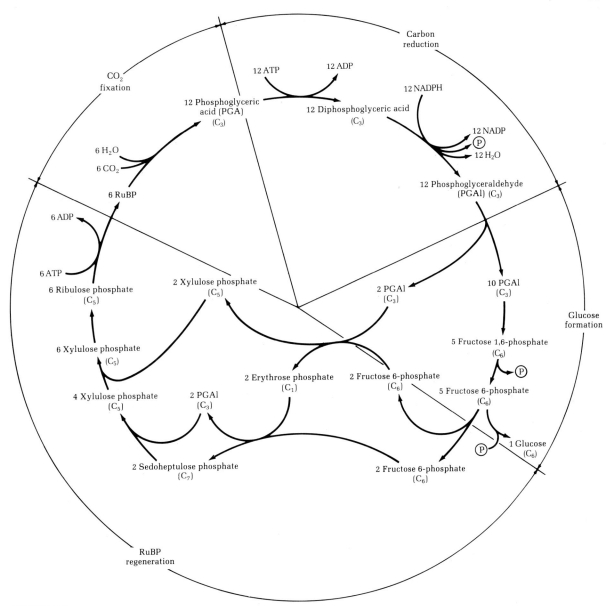

FIGURE 5-9
Net production of one glucose molecule by the Calvin cycle (dark reactions). Carbon dioxide
(CO_2) is fixed by the action of RuBP-carboxylase. The phosphoglyceric acid (PGA) formed
following CO_2 fixation is further metabolized and chemically reduced by NADPH to phosphoglyceraldehyde (PGAl). The pool of PGAl is used to regenerate RuBP and excess carbon can
be used to make simple sugars such as glucose.

Photorespiration

An efficiency-robbing process occurs in many plants as the result of a second reaction catalyzed by RuBP carboxylase. Instead of combining CO_2 and RuBP to produce two molecules of PGA in the Calvin cycle (Fig. 5-9), RuBP combines with O_2 (Fig. 5-11). (Because it can also act as an oxygenase, RuBP carboxylase is more properly called RuBP carboxylase/oxygenase or *Rubisco* for short.) The products of Rubisco's oxygenase activity are one molecule of PGA and the 2-carbon compound glycolic acid (or glycolate). The PGA generated by the addition of O_2 to RuBP can continue through the Calvin cycle, but glycolate cannot be metabolized by the chloroplast any further. Instead, it is secreted out of the chloroplast into the cytoplasm. Each time that O_2 instead of CO_2 reacts with RuBP, two carbon atoms are lost from the Calvin cycle and cannot be used in making glucose or more RuBP.

The two RuBP reactions, CO_2 and O_2 additions, are catalyzed by the same Rubisco and appear to compete with each other. If O_2 is abundant and CO_2 scarce, the oxygenation reaction will be favored. Since bright light and higher temperatures also favor

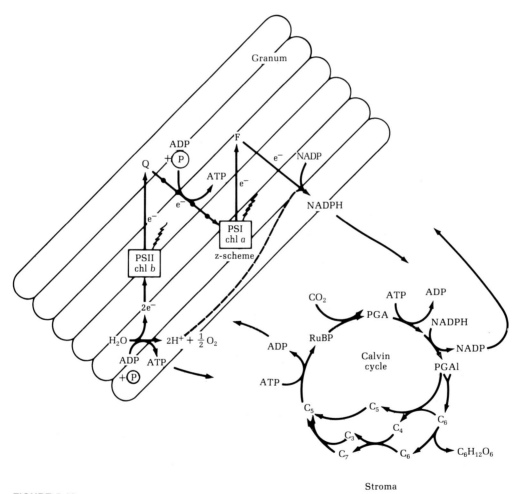

FIGURE 5-10

The C_3 photosynthetic pathway. Photosynthesis in C_3 plants involves the generation of ATP and NADPH in the light-driven reactions occurring in the grana membranes and the subsequent use of ATP and NADPH for the fixation and reduction of CO_2 in the Calvin cycle.

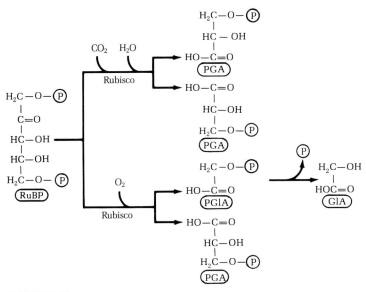

FIGURE 5-11

The competing reactions of RuBP-carboxylase/oxygenase (Rubisco). The same enzyme which catalyzes CO_2 fixation in the Calvin cycle can also cause O_2 to react with RuBP. The result of oxygenation is one molecule of PGA (versus two PGA when CO_2 is fixed) and one molecule of phosphoglycolic acid (PGIA). Normally PGIA is rapidly converted into glycolic acid (GIA).

the glycolate-producing oxidation process, on a bright summer day conditions in the leaf may become unfavorable for the productive carboxylation reaction: CO_2 concentrations decrease as CO_2 is used up in the Calvin cycle and O_2 concentrations increase as O_2 is produced in light reactions. In addition, the light levels and temperature may be promoting the reaction of RuBP with O_2 to produce the carbon-wasting glycolate. Under the conditions which exist commonly, the ratio of Rubisco's carboxylase to oxygenase activity, i.e., the ratio of CO_2 + RuBP to O_2 + RuBP is about 3:1, but it can be even lower.

Through the process of *photorespiration*, the glycolate generated by Rubisco's oxygenase activity and secreted by chloroplasts is taken up and metabolized in other organelles. The photorespiratory pathway consumes more O_2 (O_2 was also consumed initially to produce glycolate) and evolves CO_2. Although as much as half of the CO_2 entering the Calvin cycle can be lost as CO_2 through photorespiration, at least some of the gycolate produced in the chloroplast can be conserved through a series of reactions explained more fully in Fig. 5-12. In photorespiration, most of the carbon leaving the chloroplast as glycolate is recycled; the carbons are converted back into PGA and can reenter the Calvin cycle.

Photorespiration and the production of glycolate seem to be counterproductive; they rob the plant of carbon that could otherwise be used to make sugar. It has been suggested that in photorespiration the steps after glycolate production represent a mechanism for salvaging "accidentally" produced glycolate. This argument presumes that Rubisco operated strictly as a carboxylase in the early earth environment, which was largely devoid of O_2. Only as green plants enriched the O_2 level of the atmosphere did RuBP oxidation and glycolate production and loss become serious. (Some simple algae which lack glyoxysomes secrete large amounts of glycolate into their environment.) According to this theory, photorespiration via the glyoxysomes appeared in higher plants as a way of recycling, or salvaging carbons otherwise lost as glycolate. The later events of photorespiration certainly have the effect of reconverting glycolate into useful forms. (Actually only 3 out of 4 carbons that leave the chloroplast as glycolate reenter as PGA. The fourth carbon, lost as CO_2, can

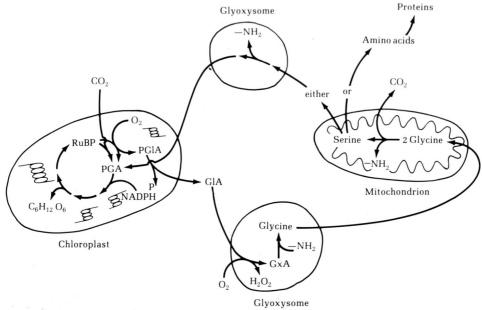

FIGURE 5-12

Photorespiration (schematic). The production of glycolic acid (GIA) in the chloroplast is usually considered the first step in the photorespiratory pathway. GIA leaves the chloroplast and enters the glyoxysome, where it is oxidized to glyoxylic acid (GxA) and further transformed to glycine. Glycine leaves the glyoxysome, enters a mitochondrion, and is converted into serine and CO_2. Serine can be converted by a series of reactions (occurring mainly in the glyoxysomes) into PGA. PGA can then reenter the Calvin cycle.

represent up to half of the CO_2 fixed in the chloroplast.)

Theoretically at least, if half the photosynthetically fixed CO_2 is lost via photorespiration, blocking Rubisco's oxidation reaction—and hence photorespiration—would increase photosynthetic productivity and plant yield by as much as 50 percent. In some experiments, researchers have chemically blocked the production of glycolate and found significant increases in photosynthesis. Although the possibility of using such chemicals on a whole field of plants is attractive, the problem of getting sufficient amounts into the chloroplasts of sprayed plants has not been solved.

Another approach to blocking photorespiration has been effective on whole plants on a modest scale. CO_2 fertilization in greenhouses artificially raises the CO_2 concentration in the atmosphere to which the plants are exposed; it has the effect of favoring the carboxylation reaction and boosting productivity

remarkably. Obviously such an approach would not be practical on a 200-ha field.

C₄ Photosynthesis

A naturally devised method of avoiding the wasteful production of glycolate and resulting photorespiration, though fairly simple, involves some redesign and retooling of the leaf and its photosynthetic machinery. The approach (to look at it from our perspective) was to reduce the O_2 concentration and increase the CO_2 concentration in the vicinity of the Rubisco enzyme so as to favor the photosynthetically productive carboxylase reaction.

Figure 5-13 shows the anatomy of a leaf redesigned to function without photorespiration. There is a separation of the O_2-producing light reactions and the Calvin cycle. The leaf has two types of chloroplast-containing cells: mesophyll cells and a distinct group of cells which ensheath the veins or vascular bundles. The chloroplasts of the mesophyll cells look typical, but the chloroplasts of the bundle sheath cells do not develop grana. Although they carry on typical Calvin cycle reactions (Fig. 5-14), by short-circuiting the z scheme, bundle sheath cells produce only ATP (not NADPH or O_2) in the light reactions. (Photosystem II and chlorophyll *b* are not present in the bundle sheath cell chloroplasts.) This has the effect of keeping the O_2 concentration in the stroma of bundle sheath cell chloroplasts lower.

While the light reactions of the bundle sheath cells are different, it is the dark reactions in the general mesophyll cells of these specially modified leaves that are atypical. The light reactions of the mesophyll cells include the usual z-scheme movement of electrons through photosystems I and II and the production of ATP, NADPH, and O_2, but there is no Calvin cycle. In the dark reactions of the mesophyll cells, CO_2 is fixed to a 3-carbon compound, phosphoenolpyruvate (PEP), producing a 4-carbon molecule. The enzyme catalyzing the addition of CO_2 to PEP is unaffected by O_2 concentrations in the mesophyll cell and is able to obtain CO_2 even when its concentration in the cells becomes very low. Because a 4-carbon compound is produced by the initial fixation of CO_2, these are called C_4 *plants* to distinguish them from the more common C_3 plants, in which only the Calvin cycle (with Rubisco) fixes CO_2 and the first stable product after CO_2 fixation is the 3-carbon compound PGA.

The 4-carbon compound produced in the mesophyll cells of C_4 plants is reduced by NADPH to form another 4-carbon compound, malic acid. After it is produced in the mesophyll cells, malic acid diffuses out of those cells, through adjacent cell walls, into a bundle sheath cell. There the CO_2 is released, leaving a 3-carbon compound again, pyruvic acid. The 3-carbon molecules PEP and pyruvic acid provide a steady one-way CO_2 "taxi service" between mesophyll and bundle sheath cells in a series of reactions called the *Hatch-Slack cycle*. The net effect of the Hatch-Slack pathway is that CO_2 is captured by the efficient CO_2-fixing enzyme in the mesophyll, transported to the bundle sheath, and concentrated in the bundle sheath chloroplasts. There it can be refixed in the Calvin cycle (with its less efficient Rubisco) in an environment enriched in CO_2, favoring the carboxylation reaction. Since O_2 concentrations are lower in the bundle sheath stroma, the photosynthetically productive reaction is favored, eliminating photorespiration almost entirely.

The C_3 photosynthetic pathway is by far the most common in terms of the number of species using it. The C_4 plants are mostly tropical grasses, but a few other plants that appear to have developed in hot and/or arid climates also have this specialized anatomy and metabolism. Their C_4 pathway and anatomy make them better adapted to the tropical conditions of high sunlight and high temperatures that would

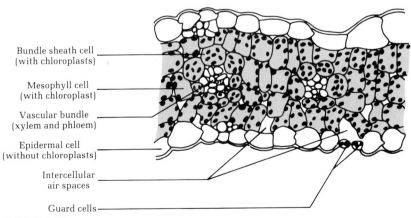

Bundle sheath cell
(with chloroplasts)

Mesophyll cell
(with chloroplast)

Vascular bundle
(xylem and phloem)

Epidermal cell
(without chloroplasts)

Intercellular
air spaces

Guard cells

FIGURE 5-13
Leaf anatomy of a C_4 plant, which separates the z scheme (light reactions) and Calvin cycle (dark reactions). The general mesophyll cells are distinctly different from the bundle sheath cells which surround the vascular bundles or veins. The differences involve the structure of chloroplasts as well: chloroplasts of the bundle sheath cells lack grana.

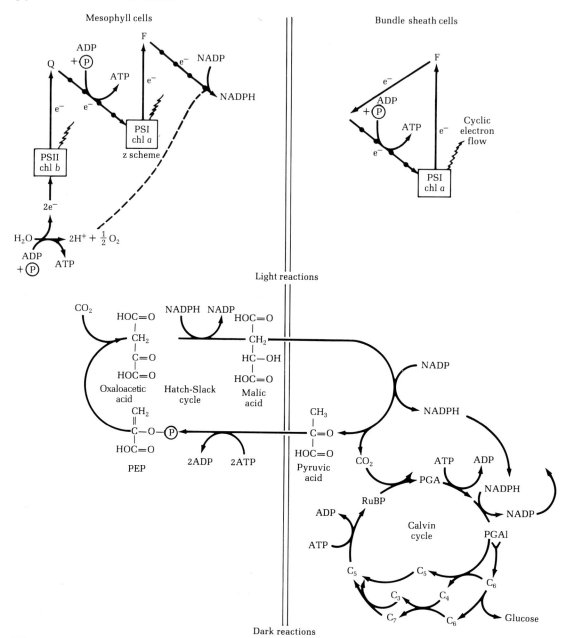

FIGURE 5-14

A general C_4 photosynthetic pathway. Each type of cell, mesophyll or bundle sheath, carries on distinct biochemical functions. The mesophyll cells exhibit the familiar light reactions, but no Calvin cycle is operating. Instead, CO_2 is fixed to phosphoenolpyruvic acid (PEP), and a 4-carbon product (malic acid in many C_4 plants) migrates out of the mesophyll cells to the bundle sheath cells. In the bundle sheath cell, light reactions produce only ATP, but the Calvin cycle operates normally. The 4-carbon compound from the mesophyll cells "drops off" CO_2 for RuBP carboxylase. The net result is a 50-fold CO_2 enrichment and O_2 reduction in the vicinity of Rubisco: conditions which favor photosynthesis and reduce photorespiration.

trigger glycolate production and photorespiration in C_3 plants. A few important crop species are C_4 plants, notably, corn, sorghum, millet, and sugarcane. Bermudagrass, a valuable pasture and hay species (significantly, only in warm climates), also has the Hatch-Slack pathway. In some places, bermudagrass, or wiregrass, as it is also known, is considered a pest, and other C_4 plants are noxious weeds, e.g., crabgrass, johnsongrass, shattercane, and pigweed (one of the nongrass C_4's). Most of our crops are C_3 plants, including such important members of the grass family as wheat, rice, barley, oats, and rye. Soybean and other valuable legumes are also C_3 plants. If the ability to use the C_4 pathway could be introduced into rice and soybean, a remarkable increase in their productivity would be predicted because of a more efficient photosynthesis under hot or dry conditions. Much effort has gone into searching for new C_4 species and a method of producing C_4 plants from C_3 species, but there have been no significant breakthroughs. The problem is obviously complex since changes both in anatomy and enzyme systems are involved. (C_3 plants may have bundle sheath cells, but they will contain no chloroplasts and no Calvin cycle.)

CAM Photosynthesis

A third variation on the theme of photosynthesis separates light and dark reactions, not into different regions of the leaf, as in C_4 plants, but according to the time of day. Plants with this *crassulacean acid metabolism* (CAM) include pineapple, many cacti, and members of the family Crassulaceae.

In CAM plants, the light reactions obviously take place in the daytime, but the initial fixation of atmospheric CO_2 occurs at night. This sequence allows the plant to keep its stomata closed during the day, conserving its water when higher temperature and lower relative humidity would normally increase the rate of water loss from the plant. Many CAM plants, including cacti, are well adapted to arid environments.

The fixation of atmospheric CO_2, which takes place when CAM plants open their stomata at night, produces malic acid (hence the "acid" in crassulacean acid metabolism). A molecule of CO_2 added to the 3-carbon PEP produces the 4-carbon malic acid. A large pool of malic acid accumulates during the night and dramatically lowers the pH of the CAM plant's tissue (Fig. 5-15). During the day, when the stomata are closed, the malic acid releases CO_2, providing a

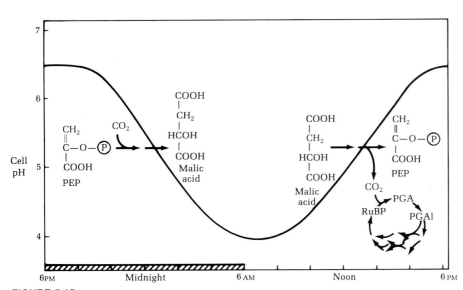

FIGURE 5-15
Typical daily pattern of pH changes in CAM plant tissues, which become acidic as malic acid is produced at night by fixing CO_2 to PEP. The acid is broken down in the daytime, releasing CO_2 for the Calvin cycle.

carbon source for the Calvin cycle. The pH of the plant's tissue rises as the acid is converted back into a less acidic 3-carbon component. Although the biochemistry is similar to that in a C_4 plant, in the CAM plant, the CO_2 provided to the Rubisco in the Calvin cycle was fixed the previous night in the same cell rather than a few moments before in an adjacent cell.

Comparison of C_3, C_4, and CAM

The CAM pathway presumably is a specific response to water limitations, while C_4 metabolism seems to be a natural way to prevent glycolate production and avoid photorespiration. But their peculiar metabolism may also enable C_4 plants to survive better than C_3 plants in arid environments. When stomata close because of water stress, the C_4 plant can still carry on some photosynthesis with its efficient mechanism for gathering and concentrating CO_2. Not so with a C_3 plant. When it wilts on a hot, dry day, its stomata close, the CO_2 concentration in its leaves drops, and the O_2 concentration increases to such an extent that photorespiration can equal or outstrip sugar production.

The greater efficiency with which C_4 plants can obtain CO_2 for photosynthesis can be demonstrated in a number of ways. Figure 5-16 shows the response of a typical C_3 plant, e.g., soybean, and a C_4 plant, e.g., corn, to increasing levels of sunlight. Many C_3 plants fix CO_2 at increasingly higher rates until the light reaches about one-half the brightness of full sun. Above that point, adding more light energy does not increase the C_3 photosynthetic rate and may actually decrease it. In other words, at higher light levels, the C_3 plant is not using the sunlight's energy efficiently. It may even use the light counterproductively, as photorespiration may increase more rapidly with increasing light levels than photosynthesis does. On the other hand, C_4 plants, which do not appear to photorespire, continue to increase photosynthetic output right up to and past the energy levels provided by full direct sunlight.

The ability of C_4 plants to outperform C_3 plants in a bright environment is dramatically demonstrated in the experiment shown in Fig. 5-17. Soybean (C_3) and corn (C_4) plants are placed in an airtight, transparent box and illuminated continuously. In a few days the soybean turns yellow and dies, but the corn continues to look healthy and grow for some time. The explanation lies in the C_4 plant's ability to carry

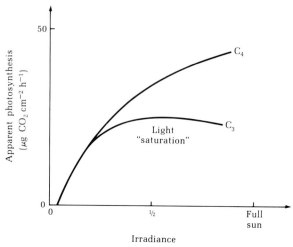

FIGURE 5-16
Generalized response of C_3 and C_4 plants to increasing irradiance. No net photosynthesis is evident until light levels reach a certain point. (The exact point depends on species and other factors.) Above that level of irradiance, photosynthesis increases steadily with increasing light until the C_3 plants begin to become light-saturated. At irradiances near that of full sun, C_3 plants may actually show decreased rates of photosynthesis.

on photosynthesis long after the CO_2 concentration in the box becomes too low for the C_3's Rubisco to pick it up. Also, the O_2 concentration increases and accelerates photorespiration in the C_3 plant. The corn, with its highly effective CO_2-fixing enzyme in the Hatch-Slack pathway, keeps the CO_2 concentration high in the bundle sheath chloroplast stroma, where its Rubisco is operating. In effect, the corn plant has killed the soybean plant by reducing the level of CO_2 below the point where the C_3 plant can obtain and fix it photosynthetically. In the final stages, the C_3 plant actually converts sugars it has already made back into CO_2 thus supporting the growth of the C_4 plant.

We should not leave the impression that C_3 plants are hopelessly outclassed by C_4 or CAM plants. In fact, CAM plants are generally slow-growing species; but in the inhospitable environments where they grow, they *can* grow. They are well suited metabolically to those conditions. There are also situations in which C_4 plants will clearly be superior to C_3 plants, but in temperate areas, some C_3 crops can be as productive as the C_4's or more so. Therefore, to the question "Which is best, C_3, C_4, or CAM?" the answer would have to be "It depends." In the hot, dry volcanic soils of Hawaii, CAM works well for pineapple. During the hot summers of the Midwest, corn and sorghum perform admirably with their C_4 pathway. But to produce grain in cool, wet England, wheat (C_3) is a better bet than corn.

RESPIRATION

The energy a plant or animal uses to generate and maintain itself comes ultimately from the sun, but it passes through several intermediate storage forms first. We have shown how some of the energy in a photon is transferred to electrons taken from water and passed with them through a number of intermediates in the z scheme to NADPH. The energy in NADPH is partially transferred, along with the electrons, to other molecules in the Calvin cycle for the production of glucose. Thus the energy in glucose is energy from the sun.

But the saga of those electrons and their energy is far from ended when they leave the chloroplast as a sugar. In fact, in terms of providing their major energy-producing service, they have not yet begun

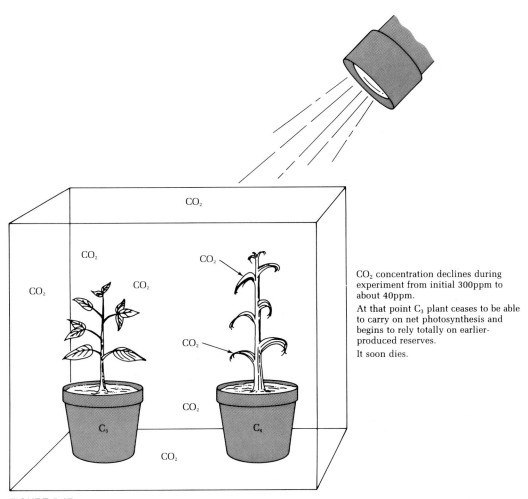

CO_2 concentration declines during experiment from initial 300ppm to about 40ppm.

At that point C_3 plant ceases to be able to carry on net photosynthesis and begins to rely totally on earlier-produced reserves.

It soon dies.

FIGURE 5-17
Demonstration of the unequal competition for CO_2 between C_3 and C_4 plants.

to work. When glucose is broken down chemically, its energy is released and the electrons revert to the low-energy state they once occupied in the water molecule. The energy released by the electrons as they move back to water may be partially conserved in coupled, energy-transferring reactions. In other words, as the sugar is stripped of its energy, the released energy can be used to increase the energy and organization of other molecules. At least this can happen if the sugar breakdown takes place in the cell, where enzymes are capable of controlling reactions and transferring the energy. When sugar is oxidized (burned) outside of a living system, the net reaction can be the same

$$C_6H_{12}O_6 + 6O_2 \longrightarrow 6CO_2 + 6H_2O + energy$$

but the energy (all 686 kcal mol^{-1}) will be liberated into the environment as heat and light. The energy is not destroyed, but it is transformed from a potentially useful chemical-energy form into a biologically unavailable, unusable form. Our discussion of respiration, one phase of metabolism, will emphasize the beauty of a system in which the sun's energy stored in glucose is used to power other metabolic events. If photosynthesis is the metabolic key that winds the mainspring of life, respiration may be called the balance wheel that governs the wound spring.

Glycolysis

Since a molecule of glucose contains more energy than can be used in any single metabolic event in the plant, the energy must be converted into smaller, more useful units. (A pocket full of silver dollars is no good when you need a dime for the parking meter.) The first steps in converting the energy of glucose into smaller, more usable packets take place in the cell cytoplasm and are catalyzed by enzymes floating around in that "living soup." This *glycolysis* (literally, sugar splitting) involves the sequential reactions shown in Fig. 5-18. Notice that some ATP is expended initially but that there is a net production of two ATP molecules and two NADH molecules for each glucose molecule undergoing glycolysis to pyruvic acid. Nicotinamide adenine dinucleotide (NAD), which resembles NADP in structure and function, is able to pick up two energy-rich electrons and pass them on later to other molecules. The negatively charged electrons attract positively charged protons, just as occurred with NADPH, so that the electron-

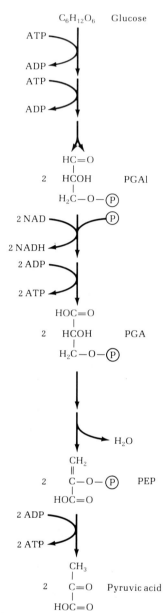

FIGURE 5-18

Glycolysis. One molecule of glucose is split into two molecules of phosphoglyceraldehyde (PGAl) following the addition of two phosphate groups (P). PGAl is oxidized, NAD is reduced, and ATP is made as phosphoglyceric acid (PGA) is formed. PGA is converted into phosphoenolpyruvate (PEP) and then to pyruvic acid with the production of more ATP. Since for each glucose molecule there will be two molecules each of PGAl, PGA, PEP, and pyruvic acid, two NADH and four ATP per glucose will be made. Because two ATP are consumed initially, two net ATP are produced per glucose undergoing glycolysis.

carrying (reduced) form of NAD is usually written NADH.

The two ATP and two NADH molecules produced during the conversion of glucose to pyruvic acid represent less than 20 percent of the energy available in a glucose molecule. Breaking one C—C bond and passing on only four electrons to NAD still leaves most of the energy in pyruvic acid. If glycolysis were the only biochemical method for converting the energy of glucose into smaller units, most of that energy would go to waste.

Fortunately, pyruvate can be metabolized further and its energy passed on to important energy-storing molecules. When mitochondria and O_2 are present, as in higher plants and animals, the pyruvate is passed along to those important organelles. But an organism without mitochondria or O_2 has other important energy-conserving pathways for pyruvate.

The most important alternative to respiration of pyruvate in mitochondria is *fermentation*, in which the 3-carbon pyruvate is converted into a 2-carbon compound, ethanol (ethyl alcohol or grain alcohol), and CO_2 is released (Fig. 5-19). Of course, fermentation has been used and abused since prehistoric times. In beverage or bread production, the organisms that carry on the fermentation are usually fungi. Yeasts, one type of fungi, can ferment glucose from sources as diverse as wheat, rice, potato, corn, and honey. In the absence of O_2, yeasts break glucose down into pyruvate and the pyruvate to ethanol. The CO_2 given off in the process may be important—as carbonation in beer or champagne or what causes bread to rise. *Anaerobic* metabolism, which takes place without O_2, is also important in making silage (Chap. 23).

Lower plants, such as yeast, are not the only organisms which can carry on fermentation following glycolysis. Many higher plants have at least a limited ability to produce ethanol. Apparently most plants have the enzymes necessary to produce ethanol, but the alcohol is so toxic that it can destroy the tissues producing it. An exception is rice seedlings growing submerged in O_2-depleted water; there the production of ethanol results in the regeneration of NAD from NADH and keeps the glycolytic process from halting due to depletion of the supply of NAD (Fig. 5-19).

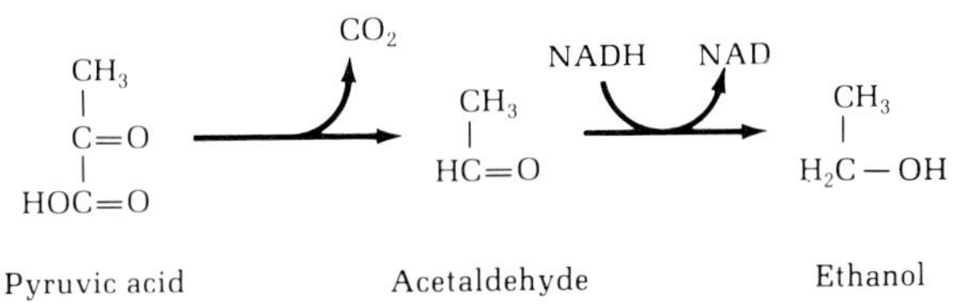

The Krebs Cycle

Pyruvate, the usual product of glycolysis, retains much of the sun's energy originally incorporated into glucose. If O_2 is present, the 3-carbon molecule can be metabolized further and stripped of its energy-rich electrons. All *aerobic* (with O_2) metabolism occurs in mitochondria. Like the Calvin cycle in the chloroplast stroma, the series of reactions is cyclical; it is named the *Krebs cycle* for the pioneering biochemist who first described it. The steps are shown schematically in Fig. 5-20. A key reaction is the addition of 2 carbons from pyruvate to a 4-carbon molecule. In the process of generating the 2-carbon group, pyruvate loses 1 carbon atom as CO_2. The 6-carbon molecule in the Krebs cycle is sequentially reduced to 5 and then 4 carbons with the simultaneous production of the intermediates NADH, ATP, and another energy-rich storage form, FADH. Although flavin adenine dinucleotide (FAD) resembles NAD in structure and ability to act as a carrier for energetic electrons and associated protons, the difference in structure means that those electrons have lower energy than the electrons of NADH.

After the second CO_2 molecule has been released in the Krebs cycle, the 4-carbon compound that remains still has more energy than the original acceptor of the 2-carbon unit. The 4-carbon molecule then undergoes further reactions involving the rearrangement of bonds and passage of electrons to NAD. The result is the regeneration of the original acceptor of the 2-carbon unit and completion of the cycle. During each turn of the cycle, three molecules of NADH, one of FADH, and one of ATP are produced. There are two pyruvate molecules produced per glucose molecule coming through glycolysis and therefore two turns of the Krebs cycle. Thus the Krebs cycle is able to convert much of the energy originally in glucose into other more immediately useful forms. The processes, which are all under enzymatic control and occur in the mitochondrial

FIGURE 5-19
Alcoholic fermentation. Pyruvic acid produced in glycolysis is converted into ethanol. In the process, 1 molecule of NADH is oxidized and CO_2 is evolved.

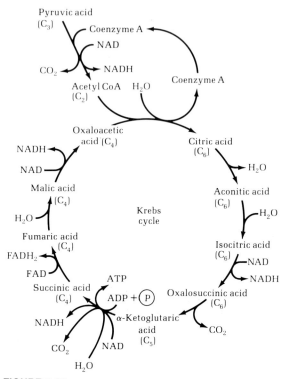

FIGURE 5-20
The Krebs cycle. Pyruvic acid loses one of its carbons as CO_2 and surrenders electrons to NAD. The resulting 2-carbon structure is fed into the Krebs cycle. In one turn of the cycle, 2 carbons are lost as CO_2 and important energy-rich intermediates (NADH, FADH, and ATP) are produced.

stroma, return the carbon of sugar to its original prephotosynthetic form, CO_2.

Electron Transport

The NADH and FADH produced during the Krebs cycle have more energy than is needed for many cellular processes, and ATP represents an energy packet of more usable size. The mitochondrion contains enzymes that can convert the higher-energy forms into the more immediately useful ATP. These enzymes are embedded in the inner membrane of the mitochondrion in folded, fingerlike projections called *cristae*. Remember that it is the presence of certain electrons that give NADH and FADH their metabolically available energy. When these electrons are removed and passed through a series of reactions in the membrane, their energy is progressively low-

ered; some of the energy released may be transferred to ATP.

A scheme of the mitochondrial *electron transport system* is shown in Fig. 5-21. NADH can pass its electrons to the low-energy form of FAD. This process, which takes place enzymatically and produces FADH, releases enough energy to produce one molecule of ATP. The FADH thus produced (or coming from the Krebs cycle) can then pass the electrons along to other molecules. The most important electron carriers, the cytochrome-containing enzymes also have a subunit that can transport electrons and occur in at least three slightly different types. The last cytochrome passes the transported electrons to O_2, producing water when protons become associated with it. In passing along the cytochromes, a pair of electrons releases enough energy to produce two molecules of ATP.

A quick review of Fig. 5-21 and the discussion above shows that three ATP molecules can be made per NADH and two per FADH. Adding up the NADH molecules (10), FADH molecules (2), and ATP molecules (4) produced directly in glycolysis and the Krebs cycle and then converting NADH and FADH into ATP at the rate of 1:3 and 1:2, we find a net of 38 ATP molecules per molecule of glucose broken down. Since the energy in 1 mol of ATP is approximately 8 kcal, only about 300 kcal of the 686 in a glucose molecule is accounted for. The rest is primarily lost as heat—lost not in the sense that it no longer exists but lost because it is converted into a form no longer available to the plant for metabolism.

The electron transport system can work only as long as O_2 is present to accept the electrons that come down the cytochrome path. When O_2 is withheld, the system backs up and the cell's NADH and FADH accumulate in the reduced form. Without the oxidized NAD and FAD forms, the Krebs cycle comes to a halt and the organism is forced to rely on glycolysis or fermentation for its ATP supply. As we indicated earlier, this is not enough for most plants.

What does the cell do with all the NADH and ATP it makes in the respiration of glucose? It uses it for just about everything a cell must do. The energy needed to drive many endergonic metabolic processes comes from the exergonic use of ATP and NADH. Living matter is assembled at the expense of the energy and organization of ATP and NADH. Thus respiration consumes glucose and partially converts its energy into the more immediately useful ATP,

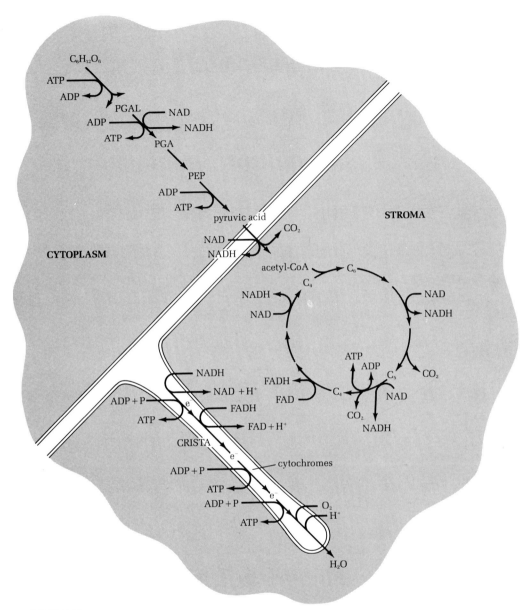

FIGURE 5-21

Glycolysis, Krebs cycle, and the electron transport system. NADH and FADH produced in the stroma move into the crista membranes and pass their electrons through a series of steps to oxygen. ATP is made at various points along the way. The last carriers in the electron transport system, the cytochromes, pass two electrons to oxygen, which then associates with two hydrogen ions to produce water.

NADH, and FADH. Since the ATP and NADH power growth directly, anything that reduces the respiratory capacity of the plant will slow down its growth.

Two or three conditions that slow respiration, thereby retarding plant growth, may become significant in an agricultural situation. Low temperatures slow down all metabolism but especially respiration. The reduced growth rate of many species during the winter is directly attributable to the overall depression of respiration. Reduced O_2 concentrations also slow respiration. Plants growing in flooded conditions suffer primarily because of the absence of sufficient O_2 to carry on ATP production in the cristae of mitochondria. Even relatively brief periods of flooding may be fatal because the plant has been deprived of ATP.

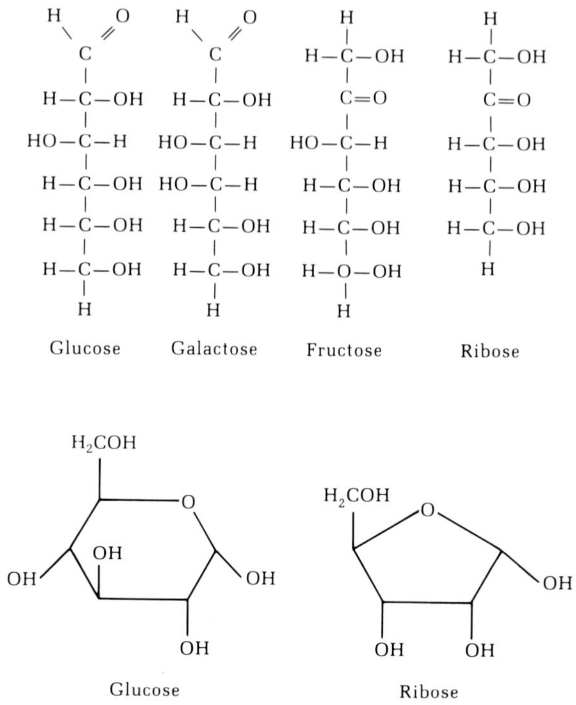

CARBOHYDRATE METABOLISM

Carbohydrates contain carbon, hydrogen, and oxygen in the ratio of 1:2:1. Glucose, one of the most common carbohydrates, is generally considered an end product of photosynthesis and the starting point for glycolysis and respiration. It is also the basic component for cellulose, the world's most abundant biochemical, and for starch, another abundant plant component.

The processes of photosynthesis and respiration involve the basic energy transformations associated with carbohydrates as some of the energy of photons is passed (via NADPH and ATP) to glucose and then some of the energy in glucose is transferred to other substances, especially NADH, FADH, and ATP. In the process, carbon-containing compounds become first more organized and then more disorganized

$$\text{light} + 6CO_2 + 6H_2O \longrightarrow$$
$$C_6H_{12}O_6 + 6O_2 \longrightarrow$$
$$6CO_2 + 6H_2O + 38ATP$$

When the carbohydrates produced in photosynthesis are not used immediately in the cellular machinery, as a source either of energy or of carbon for making other components, they must be stored. Storage as individual sugar molecules, or *monosaccharides* (Fig. 5-22) is undesirable because the necessary amounts would make the intracellular solution extremely viscous and osmotically unfavorable. Instead, monosaccharides are combined into a much larger *polysaccharide*, or molecule of starch. Because

FIGURE 5-22
Some simple sugars, or monosaccharides. The sugar molecule may be depicted as a straight chain or as a ring. Glucose, galactose, and fructose have the same molecular formula $(C_6H_{12}O_6)$.

osmotic concentrations depend on the number of molecules in solution rather than their size, a single large starch molecule has essentially the same effect on osmotic potential as the much smaller glucose molecule. (Sometimes the starch crystallizes and is completely inactive osmotically.) Glucose thus stored poses no osmotic problem and can readily be released enzymatically from the starch for respiration or other use. Some plants store relatively large amounts of carbohydrates in simple (nonstarch) forms. In both sugarbeet and sugarcane, the carbohydrate is sucrose, common table sugar composed of one molecule each of glucose and fructose.

Starch occurs in two major forms. *Amylose*, made by linking glucose molecules end to end in a long unbranched chain, has 20 to 800 individual sugar units (Fig. 5-23a). These large molecules may crystallize in the cytoplasm or chloroplast and act as a storage site for photosynthate that is osmotically inactive but readily available for use in providing energy or building materials for the cell. The second form of starch is made in the cell by joining many glucose molecules into one much larger molecule, or polymer, with a highly branched structure instead of one long chain. The *amylopectin* molecule may contain as many as 3000 or more glucose units.

Starch is an important way to store energy and carbon in many seeds or other propagative structures. Potato tubers, swollen underground stems that are stocked with starch, have dormant growing points (*eyes*) which draw on the energy and carbon of the starch reserves when they begin to grow. Potatoes are rich in amylose.

Seeds must store enough energy and raw materials for growth of the germinating seedling. The young plant calls upon these reserves until it emerges into the sunlight and can become photosynthetically independent. The seeds of many members of the grass family, including wheat, rice, and corn, are starchy, having large carbohydrate reserves. Corn

seeds are larger and have more starch than most of the other important crops. Kernels of corn have varying quantities of amylose and amylopectin. Corn types are divided into their major groups on the basis of the abundance and distribution of the two forms of starch (see Chap. 19).

Cell walls are made of several different kinds of carbohydrates. The most abundant form is *cellulose*, an unbranched chain of glucose molecules, up to 10,000 glucose subunits long, linked end to end. Cellulose looks much like amylose but has important chemical differences resulting from the slightly different method of linkage (Fig. 5-23b). Cellulose and amylose have different physical properties; amylose forms starchy or chalky grains, while cellulose tends to form tough interlocking fibers that give the cell wall its strength.

A second important difference between starch and cellulose lies in the relative abundance and distribution patterns of enzymes that break them back down into glucose. Essentially all higher organisms

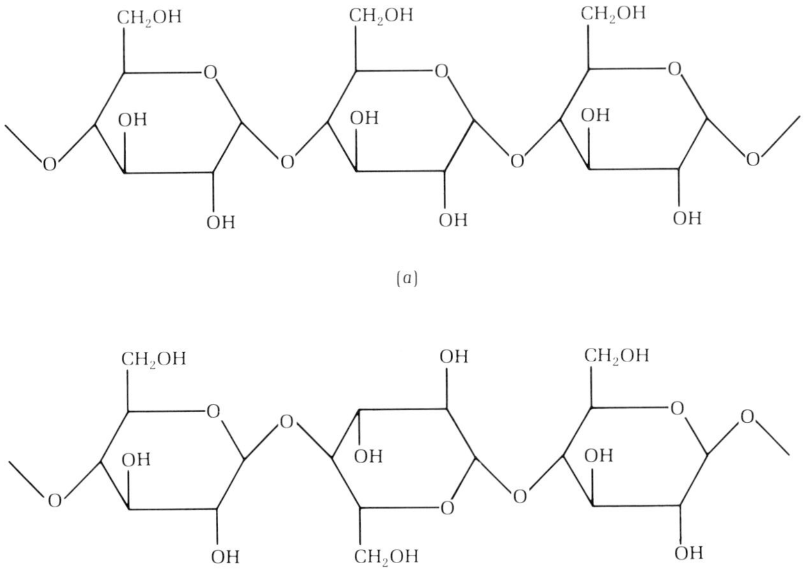

(a)

(b)

FIGURE 5-23
Two polysaccharides made from glucose. A portion of an amylose chain (a) shown along with one of cellulose (b). The difference between the two might seem slight, but it looms very large when comparing the physical and chemical properties of the two molecules.

can digest amylose or amylopectin enzymatically and use the glucose residues as needed in respiration or other processes; relatively few organisms, however, possess the enzymes to break the linkages between the glucose units of cellulose. These organisms include some bacteria and fungi and other microorganisms. No higher animals can digest cellulose. This is crucial when one considers that much of the photosynthate made in a plant is converted not into digestible starch but into cellulose.

Nevertheless, many animals eat and utilize cellulose, especially the ruminants (animals with multiple stomachs, e.g., cattle and sheep). Although people ingest cellulose in many plant foods (lettuce, broccoli, apples, etc.), they cannot get any energy from it. Termites and ruminants can use cellulose without producing cellulose-digesting enzymes themselves because microorganisms living in their digestive tract digest the cellulose and release the glucose units into the animal's gut. Since sheep and cattle can digest and utilize much more from grass than just its starches, they are important converters of cellulose into forms directly useful to people as food: milk, beef, mutton. Land and climates incapable of supporting food crops such as corn, wheat, or rice can often produce an abundance of forage for livestock. Put to this use, the land can contribute significantly as a solar energy collector and converter, making feed largely as cellulose that animals can convert into a form suitable for human consumption.

We must pass over the other important carbohydrates which function as energy-storage forms or as structural components in the cell wall. But we must say something about the third important function of carbohydrates—providing the carbon skeleton for all other biochemicals. Sugars are the necessary starting point for essentially every component in the cell. The basic 1:2:1 ratio of C:H:O may be radically altered and a number of other elements added, but the backbone of most biochemicals is the C—C bond formed in the Calvin cycle. Sugars are the feedstock used in the formation of amino acids (which give rise to proteins), lipids, nucleotides, hormones, pigments, alkaloids, and more.

We shall discuss briefly several of these biochemical types and their origin, function, and structure. Remember that each is largely carbon and that carbon was ultimately derived from carbohydrates produced in the chloroplast from CO_2, H_2O, and the sun's energy.

LIPID METABOLISM

Lipids are a diverse group of biochemicals with a limited ability to dissolve in water and a much greater ability to dissolve in organic solvents. *Triglycerides*, the largest group of lipids and one of the most common in plants, are composed of the 3-carbon molecule *glycerol* and one, two, or three molecules of fatty acids (Fig. 5-24). In plants, *fatty acids* are molecules containing mostly carbon and hydrogen; they are generally 18 to 20 carbons long and noted for their *unsaturation*. The common form in animals, *saturated fatty acids*, are so named because they contain as many hydrogen atoms as their carbons can hold; i.e., each carbon except those on the ends is bonded to two hydrogen atoms. Such fatty acids and the lipids containing them are usually solid animal fats. The fatty acids and fatty acid-containing lipids in plants are unsaturated because some of their carbons bond to less than the maximum possible two hydrogen atoms. Unsaturated fatty acids and their lipids, which are usually liquid (oils) at room temperature are considered superior to the saturated lipids for use in the human diet. Margarine is made from vegetable oil that has been *hydrogenated* to increase lipid saturation so that it will solidify.

Lipids serve several functions in the cell, one of the most important being in membranes. The *phospholipids* of membranes resemble triglycerides except that one fatty acid has been replaced by a phosphate group. The solubility of the phosphate unit gives phospholipids the peculiar characteristic of being soluble in water (*hydrophilic*) at that end of the molecule and insoluble (*hydrophobic*) at the other, the fatty acid end (Fig. 5-25). This feature is crucial in the development of the membrane's structure and properties. Some plants appear able to tolerate cold temperatures in part because they have highly unsaturated fatty acids, which remain liquid at temperatures that would cause more saturated lipids to solidify and the membranes containing them to become nonfunctional.

Lipids also function as energy and carbon reserves. Many seeds store lipids rather than carbohydrates as a source of energy and carbon during seedling germination and growth. Of course man and beast have learned to use such oil seeds for their own ends also.

By virtue of its solubility characteristics, *chlorophyll*, the green pigment so important in photosyn-

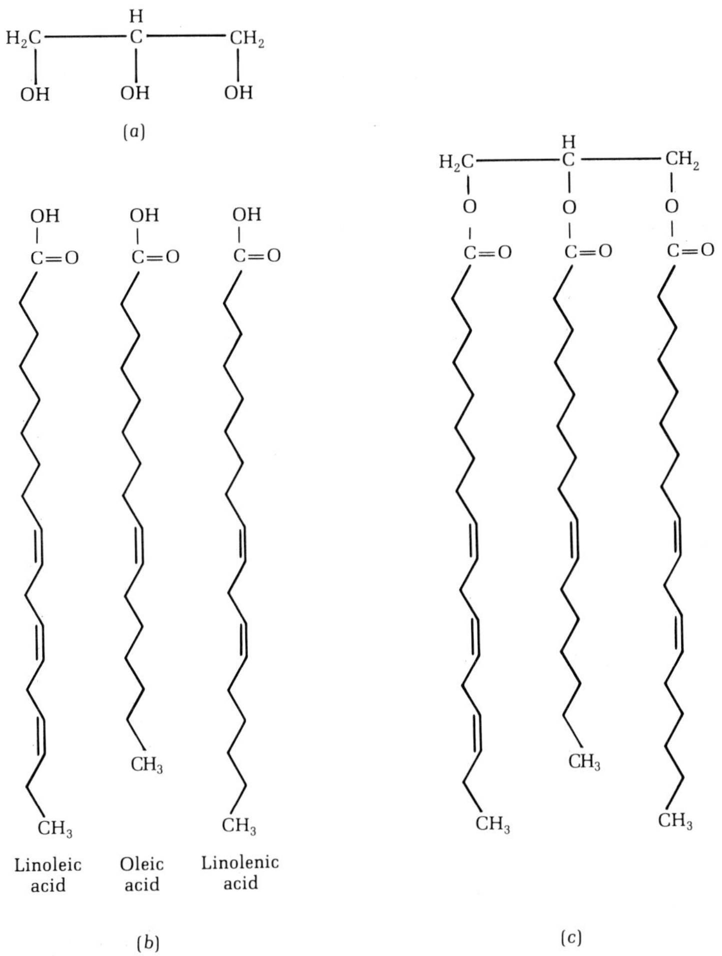

FIGURE 5-24
Glycerol (a), fatty acids (b), and a triglyceride (c). Fatty acids may have various degrees of unsaturation. Triglycerides (glycerol combined with three fatty acids) can have mixtures of saturated and unsaturated fatty acids. Oleic, linolenic, and linoleic acids have one, two, and three unsaturated bonds, respectively.

thesis, is a lipid although it is otherwise quite different from the fatty acid-containing lipids (Fig. 5-6) and serves no purpose in the cell other than gathering light. Carotenoids and carotene are lipid pigments important in plants and as A vitamins in the animal diet (Fig. 5-26). Vitamin E, another critical nutrient in animal diets, is widespread in plant oils, large amounts being found in wheat germ and corn oils. In plants or seeds, vitamin E appears to help protect the unsaturated lipids from oxidative destruc-

tion. Vitamin E is commonly lost while processing vegetable oils but replaced with artificial preservatives that are chemically much like it and appear to serve the same function.

PROTEIN METABOLISM

Biological proteins are assemblages of some 20 different kinds of amino acids. All amino acids have

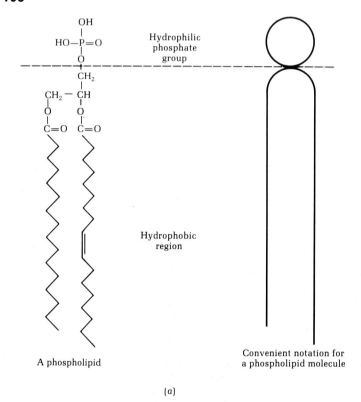

A phospholipid

Convenient notation for
a phospholipid molecule

(a)

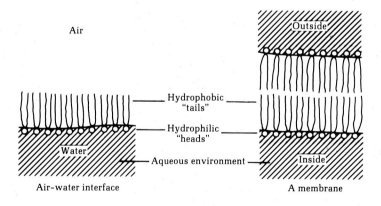

Air-water interface

A membrane

(b)

FIGURE 5-25

Phospholipids (a) and membranes. Because of the solubility of phosphate in water and the hydrophobic (literally, water-fearing) fatty acid hydrocarbon chains, there is a tendency for phospholipids to orient in a specific way when brought in contact with water. In a membrane, water is excluded from the interior and the hydrophilic (water-loving) head groups are exposed to the aqueous environment. (*From D. T. Plummer, An Introduction to Practical Biochemistry, McGraw-Hill, New York, 1978.*)

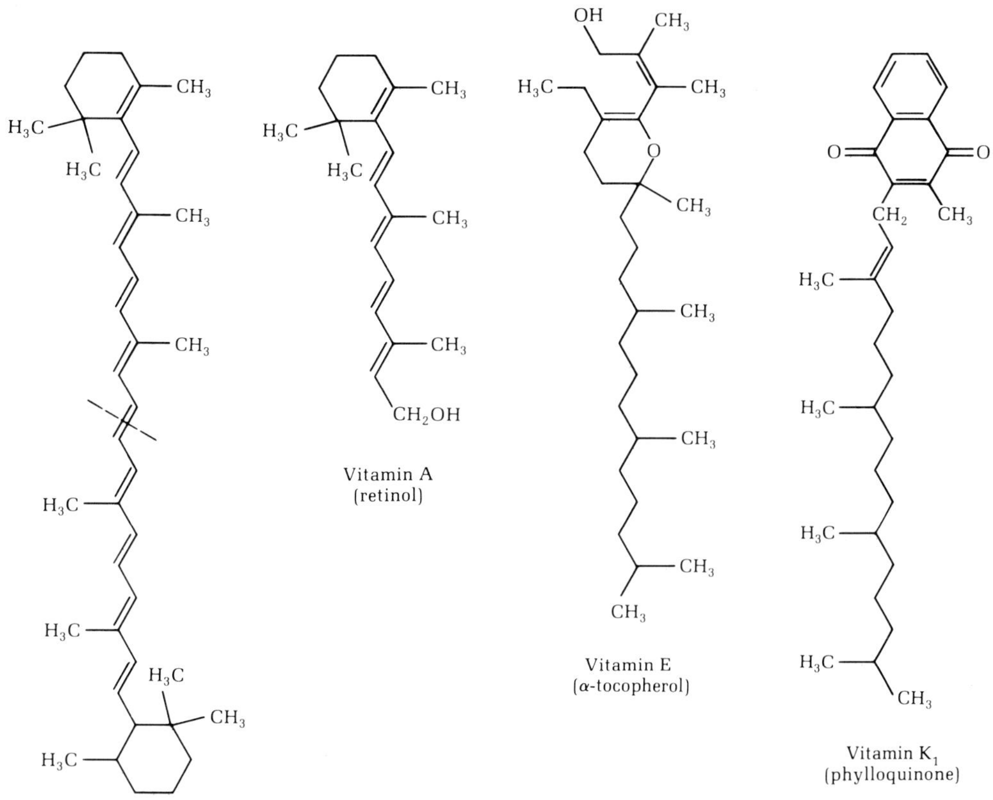

β-Carotene

Vitamin A (retinol)

Vitamin E (α-tocopherol)

Vitamin K_1 (phylloquinone)

FIGURE 5-26
Some important lipids derived from plants. Two molecules of vitamin A are derived from a single molecule of carotene. Vitamins E and K are also available in (produced by) plants. Animals, including man, must depend upon plants for the production of these critical dietary factors since our bodies cannot make them.

in common a nitrogen-containing component (the amino group) and two carbon atoms, one of which is in the acid group (carboxyl group). The carbon that lies between the amino group and acid group may have any one of 20 different components bonded to it (Fig. 5-27). Two amino acids can become bonded to each other through the formation of a *peptide* bond. As more amino acids are joined by peptide bonds, a long chain forms, with —C—C—N—C—C—N— repeating as the backbone of the molecule (see Fig. 5-2). Projecting from the backbone is the unique portion of each type of amino acid. These projecting portions which have the particular chemical and physical properties responsible for folding of the protein chain, may also become involved in catalytic activity (Fig. 5-3).

The synthesis of proteins involves a number of steps and ensures that each amino acid in the protein will be precisely the one needed at that point in the chain for proper folding or catalysis. This is a feat which sophisticated laboratory facilities cannot fully duplicate. Only in the last few years have we begun to understand how a cell accomplishes it. The next chapter discusses some of what is now known.

Of course, one of the most important functions of proteins is to be enzymes, i.e., to act as catalysts of biochemical events. Essentially everything that happens chemically in a cell is under the direction and

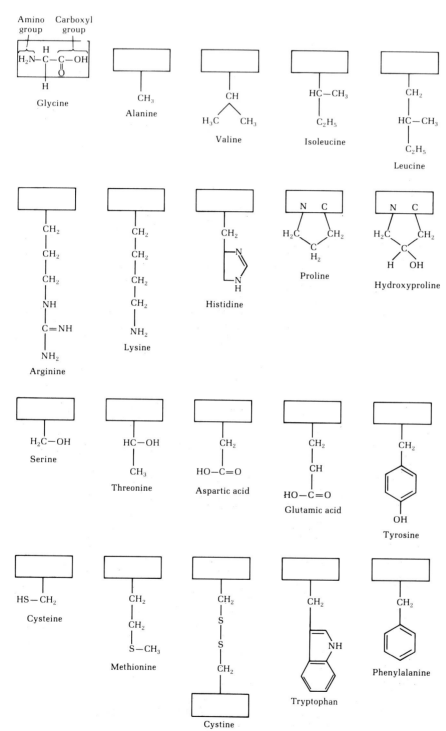

FIGURE 5-27
Twenty common biological amino acids. The amino group (—NH₂) and the carboxyl group (—COOH) are common to each amino acid. (*After M. Pyke, Man and Food, McGraw-Hill, New York, 1970.*)

regulation of the protein catalysts—the synthesis, breakdown, and conversion of all biochemicals, including the proteins themselves.

Proteins are involved in a number of other functions in the plant. Some proteins apparently inserted into membranes or embedded in them provide important regulatory functions in determining what gets into and out of a cell or organelle. In addition to creating a hole, or pore, of particular dimensions or chemical characteristics, such proteins may also be able to act catalytically in importing or exporting particular ions or molecules selectively.

Proteins may also provide a structural or storage role. In cell walls, proteins may help make cellulose fibers rigid. Some proteins, particularly in seeds, occur in structures that appear to be merely storage forms. The seeds of legumes are noted for the presence of *protein bodies*, which provide an amino acid source for the developing seedling. (The protein body is digested by enzymatically breaking peptide bonds to release amino acids, just as starch can be broken down to individual sugar molecules.) Like starchy or oily seeds, protein-rich seeds can be a valuable addition to the human and animal diet. Such legumes as soybean, chickpea, and lentil are particularly important in the protein-deficient diets of people in underdeveloped countries. Plant protein (as seeds) is relatively cheap, direct (requiring no animals for conversion), and more easily preserved than animal protein (meat).

NUCLEIC ACIDS

While carbohydrates, lipids, and proteins constitute the major portion of the cell, other, less abundant components are just as critical. The *nucleic acids* are essential as carriers of genetic information from one generation to the next and for making proteins. The two types needed for making proteins are *deoxyribonucleic acid* (DNA) and *ribonucleic acid* (RNA). Each is made up of subunits called *nucleo-*

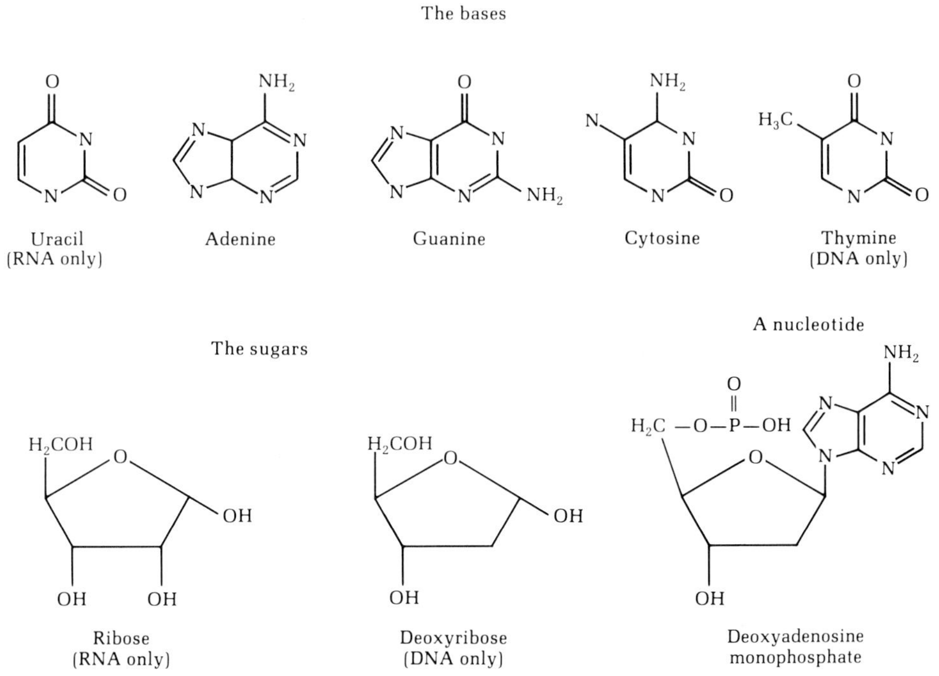

FIGURE 5-28

The components of nucleic acids. The basic building blocks of nucleic acids are nucleotides, which are composed of a phosphate group, a sugar (ribose for RNA, deoxyribose for DNA), and a base. Three bases occur with either ribose (in RNA) or deoxyribose (in DNA). The other two bases are unique (uracil for RNA nucleotides and thymine for DNA).

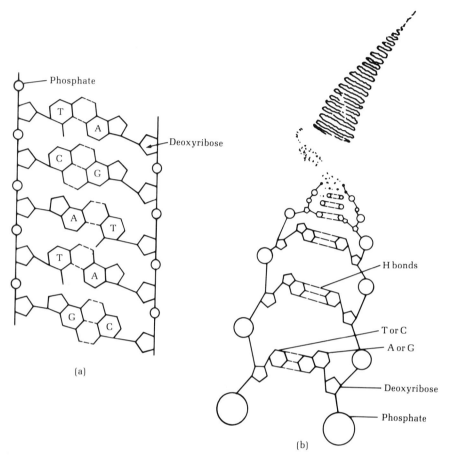

(a)

(b)

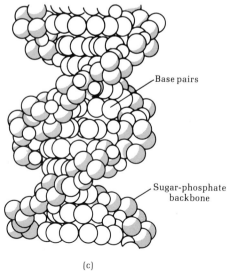

(c)

FIGURE 5-29
The organization of nucleotides into the DNA molecule.
Adjacent nucleotides are bonded to each other, creating a
sugar-phosphate backbone. (*a*) Because of interatomic attractions (hydrogen bonds), certain bases pair with each other
on opposite sides on the DNA molecule, forming a doubled
strand of nucleotides. (*b*) The double-stranded molecule can
be a million or more nucleotide pairs long. (*c*) The DNA
molecule is slightly twisted, producing a double helix. (*From
D. O. Woodward and V. W. Woodward, Concepts of Molecular
Genetics, McGraw-Hill, New York, 1977.*)

tides joined together in long chains, much like the sugars in cellulose or the amino acids in proteins. In DNA these nucleotides are formed from three components: the 5-carbon sugar deoxyribose, a phosphate group, and one of four different bases (Fig. 5-28)—adenine, guanine, cytosine, and thymine. With four different bases, there are four different nucleotides.

The DNA molecule is formed when bonding occurs between the sugar and phosphate groups of adjacent nucleotides. The bases project outward from the sugar-phosphate backbone (Fig. 5-29). It is a peculiarity of the DNA molecule that it is formed not of one such chain with the sugar-phosphate backbone but two, which lie side by side. The bases, which extend outward from each chain, pair up because of interatomic attractions with the bases of the other chain. This pairing is far from random, however, because guanine is attracted only by cytosine and thymine only by adenine. That means that if one knows the nucleotide sequence in one chain, one can predict it in the other. The significance and functionality of this arrangement will be discussed in Chap. 6.

The nucleotides of RNA differ from those of DNA in two significant ways: the sugar is ribose, not deoxyribose, and the thymidine base is replaced by uracil. This still leaves only four possible nucleotides to make up the RNA chain, which may be several hundred nucleotides long. RNA is a single strand; no double chain is formed. The chain of RNA can take at least three different forms and can provide three different functions, all important in protein synthesis (see Chap. 6).

OTHER BIOCHEMICALS

Compounds in an amazing number and complexity are produced in roots, stems, and leaves. More than 12,000 different naturally occurring chemicals have already been isolated and identified from various plants, and we have barely begun to survey the plant

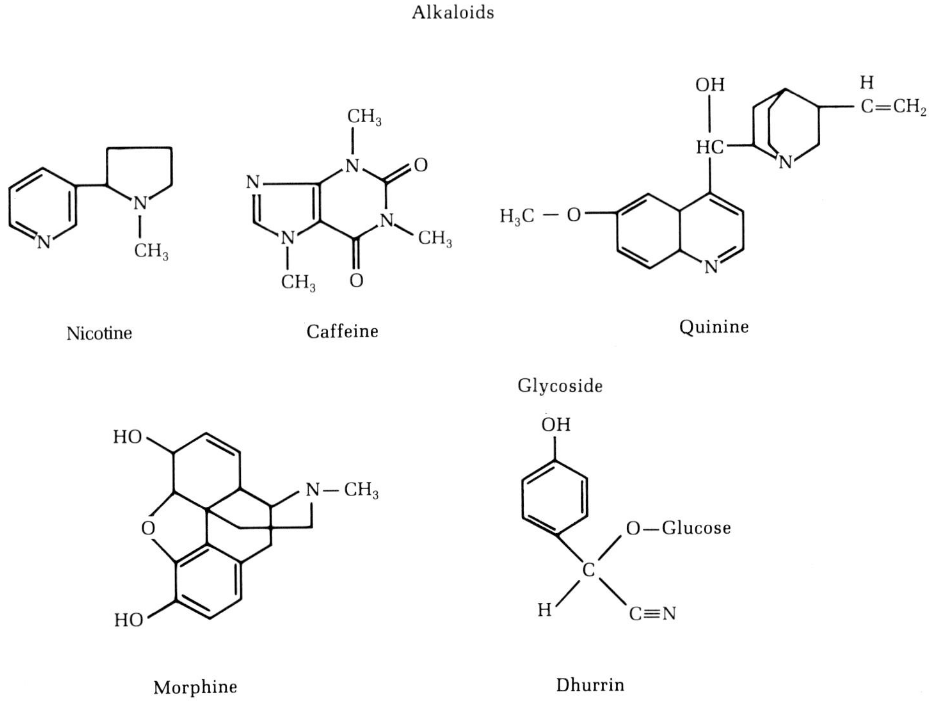

Alkaloids

Nicotine   Caffeine   Quinine

Glycoside

Morphine   Dhurrin

FIGURE 5-30
Some alkaloids and a glycoside.

kingdom chemically. Some of the important chemicals will be mentioned in other chapters. In Chap. 4 special attention was paid to the plant hormones as regulators of growth.

While plant hormones have definite and crucial roles in the plant, two other diverse groups of biochemicals, the *alkaloids* and *glycosides* (Fig. 5-30), appear to be simply by-products. If the plant does have a use for them, it is frequently not apparent. Alkaloids are alkaline compounds with a ring of carbon atoms and some nitrogen atoms; they are notable not for what they do to the plant but for what they do to people when taken internally. Some, e.g., nicotine and caffeine, are mild. Others are powerful, addictive drugs, e.g., cocaine and opium. Tetrahydrocannabinol, the active ingredient in marijuana, is an alkaloid. Some alkaloids are extremely poisonous, capable of killing human beings or animals that swallow them, suggesting at least one selective advantage: reducing predation by herbivores.

The glycosides are part sugar and part second component—a lipid, another complex carbon-containing compound, or something as unlikely as cyanide. Many glycosides are fatally toxic to people and animals, even in small amounts. Crops that are important because they contain alkaloids or glycosides will be mentioned in later chapters.

CHAPTER SIX

GENETICS AND REPRODUCTION

The most amazing attribute of any living thing is its ability to reproduce, to make a living copy of itself. Not even the most sophisticated computer-controlled device can produce an equally sophisticated duplicate. But that is exactly what all living things from the smallest bacteria to giant redwood trees and great blue whales have been doing for eons. This does not mean that reproduction is incomprehensible: it is not. The basics of how cells divide to produce new cells and how plants and animals reproduce are well known. But the ability to make copies that can make copies—that can make copies ad infinitum—involves a scheme that is both simple and profound.

CELLULAR REPRODUCTION

Most living things begin life as a single cell. The human body contains some 50 trillion cells, all derived ultimately from one cell, the fertilized egg. The billions and trillions of cells in a tree (Fig. 6-1) or a soybean plant can likewise be traced back to a single cell. Thus the process of cellular division is fundamental to all multicellular organisms; understanding it is a keystone to intelligent use of the plant world.

From Blueprints to Tools
A "mad scientist" who decided to make a cell from the atoms and molecules lying around in the laboratory would quickly discover two insurmountable

difficulties: there are no blueprints for the molecular assembly of a living cell, and it would take some impossibly tiny tools to assemble atoms and molecules into a cellsized structure. Our mad scientist is doomed to failure.

Any cell that is going to make a copy of itself must overcome the same two problems: it must have a blueprint and it must have tools. By multiplying, cells prove that they have both the pattern for cellular construction and the machinery for realizing it.

The "blueprint" for the cell's amazing act of self-duplication is a set of *genes*. For a cell or organism this set is often called its *genome* or *germplasm*. Genes are found on the chromosomes inside the nucleus of each cell and are composed principally of *deoxyribonucleic acid* (DNA). The unique chemical and physical properties of DNA give it the ability to carry (in encoded form) complete assembly and operational instructions for the cell. Genetic DNA controls all cellular activity, specifying which, when, and how many *enzymes* are to be made. These enzymes are the tools of the cells, the catalysts involved in making and then assembling molecules in precise ways. The average cell contains thousands of different enzymes, each with a specific function to perform in cellular metabolism, development, and division.

The Chemistry of Nucleic Acids
Genes and the chromosomes on which they reside are composed of proteins and nucleic acids. Each

FIGURE 6-1
Trees begin their lives as a single cell, the fertilized egg. From that single cell comes the trillions needed to form the mature organism.

chemical component is important in the overall control of cellular activity by genes. Two kinds of nucleic acids are present: DNA and *ribonucleic acid* (RNA), each with a key role in the production of enzymes.

DNA and RNA molecules are made up of repeating units, called *nucleotides*, that are joined together in long chains (Fig. 5-28). The nucleotides of DNA and RNA have three subcomponents: a 5-carbon *sugar* (deoxyribose in DNA, ribose in RNA), *phosphate*, and a *nitrogenous* (nitrogen-containing) *base*. Nucleotides can join chemically between the sugar of one nucleotide and the phosphate of another.

Each type of nucleic acid has two different kinds of nitrogenous bases: two double-ring *purines* and two single-ring *pyrimidines*. The pyrimidines in DNA nucleotides are *cytosine* and *thymine,* and the purines are *adenine* and *guanine.* In RNA, the pyrimidine *uracil* is present instead of thymine; the other three nitrogenous bases are the same as those of DNA.

DNA: The Double Helix

The most striking structural difference between DNA and RNA is not the nature of the 5-carbon sugar present in their nucleotides or the slight difference between uracil and thymine but the form assumed by the assembled chain. Besides their length (often millions of nucleotide units long) DNA molecules are distinctive for their doubled structure. Each chain of nucleotides is paired with a second chain to form the DNA molecule (Fig. 5-29). Typical RNA molecules have only a single strand of nucleotides.

The doubled structure of the DNA molecule is a natural consequence of the physical chemistry of its nucleotides. The purine guanine and the pyrimidine cytosine attract each other electrophysically. There is a similar attraction between adenine and thymine. In a single strand, the nitrogenous bases of a nucleotide are exposed. In this exposed position, the complementary purines and pyrimidines pair up, guanine to cytosine and adenine to thymine, forming a second parallel chain. This pairing of bases occurs when DNA is involved in its critical function of duplication.

Double-stranded DNA, with its two long parallel chains of paired nucleotides, might be envisioned as a minuscule twisted ladder with millions of rungs. The sugar-phosphate backbones of the nucleotide chain form the two vertical supports, and the paired bases, adenine to thymine and guanine to cytosine, form the horizontal rungs. The twist in the molecule results in a *double helix*, in which the helixes are the sugar-phosphate backbones. The purine-pyrimidine pairs extending between the helixes look somewhat like steps in a spiral staircase.

The organization of the DNA molecule was discovered in 1953 by James Watson, Francis Crick, and Maurice Wilkins, who received a Nobel prize for their effort. Their findings have enriched biology as much as any discovery in the history of that science. Not even the microscope, the cell theory, or Darwin had a greater impact on our understanding of biology; almost as soon as the structure of DNA was known, its basic method of storing genetic information and directing enzyme production became clear.

DNA to RNA to Enzyme

The sequence of nucleotides in a DNA molecule provides the blueprint of a cell. Each of the several thousand different enzymes needed to form and operate a cell is represented somewhere in the sequence of nucleotides of a cell's DNA. The sequencing of *amino acids* in an enzyme is determined by the sequence of nucleotides in DNA. How can only four different nucleotides (adenine, thymine, guanine, and cytosine) establish the sequence of the 20 different amino acids found in enzymes? And how can the nucleotide sequences of DNA on chromosomes in the nucleus direct the precise sequencing of amino acids in enzymes being made in the cytoplasm?

If only four different amino acids were inserted into enzymes, each of the four DNA nucleotides could direct the insertion of a specific amino acid. It is obvious, however, that there cannot be a one-for-one system of sequencing with 20 different amino acids. How about a two-for-one code? Some quick calculations show that there are only 16 possible combinations of the four nucleotides if they are grouped in twos (AA, AG, AC, AT, GA, GG, GC, GT, CA, CG, CC, CT, TA, TG, TC, and TT; $4^2 = 16$). But if three adjacent nucleotides (three joined along the sugar-phosphate backbone) direct the sequencing of a specific amino acid, there are more than enough possible nucleotide triplets to provide a unique

triplet for each of the 20 amino acids. In fact, there are $4^3 = 64$ different possible combinations of adenine, thymine, guanine, and cytosine when they are grouped in threes.

Coding for amino acids resembles the way a series of dots and dashes creates a telegraphic code for letters. A triplet of nucleotides that codes for a particular amino acid is called a *codon*. The whole system of codons together with the amino acids they encode is called the *genetic code*. With 64 possible codons and only 20 possible amino acids, one would expect some triplet sequences to remain unused, but when the genetic code was finally broken, some amino acids were found to be coded for by two, four, or even six different triplets (Fig. 6-2).

Thus, the giant molecule DNA has sequences of nucleotides that in groups of three prescribe the sequence of amino acids in enzymes. DNA is the blueprint by which enzymes can be specifically made. Enzymes in turn are the tools with which cells are assembled and operated.

How does a sequence of nucleotides on the chromosomes in the nucleus of a cell give rise to a sequence of amino acids being assembled into proteins elsewhere in the cell? DNA does not leave the nucleus to interact with amino acids in the cyto-

First letter	Second letter U	Second letter C	Second letter A	Second letter G	Third letter
U	UUU } Phe UUC } Phe UUA } Leu UUG } Leu	UCU UCC UCA } Ser UCG	UAU } Tyr UAC } Tyr UAA } STOP UAG } STOP	UGU } Cys UGC } Cys UGA STOP UGG Try	U C A G
C	CUU CUC CUA } Leu CUG	CCU CCC CCA } Pro CCG	CAU } His CAC } His CAA } Gln CAG } Gln	CGU CGC CGA } Arg CGG	U C A G
A	AUU AUC } Ile AUA AUG Met	ACU ACC ACA } Thr ACG	AAU } Asn AAC } Asn AAA } Lys AAG } Lys	AGU } Ser AGC } Ser AGA } Arg AGG } Arg	U C A G
G	GUU GUC GUA } Val GUG	GCU GCC GCA } Ala GCG	GAU } Asp GAC } Asp GAA } Glu GAG } Glu	GGU GGC GGA } Gly GGG	U C A G

FIGURE 6-2

The genetic code. Each triplet (codon) of an mRNA is shown with the corresponding amino acid (see Fig. 5-27). All but three base sequences code for a specific amino acid. All amino acids except *try*ptophan and *met*hionine are coded for by two or more triplets. (*From A. Roller. Discovering the Basis of Life. McGraw-Hill, New York, 1974.*)

plasm; but it releases into the cytoplasm representatives, or emissaries, that become directly involved in the synthesis of enzymes. The emissaries are RNA, a single-stranded nucleic acid that can take several different forms, three of which are directly involved in sequencing amino acids during enzyme production.

Messenger RNA (mRNA) is a relatively straight chain of nucleotides that is a faithful copy of some portion of one strand of DNA (except that uracil appears in RNA wherever thyamine appears in DNA). An mRNA molecule is copied from DNA by *transcription* (Fig. 6-3), and then it moves to the cytoplasm, where its nucleotides will act as codons that can be *translated*, using the genetic code, into amino acid sequences (Fig. 6-4). *Transfer RNA* (tRNA), also made in the nucleus, is involved in the enzyme assembly process as well. It brings the proper amino acid to the assembly site when a particular codon is being translated. *Ribosomal RNA* (rRNA) is the third type of RNA needed in the translation process. After being produced by transcription in the nucleus, rRNA moves into the cytoplasm, where it combines with a roughly equal amount of protein to form the *ribosomes*. Ribosomes provide the unique environment in which the three RNA types and certain enzymes can assemble proteins. These newly made proteins, or enzymes, then leave the ribosome to perform their vital cellular functions.

The translation of the message in an mRNA molecule is performed one codon at a time. As the mRNA moves through the ribosome, each codon is matched via tRNA with the appropriately coded amino acid. In this way, specific amino acids are brought one at a time to the ribosome and hooked together in a growing chain of amino acids. This growing protein eventually becomes a highly specific enzyme. The important question of when and how to stop was answered by the discovery that one of the mRNA codons represents a "stop" signal that terminates the translation process at the appropriate time and cuts the newly formed enzyme loose from the ribosome.

In the process of translation, the genetic code is translated from an mRNA language, in which the letters are nucleotide triplets, into a protein language of amino acid letters. Previously, in transcription, the mRNA messege was simply copied in a slightly altered form from a segment of DNA. Transcription is analogous to copying something out of a book but writing it in cursive form. In a succeeding translation, the handwritten words are changed to another language. The final translation is very much the product of the originally printed form, even if it was not translated directly. In the same way, the critical amino acid sequence in enzymes is the result of the nucleotide sequence in DNA, even though mRNA nucleotide sequences are used directly in the translation or decoding process.

A *gene* is the particular portion of a DNA molecule giving rise to a particular mRNA that in turn gives rise to a particular enzyme. Genes are the DNA sequence that, when translated, produce specific

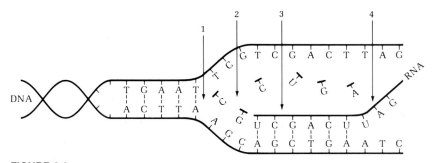

FIGURE 6-3

Transcription is the process by which an RNA molecule is copied or transcribed from DNA. The DNA double strand separates (1), and nucleotides for the growing RNA match up with complementary bases (2). The nucleotides are added sequentially and chemically bonded to form the RNA (3). The new RNA molecule can then move away (4), and the double helix can form again. (*Adapted from D. O. Woodward and V. W. Woodward. Concepts of Molecular Genetics. McGraw-Hill, New York, 1977.*)

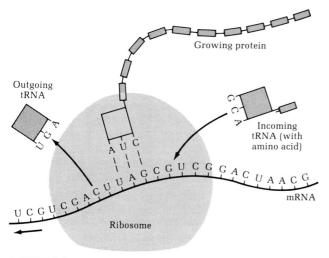

FIGURE 6-4
In translation, proteins are formed or translated from an mRNA nucleotide sequence into an amino acid sequence. Specific amino acids are brought to the assembly site in the ribosome by a tRNA. The tRNAs match with the appropriate mRNA codon and release their amino acid to be attached to the growing protein chain. (*Adapted from D. O. Woodward and V. W. Woodward. Concepts of Molecular Genetics. McGraw-Hill, New York, 1977.*)

enzymes. To revert to our earlier analogy, a gene is a portion of the cellular blueprint that gives rise to a specific enzyme tool.

Blueprints and Chromosomes

Chromosomes, the carriers of genes, are enigmatic organelles that can take on two different forms. In one, they have a distinct size and shape and become microscopically visible, discrete units. Most of the time, however, chromosomes cannot be distinguished as separate entities. Instead they are seen as a shadowy *chromatin material* dispersed throughout the nucleus. We usually describe a chromosome in its readily visualized form, even though it represents the relatively brief portion of the chromosome's existence during cell division, when the genes are passed to the next cellular generation.

Each chromosome is thought to consist of one DNA molecule together with an assortment of RNAs and proteins. Other ideas have been suggested for the packaging of DNA in chromosomes, but one DNA molecule per chromosome is a favored theory. In that theory, genes would be distinct sequences of nucleotides in the long DNA molecule or chromosome. Many hundreds of genes are located along the

length of one chromosome, and many chromosomes are typically present in a nucleus.

The specific number of chromosomes in a genome is constant and characteristic for any particular plant or animal. For plants, the characteristic number of chromosomes for the species may vary from as few as 4 to more than 500 (Fig. 6-5). There is no direct relationship between chromosome number and complexity of the plant: some primitive ferns have the highest known chromosome counts. Some closely related plants have two-, three-, or four-fold differences in chromosomal number. The numbers for some common species are barley, 14; maize, 20; wheat, 42; tomato, 24; and many oak species, 24.

With so many chromosomes in a typical plant cell and several hundred genes on each chromosome, the total number of genes is obviously high. The large number of genes is needed because a gene must be present for each different enzyme in the cell. Perhaps as many as 100,000 different enzymes are needed by various plant (or animal) cells, and the genes for each of those 100,000 must be encoded in DNA at least once somewhere in the cell.

Is there enough DNA in a typical cell's chromosomes to form 100,000 genes to produce 100,000

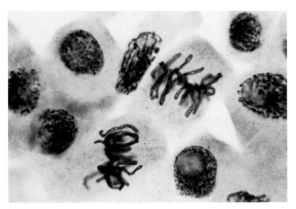

FIGURE 6-5
Photomicrographs of chromosomes during cellular division. Each chromosome is evident as a double structure (with two chromatids) joined at a common point (the centromere).

different kinds of mRNA? Yes, and then some. By some estimates, there is enough DNA in *each* cell in the human body to represent hundreds of thousands of genes, and human genomes are much smaller than those of many flowering plants.

Mitosis

Cells reproduce themselves by a complex process of nuclear division called *mitosis*. Nuclear division is usually but not necessarily followed by cytoplasmic division, or *cytokinesis*. Mitosis results in the production of two cells from an initial cell. These *daughter cells* are identical to each other and to the original cell in their chromosomal content and in the genetic contents of their chromosomes.

As the cell prepares to divide, two critical events occur. First, there is a buildup of energy. Cell division involves much metabolic work and therefore requires energy. Before mitosis and cytokinesis, energy-rich materials in the cell are metabolized to forms of more readily available energy, especially ATP. Second, before the onset of visible division, an event of critical significance and amazing precision occurs: chromosomes are duplicated. The best understood yet most profound part of chromosome duplication involves DNA *replication* (Fig. 6-6). Studies using radioactive isotopes of nucleotides have shown clearly how replication of the giant DNA molecule occurs. Recall the ladderlike structure of the DNA double helix and the fact that each rung is composed of a pair of nucleotide bases, adenine and thymine or guanine and cytosine. During replication, the weak bonds between the complementary nitrogen bases of the double helix are broken and a new helix is formed on each separated half. The result of this precise DNA replication process is two identical double helixes. The chromosome, which was a single DNA molecule before replication, now consists of two strands called *chromatids*. Each chromatid contains identical DNA material because each has as its core a double helix, one-half of which is the original, unreplicated molecule and the other half of which is an exact complementary copy of the original.

The microscopically observable processes of nuclear division can be described as a series of distinct events (Fig. 6-7), but this is a matter of convenience or simplification, since mitosis is a continuous process of stages blending from one to the next.

In the life history of a cell, i.e., from a cell's formation by cell division through the next division,

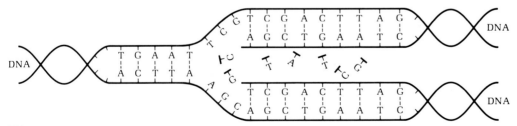

FIGURE 6-6
Replication of DNA. During the DNA-copying process, the double strand of the original molecule separates, and new nucleotides arriving to match up with each strand are then chemically joined. The two new molecules of DNA are actually only half new: one strand of each is from the original double helix. (*Adapted from D. O. Woodward and V. W. Woodward. Concepts of Molecular Genetics. McGraw-Hill, New York, 1977.*)

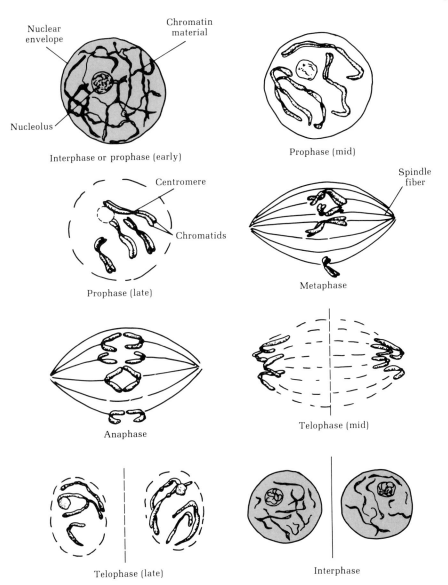

Nuclear envelope

Chromatin material

Nucleolus

Interphase or prophase (early)

Prophase (mid)

Centromere

Chromatids

Prophase (late)

Spindle fiber

Metaphase

Anaphase

Telophase (mid)

Telophase (late)

Interphase

FIGURE 6-7
In mitosis after the chromosomes are aligned along the equator of the nucleus, their chromatids are equally divided into two new nuclei. Only four chromosomes are present in this nucleus. Most plants have many more chromosomes.

the longest phase is the resting stage, or *interphase,* but the cell is resting only in the sense that the nucleus is not undergoing visible changes. During interphase, the cell is carrying out the activities typical of its normal function. All cellular organelles are working, and the chromosomes are greatly extended and dispersed as chromatin throughout the nucleus. Some time toward the end of this phase, as the energy level is built up, DNA replication and chromosomal duplication occur.

The first microscopically evident phase of the division cycle is *prophase*, during which the chromosomes shorten and thicken as a result of internal twisting or coiling. The two chromatids of each condensing chromosome begin to become visible, the nuclear envelope and nucleolus disappear, and the mitotic *spindle apparatus* begins to form. The spindle apparatus is a football-shaped "web" of protein fibers critical to chromosome division.

In *metaphase*, the chromosomes line up in the center of the spindle apparatus along its equator. Individual chromosomes, each now composed of two chromatids, are microscopically visible and identifiable, having shortened and thickened further as a result of additional coiling. The spindle apparatus, which began to form near the end of prophase, now stretches across the cell. Fibrils of the spindle attach to each chromosome at a special region, the *centromere*. The fibrils apparently help align chromosomes on the equator of the spindle apparatus.

In *anaphase*, the sister chromatids separate and become independent chromosomes. As the centromere region of a double chromosome splits, the fibrils from the mitotic apparatus contract and separate the sister chromatids, drawing them toward opposite ends of the cell.

At *telophase*, the two sets of chromosomes have moved to opposite sides of the old cell and organized into new nuclear centers. The chromosomes begin to lengthen and disperse to reform the netlike matrix of chromatin. A nuclear envelope reforms around each chromatin mass, and the nucleoli reappear. Two interphase nuclei are the end result.

In summary, DNA replication during interphase duplicates chromosomal material, temporarily doubling the size of the genome. Each chromosome entering mitosis is composed of two identical sister chromatids. During mitosis, the sister chromatids are separated into the daughter nuclei, reducing the genome to its former size. As a result of mitosis, the two daughter nuclei are genetically identical to each other and to the mother nucleus.

Cytokinesis, or cell division, usually follows telophase immediately. If the cell is to divide, an intervening membrane and cell wall must be laid down. Membrane components, cellulose fibrils, and the other cell-wall materials are synthesized and deposited to produce the separation between the two new cells.

ASEXUAL REPRODUCTION

Totipotency

Mitosis produces cells that are genetically identical; the nucleotide sequences of their DNA are exact copies. The trillions of cells in the roots, stems, and leaves of a giant redwood all contain exactly the same genome. To produce trillions of identical sets of more than 100,000 different genes is a remarkable feat, one that would be quite improbable if it were not that every nucleus is faithfully reproduced by mitosis and all are derived from a single original nucleus, the fertilized egg. In mitosis, the germplasm of the cell is passed on in its exact entirety, division after division.

A surprising but logical extension of the common origin and genome of cells in a plant is that each nucleus contains *all* the genetic information for producing not just a particular kind of cell but an entire plant. Each mitotically produced cell is genetically identical, regardless of whether it is in a root, stem, or leaf and whether it is epidermis, mesophyll, or cortex. Since three different cells with identical nuclei (and genomes) can become, say, root hairs, phloem, or guard cells, depending on where they are found, each nucleus must contain the germ plasm to be *any* kind of cell. The idea may not seem so strange when you consider that all the genetic information for making a plant had to be present in the original cell (the fertilized egg), which gave rise via successive mitotic divisions to the billions or trillions of cells in the plant.

There is proof that each nucleus in a plant contains all the genes needed to make a whole plant. It is possible to remove a group of cells or even a single cell from a plant and cause it to produce an entire new plant. *Cloning*, the process of producing new individuals by mitotically culturing cells from a mature organism, depends entirely upon the fact that all cells in an organism are genetically alike and *totipotent*. Totipotency refers to the ability of a mature cell to return to an embryonic state and to produce a whole new individual. Plant cells are particularly noted for the relative ease with which their totipotency can be expressed. Animal cells are theoretically totipotent in that each cell is genetically like the fertilized egg, but getting them to regenerate whole new individuals requires some special manipulation, particularly for the higher vertebrates.

Cloning plants in the laboratory is now relatively simple. We have learned how to trigger expression of the totipotency of plant cells and how to take economic advantage of it. But even before we learned how to manipulate cellular totipotency in the laboratory, we were taking advantage of nature's own use of the system. As we shall see, many plants also use the totipotency of their cells to generate new individuals.

Vegetative Propagation

When a strawberry plant puts out a runner, that stem grows along the soil surface for some distance and then produces a new plantlet. At first dependent upon the parent plant, in time the plantlet develops its own root system and the connecting stolon disappears. The result is a fully independent plant genetically identical to the original. (Only mitotic divisions were occurring in the development of the stolon and the plantlet.) Many other valuable species reproduce naturally by such *asexual* or *vegetative processes* not involving flowers or the fusion of egg and sperm cells.

The bulbs produced by many flowering plants, such as tulips and daffodils, are another natural means of asexual reproduction. Because only mitotic divisions are involved in the formation of bulbs, flowers grown from the bulbs will be genetically identical to the plant that made them. Red tulips or white daffodils will generate bulbs that produce only red tulips or white daffodils because the genes for flower color are being passed on faithfully in the mitotic divisions that form the bulbs. Many other plants can reproduce asexually and produce large *clones* of genetically identical plants (Fig. 6-8). This natural tendency for many plants to reproduce asexually is often used to *propagate* them. (Many asexually propagated crops will be cited in Chap. 15.)

The fact that vegetatively propagated plants will all be genetically identical to the plants from which they were produced has great commercial and practical significance. For example, someone finding a strawberry plant that yields bigger, sweeter berries than other plants or a tulip with a particularly beautiful flower might want to produce many such plants. For reasons that will become clear when we discuss sexual reproduction, the seeds of a strawberry or tulip will not consistently produce identical offspring, because the particular combination of genes

FIGURE 6-8
Clones. These sumac plants all arose from one original plant that spread vegetatively.

responsible for the desirable berries or flowers will not be preserved in sexual reproduction. Only by using vegetative propagation, which normally passes the germ plasm of a plant on in its entirety, can the developer of a champion berry or tulip mass-produce genetic copies of the original and guarantee that it will produce certain characteristics.

In some cases, vegetative propagation of plants to maintain a certain highly desirable germ plasm requires some time-consuming artificial steps to be taken, i.e., steps that bear little relation to natural plant processes. An example is grafting apple buds onto trees of a less desirable type (see Chap. 15).

One of the most spectacular examples of artificial vegetative propagation involves plant *tissue culture*. In this procedure a group of cells or even a single cell is removed from a plant and put in a solution of nutrients, vitamins, plant hormones, and sugar. The solution acts upon the totipotent cells, which begin to divide and form new cells. Manipulation of the chemicals, especially growth regulators, and other conditions can ultimately cause the mass of cells to develop roots, stems, and leaves. A single cell produces a whole new plant that is genetically identical to the original. Some commercial producers of ornamental flowers have developed their tissue-culture technology to the point that they can produce flowers by the thousands. Genetic combinations that would be lost by sexual reproduction and seed propagation are now successfully transmitted to artificially cloned plants.

Specific techniques of vegetative propagation will be discussed in Chap. 15; for now we emphasize the point that plants reproducing naturally by asexual means or propagated vegetatively with human help will be genetically identical to the plant from which they were produced. Asexual reproduction faithfully maintains the germ plasm of plants. In many agricultural situations, the constancy of genetic characteristics is absolutely essential for consistently producing plants with the desired economic or marketable value.

Some Important Variations

Although we have discussed mitosis and asexual reproduction as error-free processes and have stressed an absolutely faithful copying of the germ plasm from cell to cell or clone to clone, this is an overstatement. Faithful reproduction is the norm but exceptions or errors do occur and may figure prominently in some situations.

Replication of DNA usually produces identical DNA molecules and genetically identical sister chromatids. The sister chromatids are usually aligned at the equator and then divided between the newly developing nuclei. The operative word here is "usually." Errors in replication occur from time to time, and mix-ups in mitosis happen occasionally. Such genetic accidents, or *mutations*, produce daughter cells that are genetically different. The difference may be slight, e.g., a single nucleotide inserted in the wrong place, or more serious, e.g., a whole chromosome deleted or doubled. In all such cases we are dealing with mutations that are called *somatic* because they occur in the somatic rather than the reproductive cells.

The chances of a mutation occurring during a particular mitotic division are quite low; the mitotic process is reliable if not absolutely so. A brief mathematical exercise will demonstrate, however, that even with a very low rate of error the mutants may become significant. Take a hypothetical plant with only 1024 cells at maturity. Those 1024 cells were produced by mitotic doubling of 512 cells. Continuing to reverse this doubling, we calculate 256 cells, then 128, then 64, then 32, then 16, then 8, then 4, then 2, and finally the fertilized egg. To produce the 1024 final somatic cells there had to be

$$512 + 256 + 128 + 64 + 32 + 16 + 8 + 4 + 2 + 1 = 1023$$

separate mitoses. By the same process we calculate that a plant with 1 million cells is produced by $10^6 - 1 = 999,999$ mitoses and a plant with 1 trillion cells by $10^{12} - 1$ mitoses. If a mix-up occurred during only one of the 1023 mitoses in our hypothetical 1024-celled plant, could it be significant? It depends upon the timing. If it occurs in one of the divisions producing 2 of the final 1024 somatic cells, only 2 cells will be affected, a mere 0.2 percent of the whole; but if the single mutation occurs when there are only 16 cells, all the cells produced by successive division of the mutant cells will carry the mutation as well. As a result, 64, or about 6 percent, of the final 1024 cells would differ from the others. Would that be significant? We obviously do not have enough information to answer, but it is possible that the deviation in normal development of only 0.2 percent of the cells could be significant if it were drastic enough or occurred at a critical point.

Not only can mutations occur in the cells of a plant but they are likely, given the tremendous number of opportunities. Some cells may be genetically different from others although they are derived from a common cell, the fertilized egg. Depending on the type of mitotic mix-up and where and when it occurs, the consequences may be drastic or undetectable.

Consider the possible effects of a mutation that interferes with the ability of cells to make chlorophyll. If the mutant cell is located in a root, it will not be noticed or be of any consequence; but if it is in a shoot apical meristem, the consequences may be dramatic: all the leaves on the shoot may lack green color. (The branch may still be able to survive by drawing photosynthates from other parts of the plant.) This is not a strictly hypothetical situation. Such anomalies do arise. The name *chimera* for a plant or plant part that differs genetically (and usually visibly) from the rest of the plant (Fig. 6-9) comes from the name of the mythological creature with head, body, and tail of three different animals. Some apparent chimeras are actually the result of diseases that cause localized autonomous growth.

Although some plant chimeras are mere curiosities, others are horticulturally important. Some desirable apple cultivars first appeared as chimeras, or *sports*, on trees bearing less desirable fruits. The mutant

FIGURE 6-9
Chimera. Part of the tissue in this apple is genetically different from the rest, as evidenced by its color.

somatic cells contained genes (or combinations of them) that produced an apple in some way superior to the rest. The desirable germ plasm was maintained by propagating the sport asexually. Other chimeras are valued for the distinctive appearance they lend ornamental plants.

Other situations can lead to genetic dissimilarities within a plant. Sometimes groups of cells double or quadruple their chromosome number. The cells apparently multiply their genome by replicating DNA, duplicating chromosomes as if they were going to undergo mitosis, and failing to divide. The resultant *polyploid* condition may be fairly common in the cells of mature plant parts. Polyploidy is apparently uncommon in young, actively dividing cells. Plant scientists are not sure why cells become polyploid; the cells remain active metabolically but forego division.

Before concentrating on other methods plants use to generate differing germ plasm, we stress again that the germ plasm of a cell must be passed carefully to daughter cells if the resulting organism is to function according to the specifications in the blueprint. And, despite the occurrence of somatic mutations and chimeras, mitosis is a very reliable method for producing cells that are exact genetic copies.

SEXUAL REPRODUCTION

If organisms reproduced only by asexual means, all the plants or animals descended asexually from a common ancestor would be genetically identical, or they would change very slowly as mutations occurred in asexual reproductive parts. The genetic makeup of an individual, often called its *genotype*, would be identical in any group of vegetatively propagated plants derived from the same plant.

What would be the consequences of a world full of oak trees, grape vines, or soybean plants that are genotypically uniform? Perhaps the most striking would be a drastic reduction in the distribution of oaks, grapes, and soybeans. Instead of being found in diverse locations almost all over the globe, each species might be restricted to specific habitats in distinct geographical areas. Efforts to grow the plants in other areas would probably fail. To understand why a fixed genotype can be a serious limitation we need to learn what genotypes do.

We know that the genome of an organism determines its physical and chemical makeup. The genes in the genotype produce enzymes that give rise to particular physical and physiological characteristics. Since the physical appearance of a plant, called its *phenotype*, is the physical expression of the genotype, plants of the same genotype generally have similar phenotypes. Vegetatively propagated plants may not look exactly like each other or the plant from which they were propagated because the environment affects phenotypic expression of the genotype. Plants in hotter, drier, brighter, or more acid environments may look different from those produced from the same genotype in cooler, wetter, less sunny, or more alkaline environments (Fig. 6-10).

The ability of a genotype to produce a viable phenotype in an environment is an indication of its *adaptation* to that environment. When a plant can grow and reproduce successfully in a particular environment, it is said to have an *adapted germ plasm*; i.e., its genotype produces a phenotype that fits the environment it is placed in. With only a slight change in environment the genotype may be unable to produce a viable phenotype. If conditions get slightly warmer, drier, brighter, or more acid, the plant produced by the genotype may not be able to cope. This explains why asexually reproduced plants may be so restricted in their distribution; their germ plasm may not be adapted to more diverse locations.

If all plants of a species were asexually reproduced, genotypically identical, and adapted only to particular environments, and if the environment were constant over eons, it might be feasible for all mem-

FIGURE 6-10
Environment and phenotype. Plants that are genetically identical may differ as a result of exposure to different environmental conditions. These sorghum plants grew (rear) with, (front) without adequate fertilization.

bers of an adapted species to be genotypically uniform, to have a common germ plasm. Of course stability is not a characteristic of environments, even in the short run. Climatic patterns shift, and local conditions can change quickly. A species locked into a genotype because it reproduces only asexually is in a delicate and almost inevitably disastrous situation.

On the other hand, some method by which a species could create variation in its genotypes would increase the potential adaptability of a species. While any one genotype might not be widely adapted, the existence of several even slightly different genotypes would greatly increase the chance of some plants surviving environmental change or transfer to a different environment.

Gene Variation

Living organisms have such a method for generating new genotypes every time a new individual is formed. Sexual reproduction offers the opportunity to create unique gene combinations in each succeeding generation. Before discussing the specific processes involved in generating new genotypes, we need to see how a gene, which controls production of an enzyme, can be variable.

It would be reasonable (but incorrect) to assume that there is only one nucleotide sequence for any gene. We have argued that DNA is copied very faithfully in the replication process. The nucleotide sequence of a gene prescribing the amino acid sequence of a particular enzyme should be unchanging and the gene should therefore produce an unchanging enzyme responsible for a particular phenotypic characteristic generation after generation.

DNA is not always copied perfectly: DNA replication does not always produce two identical double helixes. There can be mistakes of different kinds, e.g., one nucleotide inserted in place of another during complementary pairing. Accidents like this become part of the genetic blueprint. When the chromatid with the miscopied gene goes to another cell following nuclear division, the *mutant* gene may cause the production of a slightly different enzyme, differing from the original, or *wild type*, by only one amino acid. The slightly altered enzyme may still be able to perform the original function, but the performance may be altered enough to cause some phenotypic difference. The difference becomes noticeable if the mutant gene is transferred to a cell that gives rise to a new individual. In the new plant, the altered gene may result in a different leaf shape, a unique petal color, the disappearance of some feature, or a change in the plant's ability to tolerate certain environmental conditions. The altered phenotype may be for the better or the worse as far as the adaptation of the plant is concerned, or it may have no significant effect at all on adaptability. In any event, the new nucleotide sequence represents variation in the germ plasm of the species.

Each phenotypically detectable variation of a particular gene is called an *allele*. Alleles result from variations (mutations) in the nucleotide sequences for a gene and there are potentially many different alleles for any one gene. (If a gene has 1000 nucleotide pairs, there are 1000 different points at which a mutation can occur.) Nevertheless only a relatively small number of alleles are seen for any gene, in part because many different mutant nucleotide sequences for a gene look the same phenotypically. For example, consider a gene that regulates synthesis of the enzyme controlling the production of a red pigment in flower petals. A slightly altered enzyme might produce less pigment, so that the petal is pink, or it might produce a darker pigment; but the many different alterations in the enzyme that result in no color at all produce the same phenotype (white petals).

Genotypic Variation

Although the variation that arises by mutation in a gene must be multiplied by the 100,000 or so different genes in an organism to give an idea of the total variation possible in a genotype, even that does not reveal the potential for different genotypes in a species. Consider the possible combinations for just two genes, A and B, with four alleles each (A_1, A_2, A_3, A_4, B_1, B_2, B_3, and B_4). There are $4^2 = 16$ unique pairings of A_x and B_x. With 100,000 genes of only four possible alleles each, there would be $4^{100,000}$ different possible unique genotypes, a number much greater than the number of individuals in any species.

By randomizing genes and creating new genomes, sexual reproduction makes each organism potentially different. Without sexual reproduction to make different genotypes with different phenotypes, organisms would be boringly similar. With only asexual reproduction, horse races would not be worth a wager.

Species

There are no wild swings in phenotype from generation to generation. Sexual reproduction allows for change in a marvelous and significant way, but many alterations are generally so unassuming or infinitesimal that they go unnoticed. There is no way that wheat, for example, can turn into something looking more like corn. The 100,000 or so genes in a wheat plant are wheat genes. They will produce only a wheat plant, even if a different allele is substituted for every gene; the alleles are all wheat alleles, all variations of wheat genes. It is the particular set of wheat genes, together with their alleles, that form the *species*.

In a genetic sense, a species is a group of organisms that has a common set of genes and a pool of alleles. This technical definition usually implies the ability of such organisms to mix and create new combinations of alleles during sexual reproduction, i.e., to *interbreed*. Wheat cannot mix its genes with soybean and produce either wheat or soybean (or anything else) because the two species have different sets of genes. Mixing wheat and soybean genes to produce a viable organism would be as likely as shuffling together blueprints of a skyscraper and a football stadium to produce a superdome.

Each species is genetically distinctive. With relatively few exceptions, plants of one species cannot mix their genes with another species, and most of the exceptions have arisen from human manipulation and disruption of the normal reproductive patterns. A few exceptions that are important in agriculture will be considered in Chap. 14.

Chromosomes and Sexual Reproduction

In plants, as in animals, each new generation begins with the fusion of *gametes* (egg and sperm) to create a *zygote* with a new, potentially unique genotype. The zygote divides mitotically and passes on its genome faithfully so that all cells produced from it are genotypically identical. In time, those mitotically formed cells become a mature plant that can produce new gametes, and the cycle starts over.

Obviously there must be a way to maintain chromosome number (genome size) in successive generations. The set of genes forming the germ plasm of a barley plant is distributed over 14 chromosomes; and exactly 14 gene carriers are maintained from generation to generation. Since there is no doubling or redoubling, somewhere along the succession of cell divisions the chromosomal number must be halved. That critical event occurs during the formation of gametes. In barley, for example, each gamete brings to the zygote 7 chromosomes, so that the correct number of 14 is retained in succeeding generations.

The 14 chromosomes of the barley zygote and subsequent somatic cells are actually two sets of 7, each of which carries all 100,000 or so genes needed to make a barley plant. Although it seems strange, having a double genome in each somatic cell is part of a system that prevents chromosomal and genetic chaos. (Wheat, even more peculiar, has a sixfold genome; the 42 chromosomes represent six sets of 7-chromosome genomes.)

Each chromosome in the nucleus of a somatic cell has a chromosomal twin (7 pairs of twins in barley, 10 in corn, and 21 in wheat). These twins are called *homologous pairs* (Fig. 6-11). Each chromosome of a homologous pair looks just like the other and carries the same genes. If one chromosome in the pair carries a gene coding for an enzyme involved in chlorophyll production, the homologous chromosome will also carry that gene at the same location, or *locus*. Each chromosome will have identically placed loci for each of 100,000 or more genes. But the alleles at a given locus may not be the same. For

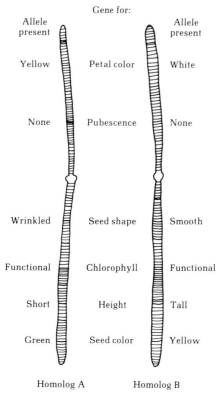

FIGURE 6-11
Homologous chromosomes. These two chromosomes are alike in containing genes to control the same traits. Although each gene is found at the same locus on each homolog, it may have different alleles.

example, the homologous loci for a gene coding for petal color may have an allele for red petals on one chromosome and an allele for white petals on the other. Such alleles have several important consequences, but first we focus on their effect on gamete formation.

Meiosis

The zygote, which contains homologous pairs of chromosomes, is said to be *diploid*. Diploid cells, including all cells produced mitotically from the zygote, contain the homologous sets of chromosomes. Before fusing to form the zygote, the nuclei of the gametes must have only one-half the diploid number of chromosomes if chromosomal number is to be maintained from generation to generation. Each such

gamete is *haploid*, containing a single set of unpaired chromosomes. How is a haploid cell formed? The type of nuclear division to turn a diploid cell into a haploid one must be different from mitosis, in which genome size and chromosomal number are preserved. In addition to halving the genome, this *reduction division*, or *meiosis*, has far-reaching genetic effects.

Despite different outcomes and different intermediate events, meiosis and mitosis have several features in common. The first steps in both involve DNA replication and chromosomal duplication. Cells about to undergo meiosis have chromosomes composed of two chromatids, each of which will be assigned to different daughter cells, as in mitosis. How does a cell double its chromosomal material and then divide into cells with only half the original chromosome number? We know that $2X \div 2 = X$, not $\frac{1}{2}X$. The solution lies in the fact that the nucleus undergoing meiosis divides twice to form four cells; from a single diploid cell, four haploid cells are produced. Mathematically, $2X \div 4 = \frac{1}{2}X$.

As in mitosis, the first events of meiosis are not visible through the microscope. During the premeiotic interphase, the DNA molecules of each chromosome are replicated and the rest of the chromosomal components are also duplicated. When the chromosomes become visible in subsequent stages, each is clearly made up of two chromatids joined at their common centromere. As in mitosis, each chromatid of a chromosome is normally identical at every gene locus. Alleles, the particular nucleotide sequences at particular loci of the DNA molecule, are faithfully reproduced in the replication process with few exceptions. The exceptions, or mutations, can become significant when they find their way into a newly created genotype but they are not frequent.

The microscopically observable events of meiosis are generally similar to those of mitosis with the important difference that they happen twice (Fig. 6-12). The two sequential steps of reduction division are typically numbered I and II.

Many critical events occur during prophase I. As in the mitotic prophase, the chromosomes condense, becoming shortened and thickened. Late in this phase, the nuclear membrane disappears, but as chromosomes become visible, the great difference between meiosis and mitosis becomes evident: homologous chromosomes are physically paired. The pairing is precise; homologous chromosomes lie alongside each other and pair up gene for gene, locus

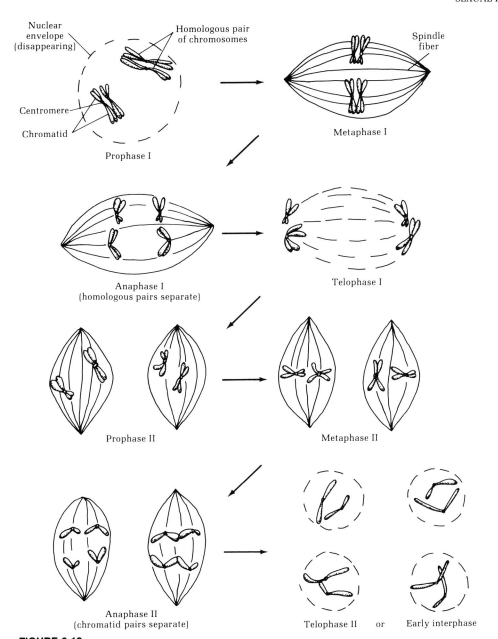

Nuclear envelope (disappearing)

Homologous pair of chromosomes

Spindle fiber

Centromere

Chromatid

Prophase I

Metaphase I

Anaphase I
(homologous pairs separate)

Telophase I

Prophase II

Metaphase II

Anaphase II
(chromatid pairs separate)

Telophase II   or   Early interphase

FIGURE 6-12
In meiosis the chromosomes are aligned in homologous pairs along the equator of the nucleus, and homologs are separated into two nuclear centers. The chromosomes are realigned along the equator of each nucleus, and the chromatids separate equally into what become four nuclei.

for locus. This pairing frequently gives rise to *chiasmata*, points where chromatids from two separate but homologous chromosomes appear to overlap (Fig. 6-13). Exchanges of whole segments of a chromatid can occur at a chiasma. The tip of one chromatid may *cross over*, or be exchanged with, the tip of its homolog. The genetic consequences of crossing-over are discussed below.

Toward the end of prophase I, a spindle apparatus is formed, and the chromosomes, still paired, move toward the center of the cell. Metaphase I is defined as the stage when the pairs of chromosomes are lined up on the equator of the spindle apparatus. In barley, for example, there would be 7 *pairs* of chromosomes aligned on the equator. (During metaphase of mitosis in barley, there would be 14 individual chromosomes

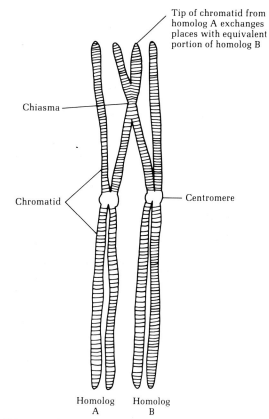

FIGURE 6-13
A chiasma forms when arms of chromatids from homologous pairs of chromosomes become entangled or otherwise exchange portions.

lined up with no positional relation to their homolog.)

At anaphase I, the first separation phase, the members of each homologous pair of chromosomes begin to move apart. Each chromosome is still composed of two chromatids, but since crossing over may have taken place, the chromatids in a chromosome are not necessarily identical. At telophase I, the segregated chromosomes congregate briefly. Each cluster has one-half the number of chromosomes of the original cell.

The period between telophase I and prophase II may be brief or extended. During prophase II, the two chromosome clusters (nuclei) reorganize, and two spindle apparatuses form. Metaphase II resembles metaphase of mitosis in that the chromosomes, still composed of two chromatids, move to the center of the cell. Anaphase II is the stage at which chromatids separate and begin to migrate in opposite directions. In telophase II, a new membrane is formed and cytokinesis occurs. The result of the two sequential divisions is four nuclei, each with the haploid number of chromosomes. Thanks to the physical pairing and crossing-over in prophase I, the genotypes may be different in all four nuclei.

From Meiosis to Gametes

In animals, the products of meiosis usually develop directly into gametes, the cells that fuse in fertilization. In plants, the initial haploid products of meiosis, called *meiospores*, undergo mitotic divisions and further development to produce the gametes. The hundreds of meiospores in the anthers of most flowers develop into the haploid male reproductive units, the *pollen grains* (Fig. 6-14). Each pollen grain contains two or three haploid nuclei at maturity. One, the *pollen-tube nucleus*, plays an important role during prefertilization. The second, when there are only two, is the *generative nucleus*. In flowering plants, it gives rise at some point to two *sperm nuclei*, the gametes that play a direct role in fertilization.

In the female parts of a flower, the meiospores develop into the haploid female reproductive unit, the *embryo sac* (Fig. 6-15), produced in an ovule after a single diploid *meiospore mother cell* in the ovule undergoes meiosis. Normally, only one of the four meiospores produced by meiosis in an ovule becomes an embryo sac; the other three disintegrate. The nucleus of the nondisintegrating haploid cell

Labels in figure:
Tip of chromatid from homolog A exchanges places with equivalent portion of homolog B
Chiasma
Chromatid
Centromere
Homolog A
Homolog B

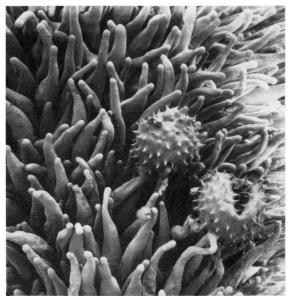

FIGURE 6-14
Pollen grains of cotton following pollination. The stigma is obviously suitable for trapping the spiny pollen grains. (*From J. Troughton and L. A. Donaldson. Probing Plant Structure. McGraw-Hill, New York, 1972.*)

divides mitotically three times to produce eight haploid nuclei ($1 \times 2 \times 2 \times 2 = 8$). The cell containing the meiospore nucleus does not usually undergo cytokinesis but enlarges considerably in the ovule to produce the embryo sac.

The eight haploid nuclei produced in an embryo sac can be divided into three groups. Three nuclei, the *antipodals*, lie opposite the ovular opening or *micropyle*; their function is unclear. Two nuclei positioned in the middle of the embryo sac and called *polar nuclei* will be actively involved in the fertilization process. The other three nuclei lie near the micropyle of the ovule. Two of them, called *synergids*, apparently play an indirect role in fertilization of the *egg*, the last of the eight nuclei and the actual female gamete.

Pollination and Double Fertilization

Following meiosis and development of the gametes, pollination must occur before sexual reproduction can continue. *Pollination* is simply the transfer of pollen from an anther to a stigma. Different methods of pollen transfer are used by different plants. When

pollen lands on a stigma, the flower's receptive surface, the pollen grain normally germinates. The pollen-tube nucleus directs the formation of a tube that penetrates the stigmatic surface, grows down the style, and into the ovary (Fig. 6-16). Eventually, it makes its way to the micropyle of an ovule. The synergids in the embryo sac may help by chemically directing the growth of the tube toward the micropyle.

The mitotic division of the generative nucleus into two sperm nuclei may take place before the pollen tube begins to grow or after the generative nucleus enters the tube; the timing depends on the species. In most members of the grass family, for example, the division occurs in the pollen tube. When the tip of the pollen tube arrives at the ovule, the two haploid sperm nuclei move into the embryo sac. One sperm nucleus fertilizes the haploid egg nucleus to form the diploid zygote, which will develop into the embryo. The second sperm nucleus fuses with the two haploid polar nuclei in the middle of the embryo sac to form a *triploid* nucleus, called the *endosperm* nucleus. Thus, five separate haploid nuclei fuse to form two nuclei. Each fusion or fertilization event is critical to further development. *Double fertilization*, as it is called, is unique to the flowering plants (angiosperms). Conifers (gymnosperms), ferns, mosses, and other non-flower-producing species have only a single fertilization event, fusion of one sperm and one egg.

The endosperm nucleus formed by fertilization of one sperm plus two polar nuclei normally begins to undergo mitotic divisions shortly after it is formed from the fusion of the three nuclei. The triple set of chromosomes is faithfully replicated in each succeeding division so that triploid endosperm tissue is produced. The endosperm provides a nutritive tissue in which the embryo can develop. The embryo forms by mitotic divisions of the diploid zygote. After several divisions, the diploid cells begin to organize into a recognizable embryonic root and shoot.

The fate of the endosperm tissue in a developing seed depends on the type of plant. In members of the grass family, a large mass of endosperm tissue persists through initial seed development and serves as a source of energy to sustain the young seedling after germination. The bulk of a grain in most grasses is endosperm tissue. Common crops such as wheat and corn are grown for their endosperm tissue, which can be milled into flour or consumed directly. The triploid endosperm tissue of the grain crops is high in starch, hence high in energy for young seedlings or for those who eat the grain. In other seeds, the endosperm is completely consumed by the developing embryo, so that the mature seed consists of nothing else. Soybeans and other legumes are examples of *exalbuminous* seeds (without endosperm). Many such seeds remain important for food and feed value, their stored food materials usually being found in fleshy cotyledons.

GENETICS

Genetics from a Monastery

In the middle of the nineteenth century, Gregor Mendel, an Austrian monk, used acute powers of observation, painstaking research, and keen logic to start a biological revolution. Mendel's ideas were so far ahead of his time that his work, published in 1866, was scarcely noticed until the turn of the century, many years after his death. His work clearly demonstrated that many observable plant traits are passed from one generation to the next in a mathematically predictable way. Mendel observed the *inheritance* of plant characteristics to occur in patterns that suggested something about the underlying genetic and chromosomal processes. Although he did not talk about genes and never saw or heard of chromosomes, his work pointed to the discovery of both.

Mendel worked with many different plants in his small garden, but he is remembered primarily for his studies with peas. He *crossed* pea plants that had certain heritable (and variable) traits and meticu-

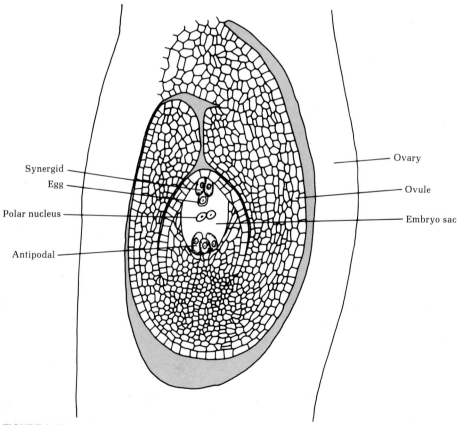

FIGURE 6-15
An embryo sac in the ovule of a typical flower. (*After A. W. Haupt. Introduction to Botany.* McGraw-Hill, New York, 1956.)

Synergid

Egg

Polar nucleus

Antipodal

Ovary

Ovule

Embryo sac

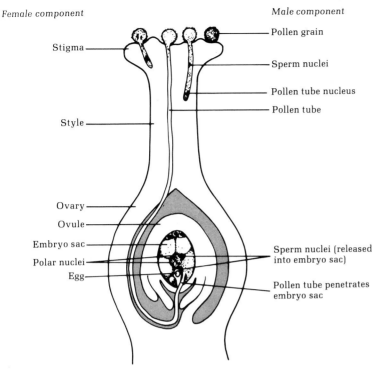

Female component   *Male component*

Stigma

Style

Ovary

Ovule

Embryo sac

Polar nuclei

Egg

Pollen grain

Sperm nuclei

Pollen tube nucleus

Pollen tube

Sperm nuclei (released into embryo sac)

Pollen tube penetrates embryo sac

FIGURE 6-16
Double fertilization occurs when two sperm nuclei meet and fuse with three nuclei within the embryo sac. One fusion occurs between a sperm and the egg to form the zygote. A second event is the fusion of a sperm and two polar nuclei to form the endosperm nucleus. (*After A. W. Haupt. Introduction to Botany. McGraw-Hill, New York, 1956.*)

lously observed the outcome. Crosses were made by carefully introducing the pollen from the flower of one plant onto the stigma of another. He collected all the seeds produced from each cross, planted them, and observed their mature plant characteristics. Mendel kept thorough records of the traits he observed and the numbers of offspring plants with each variation of the trait. This quantitative approach permitted him to see the consistency with which certain traits appeared, disappeared, and/or reappeared in successive generations. As the mathematical regularity and predictability of inheritance became apparent, Mendel formulated "laws" of inheritance, which became the basis for the modern discipline of genetics.

In his pioneering work with peas, Mendel studied seven different pairs of traits, e.g., tall versus short plants: the vines of certain plants grew 2 m or more (Fig. 4-9), while others grew no more than a few

tenths of a meter tall. Other contrasting character pairs included seed color and seed shape.

If Mendel crossed a tall plant without any short plants in its ancestry with a dwarf plant, the offspring (called the F_1, or first *filial*, generation) would all be tall, as tall as the tall parent (Fig. 6-17). When Mendel crossed F_1 plants with each other, the result, the F_2 generation, was a tall-dwarf mixture of plants in a ratio of 3:1. Mendel then allowed the F_2 plants to self-pollinate (the pollen is placed on the stigma of the same plant). When Mendel grew the seeds from the self-pollination of the F_2, now three generations removed from the original cross, all the short plants produced only short plants but two-thirds of the tall plants produced some short plants.

With these results before him, Mendel began to consider possible explanations for the observations. Several conclusions were immediately evident. The dwarfing *character*, or *factor*, though not evident in

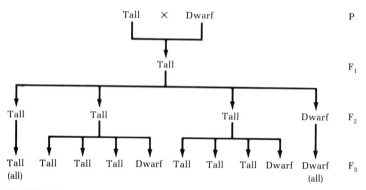

FIGURE 6-17
Mendel's crosses of tall and short peas.

the F_1 generation, must have been present because it reappeared in the F_2 generation. Mendel reasoned (and here is a great leap forward) that the F_1 plants (and in fact all the plants) must be carrying *two* factors (which we call genes) for height. The original parents of the study were envisioned as possessing two factors for tallness (tall plants) or two factors for shortness (dwarf plants). In the F_1 generation, all the plants contained one tall factor and one short factor, but only the tall factor was expressed.

If two different alleles for one trait were present but only one was expressed, the hidden factor was said by Mendel to be *recessive*; the character that did appear was called *dominant*. This was his *law of dominance*. We use the same terminology today to describe both the visible characters and the genes that control them. When a dominant gene (or, more properly, allele) is present, it will be expressed even if the alternative gene is also present in the double set.

Mendel also argued that if genes are present in a double set in each generation, there must be some process operating during sexual reproduction by which each parent can *segregate* a gene pair and contribute one gene for each trait to an offspring. Knowing what we do about meiosis, fertilization, alleles, haploidy, and diploidy, Mendel's *law of segregation* seems obvious, but he was unaware of the cellular mechanisms by which genes are transferred from generation to generation. His theories have proved remarkably accurate.

Mendel performed many more studies and worked with species other than peas. His other findings were primarily extensions of the basic pattern he disclosed

for inheritance of a particular trait with just two contrasting characters and one character dominant. This pattern, which recurs in genetic studies, is called *simple Mendelian inheritance*.

Segregation and Linkage

Since homologous chromosomes are segregated into opposite ends of the cell during meiosis, the two alleles for a gene, each carried on one chromosome of a homologous pair, are segregated into different gametes. For the F_1 plants above, half of the gametes will carry tall alleles, and half will carry short ones. The F_2 plants will then receive either two talls, two shorts, or one of each. Only when two different alleles are present in a parent will segregation be evident. (Parents with two talls or two shorts will produce only tall or short gametes, but a tall-short parent produces gametes of both types. In other words, it is still segregating for height.)

When two different traits are coded for by genes on two different (nonhomologous) chromosomes, the traits will segregate independently. For example, the gene controlling height of pea (tall versus short) and the gene controlling seed color (yellow versus green) are on different chromosomes. Therefore, they *segregate independently*; each height allele moves into a newly forming gamete during meiosis without any regard to what seed-color allele is moving along with it. That means a plant with both tall and short height alleles and both yellow and green seed-color alleles can produce gametes of four genotypes for the traits considered: tall-green, tall-yellow, short-green, or short-yellow. Each allele pair is segregating independently.

On the other hand, if genes for two different traits are on the same chromosome, they are said to be *linked* and cannot segregate independently. For example, if tall alleles were on the same chromosome with green and short were on the same chromosome with yellow, only two kinds of gametic genotypes could result. Crossing-over (the exchange of genetic material between homologous chromosomes during prophase I of meiosis) offers a way to *recombine* genes that are otherwise linked. Linkage and crossing-over create mathematical complications in genetic studies that Mendel luckily avoided because none of the seven contrasting trait pairs he studied were linked. It has been suggested that he was as lucky as he was astute. Others have suggested that Mendel may have failed to report studies that offered conflicting evidence, but this is speculative because after his death his copious notes were heaped on a bonfire.

Genotypes and Phenotypes

Chap. 14 will discuss how plant breeders manipulate genotypes in order to generate desirable phenotypes. The genome of an individual, its total set of genes, is made up of a double set of the 100,000 or so genes needed to run the cells of the species. It is a double set, because each gene is represented twice in a diploid cell, once on each chromosome of a homologous pair.

The phenotype of a plant, the sum of its physical characteristics, is the physical expression of the genotype. From Mendel's example, part of a pea genotype contains genes that may cause tallness. Tallness is the phenotypic expression of that particular portion of the genotype. The genome with its particular combination of genes gives rise to a plant with a particular combination of physical and chemical features.

The third factor in the genotype-phenotype relationship, the environment, can dramatically modify the phenotype. Even when pea plants contain alleles for tallness in their genotype, the environment in which the plants grow can keep them from becoming tall. Adverse temperatures, low light, drought, poor fertility, or diseases can cause a plant to be short in spite of a genotype otherwise programming it to be tall. Phenotype, then, is more properly explained as the physical expression of the genotype as modified by environment. Plant breeders and geneticists work with phenotypes in usually favorable environments

and manipulate their genotypes by various means in order to produce more favorable genotypes or to learn more about the genotypes themselves.

Although genotype and phenotype are used to describe the sum of the genes and their physical expressions, respectively, the words may also be used to describe the genetic situation or physical expression for a particular trait. For example, the genotype for a height-controlling gene may be represented by two tall alleles, two short alleles, or one of each. The phenotype in each case would be tall, short, and tall, respectively.

If both gene loci for a trait (one on each homologous chromosome) contain the same allele (the same nucleotide sequence), the genotype is said to be *homozygous*. Two different homozygous genotypes may occur in simple Mendelian inheritance. If both genes of the pair contain the dominant allele, the genotype is homozygous dominant. If only the recessive allele is present, the genotype for that trait is homozygous recessive. If an allele of each type is present, the genotype is *heterozygous*. Heterozygous always signifies that two different alleles are present for the gene pair.

In a heterozygous genotype, often only one of the two different alleles is expressed phenotypically. That allele shows dominance, as in the case of the tall allele for pea-stem height. In many other heterozygous genotypes, however, neither of the alleles present is clearly dominant or recessive, and some intermediate phenotype results. For example, when an allele for white petals is combined in a heterozygous genotype with an allele for red petals, the resulting phenotype may be pink. In species other than peas, a heterozygous height genotype might produce a plant of intermediate height. When one allele tends to be expressed phenotypically rather more than another, it shows *incomplete dominance*. One allele expressed without any evident phenotypic influence of a contrasting allele is said to have *complete dominance*.

Mating Systems

The *mating system* of a plant species refers to the relationship between the female parent and the source of pollen. In developing and maintaining crop species, understanding the mating system is of prime importance. Three general mating systems based on the mode of pollination are recognized: *self-pollination* (selfing), *open pollination* (outcrossing), and *mixed* (some selfing and some outcrossing). The two basic mating systems (self- and open pollination) have different genetic outcomes.

Self-pollination generally occurs when pollen from an anther lands on the pistil of the same flower. Actually, if pollen from any flower on a plant lands on any stigma in a flower of the same plant, self-pollination is accomplished, but self-pollination usually occurs in a *single* flower (Fig. 6-18). Self-pollination is often favored or ensured by special floral structures, generally in plants with *cleistogamous* flowers or florets, i.e., closed flowers which prevent the entry of foreign pollen to the pistil. Common examples are wheat, barley, and soybean. The palea and lemma, which are coverings surrounding the reproductive structures of grasses, commonly stay tightly closed during pollination, nearly ensuring self-pollination.

Outcrossing, or cross-pollination, is in a sense the opposite of selfing; the pollen of one plant is delivered to the pistil of another. Plants have a number of structures and/or characteristics to help ensure cross-pollination in some species. Obviously, only cross-pollination is possible in *dioecious plants*, those with separate male and female individuals. Among cultivated plants, dioecious species are fairly rare, occurring in some hollies, hemp, asparagus (Fig. 20-14), and a few others.

Self-incompatibility, a common feature ensuring cross-pollination or preventing self-fertilization among crops, is an inherited (genetic) phenomenon. In some cases, the mechanism to prevent self-pollination is mechanical or physical (the arrangement of anthers and stigma prevents contact or transfer). In other cases, self-incompatibility is by a chemical or metabolic mechanism, enzymes produced in the stigmas preventing germination of pollen with the same genotype. In another situation, self-fertilization is effectively prevented by retarding the growth of the plant's own pollen tubes through its style, so that it will compete poorly with other pollen.

Self-incompatibility by metabolic means prevents fertilization of a plant by an individual from the same clone or otherwise containing the same geno-

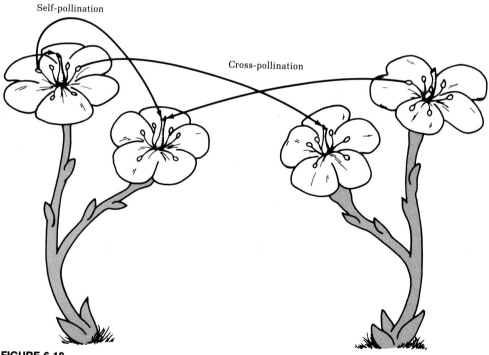

FIGURE 6-18
Mating systems.

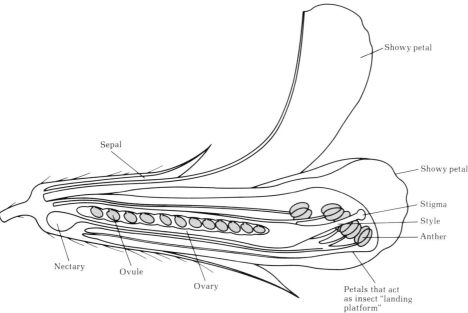

Showy petal

Sepal

Showy petal

Stigma

Style

Anther

Nectary

Ovule

Ovary

Petals that act
as insect "landing
platform"

FIGURE 6-19
An alfalfa flower (here with half cut away) is well-suited for pollination by insects. (*After a
drawing by the U.S. Department of Agriculture.*)

type. An important example is the apple. There must
be at least two cultivars of apple within bee-flying
distance to ensure cross-pollination for a good crop.
Two trees of the same cultivar, or genotype, often
will not provide adequate cross-pollination because
their exchanged pollen is recognized as if it were
self. Under laboratory conditions, plant scientists
can force self-fertilization, but under natural conditions, plants with sufficiently different genetic con-
stitutions must be present to produce compatible
pollen. Of course, without pollination by a compatible plant there will be no fertilization, and without
fertilization there will be no fruit.

Many species have structural floral adaptations
that ensure open pollination or at least heavily favor
it. Although corn can be self-pollinated, separate
staminate and pistillate flowers on the same plant
encourage cross-pollination, which in this case is
assisted by wind. The light pollen produced in the
apical tassels is blown to the exposed, elongated
pistil structures, the silks arising in profusion from
each ear. Many trees, including oaks, pecans, pines,

and palms, have separate male and female structures,
and cross-pollination is common.

In many species, flowers are adapted for insect
pollination. The arrangement of the pistil and anthers
in a flower of alfalfa together with the structure of
the petals and the presence of attracting nectar ensure
that when a bee moves down into the flower, a petal
is tripped so that the anthers brush against the bee's
body, covering it with pollen (Fig. 6-19). In some
flowers, the bee must crawl over the anthers to reach
the nectaries; in the process, its body picks up pollen
grains. In subsequent visits to flowers, the pollen the
bee carries is brushed against receptive stigmas. Some
plant species are adapted to pollination only by one
specific insect or bird. Some orchids have complex
structures which attract specific pollinators. Each of
these methods helps to promote or ensure cross-
pollination.

Sometimes cross-pollination is assured by differences in the rate of maturation of pistil and stamen.
If pollen is shed before or after maturity of the
stigmas of the same plant, selfing cannot occur. This

Generation	Genotypes	Ratio AA:Aa:aa	AA+aa:Aa	% Heterozygous
Parents	Aa X Aa			

(1/2 of all gametes [eggs and sperm] are "A", 1/2 are "a"; all are equally likely to fertilize)

Generation	Genotypes	Ratio AA:Aa:aa	AA+aa:Aa	% Heterozygous
F_1	AA 2Aa aa	1:2:1	1:1	50

(Throughout the F_1 population, 1/2 of each kind of gamete is "A", 1/2 is "a"; in an open pollination system, each sperm is equally likely to combine with any egg)

Generation	Genotypes	Ratio AA:Aa:aa	AA+aa:Aa	% Heterozygous
F_2	AA 2Aa aa	1:2:1	1:1	50
F_3	etc.			50
.				
.				
.				
F_n	etc.			50

Conclusion: Cross fertilization (open pollination) maintains heterozygosity.

FIGURE 6-20
Inheritance pattern of a single gene with two alleles (A and a) in a cross-pollinating species.

happens to some degree in corn, where the pollen of an individual plant tends to be released a few days before its silks are fully receptive.

Often the mating system of a species is not clearly cross- or self-pollinating; pollination may be mixed— some selfing and some outcrossing. Depending on environmental conditions, even cleistogamous wheat and barley florets are not totally self-pollinated. In fact, it is doubtful that any species is exclusively self- or cross-pollinated.

Generation	Genotypes	Ratio AA:Aa:aa	AA+aa:Aa	% Homozygous
Parent	Aa			0
F_1	AA 2Aa aa	1:2:1	1:1	50
F_2	4AA 2(AA + 2Aa + aa) 4aa	3:2:3	3:1	75
F_3	16AA + 8AA + 4(AA + 2Aa + aa) + 8aa + 16aa	7:2:7	7:1	87.5
F_4	etc.	15:2:15	15:1	93.75
.				
.				
.				
F_7	etc.	127:2:127	127:1	99.23

Conclusion: Self fertilization increases/maintains homozygosity.

FIGURE 6-21
Inheritance pattern of a single gene with two alleles (A and a) in a self-pollinating species. This illustration assumes a small number of offspring (four) from each plant of each genotype in each generation.

INHERITANCE PATTERNS

The genetic consequences of continuous self-pollination and continuous open pollination are quite different. When the mating system of a species favors open pollination, there is a tendency for genotypes to be highly heterozygous, i.e., for two different alleles to be present for many of the genetically determined traits. This is exactly what would be predicted if individuals of diverse genotypes are in effect swapping pollen (Fig. 6-20). Each new egg-and-sperm combination will produce many different heterozygous gene pairings. Open pollination will continue to keep genes mixed throughout the cross-pollinating population. In other words, open pollination maintains heterozygosity.

Self-pollination has a very different effect on genotypes (Fig. 6-21). Instead of maintaining heterozygosity, selfing tends to bring each gene pair to homozygosity. The mechanics of gene movement from generation to generation are such that alleles tend eventually to be paired with an identical allele in homologous pairs. In just seven generations, the offspring from a parent that is heterozygous for one trait will be 99 percent homozygous for that trait. Half of the 99 percent will be homozygous AA and the other half will be homozygous aa; but selfing is quickly leading to greater homozygosity.

Because the two radically different pollination systems are naturally built into various species, plant breeders must approach breeding and seed production of a particular species with full awareness of the normal mating system and the inherent results of each. Open pollinating species like corn, apple, or oak will normally be highly heterozygous and not readily developed to the point of genetic uniformity and consistent high phenotypic performance. On the other hand, largely self pollinating species like wheat, soybean, or tomato tend to produce individuals that are highly homozygous and produce offspring that are of similar genotypes. Each of the basic mating systems can be manipulated to the advantage of the plant breeder, but each has its own problems and limitations. These and additional genetic concepts are discussed in more detail in Chap. 14.

PART TWO

ENVIRONMENTAL FACTORS

CHAPTER SEVEN ▬▬▬▬▬▬▬

LIGHT

Sunlight is the ultimate source of almost all energy consumed on earth, especially the energy used by living things. In photosynthesis, green plants convert radiant energy into chemical forms which can be used in turn by nonphotosynthetic organisms. The basic plant process of photosynthesis, which is directly driven by radiant energy, was described in Chap. 5. The many other effects of light on plants described in this chapter include its influence on seed germination, vegetative growth, flowering, and plant morphology.

In this chapter we are concerned mainly with the direct effects of solar radiation on plants. Much of the sun's energy is converted into heat when it is absorbed, a fact of major importance to plants that will be explored in Chap. 8.

RADIANT ENERGY

Radiation is one of the ways energy can move from one point to another. In a vacuum, radiant energy travels at the rate of 3×10^8 m s^{-1}. Radiant energy behaves in part as if it were transmitted in waves and in part as if it were discrete particles. The wave characteristic is used in classifying radiation on the basis of *wavelength* (the distance between successive crests or troughs) or the *frequency* (the number of wave crests passing a point in a particular time span). Since all radiation travels at the same velocity in a

vacuum, short-wavelength radiation has a high frequency and long wavelengths correspond to low-frequency radiation. In other media, however, not only is the speed slower but the velocity of different wavelengths varies. In air, for example, blue light travels more slowly than red light.

Spectrum of Radiation

Radiant energy streaming from the sun in the *electromagnetic spectrum* (Fig. 7-1) ranges from extremely short-wave cosmic rays to very long-wave radio waves.

The unit of radiant energy is the *photon*. Its energy is greatest for the shortest wavelengths; conversely, the longer the wavelength the lower the energy per photon. For example, the energy in a photon that has a wavelength of 800 nanometers (nm) is only one-half of that of a photon at a wavelength of 400 nm. This relationship means that the shortest wavelengths (cosmic rays) have an extremely high energy level. Cosmic rays, x-rays, and other shorter-wavelength forms of radiation are extremely harmful to biological materials because of their tendency to ionize the molecules they strike by displacing electrons. The effect is similar, although not controlled or productive, to what happens when certain photons strike chlorophyll and cause electrons to be ejected. The shorter, destructive wavelengths are often referred to as *ionizing radiation*. Ultraviolet light is a mildly ionizing form that can damage biological

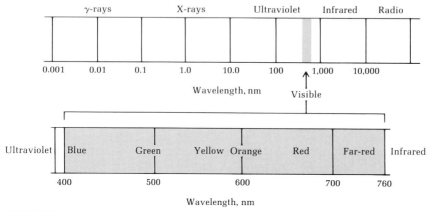

FIGURE 7-1
The electromagnetic spectrum of solar radiation. (*From H. Smith, Photochrome and Photomorphogenesis, McGraw-Hill, London, 1975.*)

tissues. (It is primarily the ultraviolet in sunlight that burns our skin.)

By definition, the visible portion of the spectrum encompasses those wavelengths to which the human eye is sensitive. Strictly speaking, only this restricted part of the spectrum from about 400 to 700 nm is *light*. In common usage, however, "light" frequently denotes a broader part of the spectrum, often including wavelengths of ultraviolet, or "black light," and infrared. The response curve of the eye varies with the individual, but the standard visual sensitivity to various wavelengths is as given in Fig. 7-2. Sensitivity of the human eye peaks at 555 nm and drops off rapidly at both longer and shorter wavelengths. The visible portion of the spectrum (400 to 700 nm) also encompasses the wavelengths active in photosynthesis and many other photochemical processes that affect plant growth and development. We shall see shortly why organisms seem to have "taken advantage" of primarily this portion of the spectrum.

Solar Radiation

Although a small amount of the energy to power civilization comes from the interior of the earth and more is contributed by atomic fission, our most abundant source of energy is the sun. "Fossil fuels" are sun/plant-derived, when photosynthetic products were converted within the earth's geochemical factory. With a diameter of approximately 1.4×10^6 km (about 100 times that of earth), the sun has a mean distance from earth of about 1.5×10^8 km. The thermonuclear processes in the sun's core keep its surface temperature at an estimated 6000°C. The nuclear activity of its interior causes the sun to emit constantly tremendous amounts of radiant energy in all directions. The earth and its atmosphere are warmed and energized by receiving only a small fraction (about 4×10^{-10}) of all the energy radiated by the sun.

All matter at a temperature above absolute zero† continuously emits radiant energy. Two important features of radiation are that the predominant wavelength of the radiation becomes shorter as the temperature of the body increases and that the total radiation per unit area from all wavelengths is proportional to T^4, where T is the absolute temperature of the radiating body. Because of its high temperature, the sun radiates much energy in the short wavelengths ranging from 100 to 4000 nm (Fig. 7-1). About 10 percent is ultraviolet (100 to 400 nm), 45 percent visible (400 to 700 nm), and 45 percent infrared (700 to 4000 nm).

The total incoming solar energy reaching the outer edge of the earth's atmosphere averages 1.94 cal cm^{-2}

---

†The absolute (or Kelvin) temperature scale is based on absolute zero, at which point matter contains no heat, and is equivalent to -273°C or -460.4°F. Ice melts at 273 K; water boils at 373 K.
Note that the kelvin (small k) is the SI name for the temperature interval on both the Kelvin and Celsius scales.

min⁻¹, a value known as a *solar constant*. We can put this in perspective by noting that in 10 days the energy arriving at the periphery of the earth's atmosphere is equal to the total known fossil fuel reserves. The solar constant varies slightly as the distance between the sun and earth changes.

Reflection by clouds can sharply reduce the amount of solar energy reaching the earth's surface. Clouds act as poor absorbers and efficient reflectors of radiant energy. This is particularly apparent in an airplane on an overcast day; when the plane breaks through the cloud cover, the light above the clouds is intense since most of the incoming energy is reflected. The proportion of energy reflected by clouds is often 50 to 90 percent, depending on cloud depth and density.

The solar radiation striking the surface of the earth, called *insolation*, can be divided into *direct radiation*, received directly from the sun, and *diffuse* (or sky) *radiation*, received indirectly because it is *scattered*, or reflected. The proportions vary depending on cloudiness, latitude, angle of the sun, and altitude. Most solar energy is received as direct radiation on a clear day and as diffuse radiation on an overcast day. Long-term averages indicate that of the total solar radiation reaching the outer edge of the earth's atmosphere, about 19 percent penetrates directly, 28 percent arrives as diffuse radiation, 19 percent is absorbed by the atmosphere, and 34 percent is reflected back into space.

Scattering by various particles in the atmosphere also alters incoming radiation. Large particles such as dust, smoke, and water droplets scatter all wavelengths nonselectively. Since white light is a mixture of all colors, the sky takes on a whitish color when there is a high concentration of such particles in the air. Smog and haze, which have become all too common, are one cause of nonselective scattering. Smog may also impart a brownish color to the air due to the presence of colored particles. Small particles, such as gas molecules, scatter the shorter wavelengths of visible light (blue) more than the longer wavelengths (red). It is for this reason that on a clear, dry day with little air pollution, the sky is very blue because of the preferential scattering of the blue portion of the spectrum. At sunset on such a day, the sky often has a reddish hue because the solar energy is passing through the equivalent of several depths of atmosphere and the shorter-wave (blue) energy is removed by scattering before the energy reaches our eyes.

As the solar energy penetrates the atmosphere, *absorption* by atmospheric components is of major consequence, especially with ultraviolet light. Oxygen and ozone absorb most of it (Fig. 7-3), a matter of importance since neither plants nor animals can stand very much ultraviolet radiation. The concern over possible damage to the ozone layer from fluorocarbons in aerosol sprays is related to an increase in the ultraviolet radiation received at the earth's surface. Water vapor and carbon dioxide also absorb some energy in the infrared region.

Of utmost importance to life on earth is the *atmospheric window effect*, which allows the relatively unrestricted penetration of visible light through the earth's atmosphere (Fig. 7-3). Ozone, carbon dioxide, and water vapor absorb and filter out specific wavelengths but freely transmit the visible region of the spectrum. Almost one-half of all radiational energy reaching the surface of the earth is in the visible wavelengths. It is not likely to be coincidental that

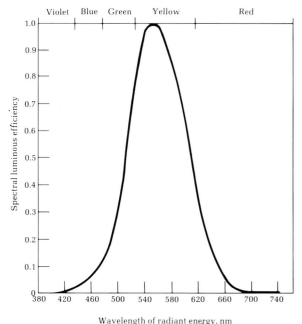

FIGURE 7-2
Standard spectral luminous efficiency curve (for photopic vision), showing the relative capacity of radiant energy of various wavelengths to produce visual sensation. [*From J. E. Kaufman (ed.), IES Lighting Handbook, 5th ed., Illuminating Engineering Society of North America, New York, 1972.*]

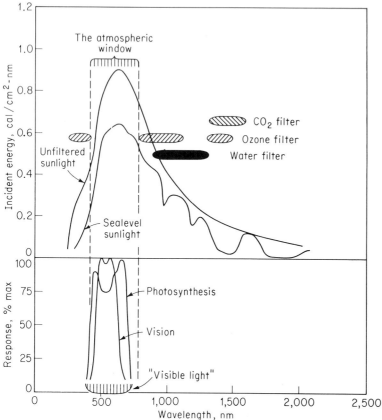

FIGURE 7-3
The radiation received by the earth from the sun is most abundant in the 400- to 700-nm range, which coincides with the wavelengths used in animal vision and plant photosynthesis. (*From A. C. Leopold and P. E. Kriedemann, Plant Growth and Development, 2d ed., McGraw-Hill, New York, 1975.*)

such biological processes as animal vision and plant photosynthesis capitalize on such a small portion of the electromagnetic spectrum. Much more likely is that the process of natural selection has favored those organisms which can use this most available range of the total spectrum.

Terrestrial Radiation

Because it has a temperature above absolute zero, the earth also radiates energy. With an average surface temperature of 14°C, its radiation is primarily in the long-wave part of the spectrum. More than 99 percent of the spectrum of *terrestrial radiation* is in the infrared region. These infrared wavelengths are strongly absorbed and reflected by water vapor, ozone, carbon dioxide, and clouds. Much of that

reflected or absorbed by the atmosphere is radiated back to earth. It is largely due to this reflection and counterradiation that the drop in temperature at night is not more severe, as it is on the moon.

Greenhouse Effect

The net effect of the atmosphere on incoming and outgoing radiation results in the *greenhouse effect*.† Incoming solar radiation is largely shortwave because

---

†The name arose by analogy with a greenhouse, but in fact although the glass in a greenhouse has some warming effect, most of the temperature increase occurs because the glass prevents the warm air from escaping and removing heat through convective mixing (Chap. 8). For this reason some authorities prefer the term *atmospheric effect*.

of the high temperature of the sun and therefore passes through the atmosphere relatively unimpeded. Terrestrial radiation, on the other hand, is long wave and passes through the atmosphere with difficulty. The net result is that the atmosphere lets in more energy than it lets out (Fig. 7-4).

FACTORS INFLUENCING INSOLATION

Daily and seasonal fluctuations in insolation are directly related to latitude, season, elevation, and time of day, all of which interact. Insolation has such a great influence on temperatures, that these factors will be discussed further in Chap. 8.

Latitude

The inclination of the earth's axis of rotation is $23\frac{1}{2}°$ to its plane of rotation around the sun (Fig. 7-5). The portion being illuminated at one time, which is only one-half of the earth, is continually changing. At our spring and fall equinoxes (March 21 and September 21, respectively), insolation reaches the earth from the north to the south pole; therefore, all points on the earth have 12 h of daylight and 12 h of darkness, and all regions of the outer atmosphere receive similar amounts of insolation. Insolation striking the earth's surface varies with latitude for two major reasons. (1) At the time of the equinoxes, the sun is directly over the equator, and the *solar angle* decreases with increasing latitude. Thus, as one moves poleward, the same amount of radiant energy is spread over an increasing surface area. This effect is easily demonstrated by shining a flashlight on a dark wall, first at a right angle and then at different angles. As the angle of incidence decreases, the same number of photons (radiant flux) is received by an increasing

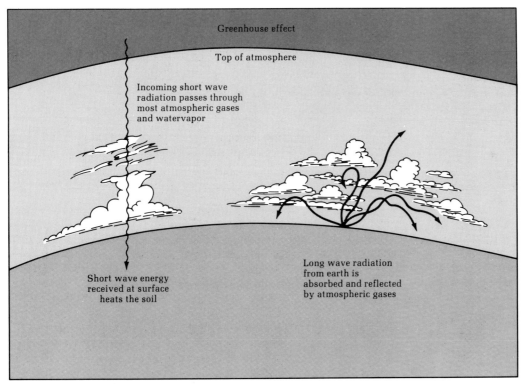

FIGURE 7-4
The greenhouse effect refers to the difference in the effect of the earth's atmosphere on incoming shortwave solar radiation and outgoing long-wave terrestrial radiation. The atmosphere is relatively transparent to shortwave solar energy but relatively opaque to long-wave radiation from the earth. (*Adapted from illustration by National Oceanic and Atmospheric Administration of the U.S. Department of Commerce.*)

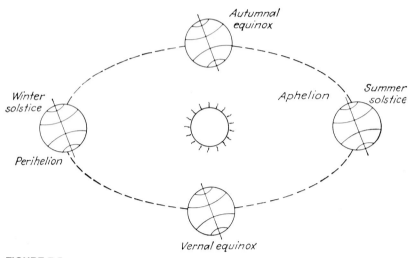

FIGURE 7-5
The relative positions of the earth in relation to the sun at the summer and winter solstices and the fall and spring equinoxes. (*From H. R. Byers, General Meteorology, 4th ed., McGraw-Hill, New York, 1974.*)

area and therefore the *radiant flux density* (number of photons received per square centimeter) decreases. (2) As the angle of the sunlight decreases, the depth of atmosphere through which the insolation has passed increases:

Solar angle	90°	60°	30°	10°
Relative depth	1	1.2	2.0	5.7

As the depth of atmosphere increases, the amount of solar energy reaching the earth decreases, owing to absorption, reflection, and scattering by the atmosphere. The same phenomenon can be seen by comparing noontime insolation with that of early morning, when the angle of the sun's rays is small.

Season
At the equator, day length is 12 h throughout the year, and solar angle is always relatively close to 90°. As one moves north or south from the equator, summer day length increases (Table 7-1), so that in spite of the lower solar angle, daily insolation is moderately high. During winter, far-northern latitudes have not only very low solar angles (actually below the horizon above the Arctic Circle) but very short days as well.

Elevation
If we compare two sites, one at sea level and one at 1500 m elevation, we find differences in incoming and outgoing radiation. It has been estimated that one-half of the total water vapor in the earth's atmosphere is in its lower 1800 m. In other words, the lower 0.2 percent of the atmosphere contains more than 50 percent of its total water vapor. Since water vapor is a much more efficient barrier to transmission of terrestrial radiation than it is to solar radiation, water vapor is of major significance in the greenhouse effect. Although a site at 1500 m receives slightly more insolation than one at sea level, its rate of radiant heat loss is much greater and the net effect is a lower temperature.

Time of Day
Insolation from the sun is of course received only between sunrise and sunset, with the maximum at *solar noon*, when the sun comes closest to the zenith. Terrestrial radiation occurs both day and night, but the rate varies with the earth's temperature (explored further in Chap. 8).

RADIATION TERMINOLOGY AND MEASUREMENT

The level of radiant energy is measured and expressed in different ways in different disciplines. Many are concerned only with the visible portion of the spectrum and its relation to such everyday activities as conventional photography and the lighting for homes, offices, shopping areas, and streets. Most of us are familiar with the use of a photographic light meter, used to determine proper exposures. Although it measures visible light, it does not provide readings in specific units, indicating instead corresponding lens apertures and shutter speeds.

For illumination engineering, more refined light meters have been developed, for which the most widely used unit of measurement used to be the *footcandle* (fc), originally defined as the illuminance (radiant energy) from a standard candle received on a 1-ft² surface that is everywhere 1 ft from the candle. (The standard candle has been replaced with a more reproducible light source.) The footcandle, also called a *lumen* (lm), is being replaced by the SI unit *lux* (lx), defined as an illuminance of 1 lm dispersed over 1 m² (10.76 ft²); thus, 1 fc equals 10.76 lx. Full sunlight at noon in summer is normally about 108 klx (10,000 fc); full moonlight is about 0.32 lx (0.03 fc). For reading, about 215 lx (20 fc) is considered adequate. Lux and footcandles are units of illumi-nance, not intensity. *Intensity* refers to a quality of the *source* of the light and is expressed in different units.

The footcandle and lux have been widely used in the past by researchers in the plant sciences, presumably because of the availability and low cost of the instruments, but unfortunately, the use of footcandle meters makes data difficult to interpret when different sources of light energy are compared, because of different spectral quality. The response curve of the sensor in a footcandle meter approximates that of the human eye (Fig. 7-2), whereas plants are most responsive to different wavelengths (see Fig. 7-8). Thus, for data to be meaningful, a researcher must provide both the illuminance and the specific light source. Typical fluorescent-lamp spectra (Fig. 7-6) are not the same as those of incandescent lamps (Fig. 7-7) or daylight (Fig. 7-3).

Instruments and Measurements

Instruments for measuring visible light, called *photometers* or simply *light meters,* are designed to match human vision and therefore are not ideal for use in plant studies. For very refined research, as with the pigment *phytochrome* described later in the chapter, *spectroradiometers* are used. They can measure the amount of radiant energy being received in specific wavelengths, so that the radiation can be

TABLE 7-1
Effect of Latitude and Season on Day Length†

Latitude, °N	Day length			
	Mar 21	June 21	Sept 21	Dec 21
0 (Equator)	12 h	12 h	12 h	12 h
10 (Caracas)	12 h	12 h 35 min	12 h	11 h 25 min
20 (Mexico City)	12 h	13 h 12 min	12 h	10 h 48 min
30 (Jacksonville, Florida)	12 h	13 h 56 min	12 h	10 h 4 min
40 (Columbus, Ohio)	12 h	14 h 52 min	12 h	9 h 8 min
50 (Winnipeg, Manitoba)	12 h	16 h 18 min	12 h	7 h 42 min
60 (Anchorage, Alaska)	12 h	18 h 27 min	12 h	5 h 33 min
70	12 h	2 months	12 h	0 h
80	12 h	4 months	12 h	0 h
90	12 h	6 months	12 h	0 h

† Adapted from A. Miller and J. C. Thompson, *Elements of Meteorology,* Merrill, Columbus, Ohio, 1970, p. 81.

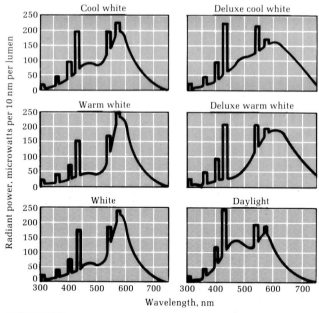

FIGURE 7-6
Spectral energy distribution curves for typical "white" fluorescent lamps. [*From J. E. Kaufman (ed.), IES Lighting Handbook, 5th ed., Illuminating Engineering Society of North America, New York, 1972.*]

described in terms of quantity (number of photons or energy content) and quality (wavelength).

A *radiometer* measures not only visible light but also the radiant energy at wavelengths both shorter and longer than visible light. Black surfaces absorb

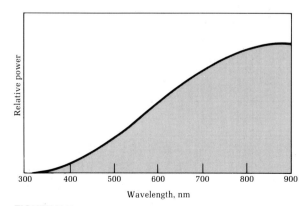

FIGURE 7-7
Spectral energy distribution for incandescent lamps, including tungsten halogen. [*From J. E. Kaufman (ed.), IES Lighting Handbook, 5th ed., Illuminating Engineering Society of North America, New York, 1972.*]

essentially all radiation striking them, whereas white or silvered surfaces reflect almost all incoming radiation. Thus, the temperature difference between a black surface and a white surface is a good measure of radiation received. A radiometer has black and white surfaces whose temperatures are compared by sensitive temperature sensors called *thermocouples*.

The trend in agricultural research is to express radiant energy in terms of *photosynthetically active radiation* (PAR), that portion of the spectrum between 400 and 700 nm. Because photochemical reactions, including photosynthesis, are driven by photons, radiant energy used by plants is best described in quantum or photon terms, a convenient unit being the einstein (E). Since a quantum sensor is generally used and

$$1 \text{ E} = 1 \text{ mol of photons} = 6 \times 10^{23} \text{ photons}$$

the measurements are expressed in microeinsteins per second per square meter (μE s^{-1} m^{-2}) between 400 and 700 nm.

Energy Receipt and Disposal
Radiant energy striking a leaf may be reflected, absorbed, or transmitted. The fate of incident radia-

tion depends on the wavelength. Most photons in the near-infrared (700 to 1100 nm) are either reflected or transmitted, greatly lessening heat buildup. Absorption of photons with wavelengths in the visible range (400 to 700 nm) may reach 60 to 80 percent, depending on specific wavelength. Solar radiation contains relatively little energy in the far-infrared wavelengths. Although most of the terrestrial radiation is in this part of the spectrum, it has little net effect on leaf temperature since leaves both absorb and radiate energy efficiently in this region.

The radiation absorbed by a leaf is disposed of in several ways, the most important being photosynthesis (Chap. 5). On a global scale, vast quantities of CO_2 are fixed by plants; estimates usually range from 45 to 136 $\times$ 10^9 metric tons per year. It is upon this solar-driven fixation of carbon in the past and present that we depend for food, fiber, and fuel.

Absorption

The two major pigments which absorb radiant energy for photosynthesis, chlorophylls *a* and *b*, are rela-

tively large organic molecules (Chap. 5). Chlorophyll *a* is present in most plants in amounts about twice those of chlorophyll *b*.

The *absorption spectrum* of a pigment indicates its tendency to absorb radiant energy of specific wavelengths. The two types of chlorophyll have slightly different absorption spectra; nevertheless, as Fig. 7-8 shows, each has an absorption peak in the blue-violet region and a second peak in the red region of the spectrum. It follows therefore that red and blue photons provide the energy to drive photosynthesis. We see a leaf as green because the red and blue wavelengths are absorbed by chlorophyll, and most of the green is transmitted or reflected. Because they are not absorbed, photons in the green portion of the spectrum are of much less use in photosynthesis.

Chlorophyll occurs only in the *chloroplast*, the organelle in which photosynthesis is carried out. In most plants, light is required for the formation of chlorophyll. Plants grown in darkness develop immature chloroplasts called *etioplasts*, which can

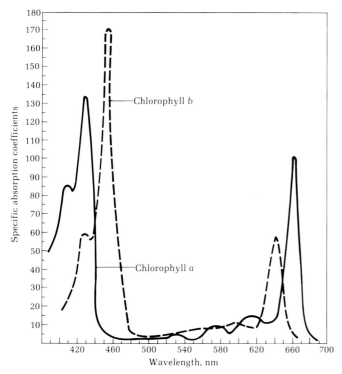

FIGURE 7-8
Absorption spectra of ether extracts of chlorophylls *a* and *b*. (*After F. P. Zscheile and C. L. Comar, Bot. Gaz. vol. 102:463–481, 1941.*)

develop into mature chloroplasts after a few hours of irradiation. This response, typical of leaves, is also apparent in the greening of potatoes exposed to light. Besides chlorophyll, chloroplasts may also contain *carotenoids*, pigments ranging in color from yellow to red. Unlike chlorophyll, carotenoids may also occur in other plastids, the chromoplasts. Though not involved in photosynthesis as directly as chlorophyll, carotenoids are thought to prevent destruction of chlorophyll at very high irradiance levels and in some cases to absorb light and pass the energy along to chlorophyll molecules.

Dissipation of Excess Energy
Only 2 to 4 percent of the radiation striking a leaf is used in photosynthesis; the remainder must be disposed of or it could literally cook the tissues. About 70 percent of the energy is dissipated through the evaporation of water from the leaf (transpiration), the remainder being given off as radiation or by conduction and convection (see Chap. 8).

PHOTOSYNTHESIS

Light Effects
The relationship between PAR level and photosynthesis depends on both environmental and genetic factors. Photosynthetic rate is a function of amount of PAR (see Fig. 5-16). In complete darkness, no photosynthesis occurs. As PAR increases, photosynthetic activity begins and increases until the *light-compensation point*, where photosynthetic fixation of CO_2 exactly equals respiratory release of CO_2 so that there is no net movement of CO_2 into or out of the leaf. As PAR increases still further, net photosynthesis (gross photosynthetic CO_2 fixation minus respiration) increases rapidly until the *light-saturation level* is approached. At light saturation, the rate of net CO_2 fixation levels off; above it net photosynthesis does not increase, either because CO_2 becomes the limiting factor or because photorespiration accelerates at the same rate as photosynthesis, or both (see Chap. 5). The light-saturation level depends on such factors as the species, CO_2 concentration, and the environment under which the leaf developed.

The light-response curve in Fig. 5-16 varies with the species. A philodendron leaf, which may be light-saturated at 100 μE s^{-1} m^{-2}, 5 percent of full sunlight, also has low rates of maximum photosynthesis. This type of light-response curve is typical of species adapted to heavily shaded habitats, e.g., the floor of a forest, a characteristic that makes these species ideal for use as foliage plants in homes and offices where PAR levels are quite low. The philodendron thrives in a low-light environment where geraniums and other plants requiring high PAR will not.

CO_2 Effects
The CO_2 level has a strong effect on photosynthetic rate. If CO_2 level rises above the normal atmospheric concentration of 340 ppm, the rate of photosynthesis at light saturation increases markedly (Fig. 7-9). Note that the PAR level required for light saturation rises with increasing CO_2 levels. From the discussion of the biochemistry of photosynthesis in Chap. 5 we know that as the PAR level increases, progressively higher concentrations of CO_2 are required to keep Rubisco active in carboxylation and use up the NADPH and ATP produced in the light reaction.

The normal atmospheric CO_2 levels of about 340 ppm may be somewhat depleted within the canopy of a crop, the amount of depletion depending on such factors as canopy density, wind speed, PAR level, and temperature. Although little can be done to modify CO_2 levels under field conditions, CO_2 enrichment has potential benefits in an enclosed structure, e.g., a greenhouse in winter. Two apparent advantages of elevated CO_2 levels are increased CO_2 uptake and decreased photorespiration, the latter the result of the greater $CO_2:O_2$ ratio. Tending to counteract these advantages is the fact that elevated CO_2 levels may cause partial closure of stomata. Another practical problem is that CO_2 enrichment is practical only during the winter, when ventilation is not needed for cooling. Nevertheless, even under the low PAR available in winter, CO_2 fertilization is useful commercially for such crops as lettuce, tomatoes, and roses.

Atmospheric CO_2
Atmospheric levels of CO_2 are of major interest today because during the past 25 years, CO_2 in the earth's atmosphere has increased by about 10 percent (from 310 to 340 ppm). The annual rate of increase is currently estimated at 1.5 to 2 percent. Most of the publicity about this change has focused on the predicted greenhouse effect on temperatures. Whereas water vapor and CO_2 are essentially transparent to

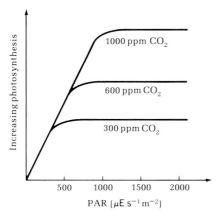

FIGURE 7-9
Schematic light response curve at three levels of CO_2. At higher CO_2 levels, both light saturation and maximum net photosynthetic rates increase.

shortwave radiation from the sun, the longer-wave infrared radiation from the earth is effectively blocked by reflection and absorption. The basic premise is that as CO_2 levels increase, the greenhouse effect will be increased and more heat will be trapped. The long-term effects of these changes cannot be foreseen with accuracy, but the doomsday predictions include sufficient melting of the polar ice masses to raise the ocean levels enough to submerge coastal areas. Also projected are major shifts in precipitation patterns.

Regardless of the possible climatic outcomes, a more definite effect from changes in CO_2 can be predicted. As CO_2 levels rise, the implications for crop productivity must be considered. The projected effects of a doubling of CO_2 levels in the atmosphere are given in Table 7-2. Even the gradual upward trend being experienced carries the potential for a major impact on crop production in coming years. The positive effects of the phenomenon have received little attention in the popular press.

Light and Leaf Anatomy

The environment in which a plant or leaf develops can influence not only the light-saturation level and the maximum rate of net photosynthesis but leaf morphology as well. A leaf developing under high PAR is thicker, often as the result of longer and more numerous layers of palisade mesophyll cells. Associated with increased leaf thickness is a greater fresh and dry weight per unit area, *specific leaf weight* (SLW), usually expressed in milligrams per square centimeter. There is a correlation between SLW and photosynthetic potential; as SLW increases, net photosynthesis also tends to increase.

At one time it was believed that species of plants could be divided into sun and shade types. The distinction was proposed on the basis of light-satu-

TABLE 7-2
Implications of Rising Atmospheric CO_2 for Crop Productivity†

Biological process	Effect of doubling current CO_2 levels
Photosynthetic CO_2 fixation	50% increase with C_3 plants
Plant dry weight and yield	20–25% increase
Primary productivity	10–40% increase
Tuberization in potato	Possible severalfold increase
Genetic improvement, plant breeding	Development of plants for optimal growth at higher levels
Water-use efficiency	Nearly doubled for both C_3 and C_4 plants
Biological nitrogen fixation	Preferential indirect increases
Mycorrhizal activities	Preferential indirect increases
Limiting factors in growth	Partial compensation for other limiting factors
Flowering phenology	Shifted according to species
Weed-crop interactions	Weed-crop ratios shifted according to species

† From S. H. Wittwer, Rising Atmospheric CO_2 and Crop Productivity, *HortScience,* vol. 18, p. 669, 1983.

ration curves, *sun plants* being light-saturated at 500 to 800 μE s^{-1} m^{-2} and *shade plants* at 100 μE s^{-1} m^{-2} or less. More recent research has shown, however, that the distinction is not clear. Some sun species can be made to respond like shade species and vice versa by growing them under a different light regime. In other words, a "sun plant" grown under low PAR responds rather like a "shade plant" (Fig. 7-10).

As the plant canopy develops, early leaves grow on the periphery of the plant and are thus under high PAR. As the plant enlarges, however, new leaves at the periphery of the canopy shade these older leaves, and to remain productive, a leaf must adapt to decreasing light level. Obviously, as the PAR received by older leaves declines, their potential photosynthesis must also decline. Fortunately, a change which often accompanies the decline in photosynthesis is a decrease in respiration. But in spite of lowered respiration, shaded leaves in some species may use more photosynthate in respiration than they produce in photosynthesis, becoming a liability rather than an asset to the plant. On many plants, heavily shaded leaves abscise before becoming "parasitic."

In forestry, trees are often characterized as *shade-tolerant* or *shade-intolerant*. Species which will survive in the dense shade cast by large, tall trees are considered shade-tolerant. Such trees can therefore become established in an existing forest, whereas shade-intolerant species cannot.

Even more striking than the different light-response curves of species growing in sunny and shaded environments is the difference in light-response curves for C_3 and C_4 plants (see Fig. 5-16). Whereas leaves of C_3 plants are light-saturated at 100 to 800 μE s^{-1} m^{-2}, C_4 species are typically not light-saturated even at twice the PAR of full sun. Thus C_4 plants have a greater production potential because they are able to make more use of the available PAR and have higher photosynthetic rates than C_3 species.

LIGHT MANAGEMENT

Agricultural Implications

Crop production in developed countries is generally designed to maximize the output of marketable yield per unit of input. Input is usually thought of in monetary terms such as costs of land, equipment, labor, fertilizer, pesticides, and, in arid regions, water. Many of these inputs are called fixed costs, because they are tied directly to the size and nature of the operation. As per hectare production increases, output per unit of input increases, assuming that variable costs such as labor, fertilizer, pesticides, and irrigation do not rise excessively. Great progress has been made in increasing per hectare yields through the introduction of improved cultivars combined with better use of fertilizers, pesticides, growth regulators, and irrigation.

In our high-technology, fast-paced society, we sometimes lose sight of the obvious. If we are to maximize production, we must maximize the interception of light by chloroplasts. It is the process of photosynthesis upon which the yield of crop plants is totally and directly dependent.

With this background, we introduce the concept of *light management*, a term used to describe attempts to maximize the utilization of this all-important resource for crop production.

Canopy Establishment

Whether planting a peach orchard, a soybean field, a crop of poinsettias in a greenhouse, or a pine forest, one wishes to establish the canopy of leaves as

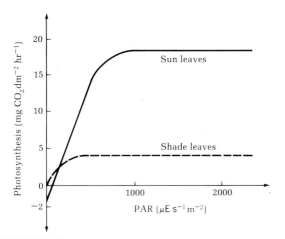

FIGURE 7-10
Sun leaves usually show higher rates of photosynthesis and dark respiration and greater light compensation points than shade leaves. The data are means for several species of sun and shade plants. (*Adapted from R. H. Bohning and C. A. Burnside, Amer. Jour. Bot. 43:557–561, 1956.*)

rapidly as possible. A view of a field of young corn plants or a newly planted citrus grove makes it apparent that most of the radiant energy is being wasted; it is striking the ground or worse is being absorbed by weeds. Only as each crop plant grows and fills its allotted space, can maximum utilization of the available radiant energy occur.

Construction, heating, and maintenance of greenhouses is very costly. To maximize cost efficiency in the production of container-grown plants such as poinsettias and chrysanthemums, young plants are often started in small pots which occupy very little bench area. As the plants grow, they are transferred to larger pots and gradually moved to wider spacing as their foliage expands. To further exploit the growth area or light-capturing potential, a partial layer of plants may be established above the main benches. This technique is often used for growing plants in hanging baskets to be sold along with bedding plants in the spring.

The process of moving plants as they enlarge is also widely adopted by growers of container-grown woody ornamentals in the field. It is much more efficient to manage large populations of newly rooted cuttings in a confined area than to spread them out according to their final space requirements.

Plant Spacing

The production of field-grown soybean, tomato, or corn obviously does not allow for changes in the plant density established at planting. For this reason, the spacing decision is vitally important. The distance between rows is frequently determined by the equipment used in planting and harvesting, but major flexibility is available in spacing within the row. The plant population per hectare is readily determined by dividing 10,000, the number of square meters in a hectare, by the product of distance between rows and distance within rows (in meters). With corn, for example, spacing between rows may be 80 cm with 25 cm between plants. Thus plant spacing is

$$0.8 \times 0.25 = 0.2 \text{ m}^2 \text{ plant}^{-1}$$
$$10,000 \text{ m}^2 \text{ ha}^{-1} \div 0.2 \text{ m}^2 \text{ plant}^{-1} = 50,000 \text{ plants ha}^{-1}$$

Although corn may be planted at a much higher population, interplant competition increases accordingly, not only for moisture and nutrients but for light as well. Therefore optimum spacing depends upon environmental as well as genetic factors.

Leaf-Area Index

One way to evaluate light competition in a crop canopy uses the concept of *leaf-area index* (LAI), defined as the leaf surface area (one side) over a unit area of soil. For example, in a soybean field if there is an average of 5 m² of leaf area per square meter of soil, the LAI is 5. The LAI varies with species, cultivar, stage of development, nutrition, available moisture, and plant population. For most crops there is a range of LAI where yields level off (Fig. 7-11). Below the critical LAI, maximum light interception is not achieved; above the critical LAI, yields may even tend to decline due to excessive shading, barren plants, and competition for water and nutrients. Total yield per hectare is the product of plant population and yield per plant. Yield per plant declines as population increases, but total yield increases as long as the percent increase in population exceeds the percent decrease in per plant yield. For example, if a 20 percent increase in plants per hectare results in only a 10 percent decrease in yield per plant, total yield will increase.

Leaf Orientation

As a general rule, an LAI between 4 and 8 is optimal. Corn, wheat, and rice, which have characteristically erect (*erectophile*) leaves, have a higher optimum LAI than species with horizontal (*planophile*) leaves. Many of the nongrass species, such as soybean and clover, have planophile leaves.

Of the PAR which strikes a leaf, about 70 percent is absorbed. Therefore, the level of PAR decreases rapidly within the plant canopy. Because of their orientation to the sun, erect leaves intercept less

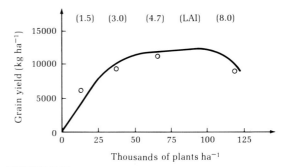

FIGURE 7-11
Relation between plant density, LAI, and grain yield for an irrigated crop of hybrid corn.

PAR than horizontal leaves. Hence more leaf area on a plant with erectophile leaves receives adequate PAR for net positive photosynthesis, and such a crop can effectively function with a higher LAI.

Keep in mind that leaves of C_3 plants are light-saturated at well below the irradiance of full sun. Therefore, leaves at the top of the canopy are often light-saturated. This does not mean that crops will peak in photosynthetic output below the irradiance of full sun; because, although upper leaves are saturated, lower leaves will become more active photosynthetically as light is added. Leaves tend to become light-saturated but the canopy does not.

Reflective Materials

In the lower portions of most plant canopies, photosynthesis is limited by low irradiance. If an inexpensive, long-lasting, reflective material can be applied to the soil, the resulting increased light levels in the lower canopy may increase crop productivity. When interpreting such data, however, we must make certain that the effects are not confounded by modification of soil or air temperature. For example, an aluminum-foil mulch reflects light into the canopy but by reflecting insolation it also lowers soil temperatures. Attempts with reflectants have not yet shown economic potential.

Shade

Sometimes crop growth or quality is reduced by direct, full sun. Artificial shade is therefore used in the production of some high-value crops. For example, in Florida, many ornamental foliage plants are shaded; growth is more luxuriant in shade, and the eventual transition to the low-irradiance environment of the home or office is facilitated.

Tobacco for cigar wrappers has been grown under shade for many years. The fields are covered with a heavy cheesecloth suspended on frames high enough to allow the movement of workers and equipment. The reduced irradiance produces the relatively large, thin leaves (lower SLW) required for this particular use.

Light Relations in Tree Canopies

In certain crops, e.g., fruit trees, the light level in the canopy is controlled by pruning. Another technique is to suppress tree size by using dwarfing rootstocks and scion cultivars. Generally in smaller trees, larger proportions of the canopy have adequate light levels. Heavy shade in fruit trees suppresses flower-bud formation, fruit set, fruit size, and fruit color. The effects of light on fruit color are apparent; the side of an apple or peach facing outward colors earlier and more completely than the side facing inward.

LIGHT STIMULI AND RESPONSES

Categorizing the multitude of effects of light on plant growth and development depends upon specific criteria—whether the light stimulus must be directional or periodic or both and whether the developmental response is directional. On the basis of these criteria, we can define three major responses to light:

Phototropism—a directional growth response to a directional light stimulus.

Photoperiodism—a nondirectional developmental response to a periodic, nondirectional light stimulus.

Photomorphogenesis—a nondirectional developmental response to a nonperiodic, nondirectional light stimulus.

Light-Receptive Pigments

The pigment systems mediating the major responses of plants to light have been under intense study for several decades. It is generally accepted that there are at least two pigment receptors of light stimuli. One is *phytochrome*, a chromoprotein with a molecular weight of about 120,000. It occurs very widely among plants and may even be universal in its occurrence. The distribution of phytochrome in plants has been confirmed in roots, stems, leaves, flowers, fruits, and seeds. (At the end of the chapter we mention another receptor pigment, riboflavin.)

Phytochrome has two forms, with different absorption peaks. The red-absorbing form (P_r) with an absorption peak at 660 nm, is readily converted into the far-red-absorbing form (P_{fr}) when plants are exposed to red light. The P_{fr} form has its absorption peak at 730 nm and is converted into the P_r form on absorption of far-red light (Fig. 7-12). Reversion of P_{fr} to the P_r form also takes place in the dark, a process much slower than the photoconversions which occur in response to specific wavelengths of

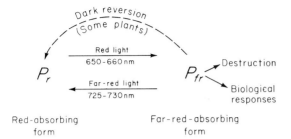

FIGURE 7-12
The red-absorbing form of phytochrome (P$_r$) is converted by red light into the far-red-absorbing form (P$_{fr}$), which may be used to induce a biological response, may be destroyed, or may be slowly reconverted into the P$_r$ form in darkness. Far-red light converts the P$_{fr}$ form back to the P$_r$ form. (*From A. C. Leopold and P. E. Kriedemann. Plant Growth and Development, 2d ed., McGraw-Hill, New York, 1975.*)

light. The active form of phytochrome is generally considered to be P$_{fr}$, although it is the subject of some debate.

Because it is richer in red than far-red wavelengths, daylight (sunlight) tends to convert the phytochrome in a plant from the P$_r$ to the P$_{fr}$ form, although the final ratio may be only slightly higher than 50:50. The influence of artificial lighting on the phytochrome balance depends on the spectral energy distribution of the source. Since fluorescent lamps are quite deficient in the far-red wavelengths (Fig. 7-6), ordinary fluorescent lighting used for growing plants is usually supplemented with some incandescent lamps to provide a better balance of wavelengths and promote more "natural" growth (in contrast to some responses noted below).

Scientists are intrigued by phytochrome and speculate almost endlessly on its mode of action. Although unproved, favored theories relate to the probable effects of phytochrome on particular hormones, membrane permeability, and enzyme systems.

Photoperiodism

Photoperiodism can be defined as the developmental responses of plants to the duration of relative periods of light and darkness. The total amount of light energy is unimportant as long as it exceeds some low minimum level required to trigger changes in phytochrome form (P$_{fr}$ and P$_r$). Responses to photo-

periodic stimuli include flowering, tuber and bulb formation, bud dormancy, and seed germination.

Flowering

The striking effect of photoperiod on flowering was discovered in 1920 by Garner and Allard while working with a type of tobacco called 'Maryland Mammoth.' Grown in the summer, this cultivar attained a height of 3 to 5 m and did not flower. In the greenhouse during the winter, however, the same cultivar would grow to a height of only 1 m, flower, set seed, and die. After trying many other treatments, they hypothesized that the response was due to the different day lengths, and they confirmed it by experiments in which they artificially altered the photoperiod. Subsequent experiments by many researchers have shown that effects of photoperiod on reproductive development are widespread among diverse plant species. Not all plants respond in the same way, and there are three more or less distinct response groups: short-day, long-day, and day-neutral plants.

Short-day plants are induced to flower only when the day length is shorter than some critical period. While the critical day length is exceeded, the plant remains in a vegetative stage. When the days become shorter than the threshold value, the plants are induced to flower (see Chap. 4). The critical day length varies with the species and cultivar. Examples of short-day plants are chrysanthemum, poinsettia, some types of tobacco, and most cultivars of soybean.

Long-day plants are triggered to flower only when a critical day length is exceeded. Until the photoperiod is longer than the threshold value, only vegetative growth occurs. As with the short-day plants, the critical day length varies both between and within species. Spinach, beet, radish, timothy, red clover, and most small grains are long-day plants.

Day-neutral plants do not respond to photoperiod and may flower under any day length. Examples are tomato, rose, African violet, and cotton.

When photoperiodism was first discovered, length of day was assumed to be the critical environmental stimulus, but subsequent research eventually showed that the length of the dark period is actually the determining factor in flowering. In other words, long-day plants respond to nights shorter than some critical period and short-day plants respond to nights longer than some critical period. An interruption of an otherwise appropriately long night with just a few minutes of light can prevent a short-day (actually long-night) plant from flowering. An interruption of the day with a few minutes of darkness has no effect at all. The night-interruption phenomenon has considerable horticultural and agronomic significance. It is much more energy-efficient to break the night (to prevent flowering of short-day plants or cause flowering of long-day plants) by a brief night interruption than to extend the day length by artificial lighting in either morning or evening. The night interruption works because the P_r (formed during the dark) is reconverted into P_{fr}.

In another interesting aspect of night interruption, the red–far-red phenomenon exhibited by phytochrome is apparent in the night interruption data (Fig. 7-13). A few minutes of red light prevent flowering; an exposure to far-red light does not. This response to either the red or the far-red light is repeatedly reversible, as is the effect on seed germination, discussed below.

In order to flower, short-day plants do not necessarily require a longer night length than a long-day plant, but for the short-day plant, the dark period must be *longer* than the critical length. For the long-day plant, the dark period must be shorter than the critical night length. A widely studied short-day plant is the cocklebur (*Xanthium*), a weed with a critical dark period of $8\frac{1}{2}$ h; when the night length is more than $8\frac{1}{2}$ h, it will flower. A long-day plant which has been widely studied is another weed, henbane (*Hyoscyamus*); it has a threshold dark pe-

riod of 13 h. When nights are shorter, it will flower. At a day length of 12 h (night length of 12 h), both the short-day cocklebur and the long-day henbane will flower.

A further breakdown of the groups can be made on the basis of whether the photoperiodic requirement is absolute or merely beneficial. Plants which must have short days to flower have an absolute requirement and are called *obligate short-day plants*. Those in which short days will accelerate flowering but often can be substituted for by temperature or other factors are termed *quantitative short-day plants*. Similar types of obligate and quantitative long-day plants also exist.

The flowering of many agronomic and horticultural crops is under photoperiodic control, but the degree of control varies widely. Most wheat cultivars are considered to be quantitative long-day plants in that longer days hasten their flowering. Eventually, however, most cultivars will flower even under short days. Soybean is obligately photoperiodic and will begin to bloom only after nights exceed some critical length. Because of the differences in day length which occur with change in latitude, soybean cultivars must be matched carefully to planting location. A cultivar that requires a night length greater than 9 h to trigger flowering should not be planted in northern Illinois, for example, since in northern latitudes it would not experience a 9-h night until too late in the growing season. On the other hand, a cultivar which will not flower until nights are shorter than 11 h might be very satisfactory in Georgia or Texas. The plant could

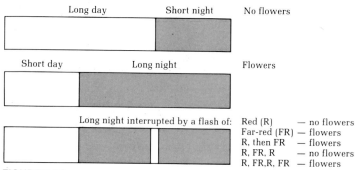

FIGURE 7-13
The influence of a short exposure of red (R) or far-red (FR) light or of sequences of R and FR during the inductive night of a short-day plant. (*From R. G. Halfacre and J. A. Barden, Horticulture, McGraw-Hill, New York, 1979.*)

grow vegetatively for much of the season and change to reproductive growth in time to produce good seed before the first killing freeze.

A particularly strong photoperiodic response is present in the short-day plant chrysanthemum. In growing young plants and producing stock plants for stem cuttings, day length must be kept above $14\frac{1}{2}$ h (or the night interrupted by a brief period of illumination) to prevent the formation of flowers. When plants reach the desired size, the day length is reduced to less than $14\frac{1}{2}$ h (night length is more than $9\frac{1}{2}$ h) and flowers are initiated. Since for full flower development night length must increase to $10\frac{1}{2}$ h or more, frequently a black cloth must be pulled over the plants daily to keep night length above the minimum.

Tuber and Bulb Induction

The photoperiodic induction of tuber and bulb formation has several similarities to the triggering of flowering. The photoperiodic stimulus is received by the leaves and somehow transmitted to the part which enlarges as a storage organ, such as tubers [potato and Jerusalem artichoke (Fig. 7-14)] and bulbs (onion).

The tuberization of potato is affected by photoperiod, but the degree of control depends on the cultivar and temperature. Some potato cultivars form tubers regardless of photoperiod but do so more readily under short days. Certain cultivars may form tubers only under short days. It has been demonstrated that the stimulus is perceived by the leaves and moves downward. Although the tuberization response to photoperiod would make the potato a short-day plant, its flowering puts it in the long-day category. The acceleration of tuber formation in the potato by short days is nullified by a night interruption, another similarity to the flowering phenomenon.

Bud Dormancy

The occurrence of bud dormancy in many species is largely a response to the shorter days of late summer and fall. In contrast, during the lengthening days of spring, bud dormancy is broken and a new season's growth begins. In many species the initiation of growth in the spring is also associated with the advent of rising temperatures. Since buds of a photoperiod-sensitive species will not break dormancy

FIGURE 7-14
Tubers of the Jerusalem artichoke, a potential source of carbohydrates for fermentation to alcohol for use as fuel.

during the short days of winter, the plant is protected against premature growth in response to a midwinter warm spell. In many deciduous species, reinitiation of growth in the late fall and early winter is prevented by a requirement for a minimum amount of hours of low temperatures. (The chilling requirement is discussed in Chap. 8.)

Lateral (axillary) bud formation usually occurs in the early summer during shoot development of temperate woody plants. In many species these buds do not grow during the season of their formation except in unusual circumstances, e.g., defoliation or severe pruning. Under normal conditions these buds do not grow, and by late summer they have gone into a state of physiological dormancy, at least partially because of the shortening days. Cooler temperatures in late summer and early fall can also contribute to the onset of bud dormancy. The photoperiodic cue leading to bud dormancy is thought to be received by the leaves in most plants. It thus seems plausible that some substance is formed in the leaves and is translocated to the buds. In many species, the onset of bud dormancy can be prevented or delayed by either an extension of day length or a night interruption. The indications are that these reactions are the result of phytochrome.

Leaf Abscission

The abscission of leaves is a normal developmental process in plants, but its timing varies with species and environment. In some plants, leaf abscission occurs gradually as the plant ages. As older leaves abscise, they are continually replaced by younger leaves at the stem apex. In contrast, a deciduous tree typically sheds all its leaves within a short period during the fall.

The shedding of leaves coincides with their senescence (Chap. 4) and may be a gradual process due to shading or result from some general environmental cue. In many deciduous plants, this cue comes from the shortening days of fall and is thus a photoperiodic response. A tree beside a street light may be slow to shed its leaves in the fall because it does not "see" short days (long nights). Eventually, the leaves abscise in spite of photoperiod because of cold temperatures.

Effect of Latitude

The striking effect of latitude on photoperiod is shown in Fig. 7-15. Since day length is essentially 12 h year-round on the equator, there is no photoperiodic cue to which equatorial plants can respond. Nevertheless it has been shown that certain tropical plants, e.g., coffee and *Streptocarpus*, grown a few degrees north or south do respond to day length.

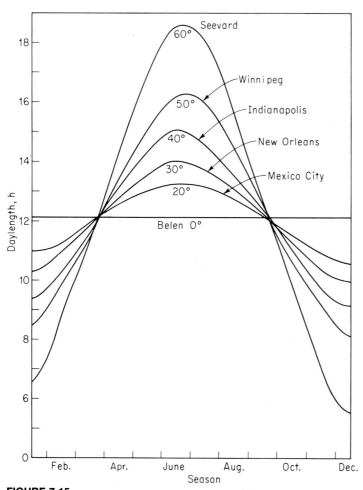

FIGURE 7-15
As latitude increases, the annual variation in day length increases dramatically. (*From A. C. Leopold and P. E. Kriedemann, Plant Growth and Development, 2d ed., McGraw-Hill, New York, 1975.*)

Flowering of some native plants in Nigeria results from seasonal day-length changes of only 15 to 20 min. Obviously, the further poleward one goes the greater the seasonal fluctuation in day length, the extremes being found above and below the Arctic and Antarctic circles, respectively. In the middle latitudes there is a sizable change in day length, the general range being from 15 h daylight in the summer months down to about 9 h in midwinter. In New Orleans, day length ranges from about 14 to 10 h; in Winnipeg it ranges from about 16 to 8 h. The changes in day length with seasons provide a dependable cue to plants. Plants can respond developmentally not only to the photoperiod of a particular day or week but (equally important) to whether the days are becoming progressively longer or shorter. The effect of photoperiod on a great many plant processes is a major factor in the natural distribution and adaptation of plants to various latitudes.

The relationship between temperature and photoperiod is depicted in Fig. 7-16. It is interesting to contemplate the tremendous range of conditions which plants at Ithaca, New York must endure, as opposed to Paramaribo, Surinam. At least part of this ability to adapt results from the cues received from photoperiod, which cause the plant to initiate various adaptive developmental responses.

Perception of the Photoperiodic Stimulus

That leaves are the organs which perceive the photoperiodic stimulus was established by J. C. Knott, who studied flowering in spinach. Some plants may require only one night of the appropriate length to be induced to flower. With some plants, e.g., cocklebur, one exposure of just one leaf to the necessary photoperiod will initiate the flowering process over the whole plant. With others, the amount of flowering increases in proportion to the total leaf surface exposed to the proper day length or the number of days of exposure.

Although the photoperiodic stimulus is received by the leaves, initiation of flowering occurs at the apex, which may be some distance away. Researchers have yet to isolate the flowering stimulus, which is apparently formed in the leaves and translocated to the apex (see Chap. 4).

Seed Germination

The seed of some species of plants require exposure to light before they will germinate. The phenomenon

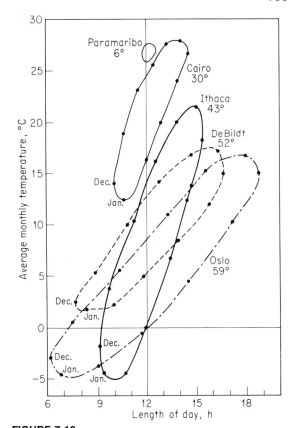

FIGURE 7-16
Seasonal changes in day length and temperature can be plotted so that by connecting the 12 points the range becomes apparent. Paramaribo has minimal seasonal changes, whereas Oslo has extreme variation. (*After J. H. A. Ferguson. Euphytica 6:97–105, 1957.*)

is common in weeds and offers them a competitive advantage because they germinate only near the soil surface or where the soil has been disturbed. An extensively studied crop species requiring light to germinate is lettuce. If freshly harvested lettuce seeds are allowed to absorb water in the dark, only a few will germinate. If the seed are exposed to red light (660 nm) after imbibing water, seed germination approaches 100 percent, but if the seed is exposed to far-red light (730 nm) after its exposure to red light, the stimulatory effect of the earlier exposure is nullified. This effect can be reversed repeatedly, like the flowering response associated with phytochrome. Once germination has actually started, it cannot be reversed.

Although light stimulates the germination of lettuce seed, certain other species are inhibited by light, e.g., the American elm and some cucurbits. Seeds of many other species are insensitive to the presence or absence of light.

Photomorphogenesis

Photomorphogenesis refers to developmental responses of plants to nonperiodic, nondirectional light. As with the several other processes associated with photoperiodism, phytochrome is the receptor pigment system, and the light dosage required is very low.

A striking photomorphogenic response of plants occurs when an *etiolated* (dark-grown) seedling is first exposed to light, converting some of the phytochrome from P_r to P_{fr}. An etiolated seedling is characterized by rapidly elongating shoots and small leaves; some also have a recurved hook. The etiolated growth can help the seedling reach light at the soil surface quickly, and the hook helps protect the tender shoot apex. Obviously, leaf growth is undesirable until the shoot reaches the light required for photosynthesis. As soon as the seedling "sees" light, the hook opens, stem growth slows, the stem thickens, and the leaves expand. Thus, phytochrome conversion triggers changes which allow the seedling to become self-sufficient. Concurrent changes in the deetiolated seedling include the formation of chloroplasts from etioplasts and the synthesis of chlorophyll and other pigments vital to the plant's continued growth.

Bean sprouts used in cooking are usually etiolated seedlings, as are the alfalfa sprouts served in salads. The seeds are sprouted in a moist, dark environment to encourage rapid germination and prevent deetiolation.

Etiolation is not necessarily an all-or-none response, however; and it is not always desirable. Although full darkness promotes maximum stem-elongation rates, at least until the energy supplies are exhausted, intermediate levels of shading likewise produce intermediate amounts of etiolated growth. Several situations in agronomy, forestry, and horticulture may produce the extra growth unintentionally with adverse results.

When plants are crowded together in a field, nursery, or greenhouse, they begin to shade each other. The close spacing creates lower levels of light for each plant than a more open arrangement. The same basic processes that help push germinating plants out of the ground and cause growth toward a light source now go to work causing the plants to grow taller. Auxin or gibberellins (or both) promote elongation of cells where light levels are low; therefore the whole stand of plants tends to have taller, thinner stems. The combination of tallness and thinness obviously can be bad structurally; such stems are much more likely to bend and break than shorter, stouter ones. Many crops show a greater tendency to *lodge*, or fall over, when planted more closely, soybean and rice being just two examples.

Perhaps you have started plants indoors in early spring for later transplanting outdoors. (Tomato, eggplant, and pepper seedlings cannot be transplanted until the soil and air warm sufficiently for protection and good growth.) Unless you have used care or ingenuity, the result may be long, thin, semiprostrate seedlings. These *leggy* plants are usually suffering the adverse effects of partial etiolation. Unless they are put in a sunny place or given additional illumination, the plants will respond to their shady environment by growing faster and thinner. In commercial production, such vegetable plants are usually grown in a greenhouse, where they get full sun.

Phototropism

When plants are irradiated from one side, they generally bend in the direction of the light, a response termed *phototropism* (see Fig. 4-8). Although phototropism has held the interest of researchers for at least 100 years, unanswered questions remain. It is now generally accepted that the curvature is due to a higher concentration of auxin on the shaded side, leading to a greater rate of cell elongation (see Chap. 4). The absorption spectrum for the perception of the phototropic stimulus indicates that the receptor pigment may be riboflavin, which absorbs light in the blue part of the spectrum.

Some plants, called *heliotropes*, "follow" the sun as it moves during the day. A common example is the sunflower, whose flowering head responds continuously to the changing position of the sun. House plants can be turned to prevent lopsided growth toward the window.

CHAPTER EIGHT

TEMPERATURE

Of the many environmental factors that affect plant life, temperature is one of the most important. Plants are able to survive in a very limited range of temperatures, compared with inanimate matter, which can exist over a range from $-273°C$ (absolute zero) to about $10,000°C$.

The ability of plant life to adapt to changing temperatures within the life range, however, is remarkable. The critical range varies widely from species to species. Banana, sweet potato, cucurbits, and many tropical plants may be seriously injured by exposure, however brief, to temperatures below $4°C$. A properly acclimated apple tree, on the other hand, seldom suffers injury at $-35°C$.

The study of temperature is complex but essential to understanding the plant world. In this chapter we explore many environmental factors influencing temperature and see how important temperature is in the world of plants.

BASIC ASPECTS OF HEAT AND TEMPERATURE

We start with some basic definitions and concepts. *Temperature* is a qualitative term that gives an indication of the *intensity* of heat in a body of matter but does not give any idea of the quantity of heat or energy present.

Heat

Heat is a form of energy. When it is transferred to (or from) a body of matter, it causes an increase (or decrease) in that body's temperature so long as the body of matter does not change state in the process. The last part of the definition is needed because a change of state, e.g., ice to water or water to water vapor, involves a transfer of heat with no change in temperature.

Units of heat in common use are the calorie and the British thermal unit (Btu).† A *calorie* (cal) is the amount of heat required to raise the temperature of 1 g of water by $1°C$.‡ A *kilocalorie* (kcal), being 1000 cal, is the amount of heat required to raise the temperature of 1 kg of water by $1°C$ (it is equivalent to the Calorie, the unit used in diet planning). In the English system, 1 Btu is the amount of heat required to raise the temperature of 1 lb of water by $1°F$; 1 Btu = 253 cal.

---

†We retain these familiar units although they are gradually being replaced by the joule (J), the SI unit.

‡Strictly speaking, this should be read "Celsius degree" when an interval is being discussed and not "degree Celsius," which is a scale reading. The SI takes note of this distinction by naming the unit of temperature difference a *kelvin* (K); it is identical on the Kelvin and Celsius scales.

Specific Heat

For the units we are using, *specific heat* is defined as the number of calories required to change the temperature of 1 g of a substance by 1°C. A high specific heat means that a relatively small change in temperature will occur when a given amount of heat energy is added.

Table 8-1 makes it apparent that water has a high specific heat, which is lowered by one-half or more when water is in the form of ice or steam. This high specific heat of water is of major importance in plant science. Since actively growing plants contain large amounts of water, its presence can act as a strong buffer against temperature change. Excessive insolation (Chap. 7) can raise tissue temperatures, leading to *sun scald* on fruits, vegetables, and tree trunks. This injury would be much worse if the tissues did not contain large amounts of high-specific-heat water.

Heat of Fusion

The *heat of fusion* is the amount of heat required to change 1 g of a substance at its melting point from the solid to the liquid state or vice versa. For water, 80 cal must be added to 1 g of ice at 0°C to get 1 g of water, still at 0°C. Similarly, 80 cal is released when water at 0°C turns to ice.

Heat of Vaporization

The amount of heat required to change 1 g of a substance at its boiling point from the liquid state into the vapor state is called the *heat of vaporization*. The same amount of heat must be removed from 1 g of the substance in the vapor state to convert it into a liquid.

Just as water has a rather high specific heat, it also has a relatively high heat of fusion and heat of

TABLE 8-1
Specific Heat of Several Substances

Substance	Specific heat, cal g^{-1} $°C^{-1}$
Water	1.00
Ice	0.50
Steam	0.48
Alcohol (ethyl)	0.58
Wood	0.42
Glass	0.20
Steel	0.11

TABLE 8-2
Heat of Fusion and Heat of Vaporization for Three Substances

Substance	Heat of fusion, cal g^{-1}	Heat of vaporization, cal g^{-1}
Alcohol (ethyl)	25.0	204
Oxygen	3.3	51
Water	80.0	540

vaporization compared with other substances (Table 8-2). These three physical properties, which make water unusual or unique, are of particular significance in plant science.

Changes of State

The addition or removal of heat to a body of matter does not always change its temperature. During a change of state no temperature change occurs until the entire mass has completed the change. Figure 8-1 plots temperature versus time. Starting with 1 kg of ice at −100°C, heat is added at a constant rate of 100 kcal min^{-1}. Since the specific heat of ice is 0.5, each kilocalorie of heat raises the temperature 2°C. From a different point of view, it would take 50 kcal to raise 1 kg of ice from −100 to 0°C. Thus, at 100 kcal min^{-1}, it takes 0.50 min to bring ice from −100 to 0°C. At 0°C, the ice's temperature rise ceases, and the temperature remains at 0°C until 80 kcal has been added, the heat of fusion of water. Thus, it takes 0.8 min at 100 kcal min^{-1} to convert 1 kg of ice at 0°C into water at 0°C. When all the ice has melted, the temperature of the 1 kg of water rises again, but now it rises at a rate of 1°C kcal^{-1}. To raise the temperature of 1 kg of water at 0°C to 100°C takes 100 kcal or 1 min. When the temperature reaches 100°C, the boiling point of water, it remains constant until 540 kcal (the heat of vaporization) has been added. It therefore takes 5.4 min to convert all the water into steam. After this conversion is complete, the temperature rises at a slightly faster rate than for ice since the specific heat of steam is only 0.48 kcal kg^{-1} °C^{-1}.

Heat Transfer

The transfer, or movement, of heat can occur by three different processes, in all of which the net

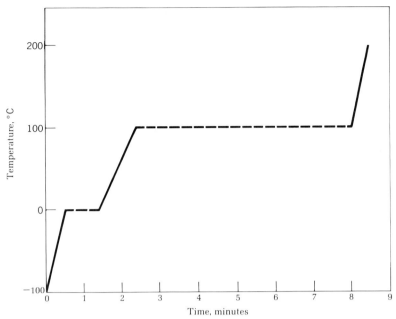

FIGURE 8-1
Plot of the temperature rise of 1 kg of ice starting at −100°C and with the addition of 100 Kcal of heat per minute. (*From R. G. Halfacre and J. A. Barden, Horticulture, McGraw-Hill, New York, 1979.*)

movement is always from the warmer to the cooler body:

Conduction is the flow of heat through a substance. The rate of flow is proportional to the cross-sectional area of the conductor and the temperature gradient and varies with the substance. For example, steel is a good conductor, wood a relatively poor conductor, and air a very poor conductor.

Convection involves the transfer of heat by a moving agent. When heat is added to a gas or a liquid, the density of the gas or liquid changes and a circulatory motion is established whereby heat is transferred. Thus, convection depends upon establishing convection currents. One can see convection currents in a pan of water being heated on a stove or in air over a very warm surface, e.g., black pavement exposed to direct sun.

Radiation is the transfer of energy without any connecting medium (Chap. 7). Radiant energy consists of electromagnetic waves traveling at the speed of light, 3×10^{10} cm s^{-1}. It covers the entire electromagnetic spectrum from gamma rays to ultraviolet,

visible light, and radio waves. Infrared is the radiant form easily detected as heat.

Temperature Measurement

In the past the United States used the Fahrenheit scale, on which 32°F is the freezing point of water and 212°F is the boiling point. Currently, we are going through a transition period of gradual conversion to the Celsius (formerly called centigrade) scale, on which the freezing point of water is 0°C and the boiling point is 100°C. The Appendix contains a conversion table.

Simple Thermometers A thermometer is a device for measuring temperature based on the principle that most substances expand as their temperature increases and contract as it decreases. A simple thermometer consists of a fluid, a glass reservoir, capillary tube, and safety chamber. The reservoir and part of the capillary tube are filled with mercury or colored alcohol. The shape of the reservoir varies from spherical to cylindrical, and the size of the tube in relation to the volume of the reservoir determines the "openness" of the scale. The remainder of the tube and the safety chamber are evacuated to allow for expansion of the liquid.

Simple thermometers can be very accurate if properly manufactured, but frequent checking is essential. A major problem develops if the temperature scale is not marked on the thermometer tube, so that if the thermometer shifts in relation to the scale, readings become inaccurate. Perhaps the easiest way to check the accuracy of a thermometer is to make an ice-and-water mixture in an insulated container and gently stir it with the thermometer. After a few minutes, the temperature will be 0°C, the melting point of ice and the freezing point of water. This provides a reproducible calibration temperature at which the thermometer can be checked and reset if necessary. The same procedure is suitable for standardizing the maximum, minimum, and maximum-minimum types described below.

Minimum Thermometers This type of thermometer both measures and records the minimum temperature reached (Fig. 8-2). Minimum thermometers contain alcohol, in which a metal index is suspended inside the tube (Fig. 8-3). Minimum thermometers are commonly used in weather shelters. If they are not securely fastened, shelter vibration due to wind gusts may cause the index to drift, especially if the thermometer is not exactly level.

Maximum Thermometers Temperature-observing stations have a maximum thermometer in combination with the minimum type described above. A maximum thermometer contains mercury but differs from the simple type in having a constriction at the base of the stem. As the temperature increases, the mercury in the reservoir expands and is forced past the constriction, but as the temperature decreases, the mercury cannot fall below the constriction (Fig. 8-3b). A similar thermometer with a limited range is the clinical thermometer used in medicine.

Maximum-Minimum Thermometers This type measures and records both maximum and minimum temperatures (Fig. 8-4). As the temperature increases, the mercury expands and moves upward on the maximum side, pushing the index immediately in front of the mercury. As the temperature drops, the mercury moves back down the tube, leaving the

FIGURE 8-2
A minimum thermometer is mounted horizontally. As the temperature decreases, the alcohol contracts, drawing the index down the tube by surface tension. As the temperature increases, the alcohol expands but bypasses the index. The distal end of the index remains at the minimum temperature experienced until it is reset by tilting. (*Taylor Instrument, Sybron Corp.*)

lower end of the index at the maximum temperature reached. As this contraction occurs, the minimum index is pushed upward on the opposite end. Thus, the minimum temperature is read at the lower end of the second index. The temperature scales are reversed on the two sides of the thermometer so that at any given time the current temperature can be read on either side. The indexes are reset on the top of the mercury columns with a magnet.

Location and Shielding of Thermometers Thermometers must be accurate, properly positioned, and shielded to prevent errors. Since we generally want to determine air temperature, the thermometer must be shaded from direct radiation from the sun, the ground, and adjacent buildings. The thermometer should not be exposed to the sky, as readings will be below air temperature because of excessive radiation from the thermometer to the cold sky. A typical weather observation station is a white louvered structure which allows free air circulation but no direct penetration of radiation (Fig. 8-5). The multiple thermometers used in crop areas require less elabo-

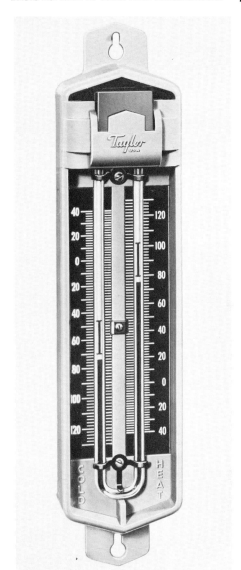

FIGURE 8-4
Maximum-minimum thermometer. The metal indexes are at or above the mercury. The maximum and minimum temperatures are read at the lower end of the appropriate index. (*Taylor Instrument, Sybron Corp.*)

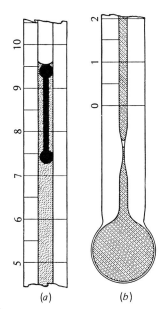

(a) (b)

FIGURE 8-3
(a) Minimum thermometer showing index. (b) Maximum thermometer showing constriction in bore of tube. (*From W. L. Donn, Meteorology, McGraw-Hill, New York, 1975.*)

FIGURE 8-5
Standard weather shelter for housing a maximum and a minimum thermometer. Note the double roof and louvered sides.

rate shelters, which provide protection primarily against the ground and sky.

FACTORS INFLUENCING TEMPERATURE

The daily and seasonal temperature fluctuations at a particular location are directly related to latitude, elevation, and any other local factors that affect the receipt and dispersion of solar energy. These factors, already considered in Chap. 7, will be discussed here as they relate directly to temperature.

Latitude

Representative temperatures for several latitudes in the northern hemisphere are given in Table 8-3. The temperature differences associated with latitude are caused by differences in insolation, which in turn depend both on day length (Fig. 7-15) and the angle of the sun.

Season

The effects of season on temperature are closely related, of course, to the influence of latitude. As Table 8-3 shows, poleward from the equator temperatures not only decline but exhibit greater seasonal fluctuation, largely from the dual effects of solar angle and day length (Table 7-1). As Table 8-3 shows, the mean temperature difference between 0° (equator) and 90°N (north pole) in July is 27°C but in January is 68°C.

Elevation

Temperatures decrease with altitude. Beyond the earth's atmosphere, where there is nothing to absorb radiation, temperatures are constant and very cold. For each 100 m increase in elevation, mean temperatures decline about 0.6°C. On the surface of the earth, a zone of permanent snow exists above 4500 m in the tropics and above 3000 m in temperate zones. A striking example of the effect of altitude on temperature is found in South America: Belem, Brazil, 19 km from the equator at an elevation of about 10 m, has a mean annual temperature of 29°C; Quito, Equador, also 19 km from the equator but at an elevation of 2835 m, has a mean annual temperature of 13°C. Similar instances can be cited in many

TABLE 8-3
Mean Celsius Temperature at Five Latitudes in the Northern Hemisphere

Latitude	Annual	Jan.	July	Range Jan.–July
90°	−26	−41	−1	40
60°	3	−16	14	30
30°	18	14	27	13
10°	26	26	27	1
0°	26	27	26	1

areas of the United States where sizable differences in elevation exist in close proximity.

The effects of altitude, and consequently temperature, on plant growth are readily apparent in any specific area. Hawaii, for example, can grow its famous tropical plants only in low areas; temperate crops flourish on higher mountain slopes.

Time of Day

The gross effect of time of day on temperature is obvious to us all. At any given moment, local temperature is determined to a very large degree by the balance of incoming versus outgoing radiation (Chap. 7). If we plot the energy received from the sun and that lost by the earth in relation to temperature, we can see the *lag effect* (Fig. 8-6), which occurs in both daily and annual trends. Thus, the maximum daily temperatures occur in midafternoon, and the daily minima occur just before sunrise. This same lag effect is observed in the annual maximum temperatures, which occur in late July and August in the northern hemisphere, well after the time of greatest insolation. Likewise, the annual minimum temperatures occur in late winter, well after the period of least insolation.

During the day, insolation passes through the atmosphere, striking various objects on the earth's surface. As insolation is absorbed, much of the radiant energy is converted into heat, elevating surface temperatures. Soon after sunrise, the temperature of exposed surfaces begins to exceed that of the adjacent air, and heat is transmitted to the air by conduction. Convection currents begin gradually, and heat is moved upward. Throughout the day, however, the warmest air is found closest to exposed surfaces, and temperatures decline with increasing distance from the earth's surface.

After sunset, even though the earth no longer receives solar radiation, outgoing radiation continues through the night. The radiation absorbed by an object during the day is reradiated back to the sky as infrared or heat waves.

Because of the radiative cooling of objects, the temperature of many exposed surfaces can drop below that of the adjacent air. This is particularly true of such efficient radiating surfaces as leaves, which have a large area in relation to their volume. When these surfaces become cooler than the surrounding air, heat is conducted from the air to the leaf—the reverse of the heat flow during the day. Since heat is radiated more rapidly than it is absorbed, the flow of heat from the air to cold surfaces continues all night. High levels of moisture in the air suppress radiant heat loss; nighttime temperatures

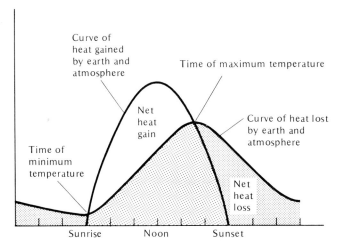

FIGURE 8-6
Variation of insolation and temperature during the day and night. The time of maximum temperature lags behind the time of maximum insolation. Similarly, maximum annual temperatures lag behind maximum solar radiation received. (*From W. L. Donn, Meteorology, McGraw-Hill, New York, 1975.*)

fall much more rapidly in deserts than in humid areas.

Temperature Inversions

During the night, as heat conduction from the air to surfaces continues, the air near the ground becomes cooler than the air above it and since cool air is denser than the warm air above it, no convection currents are formed. If there is no wind to cause mixing within these lower, cooler levels of air, a *temperature inversion* results, i.e., reversal of the normal situation present during the day (Fig. 8-7). Conditions conducive to temperature inversions in-

clude long nights, clear skies, and dry air. Long nights facilitate more complete heat loss through radiation. Since clouds are effective barriers to terrestrial radiation (see Chap. 7), temperatures drop much more slowly on cloudy nights. Calm conditions are also necessary for a temperature inversion to develop.

Local Factors

Oceans and Continents Since water absorbs heat to greater depths and has a higher thermal conductivity and specific heat than soil, it warms and cools

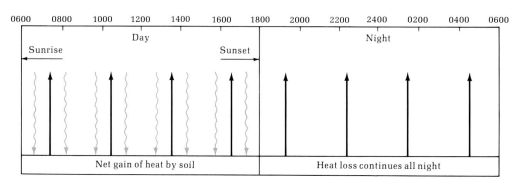

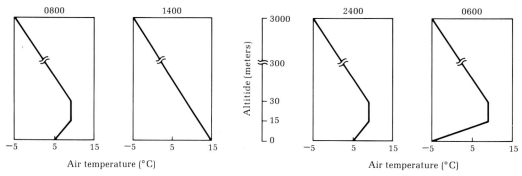

FIGURE 8-7

(*Top*) During the day, since incoming radiation exceeds outgoing radiation, soil and other surfaces experience a net gain in heat. At night, terrestrial radiation continues, and surface temperatures fall below air temperatures. (*Bottom*) Diurnal temperature variation differs markedly with altitude. Major changes occur near the earth's surface. The normal situation during the day is shown by the graph at 1400 h. The temperature inversion occurs in the lower 15 m, seen in the three other graphs. (*Adapted from graphs by National Oceanic and Atmospheric Administration of the U.S. Department of Commerce.*)

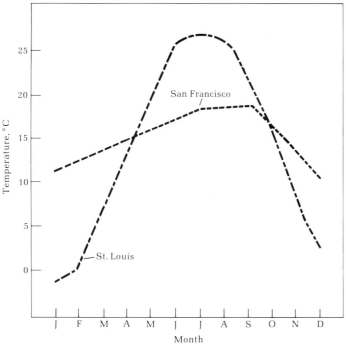

FIGURE 8-8
Monthly temperature means for St. Louis and San Francisco, which have the same mean annual temperature but different climates. (*From R. G. Halfacre and J. A. Barden, Horticulture, McGraw-Hill, New York, 1979.*)

much more slowly than land masses. The moderating effect of oceans on land, called the *oceanic effect*, is more apparent on the west coast of the United States than on the east because of prevailing westerly winds. The effect of the Pacific Ocean is obvious from Fig. 8-8, which compares temperatures in San Francisco, California (oceanic climate) with those in St. Louis, Missouri (continental climate). Both cities lie at about 38°N. This graph demonstrates why mean annual temperature is a poor index of climate, as both cities have a mean of about 13°C. The monthly means for San Francisco range from about 8 to 18°C, while those for St. Louis range from −1 to 26°C. The effects of these temperature differences on plants are dramatic. To survive in St. Louis, a perennial plant must be more resistant to both high and low temperatures. The two cities also have very different precipitation patterns (Chap. 9).

The buffering action of smaller bodies of water is apparent on a lesser scale in many areas. The moderating effect of Lake Michigan on temperatures east of the lake accounts for the concentration of fruit production in southwestern Michigan. The southeastern shore of Lake Erie and the Finger Lakes region in New York are major grape-producing areas for the very same reason.

Aspect What direction a slope faces can alter temperatures sufficiently to affect some crops. In the northern hemisphere, similar slopes rank from warmest to coolest, as follows: south, west, east, north. The south slope receives more insolation than any of the other three because it is exposed to sun almost all day; even in winter, when the sun is low in the sky, the sun's rays are more nearly at 90° to the soil surface. The west slope tends to be slightly warmer than an east slope because of warming by the air all morning plus direct insolation all afternoon; therefore less insolation is used in melting ice or evaporating water on west-facing slopes. The effect is greatest in winter, when the sun is low in the sky, but is also apparent in higher latitudes throughout the year. Since the influence is particularly strong on soil temperature, low-growing crops, like strawberries, are more affected than apple trees.

Air Drainage At night, when air near the ground becomes colder, it also becomes denser than the warm air above. The denser cold air flows downhill and collects in low-lying areas from which it cannot drain. This *cold-air drainage* can lead to the formation of *frost pockets*. In low-lying areas, plants are subject to freeze damage on nights when the general air temperature may not actually drop below 0°C.

Crops especially sensitive to freeze damage should be planted on sites selected on the basis of good air drainage. It is not uncommon for fruit growers to plant orchards on hillsides with less than ideal soil, simply to have good elevation, slope, and associated air drainage (Fig. 8-9). A further refinement is to plant early-blossoming cultivars at the top of a slope, with the latest-blossoming cultivars at the lower levels. An apple cultivar such as 'Delicious,' which normally blooms relatively early, should be planted on the upper slope, where damage is least likely. Late-blooming cultivars, e.g., 'Rome Beauty,' which blooms 5 to 7 days later, can be more safely planted in the lower, more frost-prone locations.

Soils Factors determining how fast soils warm up in the spring include surface color, soil density, rate

FIGURE 8-9
An apple orchard planted on a sloping hillside in western Maryland. The air drainage is excellent, but the soil is quite thin, leading to frequent missing trees. (*From R. G. Halfacre and J. A. Barden, Horticulture, McGraw-Hill, New York, 1979.*)

of heat conduction, and, most important, water content.

Sandy soils, which have a very low water-holding capacity and tend to warm up faster in the spring, are often referred to as *early soils*. Organic, or muck, soils, with high water-holding capacity are *late soils* and the last to warm up. All other factors being equal, loam, silt, or clay soils are intermediate between the two.

HOW PLANTS ARE AFFECTED BY TEMPERATURE

In relating temperature to plant growth, *cardinal temperatures* include:

> The Optimum—at which a plant functions best
> The Minimum—below which a plant cannot grow
> The Maximum—above which the plant cannot grow

Cardinal temperatures, especially the maxima and minima, vary greatly with species, although the extremes lie outside the traditional crop plants. The assignment of cardinal temperature becomes complicated because the critical temperatures for a particular species vary with cultivar, stage of development, tissue involved, length of exposure, and other environmental factors such as insolation and water stress.

Plant Growth

Temperature is an enormously important factor in total plant growth and development. Whereas light directly affects the aboveground, or photosynthesizing, portions of a plant, the whole plant is influenced by temperature. Particularly striking is the response of the plant's diverse biochemical reactions. The effect of temperature on physical or chemical processes is quantified as the Q_{10} of the reaction. A Q_{10} of 2 indicates that the rate of a reaction doubles for each 10°C increase in temperature; if the Q_{10} is 3, the reaction rate is tripled by a 10°C rise. Nonenzymatic reactions usually have a Q_{10} of about 1.2, whereas many enzyme-catalyzed processes have a Q_{10} of 2 or more. Between 0 and 25°C, respiration of many plant tissues has a Q_{10} between 2 and 3. From 25 to 35°C, the Q_{10} may decrease to nearly 1 because insufficient oxygen is available in the tissues for the rate to increase further. Above 35°C, respiration rates often decline due to high-temperature inactivation of the enzymes.

The influence of temperature on photosynthesis is complex because both physical and biochemical effects are involved. Although the light reactions are not strongly temperature-dependent ($Q_{10} \approx 1$) the dark reactions have a Q_{10} of about 2. At low temperatures, increasing the temperature will increase photosynthetic rates because the dark reactions can speed up. At higher temperatures, however, increasing the temperature may have no effect because the light reactions cannot supply energy to the Calvin cycle any faster. Net photosynthesis (the difference between gross photosynthesis, on the one hand, and photorespiration and dark respiration, on the other) may actually decline as temperatures increase above some point because the Q_{10} of photorespiration and dark respiration is higher than the overall Q_{10} of photosynthesis.

Crop Development

For many decades, researchers have attempted to correlate crop development with temperature. If this were possible, one could not only predict the maturation date for a specific crop accurately, but schedule successive plantings to give a continuous, orderly harvest. Large vegetable processors often contract with growers for such crops as bean, pea, and corn. For peak efficiency of their facilities, the processors need a continuous supply of each crop, but because of increasing temperatures as the season progresses, weekly plantings of corn or other vegetables will mature at intervals shorter than a week.

Perhaps the most widely used method of temperature summation is the *growing degree-day*, each crop needing a certain number of growing degree-days to reach maturity. A base temperature is established for each crop, below which growth and development are minimal. Every day, the base temperature is subtracted from the mean for that day to determine the degree-days. For example, with pea, a base temperature of 40°F (4.4°C) is used; on a day with a mean temperature of 50°F (10°C), 10 growing degree-days would be accumulated. (If the daily mean is lower than the base temperature, no growing degree-days are added or subtracted.) Using long-term averages, we can approximate the number of growing degree-days needed to reach maturity. An average figure of 2000 growing degree-days is common for many cultivars of pea. For corn, a warm-season crop, a base temperature of 50°F is used and the number of growing degree-days to maturity is often 3000 or more. One can match the field corn hybrid to the length of the growing season. Early-maturing cultivars may require 90 typical days, whereas late hybrids may mature in 200 days. The growing degree-days required for these two extremes vary accordingly. Although it is a fairly good method of approximating a harvest time, the system is not totally accurate because it ignores the nonlinear Q_{10} response of plants and such important factors as irradiance, day length, nutrition, disease and insect problems, and water relations.

TEMPERATURE STIMULI AND DEVELOPMENTAL RESPONSES

Dormancy

Of prime importance in the physiology of many plants' survival is *dormancy*, a state of growth cessation due to internal or external factors. A plant develops hardiness during the summer and fall but does not reach maximum hardiness until it has been exposed to progressively colder temperatures. Under normal conditions, growth declines and ceases in late summer, leaves are shed in the fall, and as the temperature and day length decline, hardiness increases. Thus, dormancy is a prerequisite for the development of maximum hardiness in deciduous plants. Many perennials, e.g., alfalfa, crownvetch, red clover, and asparagus, die back to the ground each fall, a mechanism of tremendous benefit in surviving cold winters.

In alfalfa, three distinct winter hardiness types are recognized: nonhardy, hardy, and very hardy. The difference in hardiness is reflected in the amount of growth the plants make with the onset of short days and decreasing temperatures. Nonhardy types continue to grow until killed by freezing temperatures, while hardy types stop growing as temperatures and day length decline. The nonhardy types offer an advantage to the grower in that they recover sooner after cutting than hardy types. The combination of quicker recovery and extended growth period of nonhardy types permits extra cuttings in the south. In northern states and Canada, only two to three cuttings can be made from the hardy types, and the last cutting should be at least 6 weeks before the first freeze. The subsequent growth provides the reserves required for winter survival of the roots.

The onset of dormancy and emergence from it are gradual processes that vary in different portions of a plant. For example, growing points of the axillary buds on a current year's apple shoot are fully dormant in mid- to late summer, even though the leaves are still fully functional photosynthetically and fruit growth is proceeding actively. In the fall, the entire aboveground portion of the plant gradually becomes dormant. The roots do not; they continue to grow as long as the soil temperature is above a minimum of about 4°C. In part because they lack dormancy and in part because they are not exposed to temperatures as low, the actual cold tolerance of the roots is considerably less than that of the top of a plant.

Bud Dormancy During the vegetative growth of shoots on a deciduous perennial plant, buds are formed in the axils of the leaves. Most of these axillary buds do not develop into shoots during the early summer because they are suppressed by the apical meristem of the main branch. This apical dominance was discussed in Chap. 4.

Bud dormancy is brought about by a variety of environmental factors, including shortening photoperiods, lowered temperatures, and severe drought. In some species, growth resumes as soon as the environment becomes favorable; in others, however, growth is not initiated until bud dormancy has been broken, usually by exposure to cold temperatures. In other words, the buds remain dormant even in warmer weather if they have not experienced enough cold to break their dormancy. This type of dormancy is often called *rest*. The need for exposure to cold temperatures, termed the *chilling requirement*, is usually expressed as the number of hours below 7°C required to break bud dormancy. Among deciduous fruit trees, the chilling requirement is of major importance in commercial production. For example, the southern limit of peach production has been set by a chilling requirement of 600 to 1000 h for most cultivars. This simply means that they must experience 600 to 1000 h at temperatures of 7°C or lower before their buds can begin to grow again. Breeding peach cultivars with chilling requirements of 250 h or less has recently extended the commercial peach production into Florida, where there is not enough cold weather to break rest of previously available cultivars.

Until its specific chilling requirement has been met, a plant will not grow and develop normally. In many flowering trees, such as the peach, flower buds have a slightly shorter chilling requirement than vegetative buds. After an unusually mild winter the tree may flower normally, but vegetative budbreak will be sporadic. If the winter is very mild, even the flower buds will open unevenly, with the result that the bloom period is greatly extended. This is devastating for the commercial peach grower.

Part of the chilling requirement can be satisfied or replaced by special treatment. Exposure to high but sublethal temperatures and to certain chemicals will induce buds to break as long as most of the necessary chilling has already been received. Such treatment may have commercial application in regions where the growing season is suitable except that the winters are too mild, e.g., the southern United States and areas south of the Mediterranean.

Vernalization

In many plants, temperature plays a vital role in flower initiation and development, particularly in winter-annual and biennial plants which produce only vegetative growth during the first growing season. An extended cold period during the winter induces flowering in the second season. This induction of flowering by exposure to cool temperatures is called *vernalization*. Many plants would remain vegetative indefinitely without a cold period. Cool-season grasses (bluegrass, orchardgrass, tall fescue, etc.) are also triggered into flowering by cumulative extended exposure to cool temperatures.

Winter wheat is planted in late fall; it germinates before the onset of freezing temperatures. Exposure to the extended cool temperatures of winter as a young vegetative plant vernalizes the wheat, which flowers in the spring and is usually ready for harvest by early summer. Winter wheat planted in the spring remains in a short vegetative state (rosette) and does not flower until much later (if it can survive the heat of summer). As the name implies, spring wheat, which is genetically and physiologically different, is planted in the spring and is harvested in the early summer with no overwintering.

Stratification

Many seeds go through a dormant state in which germination is not possible; and for some seeds the dormancy can be broken only by cold treatment. The dormancy-breaking process is called *stratification*. With many seeds, there is a direct relationship between the requirements for breaking dormancy and the survival of the young seedling. Red oak, apple, and peach seeds must be exposed to an extended period of damp cold before they will germinate. This low-temperature requirement has led to the practice of storing such seeds at 2 to 4°C under moist conditions for several weeks or months before planting. The requirement for cold-temperature exposure before germination effectively prevents natural germination in the fall or early winter and has obvious survival value.

Tuberization

The potato thrives under cool temperatures and grows best at a mean temperature of about 20°C. The optimum range for the production of tubers is 15 to 18°C, and tuber production is suppressed at temperatures above 20°C. Above about 30°C, tuber growth is completely inhibited, presumably because carbohydrate utilization by respiration of the plant as a whole equals or exceeds carbohydrate production by photosynthesis. Since the potato is productive only under relatively cool temperatures, it grows best in the northern parts of the United States and Europe. Potatoes are produced in the middle latitudes of the United States by planting short-season cultivars early enough in the spring for tuberization to occur before the heat of midsummer. In Florida, potatoes are grown during the winter, when temperatures are suitable.

Thermoperiodicity

Under natural conditions, plants experience a diurnal fluctuation in temperature. With few exceptions, night temperatures are lower than those of the day, and the difference varies with season, latitude, and weather. The optimum day temperatures for plant growth are higher than the optimum night temperatures. *Thermoperiodicity*, the fact that many plant responses occur under fluctuating rather than constant temperatures, is much less apparent in the tropics than in the temperate zones, where the day-night fluctuations are much greater.

Classical experiments by F. W. Went showed that tomato plants grew fastest when day temperatures were 26°C and night temperatures 18°C. In this environment they grew about 20 percent more rapidly than comparable plants held at constant temperatures of 26°C and 33 percent more rapidly than those held at 18°C. As the plant approached maturity, the optimum day and especially night temperatures decreased.

Although in the field we have little control of day or night temperatures, greenhouse crops are usually grown with day temperatures about 5 to 10°C higher than night temperatures. There is now considerable interest in using lower than normal night temperatures in greenhouses to reduce heating costs.

Although the complete physiological explanation for thermoperiodicity remains a mystery, we can offer a partial explanation. Since plant growth depends on the daytime products of photosynthesis, optimum day temperatures are those associated with maximum net photosynthesis. At night, as plants continue to respire, the products of photosynthesis are used. Since respiration has a Q_{10} of at least 2, a low night temperature decreases respiratory losses. One can think of thermoperiodicity as providing the optimum balance between production and loss of photosynthates, though this is an oversimplification.

Length of Growing Season

Probably the first, and quickest, assessment of the crop production potential for an area is based on the length of the *growing season*. It is often expressed as the average number of consecutive freeze-free days or the period from the last freeze in the spring to the first in the fall. For tender crops (killed by 0°C), e.g., tomato or soybean, the freeze-free period is the maximum effective growing season; but for more hardy plants, e.g., cabbage and wheat, the effective growing season is extended on one or both ends because of their tolerance of subfreezing temperatures. Length of the freeze-free growing season varies from more than 320 days in tropical and subtropical areas (365 days in Hawaii) to 100 to 120 days in northern sections of the United States down to only 80 or 90 days in the interior regions of Alaska. Although the length of the growing season decreases as one moves northward, the greater day length in summer partially compensates for the short season by accelerating crop growth. Species and cultivars that will mature under the local conditions must be chosen.

In any given region, the length of the growing season is markedly affected by elevation and proximity to large bodies of water. For example, the average number of freeze-free days in coastal Virginia is 225, whereas at the same latitude the mountainous western regions of the state have only 165 days.

The length of the growing season is often used to indicate what *cannot* be grown but does not tell us what *can*. Even though the growing season may be long enough from last to first freeze, if midseason temperatures are not high enough, the soil adverse, or rainfall insufficient for the crop in question, its production is not practical.

HIGH-TEMPERATURE STRESS

Plants can suffer stress from either low or high temperatures. Although the nature of the metabolic and physical damage differs between low- and high-temperature stress, injury or death may result from either.

The effects of high temperature are quite diverse but all result from an excessive heat load. As a point of reference, the amount of insolation received in the middle latitudes (20 to 40° north or south of the equator) in midsummer is between 400 and 500 cal cm^{-2} daily. Less than 5 percent of the radiant energy is used in photosynthesis, and much is converted into heat energy. Recall that 1 cal of heat can raise the temperature of 1 g of water 1°C. The fact that 1 cm^2 of leaf has much less than 1 g of water suggests the potential for an excessive heat load.

Most crops experience heat stress sometime during the growing season. The stress is often temporary or mild and therefore has only minor effects on growth and yield. In other situations, however, severe or long-term heat stress can markedly reduce yield and even kill the crop.

Stress Mechanisms

Excessively high temperatures cause injury by dehydration, denaturation of proteins, or metabolic imbalance.

Dehydration The rate of water loss from a plant or any wet body is temperature-related (Chap. 9). As leaf temperature increases, evaporation also increases, so that unless the stomata close, water loss may be injurious. Usually, as water loss from leaves exceeds the uptake of water by the roots, water stress develops, the stomata close, and evaporation slows. But with this slowing, leaf temperatures rise markedly, since less water is evaporating and dissipating heat. The heat of vaporization of water is 540 cal g^{-1}, which means that all 400 to 500 cal of insolation can be dissipated when just 1 g of water is evaporated per square centimeter of leaf. If adequate water is available, the leaf can even become cooler than its surroundings. On the other hand, under drought conditions, a leaf exposed to intense radiation from the sun may become 10 to 15°C hotter than air temperature. Temperatures as high as this lead not only to further desiccation but also to the second type of heat injury.

Denaturation of Proteins We have already seen that proteins are complex high-molecular-weight molecules, often in solution in the cell. Many plant proteins function as enzymes for biochemical processes vital to the growth and survival of the plant. One important characteristic of all proteins is that when they are heated to even moderate temperatures, they become irreversibly altered, i.e., denatured. All molecules vibrate faster as temperature increases above absolute zero, where molecular motion ceases. At even moderately warm temperatures, large, loosely arranged molecules may be shaken hard enough to become disorganized. A familiar example of protein denaturation is cooking egg white. Denaturation of some proteins begins at about 35°C, and many proteins are inactivated by 50°C. There is a critical relation between temperature and the length of exposure required to cause injury; e.g., a plant can withstand 40°C without significant injury for much less time than it can withstand 35°C.

Metabolic Imbalance This third mechanism of heat-stress injury results when photosynthetic and respiratory processes are out of balance. In C$_3$ plants, both photorespiration and dark respiration increase rapidly with temperature and at high temperatures may actually exceed photosynthesis. Thus, at high temperatures, the plant is losing ground from a metabolic standpoint. C$_4$ plants, which have no photorespiration, can thrive at much higher temperatures and are well adapted to warmer climates. The effects of metabolic imbalance on the plant are much subtler than those of dehydration or protein denaturation since there are no immediately visible effects. The long-range effects, however, can eventually cause death.

Defense Mechanisms

Only about 1 or 2 percent of the radiant energy the aerial parts of plants receive from the sun is utilized in photosynthesis. The remainder must be disposed of to prevent tissues from reaching damaging temperatures. Since most of the infrared wavelengths are either transmitted or reflected, they do not heat the leaf greatly. We are concerned with the energy in the visible region of the spectrum, which is absorbed and converted into heat and which is dissipated through conduction, convection, radiation, and evaporation. As long as leaf or tissue temperature is above the adjacent air temperature, conduction and convection of heat from the leaf to air occurs. Heat dissipation by conduction is increased greatly by wind, which keeps the air near the leaf cooler, thus increasing the temperature gradient. Dissipation of heat by radiation is important to leaves, particularly on the top of the canopy. Not only are these leaves warmest, as a result of direct insolation, but they reradiate directly to the sky and receive less radiation from the soil and other leaves in the canopy. By far the most important mechanism for heat disposal is the evaporative cooling effect, already discussed. Many estimates indicate that 75 to 80 percent of heat dissipation by leaves occurs as the direct result of energy absorption as water evaporates. For this reason, if stomata close due to water stress, greatly suppressing evaporative cooling, leaf temperatures may reach critically high levels.

The heat load on a leaf depends on other factors than level of insolation. Leaves oriented obliquely to the sun's rays absorb far less energy than those at a right angle. Vertical (*erectophile*) leaves usually receive sufficient radiation for maximum photosynthesis but are much cooler than horizontal (*planophile*) leaves. A lighter leaf and stem color can reduce potential heat load, as used to advantage by grayish desert species. Such leaves and stems can reflect far more radiation than dark-green surfaces. Thick cuticles and large numbers of leaf hairs have also been reported to increase reflection.

Summer Temperature

For crops or plants grown on a large scale, growers are obviously limited in their control of summer temperatures. There are, however, some things that can be manipulated, e.g., exposure, the use of shade to reduce leaf temperature, and the use of mulch to modify root temperatures. Most important, however, is selecting a plant or crop that is adapted to the temperatures commonly experienced in a specific area. For commercial production, we select crops that are well adapted, avoiding those with poor chances for success. Landscape plants can be used in marginal areas because they are not a source of income and because it is possible to take precautions against injury which would be impractical on a large scale.

Reducing Summer Temperatures For horticultural crops of very high value, e.g., ornamentals, certain techniques are used to reduce the unfavorable effects of high summer temperatures. Shading is a common practice in the production of nursery, flowering, and foliage plants. One type of shading is a lath house, which consists of wood lath spaced to provide varying amounts of shade (Fig. 8-10). Snow fencing is often used with the wood strips running

FIGURE 8-10
A lath house used to provide partial shade for plants unable to thrive in full sun. Spacing of the lath can be varied to provide more or less shade.

from north to south to provide a continually changing shade pattern as the sun moves from east to west. One problem with wood lath is that the supports must be substantial to hold the weight. A much lighter system is Saran shade cloth (Fig. 24-25), which has become common and is available in densities of weave to provide from minimal to 95 percent shade. Shading of these types has little direct effect on air temperature; its effect is primarily on tissue temperature. The temperature of leaves and other tissues can be markedly reduced by intercepting part of the insolation.

Excessive soil temperatures can be detrimental to many plants. When soil heating is a problem for high-value crops or plants, mulching is commonly practiced. A layer of mulch on the soil surface reduces the insolation reaching the soil and lowers summer soil temperatures considerably. Most mulches are of some organic material, the type used depending largely on cost and availability. Straw, hay, chopped corncobs, wood chips, bark, leaves, sawdust, grass clippings, and white stones—all are good insulators in that they absorb little heat and conduct it poorly. Mulches also have beneficial effects on weed control and soil moisture.

The most widely used procedure for temperature control in greenhouses, other than ventilation, is to apply a shading compound to the outer surface. This material, usually a white latex or calcium paint mixture sprayed on the glass, reflects some of the insolation, thus preventing it from entering. Shading compound gradually wears off during the summer, and any that remains in the fall can be washed off. Saran shade cloth is sometimes used on the outside of greenhouses for the same purpose. In addition to shading or where the reduced irradiance is undesirable, *fan-and-pad cooling systems* are used. Large exhaust fans mounted on one side of a greenhouse draw air through moist pads mounted on the opposite side, thus providing considerable evaporative cooling especially in low-humidity areas (Fig. 24-1). Just as it does in a leaf, water evaporating from the pad absorbs tremendous quantities of heat and cools the passing air.

In relatively small-scale field cropping situations, as well as in greenhouses, increasing use is being made of overhead sprinklers for their evaporative cooling in addition to irrigation per se. This is often called air conditioning, although the plants, the soil, and the surrounding air are all cooled by the evap-

oration of water. Because leaf temperatures are lowered, respiration is reduced. Since water stress and the resultant closing of stomata can be minimized, photosynthesis is increased. The reduction in tissue temperature and water stress is also effective in the field in increasing fruit-set in tomato and bean, reducing abscission of young fruit in grapes, apricots, and citrus, and increasing the general quality of many crops. This technique is also commonly used to reduce heat stress on golf putting greens planted with heat-sensitive C_3 species.

LOW-TEMPERATURE STRESS

Low-temperature injury can occur in spring, fall, or winter and in virtually every part of our country. Those who depend upon a crop for their living are acutely aware of the possibility of a surprise freeze, either in the midst of a balmy spring or before summer seems really past. The kinds and types of low-temperature injury are numerous.

Types of Injury

Chilling Injury Not all low-temperature injury consists of damage from extreme cold. Many tropical and subtropical plant parts are injured by temperatures above freezing (between 0 and 5°C). The browning of the skin when bananas are stored in a home refrigerator is an example of chilling injury. Since ice crystals do not form, the injury cannot be due to freezing. Some tissues suffering chilling injury show a decrease in protein level and an associated increase of amino acids. Since proteins are directly involved in enzyme activity, the normal metabolism can be disrupted. Another form of chilling injury is the apparent disruption of normal membrane function, which is vital for normal organelle and cell activity.

Freezing Injury Although freezing injury implies the formation of ice crystals distinguishing it from chilling injury, the formation of ice crystals does not always lead to injury. Damage from ice formation depends upon where, how quickly, and how much is formed.

A sudden temperature drop is apt to cause more injury than a gradual decrease. With very rapid freezing, cell sap may freeze in the cell. When intracellular ice is formed, membranes are ruptured by ice crystals, protoplasm is disorganized, organelles are destroyed, and death results because molecular and cellular order is destroyed.

When tissues freeze slowly, the effects may be quite different, although death is often the same end result. Because it contains fewer solutes than that in the cell, the water between the cells has a higher freezing point and freezes first. As this happens, water is withdrawn from the cells, leaving the contents of the cell even more concentrated and with an even lower freezing point. At least two theories have been proposed to explain the death of cells following intercellular ice formation: (1) dehydration causes the cell sap to become so concentrated as water is withdrawn that the protoplasm become disrupted; (2) as the intercellular ice masses continue to enlarge, the cells are crushed or the membranes are punctured by the ice crystals. Mushy fruit, vegetables, or leaves following a freeze are evidence of cellular disruption. Control of freeze damage will be discussed later in the chapter.

Winter Desiccation An important type of damage, especially to evergreen plants, is winter desiccation. Although often associated with cold temperatures, the injury actually results from drying out of tissues, in particular, the leaves. By midwinter, the soil becomes quite cold; in northern areas, it may freeze to a depth of 0.5 m or more, which includes a sizable part of the root system. If a warm period occurs, with sunny days and windy conditions, evergreen plants may begin to transpire quite rapidly. With the ground frozen, the roots are unable to obtain enough water to meet the transpirational demand, and desiccation injury results. Winter desiccation is particularly severe on evergreens that are not well adapted to their location. Winter desiccation, or winter burn, can also occur on certain deciduous plants; e.g., raspberry has fleshy canes that can lose water rapidly.

Damage from winter desiccation can be minimized to a certain degree. Sheltered locations and windbreaks are effective in decreasing the severity of this problem because they lower wind speed and thus transpiration. Unheated polyethylene houses are used to protect container-grown nursery stock in regions subject to winter desiccation. Mulching is also effective, both because it reduces the loss of moisture from the soil and because it slows the cooling process and may prevent the soil from freezing. Watering when the ground is not frozen also helps to minimize winter desiccation by ensuring adequate soil moisture. This is particularly important on newly transplanted evergreens. The application of antitranspirant sprays has been moderately effective in reducing the transpirational losses and there are several materials of this type commercially available. Consisting primarily of plastic or wax emulsions, they form an artificial coating on the leaves which can effectively reduce water loss from tissues.

Frost Heaving Frost heaving is a serious problem with plants that have just been planted or transplanted in the previous warm season. It is a particular problem with top-rooted plants and in soils that are poorly drained. Water seeps into crevices in the soil, and as it freezes, it exerts upward pressure. As this occurs repeatedly, plants can be pushed up several centimeters above their original position, breaking the roots and exposing them to drying winds and extreme cold. One solution is using a heavy mulch to suppress daily freeze-thaw cycles. Alfalfa planted in wet soils often suffers from frost heaving, but the problem can be alleviated by interplanting a companion crop which shades the soil and reduces the freeze-thaw cycle.

Root Injury Since the roots of a plant typically do not become as fully dormant as the aboveground portions, root injury may result as soil temperatures get very cold. Root injury is more common in the midsection of the United States than in far-northern areas, where a winterlong snow cover provides considerable insulation, reducing conductive and radiative heat losses from the soil. Snow cover is crucial in the survival of winter small-grain crops in the midwestern United States. Root injury is a major problem with container-grown ornamentals in the nursery, but inexpensive polyethylene houses mentioned above provide considerable protection.

Bark Splitting With apple trees in particular, bark splitting is all too common an occurrence (Fig. 8-11). It is often associated not with extreme cold but with a moderately cold period in the fall, before the onset of total dormancy. Bark splitting occurs with temperatures in the range of -7 to $-10°C$ in November, especially when the temperature drop is rather sudden. When the bark splits, it curls back and exposes the cambium beneath. If the loose bark can be tacked down soon enough to prevent drying

FIGURE 8-11
Bark splitting and tissue death on the trunk of a young apple tree. Freezing injury occurred more than a year before the photograph was taken. (*From R. G. Halfacre and J. A. Barden, Horticulture, McGraw-Hill, New York, 1979.*)

of the cambium, injury is often minimal. Bark splitting becomes a problem in southern landscapes, where periods of winter warmth are often followed by a sudden drop in temperature. The plants are often fooled into breaking dormancy and are thus susceptible to cold injury.

Southwest Injury Tree trunks exposed to afternoon sun in midwinter suffer a type of damage known as *southwest injury*. In winter, when air temperatures can be quite low, exposed tree trunks absorb inso-lation, raising bark temperatures above air temper-atures by many degrees. The southeast and southern sides of the trunk cool gradually as the sun moves, but the southwest side is warmed by the afternoon sun. When the sun sets, tissue temperatures can drop rapidly to air temperature. Southwest injury, the result of this rapid freezing, shows up as cracking and often death of the bark on the southwest side of the trunk.

In areas where this type of injury is prevalent it is standard practice to protect tree trunks. The trunks of ornamentals are often wrapped with burlap or plastic tree guards. On a larger scale, as in orchards, trunks are painted white to reflect part of the inso-lation and to suppress the temperature rise of the trunk (Fig. 8-12).

Temperature Categories of Plants

Plants can be classified according to their tempera-ture response or the growing-season temperatures required for optimal growth. *Cool-season plants* and *warm-season plants* are terms used especially with vegetable and agronomic crops. Those doing best in cool temperatures (18 to 24°C maximum) include spinach, lettuce, potato, pea, wheat, barley, oats, rye, Kentucky bluegrass, tall fescue, and timothy. Ex-amples of warm-season (25 to 35°C maximum) crops are tomato, corn, squash, sweet potato, melons, sorghum, rice, sugarcane, peanut, cotton, bermu-dagrass, St. Augustine grass, and centipedegrass.

FIGURE 8-12
Peach tree trunk which has been treated with white latex paint to reflect some of the insolation and thus reduce drastic temperature changes.

Others are intermediate, and some have their optimum above or below these ranges.

A second classification system is used only with vegetables, ornamentals, and flowering plants, which can be categorized as *hardy, half-hardy,* or *tender* primarily in terms of tolerance to temperatures in the range of 0°C:

Hardy plants can withstand minimum temperatures of approximately −4 to −2°C. Examples are pea, spinach, kale, turnip, cabbage, and pansy. They are usually planted in early spring, 3 weeks or more before the average last freeze for the area. In southern areas, hardy crops can be planted in the fall. Generally speaking, a hardy plant thrives in cooler temperatures and will actually fail in warmer areas.

Half-hardy crops can survive minimum temperatures of −1 to 0°C and are often planted 2 to 3 weeks before the average last freeze date. Examples include carrot, beet, lettuce, and celery.

Tender crops cannot withstand 0°C. They include bean, corn, squash, melons, cucumber, tomato, and many annual flowers. These are not usually planted or transplanted outdoors until after the danger of freezing temperatures is past unless protected by the hot caps or row covers (Fig. 8-13).

There is a correlation between hardiness and the temperatures at which the seed will germinate. Hardy species will generally germinate at relatively low soil temperatures and can therefore be planted early. Half-hardy types need somewhat higher soil temperatures for germination, whereas tender species need considerably higher temperatures.

In a particular category of plants, there may be subtle but important differences in cold tolerance.

FIGURE 8-13
A field of tomatoes in which each plant is covered by a hot cap made of translucent paper. The hot cap acts as a miniature greenhouse, providing protection against intense sun, wind, and freezing temperatures. (*U.S. Department of Agriculture.*)

For example, sunflower seedlings are far more freeze-tolerant than corn. To take advantage of the full growing season, sunflowers are preferred to corn in areas where spring freezes are likely to occur.

In spite of these distinctions, many crops are grown in areas not normally considered ideal. For example, the potato, a cool-season crop best suited to the northern tier of states in the United States, is grown as far south as Florida. This is accomplished primarily by adjusting planting time and cultivar selection to match the species with the locality. In the northeast, long-season cultivars are planted in the spring and harvested in the fall. Further south, shorter-season cultivars are planted earlier and harvested before the warmest part of the growing season occurs.

Wheat is grown from central Texas to the Canadian provinces, but both the type of wheat and time of planting vary with latitude. Winter wheat is planted in the fall, usually when soil temperatures decline to less than 13°C. After germination, it overwinters in the seedling stage and resumes growth in the spring. In more northern areas, typical winter temperatures (−40°C) kill even the most hardy cultivars; wheat is planted in the spring in South Dakota and northward. Winter wheat is preferred in regions where seedling mortality is not excessive since it generally outproduces spring wheat.

Hardening

The resistance of a plant to environmental stress depends upon a number of factors, particularly the growth status of the plant when the stress occurs. Although this is especially true at low temperatures, it is apparent with high temperatures and drought stresses as well. In fact, a plant that has developed resistance to one type of environmental stress is also often resistant to others as well. When a young greenhouse-grown plant is hardened off by gradual exposure to cooler temperature, temporary water stress, and reduced nutritional levels, it becomes more tolerant of both heat and cold as well as drought. Many changes occur during this hardening process, including slowed growth rate, increased cuticle thickness, and loss of succulence in both leaf and stem tissues, all of which help the plant survive.

Mechanisms The mechanism of hardening include important biochemical changes as well as the ana-

tomical ones. As growth slows in the late summer and fall, so does the utilization of photosynthates for new growth, leading to an accumulation of carbohydrates, pectins, nucleic acids, and various proteins in the tissues of the plant. Many plants convert starches into sugars as the temperature falls. The increased concentration of solutes in the cytoplasm acts like the antifreeze in a car. Another metabolic phenomenon associated with hardening involves changes in membrane components. As hardening progresses, plant-membrane fatty acids are replaced by more highly unsaturated fatty acids in the phospholipid fractions; the more unsaturated the fatty acids the lower their freezing point. This is comparable to replacing high-viscosity oil in an engine with low-viscosity oil for the winter. These processes are reversible, and a period of warm weather in late fall or winter will cause the plant to decline in hardiness, as it does at the onset of spring.

It has recently been learned that much of the difference in cold hardiness between plants is due to the ability of some species, once hardened off, to avoid ice formation in plant tissue. About 80 percent of deciduous trees and shrubs (including many important horticultural plants) apparently have supercooled water in the functional xylem at temperatures as low as −40°C, where one would expect ice. The floral primordia in *Rhododendron*, azaleas, *Vaccinium*, and many *Prunus* species survive by supercooling, although not often, to −40°C. As research on cold hardiness continues and our understanding of the phenomena increases, perhaps we shall learn ways of decreasing the severity of cold injury.

Factors Affecting Cold Hardiness The amount of foliage on a plant affects the hardening process. Leaf injury by insects or disease and late-summer pruning tend to make plants less hardy because of reduced leaf area and photosynthetic capacity. With fruit species, an excessively heavy crop can leave a tree with diminished carbohydrate reserves for the development of maximum hardiness. Severe drought, which shortens the normal growing season, can have a similar effect. Nutrition is also important. Excessive nitrogen levels in late summer and fall delay the normal cessation of growth and the onset of dormancy. Such plants are particularly susceptible to early-fall freeze damage. Poor nutrition during the growing season can also reduce the degree of hardiness developed, particularly to severe midwinter freezes. Potassium appears to be especially necessary for the development of maximum hardiness.

In trees, the buds and other tissues in close proximity to the leaves develop maximum hardiness first; the crotches and trunk are the last to harden. This is apparent when an early cold snap in the fall injures the trunk and major crotches but not the rest of the tree (Fig. 8-11).

FREEZES AND FROSTS

Through the years, the terms *freeze* and *frost* have been used more or less interchangeably, especially in agriculture. Meteorologists recommend the term *freeze* for any situation where the average temperature of an area drops below 0°C. In this section, we consider various terms and usages but strongly recommend *radiational freeze* and *advective freeze* as most descriptive and useful terms.

Radiational Freezes A freeze associated with calm conditions, radiational cooling, and a temperature inversion is called a radiational freeze. The likelihood and severity of a radiational freeze depend on the factors which encourage radiant heat loss, as described earlier under "Temperature Inversions."

Advective Freezes When freezing temperatures occur as the result of the invasion of a large cold-air mass, the term advective freeze is used. In much of the United States large, dry, cold-air masses move southward out of the Arctic regions and cause rather sudden, drastic temperature drops. Although radiant heat losses occur during an advective freeze, the situation is quite different from a radiational freeze. These Arctic air masses are accompanied by windy conditions; no temperature inversion is present. Much of the heat loss is directly to the cold air by conduction, and freeze damage control is much more difficult.

Freeze Damage Control
The best method of freeze damage control is good site selection. It is all too common for growers to invest thousands of dollars in freeze-protection systems for a site which should never have been planted. Time and effort spent in assessing the normal tem-

peratures experienced in an area before setting out an orchard or other long-term, freeze-sensitive crop are time and effort well spent. Such scouting can be accomplished by strategically placing accurate minimum thermometers at several locations in the early spring and keeping records. Information can sometimes be gained by talking to long-time residents of the area, who may remember a time when the proposed planting site was in crops and how comparably productive it was.

Radiational Freezes
Several methods are effective in preventing or minimizing freeze damage to crops during a radiational freeze. Some techniques involve small expense, but others require large investments in labor or equipment.

Reducing Outgoing Radiation We have already mentioned hot caps, the small tents of translucent waxed paper often placed over tender crops, e.g., tomato, pepper, and eggplant, transplanted in the field before the average last freeze date (Fig. 8-13). Put on at the time of transplanting, hot caps can also reduce the stress of intense sun and wind on young plants not yet adapted to field conditions. As the plants become established and better adapted to their new environment, the hot caps can be removed (and put back if a freeze is imminent) or gradually torn open wider. The transition from a sheltered greenhouse environment to outside conditions is smoothed and, as the season progresses, the likelihood of a freeze declines.

In certain areas of the country, large clear polyethylene *row covers* are used. They resemble hot caps but cover entire rows rather than individual plants. The sides or ends of these covers can be raised during warm days and lowered at night. Because of the expense, row covers (and to a lesser degree hot caps) are used primarily on high-value crops.

Such coverings form an effective barrier to radiant heat loss since radiation from plant and soil is absorbed or reflected by the cover. Part of what is absorbed is reradiated back down, and part is radiated outward and lost. The net effect with either type of cover is higher temperature and relative humidity for the plants, both day and night.

Cranberries are grown in low-lying bogs (Fig. 9-3), constructed so that they can be *flooded* during freeze conditions and drained after the danger is past. Most crops cannot withstand extended periods of submersion during the growing season, but cranberries show no ill effects from up to 48 h under water. Cranberry bogs are often flooded before the onset of winter and the entire bog freezes solid. It remains frozen all winter and the plants are thereby protected from the extreme winter cold.

A common method of controlling radiational freeze damage to strawberries was to cover the field with straw. The cost of labor makes this method very expensive because the straw cannot be left on the plants during the day and may have to be applied and removed several times during the blossoming period.

Adding Heat One of the most common techniques in reducing freeze damage is the generation of heat. Many different materials have been burned to generate heat, but increasingly stringent air-pollution regulations have drastically changed the situation. For example, burning used automobile tires generates considerable heat, but the accompanying foul-smelling, long-lasting smoke is no longer tolerated in most areas of the country.

For many years, the standard heater used in orchards was a 20-liter (L) can containing fuel oil. These smudge pots were set out in the area to be heated, filled with oil, and covered with a tight-fitting lid. To fire the pots, the lids were removed and enough gasoline added and lighted to ignite the oil. Because smudge pots did not burn very hot, considerable smoke was produced along with the heat. Smudge pots were therefore banned in many areas. (Smoke particles in the air are of no value in reducing radiant heat loss, and a heavy smoke layer can delay warming after sunrise because smoke particles absorb some solar radiation.) More efficient oil burners being used include hot-stack and return-stack burners (Fig. 8-14), in which the flame is confined within the stack. They burn oil more completely and hotter with less smoke than smudge pots.

To avoid air pollution and reduce labor input drastically, most modern heating systems use permanently installed pipelines to carry oil or gas (natural or propane) to individual burners. These systems are expensive to install but are much more efficient in their use of labor. The actual heaters vary greatly from system to system, being very elaborate

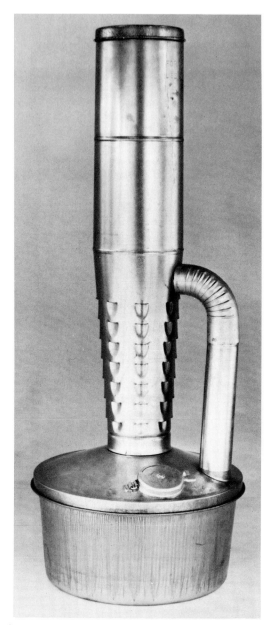

FIGURE 8-14
The Scheu Hy-Lo University Return Stack Orchard Heater. (*Scheu Products Co., Inc.*)

or homemade from locally available materials. Many oil systems of this type use a furnace nozzle to vaporize the oil for very clean burning.

Using a Temperature Inversion An entirely different principle for radiational freeze protection is the use of large fans, often referred to as wind machines (Fig. 8-15). These pivoting fans are mounted 9 to 12 m high on towers. Various models are available, including electric, gasoline, or diesel-powered. Some have a single fan; others have two fans facing in opposite directions. The idea behind using wind machines is to mix warm air at the top of a

FIGURE 8-15
A wind machine in an apple orchard. During a temperature inversion, the large fan mixes warm air above the trees with cold air near the ground to elevate temperatures around the trees. (*Orchard-Rite.*)

temperature inversion with colder air below. *Without a temperature inversion, wind machines are totally ineffective.* The degree of protection provided depends on the size of the inversion, usually expressed as the temperature difference between about 1.2 and 15 m altitude. With a large inversion (4 to 6°C), the temperature at crop level can be raised 2 to 3°C, whereas with an inversion of only 2 to 3°C temperature increase will be minimal. The effect is greatest close to the wind machine and decreases with distance. The initial investment is large, but labor requirements are small. Rises in petroleum prices make wind machines increasingly popular because they require much less fuel than heaters.

A recent innovation, similar in principle to wind machines, is the use of helicopters. In certain areas it is possible to contract for helicopters to fly at low altitudes over an area and provide mixing of air. The cost per hour is high, but there is no capital investment and they are used only when needed. By using thermometers the pilot can seek out, and fly in the warmest layer of the inversion, thereby producing maximal temperature increase at crop level. This flexibility gives helicopters an advantage over wind machines, which are fixed at a particular height.

For maximal protection, heaters are sometimes used with either wind machines or helicopters. Thus, growers are not only using the warm air atop the inversion but adding heat as well.

Using Overhead Irrigation Under the proper conditions, overhead irrigation is an effective system for freeze protection. As water freezes, the heat of fusion (80 cal g^{-1}) is liberated, and in this case much of it is absorbed by the plant and surrounding air. For this procedure to be of value, enough water must be added to maintain a constant film of water over whatever ice has formed. As long as liquid water is present, the temperature of the plant and ice will remain at about 0°C, which is not injurious to most crops. This type of system applies water continuously during the danger period, starting with a low rate of application and increasing it as necessary. The continued formation of icicles indicates that enough water is being applied. On the other hand, if the water freezes immediately on contact, the temperature is below 0°C and damage is likely to occur. An average rate of irrigation would be about 0.25 cm h^{-1}. To be practical the system must meet these requirements: a plentiful water supply, sufficient pipe and nozzles to irrigate the entire area, and good soil drainage to avoid waterlogging the soil. Since even a moderate wind will accelerate evaporation, the system is most effective under calm conditions. This overhead irrigation technique is more effective in humid than in arid climates because the heat of vaporization of water is 540 cal g⁻¹, so that for every gram that evaporates, almost 7 g must freeze to compensate.

Once started, irrigation should be continued all night and long enough after sunrise for the ice to melt. If the system is shut off too soon, evaporative cooling may pull the tissue temperature below freezing and cause serious damage. Another consideration is that evergreen trees, such as citrus, must support the tremendous ice load which can develop; therefore irrigation is frequently limited to low-growing crops or deciduous fruit trees which have only limited leaf surface during the period of freeze danger.

Advective Freezes

The windy conditions that accompany an advective freeze make most of the above protective measures either much less effective or totally useless. Heaters will do very little good, wind machines and helicopters are of no value, and overhead irrigation would be much less effective. Hot caps, row covers, and straw mulch will help, but perhaps the only completely effective method would be flooding, as used in cranberry bogs.

Trunks of young citrus trees are protected from cold injury by mounding soil or, more recently, by insulated wraps (Fig. 8-16). Even if the branches of the tree are killed by a freeze, a new top can be grown much faster than if the whole tree had to be replaced.

WINTER TEMPERATURE

Minimum temperatures during the cool seasons have profound effects on the survival of perennial plants. Many tropical and subtropical species cannot withstand temperatures below 4°C. Their distribution is limited to true tropical and subtropical areas. Such crops grown in the United States include banana, avocado, pineapple, and mango, which are grown

FIGURE 8-16
Young citrus tree with trunk wrap in place to protect against freeze injury. If the limbs above the wrap are killed, buds on the protected trunk will form new branches.

largely in California, southern parts of Florida and Texas, and Hawaii. Citrus species are slightly hardier but are limited commercially to areas which experience only infrequent, light freezes (−2 to −3°C). Many common landscape plants from the tropics or subtropics are grown only as indoor plants in the northern areas. Pittosporum, podocarpus, and camelia, for example, which southern homeowners enjoy in their gardens, are seen only indoors in the cooler parts of the United States.

Plant-hardiness maps (Fig. 8-17), used to aid in plant selection, are based on long-term averages and help predict, although with limited accuracy, the minimal temperatures to be expected in a general geographic area. When one knows the minimum temperatures a plant can withstand, one can decide on the feasibility of its use. Careful selection of specific sites will affect survival, especially in transition zones.

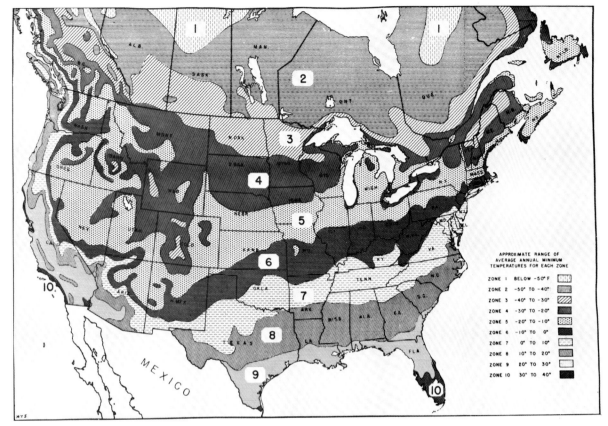

FIGURE 8-17
Plant hardiness map indicating average minimum temperatures for 10 zones. (*U.S. Department of Agriculture.*)

Unfortunately, even long-term averages are not always dependable. Certain horticultural industries have been profoundly affected by one abnormally cold period. An outstanding case in point is the 'Baldwin' cultivar of apple, which until the winter of 1933–1934 was the major cultivar grown in New York and New England. During that uncommonly cold winter, great numbers of 'Baldwin' trees were killed, and the cultivar never regained its popularity. More winter-hardy cultivars have replaced it.

CHAPTER NINE ███████

WATER

The importance of water for crop production cannot be overemphasized. Within a given temperature zone, the availability of water is the most important factor in determining which plants can grow and what their level of productivity will be. Future development of new areas of agricultural production around the world depends largely upon water supply, and it is upon production from such new land that future generations will depend for food.

Water performs many vital functions in plants:

1. It is a necessary constituent of all living plant cells and tissues.

2. It serves as a biochemical medium and solvent as nutrients from the soil and some organic compounds move in solution from their site of uptake, production, or storage to sites of utilization.

3. It is a chemical reactant or product in many metabolic processes, including photosynthesis, although relatively little water is actually consumed or produced in metabolism.

4. Without cell turgor resulting from water movement into cells, young cells would not expand.

5. Functioning of stomata and normal plant turgidity depend directly upon adequate amounts of water.

6. It is a coolant and temperature buffer.

Two of the physical properties of water are of major consequence in the temperature relations of plants. As water evaporates heat energy is absorbed at the rate of 540 cal per g of water evaporated. Plant transpiration thus serves to suppress buildup of heat in leaves (Chap. 8). The high specific heat of water means that water can absorb or release relatively large amounts of heat energy without a major increase or decrease in temperature. The high water content of plant tissues buffers them against temperature change.

Plants can be classified on the basis of their water needs: *Xerophytes* grow in dry places; *mesophytes* grow in moderately moist areas; and *hydrophytes* thrive under very moist or even flooded conditions. Xerophytes (meaning "dry plants") often have special anatomical, physiological, and biochemical mechanisms that minimize water loss, e.g., small leaf area, thick cuticles, C_4 or CAM photosynthesis, sunken stomata, or pubescent leaves. Some xerophytes, e.g., cacti, have a large water-storage capacity. Mesophytes, the plants with intermediate water requirements, include most crop plants such as pine, corn, tomato, soybean, and peach. True hydrophytes ("water plants") generally have small root systems and limited vascular tissue, but they often have specialized tissues that conduct and store air. The water lily is a true hydrophyte; rice is the major crop plant coming closest to being a hydrophyte. Sorghum and the millets come about as close as any field crop to being xerophytic, although certain range plants which provide forage for livestock are able to survive under even drier conditions.

PRECIPITATION AND CROP PRODUCTION

There is a close correlation between precipitation and world crop production. Those regions receiving adequate precipitation (particularly during the growing season) have a greater potential for crop production. Areas with very limited rainfall are highly productive only if additional water can be supplied by irrigation. Crop distribution is obviously also strongly affected by other environmental factors, such as temperature and soil.

Annual Precipitation

Average annual precipitation around the world ranges from 0.05 cm in parts of Chile to more than 1180 cm in parts of Hawaii. In the contiguous 48 states, the extremes range from less than 5 cm in Death Valley, California to more than 317 cm in northwestern Washington. When very small areas are considered, Hawaii provides some of the most amazing differences in precipitation in the entire world. For example, the island of Kauai is approximately 47 km wide from east-northeast to west-southwest, and over that distance the average annual precipitation varies from about 56 cm on the leeward slopes of this mountainous island to 1184 cm on the windward peaks (Table 9-1). As moisture-laden trade winds pass up over the island, cooling and the resultant condensation cause heavy rainfall at higher elevations. As these air masses move down the leeward side, they warm up and become relatively dry. Because of this, such islands have their wet and dry sides; the dry side is said to be in the *rain shadow* of the mountains. The same phenomenon occurs in Haiti, Sri Lanka, and other islands of similar size; although surrounded by water, they may have some very arid regions.

Monthly Precipitation

For most crops, since it is primarily rain falling during the growing season that is available for crop production, data on mean annual precipitation may be deceptive in determining a region's potential for supporting agriculture. The most useful precipitation records are on a monthly basis. For example, although San Francisco, California and Dodge City, Kansas have essentially the same mean annual precipitation (Fig. 9-1), seasonal distribution is so different that the potential for crop production is totally different in these two areas. In San Francisco, only 8 percent of the total annual precipitation falls between May and October, while in Dodge City, 73 percent of each year's precipitation falls during those months. Without irrigation, crop production in an area like San Francisco is limited to the winter season or to drought-tolerant perennials. Dryland, i.e., nonirrigated crop production in the summer obviously has greater potential in an area like Dodge City. An area in the humid east with heavier precipitation has even greater potential, not only because of more rainfall but also because of favorable seasonal distribution (Fig. 9-1). We note here that comparison of rainfall distribution around San Francisco, Dodge City, and Washington, DC does not provide the whole picture because it omits another vitally important factor, the evapotranspirational potential, discussed later in this chapter.

Type of Precipitation

In addition to the amount and monthly distribution of precipitation, the type can be of major consequence in agriculture. The amount and persistence of snowfall is important in winter survival of overwintering crops in northern areas because snow is an efficient insulator. A continuous snow cover during the cold-

TABLE 9-1

Mean Annual Precipitation for Several Locations on the Island of Kauai, Hawaii

Location	Distance from ENE coast, km	Elevation, m	Precipitation, cm
Anahola	2	55	124
Kaneha Reservation	10	253	254
Hanalei Tunnel	18	371	444
Mt. Waialeale	26	1570	1184
Puehu Ridge	42	507	91
Hukipo	45	244	64
Kehaha	47	3	56

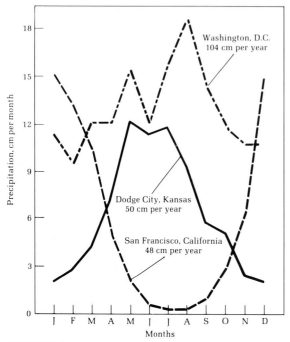

FIGURE 9-1
Precipitation by months at three locations in the United States. San Francisco and Dodge City have similar total annual precipitation, but very different seasonal distribution. (*From R. G. Halfacre and J. A. Barden, Horticulture, McGraw-Hill, New York, 1979.*)

est parts of winter in the northern tier of states allows survival of plants that would otherwise be killed. Open winters, in which there is little snowfall, are more likely to reduce stands of grains such as wheat and barley. If snow to a depth of 0.3 m or more covers unfrozen soil, little or no freezing will occur. Thus snow can provide good protection against cold injury to roots and young seedlings such as winter wheat.

Since hail can totally devastate crops in a few minutes, the average frequency of hailstorms, which varies widely from place to place, is also important in determining the agricultural value of an area.

Rate of Precipitation

Also important are the rate at which rainfall occurs and the total amount accumulating during any given rainfall period. During a light shower with a total accumulation of only 0.25 cm, very little water can penetrate the soil to reach roots because it is held on the surface of leaves and in the few surface millimeters of soil, only to be lost rapidly by evaporation. A very intense rain, typical of summer thunderstorms, can fall so fast that a large proportion is lost through *runoff*. For maximum benefit, a slow, soaking rain is best, as a higher proportion *infiltrates* into the soil. Obviously, only water that infiltrates comes in contact with the roots, the primary plant organ of absorption. Infiltration also depends on the type, texture, and structure of the soil and any covering which slows runoff and protects the soil surface against compaction by raindrops. The coarser the texture of a soil the more rapidly it will soak up water. Soils with a well-developed structure have more channels to admit water, and a soil with vegetation or a porous mulch will also take up more water than a bare soil because runoff is slowed, allowing more time for infiltration to occur. In areas of low annual rainfall, the grower may leave furrows in the soil surface to minimize runoff and to capture maximum snow in the winter months.

RELATIONS BETWEEN PLANT, SOIL, AND WATER

The water content of plants varies widely, not only with the type of plant but also with the tissue and stage of development. Many properly dried seeds contain as little as 10 percent moisture, whereas succulent fruits and vegetables may contain more than 95 percent. Most tissues have a water content somewhere between these two extremes, but actively growing plants are typically high in water content. Since the water in plant tissues is readily lost by transpiration, it must be replaced. An actively growing squash plant may absorb and then transpire about 10 times its fresh weight each day, a turnover of its entire water content approximately once every daylight hour. A full-sized corn plant may transpire as much as 4 L of water in a day, although 1 L daily would be a better average figure for its entire life. To obtain this quantity of water from the soil and avoid desiccation the absorbing and transporting mechanisms of a plant must be highly efficient.

Root System

The water needs of most plants are met almost exclusively from the soil. The notable horticultural exception being epiphytic orchids, which absorb necessary quantities of water through highly developed aerial roots. Certain other species can absorb water through modified leaf hairs, the base of leaves (bromeliads), or needles (certain conifers). In general, however, plants are dependent on water taken up through roots in the soil.

Although the type of root system varies greatly with species, plants have larger and more expansive root systems than is commonly realized. For example, a squash vine with 140 leaves and a stem length of about 9 m can have about 86 m of main and branch roots. This does not include root hairs, which usually increase the surface area of the root system by 5 to 12 times (Fig. 2-4). A single plant of winter rye has been estimated to have 14 billion root hairs with a surface area of 1310 m².

Because it is primarily the young, actively growing regions of the root with its root hairs that are active in water absorption, how fast roots grow and root hairs are produced is crucial. A rapidly growing squash vine produces daily up to 300 m of new roots plus root hairs. That represents a great increase in the volume of soil being explored for water. It is easy to visualize the significance of new root growth and consequent soil exploration by comparing irrigation frequencies of plants growing in a container with those growing in the field. A large tomato plant in a 4-L pot may require watering twice a day to keep it from wilting; a comparable plant in the field may go 1 to 2 weeks without rainfall. This difference occurs largely because the plant in the field can continue to extend roots into new soil and obtain water whereas the plant in the pot rapidly extracts the available moisture from its limited soil mass and has no additional soil to explore for water.

Root Distribution

In some plants, e.g., corn, the root system may be larger in lateral spread than the aboveground parts. Others, with taproots, e.g., alfalfa, may be more limited in lateral spread but penetrate deeply. In many plants, the area occupied by roots approximates the aboveground distribution of shoots. It is frequently easy to explain the adaptation of particular plants to their environment on the basis of their inherent rooting characteristics; cacti, for example, generally have extensive shallow roots, which allow maximum benefit from the infrequent and light rains experienced in arid areas. Crop plants vary widely

in average rooting depth; shallow-rooted species include lettuce, strawberry, and radish, and more deeply rooted types are tomato, alfalfa, cotton, and corn. Alfalfa roots up to 10 m deep have been reported. Many large perennial fruit trees, e.g., peach, apple, and pear, have about one-half their roots within 60 cm of the soil surface. Other tree crops, e.g., pecan and walnut, tend to be more deeply rooted.

Factors Limiting Root Penetration

Many factors can have a marked effect on the rooting depth of a particular species. Among them are a *high water table* (upper edge of free water in the soil), a hardpan, poor soil aeration, and soil compaction. Most plant roots are unable to grow or even survive in saturated soils because of oxygen deprivation. Problems can occur when the water table rises into the root zone after a heavy rain or a field is flooded by a stream. The severity of the problem varies not only with species but with the stage of plant growth. Although rice thrives with its roots submerged, many other plants cannot survive flooding for more than a very few days. Soybean is more sensitive to flooding than corn, but neither can live for long when the roots are under water. Apple and pear trees can withstand flooded conditions for 2 to 4 weeks in the dormant season, periods that could be fatal during the growing season.

A widely fluctuating, sometimes shallow water table is undesirable in cropland. When the water level drops, roots may penetrate into the lower soil layers, only to be killed as the water table rises once again. A high water table in the spring and early summer prevents deep root growth, possibly making the crop more drought-susceptible in the late summer than one growing over a uniformly deep water table. For crops of high cash value, a soil with a high water table but otherwise desirable characteristics can be permanently drained by one of several methods, depending on the situation. In cranberry bogs, the fields are usually trenched and the excess water is drained or pumped out of the trenches (Fig. 9-2). In more typical situations, drainage tiles or perforated pipes covered with coarse gravel can be laid in ditches at the appropriate depth. When the ditches are filled with soil and leveled, the entire field is available for cropping, with no permanent trenches to hinder normal cultural practices. The spacing of drainage lines depends on soil type; since lateral

FIGURE 9-2
A cranberry bog in New Jersey. Since cranberries are grown in swampy areas, the bogs must be artificially drained. Ditches are dug and the level of the water table is regulated by pumping water out of the ditches. To ensure even distribution of water throughout the bog, permanent-set irrigation systems are often used.

movement of water is more rapid in coarse-textured than fine-textured soils, drainage lines can be more widely spaced in the coarser soils.

A field in which a high water table is a potential problem should be evaluated, preferably in the spring when excess water is most likely, by digging holes and periodically examining the level of water in them. After a heavy rain, the water table often rises; but if drainage is adequate, it soon drops to an acceptable depth. If the level of free water stays undesirably high in the soil, the site should either be drained or not planted at all. Careful evaluation before planting is the most effective prevention against crop failure, particularly for a long-term crop like an orchard or a perennial deep-rooted crop like alfalfa.

A *hardpan* (also called traffic pan, plow pan, fragi pan, clay pan, or just pan) is a relatively impervious layer in the soil that can prevent the penetration of roots, causing a plant to be shallow-rooted. This effect is particularly apparent in dry seasons for normally deep-rooting plants, which are the first to suffer from moisture stress. The effect is similar to that described above for a container-grown plant. If there is a hardpan at 30 cm, it would be better to grow shallow-rooted crops, e.g., small grains, than deep-rooted species, e.g., alfalfa or tree fruits.

Little can be done where soil is truly shallow, with bedrock near the surface, but if a shallow hardpan

is present, its effects can be reduced by subsoiling to break up the restricting layer. A *subsoiler* (also called a slip chisel ripper, or deep plow) is a heavy subsurface implement designed to break up a hardpan (see Fig. 10-11). Subsoiling is most effective when the hardpan is relatively thin and the soil beneath it is suitable for root growth. Driving heavy equipment over soil, especially when soil moisture is high, can cause soil compaction. Obviously it is easier to avoid than to cure the problem, but here too subsoiling may be helpful. Subsoiling permits deeper penetration of roots and access to moisture (and nutrients) at the greater depths.

Crops differ sufficiently both in rooting depth and tolerance to excessive soil moisture for a producer to avoid these problems, at least in part, by judicious planning. For example, fruit growers know that deciduous fruit trees rank in increasing sensitivity to excess soil water as follows: pear, apple, peach, and cherry. Pear trees will grow and produce well on a wet site where cherry trees would die during the first year. In some species of fruit trees there is also the possibility of rootstock selection to increase tolerance to wet soils. Two forage legumes, alsike clover and birdsfoot trefoil, are capable of growing in wet soils, while alfalfa does much better in well-drained situations. Among the forage grasses, reed canarygrass is noted for being one of the most tolerant to wet conditions.

As plant roots respire, they use oxygen and liberate carbon dioxide. When soil aeration is restricted by flooding or compaction, oxygen may be depleted and carbon dioxide may accumulate as a result of root and microbial activity. The tolerance to poor aeration varies with species, but the roots of most crop plants cannot survive at oxygen levels below 2 to 5 percent. For normal root growth, oxygen levels must be in the range of 5 to 15 percent, depending on species. It seems that low oxygen is more detrimental than high carbon dioxide, although the two generally occur simultaneously. A striking exception is rice, which thrives with roots under 10 to 15 cm of water where soil conditions are nearly anaerobic. Rice's metabolic adaptation to extremely low oxygen levels, though not unique, is uncommon.

WATER POTENTIAL

The ability of plants to survive and grow requires a precise control of water relations within the plant.

The water content of a plant at any given moment reflects the balance between the uptake and loss of water over the previous few hours.

We suggested earlier that active, fully hydrated plant tissues might vary considerably in water content. In some cases, the water needs of a cell or tissue that is 90 percent water may be fully satisfied while in another cell or tissue at 98 percent it is not; the 90 percent tissue would not be able to absorb more water if available, but the 98 percent one would. Therefore, when one considers *water relations* in a cell, tissue, or plant, water content per se is obviously not very helpful in establishing the true status. *Water potential* describes water status, i.e., whether the cell, tissue, or plant is in a deficit situation and capable of taking up additional water or is filled and can absorb no more.

If a plant, tissue, or cell or other water-holding entity contains all the water it can hold, i.e., if it will take up no more when brought into contact with pure water, its water potential is zero. On the other hand, if the entity takes up pure water, it must have had a nonzero water potential, i.e., there is a potential, or tendency, for water to diffuse into it. Water potential used to be called *diffusion pressure deficit* because it describes the tendency for water to diffuse into a water-deficit area. In a sense, there is a tendency for water to be drawn or sucked into an entity or area with a nonzero water potential. This picture of what is happening explains the even older term *suction pressure*. (*Saugkraft*, which means "suction power," was the term used by German plant scientists who originated the concept of water potential in the 1930s.) The three terms suction pressure, diffusion pressure deficit, or water potential refer to the same thing, the tendency of water to move into an area such as a cell, tissue, or plant. *Water potential, ψ* (psi) is the current term.

Water potential, which indicates the tendency of water molecules to diffuse, evaporate, or be absorbed, is usually expressed in terms of pressure. The SI unit is the megapascal (MPa), equivalent to 9.87 atmospheres.

By convention, the ψ of pure water at standard temperature and pressure is zero. As solutes are added to water, ψ decreases and thus becomes negative (less than zero). This reduction in water potential occurs because the solute molecules inhibit the free random motion of water molecules, and pure water would therefore tend to move into the solution. Pressure or hydrostatic forces also affect ψ. Above

standard atmospheric pressure, ψ is increased; i.e., water will tend to move from an area of high hydrostatic pressure to an area of lower pressure; below standard pressure, ψ is decreased. The significance of these pressure effects will become apparent in the following discussion.

The concept of water potential helps us understand the uptake, movement, and loss of water. Water moves from a region of higher to a region of lower water potential. You may have already reached the logical (and correct) conclusion that, within a plant, ψ is highest in the roots and becomes progressively lower in the stems, the lowest ψ being in the leaves.

Movement of water into and out of cells does not take place strictly by unhindered diffusion. The membranes of a living plant cell are *differentially permeable* in that they allow relatively free passage of the small water molecules but markedly inhibit passage of larger molecules such as sugars, amino acids, and even ions such as K^+ and NO_3^-. The diffusion of water through a differentially permeable membrane is called *osmosis*. The moisture in a root cell is in the form of a relatively concentrated solution separated from the very dilute soil solution by a differentially permeable membrane. Therefore, water moves through the cellular membrane by osmosis, i.e., from a region of high ψ in soil solution to a region of lower ψ in the cell. The component of water potential due to solutes is called *osmotic potential* ψ_π. Since solutes lower ψ_π, it is always negative—the more concentrated the solution the more negative the ψ_π.

As a cell absorbs water by osmosis, it cannot expand indefinitely because the cytoplasm is confined by the cell wall. This containment of the expanding cell causes a turgor pressure to develop and contribute to ψ. The pressure component of ψ is called *pressure potential* denoted by ψ_p.

As indicated in the equation below, water potential is equal to the sum of osmotic potential and pressure potential:

$$\psi = \psi_\pi + \psi_p$$

As the cell absorbs water, the concentration of solutes in the cell is decreased by dilution. Because of the buildup of pressure combined with solute dilution, the rate of water uptake gradually slows after an initially rapid stage. When ψ inside the cell rises so that it is equal to ψ outside the cell, the cell is at equilibrium and no net movement of water occurs. Water molecules still move through the membrane, but they enter and leave the cell at equal rates.

The situation we have just described would apply to a single, isolated cell bathed in a relatively dilute solution. In a plant, however, cells are not isolated but in contact with adjacent cells, so that the basic phenomenon in the plant as a whole is more complex.

WATER UPTAKE

Between the root hairs and the stele of the root, water takes the path of least resistance, to a large extent bypassing the membranes and cytoplasm of the root hair and cortical cells. Since the cell walls consist largely of cellulose, they act like a wick. Water diffuses through the connecting cell walls until it reaches the suberized cell walls of the endodermis, which do not allow the passage of water. Because the water must penetrate the membrane and protoplast of the endodermal cells, they serve as an osmotic barrier between the outer and inner cells of the root. In other words, water can readily diffuse from the soil solution through the cell walls into the root as far as the endodermis. At this point, a semipermeable membrane must be passed. Because the endodermal cells contain sugars, ions, and other solutes, their ψ is more negative than the water diffusing inward from the very dilute soil solution. The endodermal layer acts much like the single cell described earlier and absorbs water.

Another aspect of water potential is important in considering relations between plant, soil, and water. Soil particles have a strong affinity for water molecules. The water molecules directly in contact with a soil particle are held very tightly; as distance from the soil particle increases, the water molecules are held less tightly. These forces contribute the major component of water potential in soil, where osmotic potential is normally quite low. When a soil contains all the water it can hold against gravity, the water potential is close to zero. As water is removed by plant roots or by evaporation, the layer of water around soil particles becomes thinner and progressively more tightly held. By the time ψ in the soil reaches about -1.5 MPa, most plants can no longer remove appreciable quantities of water. We shall return to soil water in Chap. 10.

WATER MOVEMENT

After the endodermis, water must pass through the xylem on its way to the leaves. Since xylem consists of vessels and tracheids which at maturity are non-living and therefore do not contain functional membranes, how can ψ in the xylem be more negative than in the endodermis? This is where the pressure aspect of ψ comes into play.

Transpirational Pull

The most negative ψ in the plant is in the mesophyll cells of the leaf (Fig. 9-3). Because the xylem elements are continuous from the leaves to the roots, the low ψ in the mesophyll cells draws water upward through the xylem. A rapidly transpiring plant actually has a negative pressure (or partial vacuum) in the xylem.

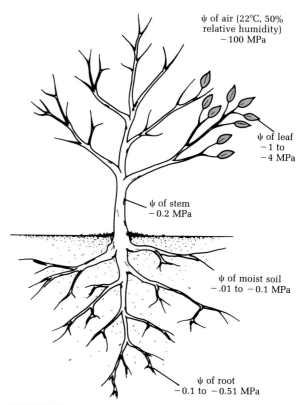

ψ of air (22°C, 50% relative humidity) -100 MPa

ψ of leaf -1 to -4 MPa

ψ of stem -0.2 MPa

ψ of moist soil $-.01$ to -0.1 MPa

ψ of root -0.1 to -0.51 MPa

FIGURE 9-3
Water potential in soil-plant-air continuum. The ψ is most negative in the atmosphere. In the plant there is a gradient, ψ being most negative in the leaves and least negative in the roots, where ψ approximates that in the soil.

As noted earlier, where the pressure becomes more negative, ψ also becomes more negative. When xylem ψ is more negative than the ψ in endodermal cells, water moves from endodermal cells into the xylem. Its tubelike structure enables the xylem to act like a pipe. The negative pressure generated in the leaves is readily transmitted downward to the roots. As water moves from the endodermal cells into the xylem as a result of the ψ gradient, ψ in the endodermis is lowered and additional water is absorbed from the soil through the outer cells of the root.

Root Pressure

Although the major driving force for the upward movement of water in plants is usually the transpirational pull just described, some specific phenomena obviously cannot be explained this way. The exudation of water from the stump of a decapitated plant gives evidence that the roots can generate a positive pressure, particularly apparent in the "bleeding" of grape vines pruned in the spring. Leaves of certain plants exude water as a liquid by a process known as *guttation*, the water being released through specialized structures called hydathodes. Most commonly this occurs during the night following a day in which transpiration has been rapid. Bleeding of xylem contents and guttation result from a positive pressure in the plant which can only reflect a force generated in living roots and is therefore attributed to root pressure.

Nutrients such as NO_3^-, K^+, and Ca^{2+} are absorbed by roots, frequently against a concentration gradient. In other words, root cells are able to absorb these ions even though their concentration is higher inside the cell than in the soil solution. Because the root cell is working against the concentration gradient, such nutrient uptake requires the expenditure of energy to "pump" the nutrients into the xylem cells. When transpiration is very slow, these ions accumulate in the xylem and lower its ψ. When the ψ in the xylem falls below the ψ in the endodermal layer, water moves osmotically from the epidermal cells into the xylem. This process is the mechanism by which root pressure develops.

TRANSPIRATION (WATER LOSS)

The evaporation of water from aerial plant parts, or transpiration, generates most of the force which causes water movement up the plant. Water evaporates from the surface of mesophyll cells in the leaf and diffuses through the stomata into the atmosphere, where ψ is often -100 MPa or even more negative. As Fig. 9-3 shows, the steepest ψ gradient by far is between the leaf and the atmosphere. As water leaves the mesophyll cells, ψ in these cells becomes more negative, causing water to move into them from the nearby xylem elements in the leaf veins. This is the origin of transpirational pull.

Stomata

The number of stomata (Fig. 9-4) is not only large but shows wide variation with species and environment. Estimates range from 2000 to 100,000 stomata cm^{-2}. Even when stomata are fully open, the total area of the pores is usually less than 3 percent of the leaf area; nevertheless, water loss through this 1 to 3 percent is approximately one-half of the amount that would be lost from the surface of a pan of water equal to the area of the leaf. It is thus apparent that the stomatal pores are far more efficient in exchange of gas (O_2, CO_2, and H_2O) than their combined surface area would indicate. The explanation lies in the fact that diffusion through pores is proportional to their perimeter rather than their area.

Factors Influencing Transpiration

The diverse factors affecting rates of transpiration include irradiance, temperature, wind speed, soil water, relative humidity, and several plant factors. Increasing irradiance normally accelerates transpiration by causing stomata to open wider. As an indirect effect of irradiance, when energy from sunlight increases, leaf temperature is elevated; therefore the water-potential gradient between the leaf and the air is also increased.

Under calm conditions, the air surrounding a transpiring leaf gains water vapor, i.e., its water potential increases and the water-potential gradient between leaf and air is therefore reduced. As wind sweeps this moisture-laden air away from stomata and replaces it with air of lower water potential, the ψ gradient is increased and transpiration accelerates. This effect is greatest as wind speed increases from 0 to 8 km h^{-1}, after which the effect increases only slowly. A confounding factor is that, if the irradiance is high, increasing wind speed may cool the leaf, lowering the internal water potential and thus lowering the gradient and transpiration.

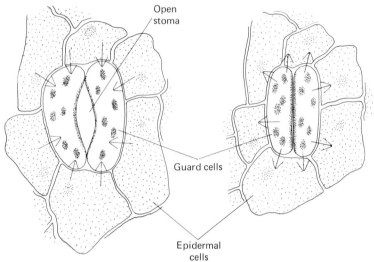

Open stoma

Guard cells

Epidermal cells

FIGURE 9-4
As water moves into guard cells, the cells become turgid and the stoma opens. As guard cells lose turgidity because of loss of water, the stoma closes. (*From R. G. Halfacre and J. A. Barden, Horticulture, McGraw-Hill, New York, 1979.*)

When soil moisture is abundant, transpiration rate is determined largely by conditions in the air and at the transpiring surface; but when water is less available to the roots, transpiration may run ahead of absorption. Then water stress can develop and limit transpiration. As water stress develops in a plant due to the inbalance between uptake and loss, the the water content of the leaf can drop below a level that will maintain full turgor and the stomata start to close. This effectively minimizes further tissue desiccation.

Certain anatomical, morphological, and physiological characteristics of plants can dramatically affect the rate of transpiration. At midday, since upright leaves intercept less radiation than horizontal leaves, they tend to be cooler and transpire less. Some plants, e.g., pineapple, have sunken stomata that are less affected by wind speed. Other plants have dense hairs surrounding stomata. CAM plants open their stomata at night rather than during the day, greatly reducing water loss. Corn leaves roll up when they wilt and reduce water loss.

Although we have discussed water absorption by roots and transpiration as interdependent processes, each can occur somewhat independently for limited periods of time. Under conditions of rapid transpir-

ation, water uptake may lag behind water loss briefly. If water loss exceeds absorption sufficiently, the plant wilts until the water balance is restored. A deficit that develops during the day is erased at night by water uptake by roots if the soil is not too dry. Guttation is an indication of excess uptake, more than satisfying any deficit that developed during the day. Plant parts severed from the root system continue to transpire, as is apparent with cut flowers, which wilt and desiccate rapidly without water. When the stems of cut flowers are placed in water, they may be able to live for extended periods. Without roots, water uptake can occur only by transpirational pull since obviously no root pressure or root osmosis develops. Blockage of xylem elements limits the life of cut flowers, but the addition of certain chemicals can significantly extend it by delaying the development of xylem blockage.

Measuring Efficiency of Water Use
One way of expressing how efficiently plants use water is the *transpiration ratio*, the amount of water transpired by the plant per amount of dry matter produced. To determine dry matter, the plant (usually only the aboveground portion is measured) is dried in an oven. The loss in weight during drying

gives the moisture content; the remainder is dry matter. To calculate the transpiration ratio, one must know the amount of water transpired by the plant during its entire existence to produce its final dry weight. Assume that total transpiration of a plant with a 0.4-kg dry weight was determined to be 200 kg of water. Its transpiration ratio is 200/0.4 = 500. In other words, the plant required 500 units of water (lost in transpiration) to make 1 unit of dry matter. Because the transpiration ratio is expressed as weight of water per weight of dry matter, the units must be the same but can be in terms of grams per gram, kilograms per kilogram, tons per ton, etc. Transpiration ratios vary from about 30 to 3000, but, crop plants generally fall in the range of 300 to 1000 (see Table 9-2).

In an agricultural context, it is most informative to know the amount of water used over an area to produce a given amount of dry matter. Although desert plants have a surprisingly high transpiration ratio (about 2000), they are able to survive in arid locations because they grow (and use water) so slowly. Estimates of cactus growth indicate an annual production of only 225 kg of dry matter per hectare. With a transpiration ratio of 2000, the amount of water used by cacti in 1 ha would be 225 kg ha^{-1} × 2000 = 450,000 kg. This is the equivalent of 450,000 L of water over the entire hectare, or 4.5 cm of rainfall.

TABLE 9-2
Transpiration Ratio of Selected Crops and Weeds

Crop	Transpiration ratio†
Alfalfa	900
Barley	800
Soybean	700
Lambsquarter	700
Oats	600
Red clover	600
Wheat	500
Potato	400
Sudangrass	400
Corn	400
Pigweed	300

† Expressed as kilograms of water transpired per kilogram of dry matter produced.

Consider a crop of peach trees producing a total of 5600 kg of dry matter (leaves, fruit, and wood) per hectare annually. The water needs will be much greater even though the transpiration ratio of peaches is much lower than for cacti. With a transpiration ratio of 500, 1 ha of peach trees would absorb 500 × 5600 = 2.8 × 10^6 kg. This is the equivalent of 2.8 million liters, or 28 cm of rainfall.

Another way to express relationships between growth and transpiration is *water use efficiency*, or the number of units of water transpired per unit of *economic yield*. If a particular cultivar of soybeans has a transpiration ratio of 600 but only 30 percent of the plant dry matter is beans, it takes 2000 kg of water to produce 1 kg of soybeans. The water use efficiency for soybeans then is 2000.

When the transpiration ratio or water use efficiency is calculated, only the water transpired by the plant is included. The amount of total precipitation or irrigation ultimately absorbed by the crop plant varies from 0 percent after very light showers, in which water is intercepted by the foliage and lost to evaporation, to somewhat over 50 percent under ideal conditions. The remainder is lost through runoff, evaporation, uptake by competing vegetation, and percolation below the root zone. A rule of thumb is that about one-third of the total water falling on cropland is absorbed by the crop. With this figure of one-third, the cactus plants would require about 4.5 cm × 3 = 13.5 cm of annual precipitation, whereas the peaches would require 28 cm × 3 = 84 cm to produce a crop.

Soil moisture, nutrition, soil aeration, temperature, irradiance, disease, plant age, plant population, atmospheric humidity, and wind speed can all influence the transpiration ratio. In general, the transpiration ratio is at its minimum when each factor is at its optimum. Thus, a crop growing in an ideal environment would be most efficient in its utilization of water.

Methods to Reduce Transpiration

Suppressing transpiration in crop plants may be an absolute necessity or merely a desirable goal, depending upon the situation. The detrimental effects of water stress on crops include suppressed growth and yield, delayed flowering, abortion of flower, pod, or grain, and a reduced grain-fill period. Many crops develop specific disorders following even temporary water stress. Some of the common problems related to water stress are blossom-end rot of tomato, poor fruit set in grape and soybean, delayed flowering in many species, the shedding of young fruits, and shriveled or shrunken grains in wheat and corn.

Leafy cuttings being rooted will not survive unless transpiration can be drastically reduced. Cuttings are usually placed in beds and misted frequently or covered with polyethylene (Chap. 15) to keep relative humidity very high. New cuttings are usually also shaded to reduce leaf temperature, thereby lowering the water-potential gradient between the leaf and air. These precautions are not necessary with dormant hardwood cuttings because they usually have started to form roots by the time leaves appear. Newly rooted cuttings of ornamentals are taken from the propagation bed and grown for a while under lath or shade cloth to permit further root development before exposing the plants to the greater transpirational demand imposed by full sun and wind. When bare-root trees or shrubs are transplanted, the tops are usually cut back by about one-third to compensate for root loss during digging. This allows the damaged root system to meet the transpirational needs of the top of the plant.

Commercially available *antitranspirants* are effective under some situations. Most of them are wax or plastic emulsions diluted with water and sprayed on the foliage to form a film when dry. They have been most widely used for the prevention of winter desiccation in horticultural plants. Although the potential of these materials on actively growing crops and transplants has been evaluated, in general they have not proved commercially practical. In reducing transpiration, the film may also suppress photosynthesis if the stomata are covered. Unfortunately, too, the film tends to have a fairly short effective life. Another potential drawback with this and other methods is that, with marked suppression of transpiration, leaf temperature rises, and the resulting increase in the water-potential gradient tends to negate the decrease in transpiration. Elevated temperatures can also have other detrimental effects (Chap. 8).

An entirely different approach to reducing transpiration has been explored in recent years. Certain chemicals are known to cause stomatal closure, and others can reduce or prevent stomatal opening. When stomata close partially, transpiration is usually decreased more than photosynthesis. A chemical that could control stomatal opening without such adverse side effects as reducing photosynthesis could be of tremendous value in agriculture. Control of transpiration could conserve great quantities of water, reducing the need for irrigation or producing a high yield on lower rainfall or both, by reducing transpiration ratio or improving water use efficiency.

Whether such chemical control of transpiration will ever become commercially practical remains to be seen, but the potential value to agriculture is enormous.

MOISTURE CONTROL DURING STORAGE AND MARKETING

Control of moisture after harvest is of major importance with many horticultural and agronomic products. Fresh fruits, vegetables, and flowers are stored at cool temperatures and high relative humidities to minimize water loss. In most storage environments for such commodities, the relative humidity is kept between 90 and 95 percent, just enough below saturation to avoid condensation, which would accentuate disease problems. Excessive water loss obviously results in shriveling of fruits and fleshy vegetables and wilting of leafy vegetables.

The natural waxes on fruit surfaces are variable, depending on the species, but usually provide considerable protection against water loss (Fig. 9-5). Many fresh vegetables are artificially waxed to maintain their crispness. Waxed cucumbers have a greasy coating; a harder wax is used on turnips. Polyethylene bags are commonly used with apples, citrus fruits, carrots, and lettuce, not only to reduce water loss but to simplify handling. When a semirigid cardboard or styrofoam base is used with an overwrap, relatively perishable items like tomatoes, blueberries, cherries, and plums can be attractively displayed and protected against mechanical damage and excessive moisture loss (Fig. 9-6).

Waxing is also used in marketing certain ornamental plants to avoid desiccation, e.g., rose bushes are handled by digging the dormant plant, putting damp peat moss around the roots, and tying a polyethylene bag over the root system. The rather fleshy nature of the stems, however, means that transpiration during storage, shipping, and marketing can lead to desiccation and death. A common procedure is to dip the stems of the plant in melted wax to provide a reasonably continuous coating. As the

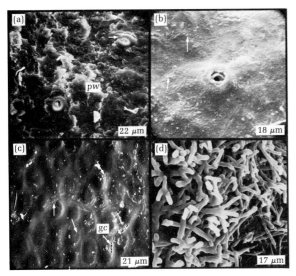

FIGURE 9-5
Scanning electron photomicrographs of variations in fruit surfaces. (a) Orange fruits are covered by platelet wax (*pw*) and have stomatal openings (*st*) which may be at least partially covered by wax. (b) Mango fruits have a thin, slightly pebbly wax coating (*black arrows*) and lenticel openings (*l*) similar to those on apples. (c) Tomato fruit have a thin but pebbly epicuticular wax coating (*white arrows*) and no natural openings. Growth cracks (*gc*) or handling injuries (*hi*) are evident on the surface. (d) Blueberry fruit have rodlet surface wax and occasionally an opening (not shown) heavily covered by rodlet wax is present. (*Courtesy of L. G. Albrigo and P. Anderson.*)

bush is planted and growth starts, the wax gradually falls off, but not before it has served its purpose.

Since most grain crops are harvested mechanically, moisture content determines when harvest can begin. Harvesting equipment does not effectively separate the grain from the plant and chaff until the moisture content is relatively low. Harvest of such crops as wheat, barley, and soybean is usually started when the moisture content of the seed declines to about 15 percent. Corn and sorghum can be harvested when the moisture content reaches about 27 to 30 percent.

For long-term storage, most grains must be dried to moisture levels somewhat below that of normal harvest. Less drying is necessary for wheat, oats, barley, and other crops that are drier at harvest than corn and sorghum. Although seeds high in oil must be dried to moisture levels of 8 to 10 percent for safe storage, starchy grain crops are generally safe at about 13 percent.

Unless grain is adequately dry when stored, microbial growth or fermentation can affect its quality. Wheat at moisture levels of 14 to 16 percent in deep bins will deplete the oxygen level, encouraging the growth of anaerobic organisms. The result is wheat with a dull appearance and high acidity, unsuitable for making bread.

Harvested grain is dried before storage by forcing air through the grain. If the relative humidity of outside air is low, unheated air can be used to lower the moisture level by 2 to 4 percent. If the humidity is high or a great deal of moisture is to be removed, heated air is much more efficient. Since higher petroleum fuel costs make heated air much more expensive, alternate fuel sources and storage methods are being evaluated.

Grain is stored in several types of enclosures, but all should be designed for ease of handling, freedom from dampness, and protection from insects and rodents. Relatively tight bins constructed of metal, wood, or masonry are the most common storage structures for grain. Controlling temperature differences and moisture in bins is the key to long-term storage of high-quality grains.

COPING WITH LIMITED MOISTURE

Having described water uptake, movement within the plant, and loss through transpiration, we now return to the availability of adequate water. In the humid eastern United States, soil-moisture shortages can be detrimental but are generally of short duration; in arid regions, limited soil moisture is the rule rather than the exception. The strategies developed by plants and growers of plants to cope with limited moisture range from passive to active, and some are both expensive and intensive.

Drought-Tolerant Species
Crop species vary considerably in their ability to withstand drought. In areas subject to frequent periods of limited soil moisture, it certainly behooves one to grow species with good drought tolerance. For example, among forage crops, millet and sorghum are two of the most resistant to drier conditions and are productive in areas where more drought-sensitive plants such as alfalfa would grow poorly. This does

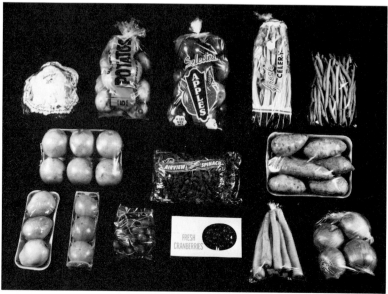

FIGURE 9-6
Produce in polyethylene bags and overwrapped trays. (*Courtesy of R. E. Hardenburg.*)

not mean, however, that sorghum and millet will perform better in a dry than a moist environment, as productivity is increased with adequate soil moisture; however, in a relatively dry environment, they suffer less than most other forage and hay species.

Perhaps the most legendary drought-tolerant crop plant is the date palm, which not only thrives under very dry conditions but requires the hot, dry climate of arid regions for proper fruit development and ripening.

Fallowing

Fallowing of land (described in more detail in Chap. 17) is used in regions with inadequate precipitation for annual cropping. The procedure involves cropping a given piece of land every other year and keeping the soil free of all vegetation, including weeds, during the noncropping year so that fallow-year precipitation accumulates in the soil. Usually bands of crops (generally wheat) alternate with uncropped or fallow bands.

No-Till Systems

A major change has recently taken place in the method of growing corn and some other crops. Traditionally, land was plowed in the fall to turn under crop residues and to control weeds. In the spring, the soil was disked to break up clods and to smooth the soil surface before planting. Now a method that decreases costs and conserves both moisture and nutrients has been developed. After harvest, the field is disked only enough to cut up crop residues and prepare a rough seedbed for a cover crop of rye, wheat, or other winter-hardy small grain. A few days before seeding corn in the spring, the cover crop is killed with a herbicide. Specially designed planters seed the corn into narrow slits made in the killed cover crop. This method not only saves labor and fuel but also conserves moisture. Leaving the residue of the cover crop on the soil surface markedly reduces surface evaporation, suppresses surface runoff, and increases infiltration.

There are many modifications of this practice and different terms to describe them—no till, minimum tillage, reduced tillage, conservation tillage, etc. The concept of soil management is being used with other agronomic crops and some vegetables.

Mulches

In Chap. 8, we briefly discussed mulching as a means of lowering soil temperature. A mulch is a layer of material applied to the surface of the soil, where it remains for weeks, months, or even years. It may consist of a wide diversity of organic and even inorganic substances applied as mulches (the mulch may be grown in place and then killed, as with no-till systems). Common organic materials include straw, leaves, sawdust, corn cobs, peanut hulls, pine bark, pine needles, wood chips, and peat moss. The mulch is often chosen on the basis of cost: in the corn belt, corn cobs may be the least expensive; in other regions, pine bark, peanut hulls, or sawdust may be cheaper. The cost of the mulch, which includes the labor of applying it, limits its use to intensively grown, high-value crops or plants.

In the eastern United States, strawberries are frequently mulched with organic materials such as straw. The benefits include moisture conservation, weed control, better conditions for the pickers, and cleaner berries. Researchers are exploring the possibility of using a living mulch for strawberries to reduce the cost by a late-summer seeding of grass in the strawberry field. After reaching adequate height, the grass would be killed with a herbicide (unless a freeze-sensitive species is used, which would be killed by the first fall freeze). The grass would fall over the strawberries, providing winter protection along with the other benefits during the next growing season.

In home and commercial landscapes, arboretums, and botanical gardens, mulches are used for both aesthetic and cultural purposes. A mulch under trees, shrubs, and flowers provides an attractive setting and reduces maintenance costs. Pebbles or small stones are occasionally used as a mulch in such situations.

Organic mulches provide a great many benefits to plants. Among these are moisture conservation, increased levels of organic matter in the soil, erosion control, improved soil structure, improved water infiltration, reduced temperature fluctuations, and improved nutrient availability. Although this chapter relates primarily to water, the other effects of mulches will also be briefly elaborated.

Moisture is conserved because an organic mulch suppresses the growth of weeds, keeps soil temperatures cooler, and protects the soil surface from both wind and rain. Organic mulches improve soil structure through the addition of organic matter, thereby increasing the infiltration rate of water dramatically. The added organic matter also improves the water-holding ability of the soil. By both protecting the

soil surface and increasing infiltration rates, mulches minimize soil erosion.

The effects and significance of mulches on soil temperatures vary both with the season of the year and the crop. In general, organic mulches stabilize soil temperatures, lowering harmful summer highs and raising harmful winter lows.

As the mulch is broken down and organic matter is added to the soil, any nutrients present are released into the soil as well. With most organic mulches, e.g., corn cobs or sawdust, nutrient content is low, but when leaves or hay are used, the quantities of nutrients released may be sizable.

The use of polyethylene film as mulch is now widespread. The film is usually clear or black and about 0.1 mm thick. With widely spaced vine crops such as cucumber and cantaloupe, the seeds or transplants are set in a single row down the center of a strip of plastic about 1 m wide. With many crops, e.g., eggplant, sweet corn, and tomato, two rows are set in each strip of plastic about one-fourth of the way in from each edge.

Plastic mulch greatly modifies both soil moisture and soil temperatures (Chap. 8). The plastic film is applied when soil moisture levels are high and allows little moisture loss aside from that transpired by the crop. It might seem that a plastic mulch would eventually lower soil moisture levels by shedding rainfall or irrigation water. This is not normally a problem because water can reach the ground through the hole around the plant stem and can also penetrate the soil between adjacent strips and move laterally into the root zone. A major consequence is the almost complete elimination of evaporative losses from the soil surface.

Terracing

On sloping land, since a large proportion of the rainfall tends to run off before it can percolate into the soil, moisture stress can occur even though the same crop on relatively flat land receiving the same amount of precipitation would not suffer. Contour planting helps reduce runoff, but terracing is even more effective. On moderate slopes, terraces are merely low, broad ridges formed along the contour. Because the flow of water down the slope is slowed by the ridge, greater amounts of water percolate into the soil. In addition to increasing soil moisture, terraces also reduce soil erosion markedly. Terracing of steep mountainsides has been practiced for centuries in many parts of the world. From a distance, the mountainside has the appearance of giant stairs— flat terraces separated by steep walls. Without such terraces it would be impossible to perform normal cultural practices on this sort of terrain.

IRRIGATION

Although irrigation has been practiced for many centuries in arid and semiarid areas, its use has recently expanded greatly even into humid climates. Increased food production to meet the needs of the world's rapidly expanding population will require not only better management of land currently used in food production but the utilization of land now too dry for agriculture.

Determining Feasibility

Insufficient water is the major limitation to cropping of vast areas of the world. In many areas, land is productive only because of irrigation, and in many others, growing crops is greatly improved by irrigation. In an arid or semiarid climate, the need for supplemental water is obvious, and without it, crop production could not even be considered. In areas that normally receive sufficient precipitation, however, the economic feasibility of irrigation becomes much more complex. As already discussed, seasonal distribution of precipitation is of prime importance. If weather records indicate that sufficient rainfall can generally be expected during the growing season, the decision to purchase irrigation equipment will be based on the anticipated frequency of insufficient rainfall or the possibility of additional yield even in good years. If a serious drought occurs on the average of once in 10 years, irrigation is much less practical than if a profit-destroying drought occurs once every 3 years. Irrigation can often be considered as insurance against dry periods, which may reduce yield or, if extreme, cause the loss of the entire crop. Since the unpredictability of weather makes this decision a gamble, probably the best one can do is to base decisions on the long-term averages for the area.

It is important to know both the amount of precipitation that can be expected during the growing season and how rapidly this water will be depleted. Probably the most useful measure of water loss, the *evapotranspiration potential*, takes into account both evaporation from the soil and transpiration by the

crop. Even though precipitation appears to be adequate, its monthly distribution should be considered in relation to the evapotranspiration potential. The rainfall advantage Dodge City has over San Francisco (Fig. 9-1) is reduced by its higher average temperatures and lower relative humidities, which increase the evapotranspiration potential. As Fig. 9-7 shows, the demand for moisture frequently exceeds the supply during the growing season. If this imbalance of current supply and demand cannot be met from reserves in the soil, moisture stress will develop and irrigation may be economically advantageous.

Methods used to estimate evapotranspiration potential include elaborate calculations of environmental parameters and the use of a *lysimeter*, a large container of soil in which a crop is grown so that the water loss can be determined by weight loss of the entire container (after correcting for dry-matter gain by the plants).

Perhaps the most common method of determining evapotranspirational potential uses available data such as temperature, relative humidity, wind speed, and cloudiness. The rate of evapotranspiration depends on a complex interaction of soil type, crop, stage of development, plant population, and environment. Since a crop in Texas exposed to relatively high temperatures, low humidity, and intense sunshine has a higher rate of evapotranspiration than a

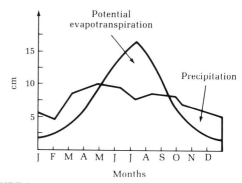

FIGURE 9-7
Annual patterns of precipitation and evapotranspirational potential. Although total annual precipitation is adequate for the crop, moisture is in excess from September to May but inadequate from May through August. Some of the excess is stored in the soil but irrigation is often beneficial during periods of peak evapotranspirational potential. (*From M. D. Thorne, Moisture Management, Crops and Soils Magazine, December 1979.*)

comparable crop in Illinois, more water is needed in Texas than in Illinois to produce an equivalent yield of the crop.

Rooting Depth Important morphological and anatomical differences between crops affect the decision to install irrigation. Depth of rooting can markedly affect how long the crop can thrive without additional water, either as rain or from irrigation. Lettuce, bentgrass, and peanut, being relatively shallow-rooted, deplete the *root zone* of moisture more rapidly than such deep-rooted crops as alfalfa, tomato, and trees. Hence irrigation is more necessary for the shallow- than deep-rooted crops in drought-prone areas.

Soil Type and Depth The ability of a soil to retain water for plant use is related to its texture. A coarse, sandy soil has little water-holding capacity, while a finer clay-loam has a much greater potential to hold water. Since as a general rule, a clay-loam will hold twice as much available water as an equal volume of sandy loam, all things being equal, irrigation is much more likely to be necessary in sandy soils than in those of a finer texture and on shallow soils than on deep soils with no rooting barrier. Soil moisture is covered more thoroughly in Chap. 10.

Stage of Growth The water needs of a crop change markedly through its life cycle. A field of young corn plants obviously needs less water than the same field when plants have reached their maximum height. Another vital consideration is that the effect of moisture stress is much more severe at some stages than at others. For example, Fig. 9-8 shows how moisture stress at different stages in the life cycle affects the yield of corn. Moisture stress during the early reproductive phases (tassel and silk emergence) has a much more detrimental effect than an equal period of stress earlier or later in the growth cycle. If irrigation water is limited, its use will be most beneficial during silk emergence.

Growth Pattern The growth pattern of a crop is of major importance in deciding when, whether, or how much to irrigate. An entire crop of strawberries can be lost during a 3- to 4-week drought in the spring when the fruits are developing. A similar period of dry weather could also reduce the yield of long-season crops, such as apples, but the net effect

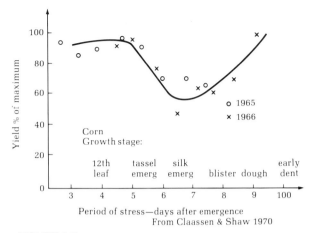

FIGURE 9-8
The effect on yield of 4 days of moisture stress on a corn crop depends on the stage of growth when the stress occurred. (*From M. D. Thorne, Moisture Management, Crops and Soils Magazine, December 1979.*)

would be much less drastic because of the longer period of fruit development. Another example can be drawn from a comparison of apples and peaches, which have different fruit-growth patterns. Apples enlarge at a more or less linear rate from full bloom to harvest. Peaches, however, have a rapid growth phase, or *final swell*, during the latter part of the season. Inadequate soil moisture during the final swell of peaches is much more harmful than a dry period of comparable length occurring at any time during the growth of apples.

Crops also can suffer when the availability of moisture fluctuates. As soil moisture is depleted, growth slows down. When water becomes available again, especially if it comes in large quantities, growth resumes rapidly. This rapid growth can cause splitting, as in 'Stayman' apples (Fig. 9-9) and cabbage, for both of which the crop can be essentially destroyed in a few hours. Something similar can happen to the seeds of soybean. By maintaining an even growth rate, irrigation can minimize such problems.

Balance of Cost and Benefit The potential returns from an irrigation system depend on the crop, how often the system will be used, and the increased value to be realized. The costs are variable, depending on the system chosen, the frequency of use, and the

FIGURE 9-9
Cracked 'Stayman' apple resulting from a dry period followed by heavy rains in the early fall. Internal pressure due to rapid water uptake caused the fruit to split. (*From R. G. Halfacre and J. A. Barden, Horticulture, McGraw-Hill, New York, 1979.*)

cost of water. Water costs depend on location. In large irrigated areas in the western United States, the cost of water varies by a factor of 10 or more. Some producers can use lakes and rivers, but others must use deep wells, adding considerably to water costs. The high cost of water (which reflects its availability) may be a major reason for selecting a more efficient irrigation system.

Water Rights and Availability Another concern faced by many growers considering irrigation is water rights. The laws regulating the use of water vary widely in different parts of the United States but are important everywhere. Increasing demands being placed on our finite water supply lead to many lawsuits. It is not easy to decide who is entitled to water when agricultural producers, cities, industries, and those interested primarily in recreation are all competing for it. It certainly behooves anyone contemplating installation of an irrigation system to be sure of the water rights before purchasing the system. Chap. 13 will look more closely at the concern of continuing availability of water for irrigation.

Systems of Irrigation
The four basic irrigation systems used today are surface, aerial (sprinkler), trickle, and subsurface. Each category offers a variety of types designed to meet particular situations.

Surface Irrigation
Because of the water-distribution systems involved, surface irrigation, which applies water directly to the soil surface, is suited only to quite flat or level land. One reason for its widespread use in arid and semiarid regions is low evaporational loss, since the water does not pass through the air. More important in such areas is the fact that irrigation water is often supplied to the field in ditches or large pipes; using gravity flow for surface distribution eliminates the need for power to drive pumps. Surface irrigation has another advantage: since water does not wet the foliage, problems with disease are reduced. Surface application is of little or no value, however, in the freeze protection or "air conditioning" discussed in Chap. 8.

Flood Irrigation Flood irrigation is used primarily with those crops that can tolerate excess water. Since it usually involves flooding an entire field, this form of irrigation is limited to essentially level land. Cranberries and rice are commonly grown using this method. With cranberries, flooding is also useful for freeze protection, insect control, and harvest. With rice, two additional advantages of flooding are weed control and nitrogen fixation by aquatic algae.

Basin Irrigation This modified system of flooding is used in some orchards (Fig. 9-10) and also applied to pastures and such crops as small grains and alfalfa. The land must be relatively level, but small differences in elevation can be accommodated. Ridges or

FIGURE 9-10
Basin irrigation of a citrus grove in Texas. Ridges are raised by machine and each elevation is flooded separately.

levees are raised along contour lines, and water is introduced into individual basins, which are essentially level terraces. The levees may be permanent and seeded to grass or built before each irrigation.

Furrow Irrigation Perhaps the most widely used type of surface irrigation for crops is furrow irrigation. Typically used with row crops, such as vegetables, sugarbeet, and corn, it offers more even distribution and more efficient use of water than flood or basin irrigation. The water is delivered in a head ditch or gated pipe to the high end of the field, from which it flows into individual furrows (Fig. 9-11). Depending on the planting system, the irrigation furrows may be between single rows of plants or two rows of the crop may be in the bed between furrows.

One disadvantage of the method is that ensuring a sufficient supply of water at the far end of the field requires giving excess water to the end closer to the source. Perhaps the greatest disadvantage, however, is the high labor requirement since furrow irrigation requires constant supervision and maintenance to regulate the amount of water. If water is applied too fast, individual furrows may merge and form channels, leading to soil erosion; if it is applied too slowly, the distant end of the furrow will not receive enough water.

Aerial (Sprinkler) Irrigation

One of the most widely used types of irrigation, the use of sprinklers, comprises a vast array of systems, all delivering water through the air to the crop and soil. The basic parts of any sprinkler system are a pump, main delivery pipe, lateral pipes, and sprinklers.

Permanent Systems Permanent systems are installed with no expectation of moving them. Most permanent systems are installed with both the main line and laterals underground and only the risers and sprinklers aboveground (Fig. 9-2). The main lines and laterals are often made of polyvinyl chloride plastic (PVC), although some steel pipe is also used. These systems are often used for such high-value crops as fruit trees, grape, tobacco, and nursery crops. Applications where convenience, appearance, and low maintenance become more important than cost include lawns, golf courses, and landscapes.

Major disadvantages of permanent systems are the very high initial cost and lack of flexibility; their

FIGURE 9-11
Onion field in Colorado with furrow irrigation. Water is siphoned into each furrow from the main ditch in the foreground; two rows of onions are in each bed.

chief advantage is the minimal labor requirement once they are in place.

Portable Pipe The availability of aluminum pipe with quick-coupler connections gave tremendous impetus to sprinkler irrigation. In many situations where a farmer cannot justify the much greater expense of a permanent installation, the use of a *portable pipe* is practical. Assume that a farmer has 50 ha of land under cultivation and needs to be able to irrigate once every 10 days for 8 h. If only one setting is used per day, a system covering 5 ha is needed. If it is possible to irrigate two areas per day, only 2.5 ha would have to be done at one time but two moves per day would be required.

In conventional portable-pipe systems, the laterals are spaced 12 to 27 m apart, with exact spacing based on sprinkler pattern (Fig. 9-12).

A drawback of portable systems is the high labor requirement. Although aluminum pipe is relatively light, considerable labor is involved in each move. The increasing cost of labor has led to the development of the systems described next, which have much lower labor requirements.

Self-Propelled Systems In the self-propelled *power roll* system, the lateral pipe is mounted as the axle in large, lightweight wheels spaced at 18 to 30 m (Fig. 9-13). Sprinklers mounted at intervals on the pipe operate when the system is stationary and the sprinklers are upright. The main line is of flexible

FIGURE 9-12
A portable pipe sprinkler irrigation system in operation. (*U.S. Department of Agriculture.*)

hose, or else couplings are made at proper places in a portable main line of hand-moved aluminum pipe. The system is driven to each new setting by a small engine mounted in the center (or in some cases at one end with a drive shaft going to the center, where the drive wheel is located). Power-roll systems may be up to 500 m long or considerably longer.

FIGURE 9-13
A power roll system. Drive wheels are in the center.

The completely mechanized *center-pivot* system is designed to operate with minimal labor. The pipe (usually steel) and sprinklers are mounted on A frames and wheels or skids, and the entire system rotates around a swivel joint at the pivot point, often a deep well (Fig. 9-14). The couplings are strong but flexible enough to allow for variation in terrain along the A-frame towers. It is fascinating to fly over an arid or semiarid region and see round, beautifully green sections in the otherwise brown landscape (Fig. 9-15). These areas may range from 8 to 80 ha or more, depending on the length of the lateral pipe, which often is 150 to 450 m. The A frames are about 30 m apart, and cables or trusses are used to support the pipe between them. Because each tower must travel at a different speed to keep the lateral pipe straight, each has a separate drive mechanism (electric or hydraulic motor), which can be adjusted individually. As distance from the center increases, the area covered by each section of lateral pipe increases, so that for uniform water distribution sprinkler size must be greater or the spacing between sprinklers less. Not only does this system require less labor but the lateral pipe can be high enough to

FIGURE 9-14
Center pivot systems rotate around a central source of water, usually a well. (*Valmont Industries, Inc.*)

FIGURE 9-15
Fields irrigated by center-pivot system are usually round. (*Valmont Industries, Inc.*)

clear a crop up to 2.5 to 3 m tall, compared with the power-roll type, which is normally designed to clear only about 1.2 m. Some systems have additional sprinklers that operate in the corners so that rectangular fields can be irrigated. Newer systems are available to irrigate fields of almost any shape.

Another approach for square fields is the use of *self-propelled side-move systems* with multiple sprinklers. They operate like the center-pivot system in that each A-frame tower is powered, but it travels in a straight line instead of rotating around a central pivot. The main line is a flexible hose, which is gradually straightened out as the lateral moves away from the water source or looped as the lateral moves toward it. In other types, the water is pumped from a main ditch running through the field.

Major disadvantages of the center-pivot and side-move systems are high initial investment, high energy cost for operation, and high evaporative losses.

Single-Sprinkler Systems Single-sprinkler, or gun, systems have one large nozzle instead of the many small sprinklers used in the types described above. At one setting, a gun sprinkler can irrigate 0.4 to 2.4 ha, depending on the nozzle size and water pressure. Hand-moved and tractor-pulled models are moved from place to place like the portable-pipe systems described earlier. The popular self-propelled traveling gun moves through the field while irrigating (Figs. 9-16 and 9-17) and may be driven by water power or an engine. The flexible water-supply hose is similar to that described for the self-propelled side-move type. With the main line in the center of the field, the traveling gun is started at one side with the supply line extended. It then moves to the center, looping the supply line, which is extended again as the gun moves to the opposite side of the field. The gun is moved over to the next area and the direction of travel is reversed.

Trickle, Drip, or Daily-Flow Irrigation

Although field application of trickle irrigation is relatively new, watering systems of this general type have been used in greenhouses and container nurseries for many years.

For the production of greenhouse crops in benches, i.e., in a bed of soil rather than in individual pots, the two most widely used trickle systems are perforated plastic pipes and soaker hoses. When the water is turned on, the perforated pipes lying on the

FIGURE 9-16
Traveling gun irrigation system, which moves by a cable and winch. Water is supplied by a flexible hose. (*From R. G. Halfacre and J. A. Barden, Horticulture, McGraw-Hill, New York, 1979.*)

FIGURE 9-17
Traveling gun irrigation of a corn field in Nebraska.

soil emit small streams of water. A soaker hose is made of porous material designed to allow water to seep out and drip onto the soil.

For container-grown plants, small tubes from a plastic supply line lying among the rows of pots deliver water to the individual containers. The delivery tubes are so small that the name *spaghetti tubing* is perfectly appropriate. A lead weight is usually attached to the distal end of the delivery tube to hold it in place.

Whether plants are grown in containers or greenhouse benches, the soil volume is limited. At each watering, the whole soil mass is thoroughly wetted; during periods of high transpiration, watering may be required 2 to 3 times a day.

With trickle irrigation systems for field conditions, the goal is to add enough water to wet the soil containing one-third to one-half of the plant roots. Instead of operating 2 or 3 times a day for a few minutes, as the greenhouse system does, water is applied continuously but slowly for several hours every day. If one-third to one-half of the root system is continuously supplied with adequate moisture, the water needs of the entire plant can be met.

Water is applied through outlets called *emitters*, which are either simple tubes inserted into a plastic pipe or something much more elaborate. One emitter per plant is adequate for tomatoes and other small plants; two emitters are used for large plants such as fruit trees.

Probably the most important advantage of trickle irrigation relates to water savings. Since water is applied at point sources and only a small portion of the soil surface is wetted, surface evaporative losses are reduced dramatically. The potential water savings from trickle irrigation compared with surface or aerial types range from 20 to 50 percent. Pumps and power requirements are also much lower because they are operated at much lower pressures and handle far less water per hour.

Subsurface Irrigation

This type of irrigation raises the water table to within 0.3 to 0.6 m of the root system of the crop, and the water moves up by capillarity (Chap. 10) into the root zone. Application is much more limited than that of other systems because certain ideal conditions must be present for it to be practical. If the water table is to be raised artificially, it must already be relatively near the surface or there must be an impervious layer that will restrict downward movement of water. These restrictions mean that only sites which require drainage of excess water in the wet season will be suitable for subsurface irrigation in the dry season. Fortunately, the system developed can serve both the drainage of excess water and the supply of needed water.

There are two basic systems of subsurface irrigation. One consists of digging open-ditch laterals, which are connected to a head or main ditch. The other involves subsurface laterals of perforated pipe, jointed concrete, or drain tiles, which in turn are connected to a main line. The desired level of water is maintained by pumping water in or out.

CHAPTER TEN

SOILS

Although plants can be grown *hydroponically*, i.e., in aerated water instead of soil, soil is essential for the large-scale economical production of the major crop species on which human life depends. In the context of world food production, soil is often equated with land. The surface of the earth is about 30 percent land, but only about 10 percent of the land area is *arable*, i.e. suitable for cultivation. At present about 50 percent of the arable land is under cultivation, and much of what remains is less well suited to crop production. As we look toward the twenty-first century, we are faced with the need to bring new land areas into food production and to improve our stewardship of the land area presently being used. We return to these issues in Chap. 13.

Soil can be defined as the outer, weathered layer of the earth's crust which has the potential to support plant life. As a growing medium for plants, soil provides physical support, water, and nutrients. We shall discuss each of these functions of soil in supporting plant growth. Chapter 9 has already dealt with soil water to some extent, and Chap. 11 will provide additional information on soils as a source of plant nutrients.

COMPONENTS

The five components of a typical soil are mineral particles, organic matter, water, air, and organisms.

The mineral particles originate from the breakdown of parent material, and the organic matter comes from decomposition of plant and animal life. The solid components are interspersed with *pore spaces* of innumerable sizes and shapes, which take up about 50 percent of the soil's volume and are occupied by water and air in various proportions.

Mineral Particles

Mineral particles include a broad spectrum of physically and chemically diverse discrete crystalline fragments classified according to diameter:

Sand—0.05 to 2 mm
Silt—0.002 to 0.05 mm
Clay—<0.002 mm

Material from 2 to 76 mm is *gravel,* and pieces over 76 mm are *stones.* The larger particles, such as sand, are essentially inert chemically and are called *primary minerals.* The smaller particles, especially those the size of clay, are chemically active and called *secondary minerals.* Part of this activity relates to surface area, which is vast in a volume of clay particles because of their small size and platelike structure. This activity will be discussed below, when we consider some chemical properties of soils.

Organic Matter

Although the organic fraction of a *mineral soil,* i.e., one predominating in mineral particles, constitutes only about 1 to 6 percent, its importance far out-weighs its proportion. The organic component consists of plant and animal residues in various stages of decomposition. The action of microorganisms and chemical reactions cause decay, a continuous formation of a dark, amorphous organic material called *humus*. Its composition cannot be described in chemical terms but humus is rather resistant to further breakdown. When it does ultimately occur, with the release of carbon dioxide, water, and mineral components, it is called *mineralization*.

Water

The importance of water to crop production is obvious. The moisture-holding capacity of soil increases as levels of organic matter rise and as the size of mineral particles decreases. These same factors are of prime importance in water infiltration and movement in soils.

After water has been added to a soil, the smaller pore spaces may remain full of water whereas the larger ones tend to drain and refill with air. A rule of thumb for a good soil is that its pore space should be filled about equally with air and water (Fig. 10-1). The proportion of the pore space occupied by air and water constantly changes as water enters or leaves the soil.

It is important to keep in mind that the water in soil is not pure water but a solution of various substances, many of which are nutrients needed by plants. The composition of the soil solution is continually altered by the addition or removal of water and chemical components, e.g., fertilizers. We shall explore soil solution as an ever-changing constituent of the soil later in this chapter.

Atmosphere

Since roots must have oxygen in order to respire and function normally, adequate soil aeration is of prime importance in maintaining optimum growth and production. The large pore spaces of some soils allow relatively free gas exchange between the air in the soil pore space and the atmosphere. In soils containing high amounts of clay, however, poor aeration can result because the pores are very small and a high proportion of the existing pore space is occupied by water. Normal concentrations of oxygen and carbon dioxide in the general atmosphere are about 20 and 0.03 percent, respectively. When conditions limit gas exchange between the soil atmosphere and

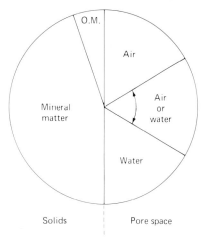

FIGURE 10-1
Relative volume of solids, air, and water in a typical topsoil (*OM* = organic matter). (*From L. M. Thompson and F. R. Troeh, Soils and Soil Fertility, 4th ed., McGraw-Hill, New York, 1978.*)

the general atmosphere, oxygen may be depleted and carbon dioxide may accumulate as the result of respiration by both roots and microorganisms. Low oxygen levels and high carbon dioxide levels are both detrimental to roots.

Organisms

Besides visible plant roots and animals, soils contain tremendous quantities of microscopic organisms, especially in surface layers. Representatives of essentially every major animal group, including birds, spend part of their lives in the soil. These living creatures, especially the microorganisms, are of utmost importance in soil formation, influencing soil properties and the synthesis and degradation of the organic fraction.

Animals The activities of larger animals, e.g., woodchuck, prairie dog, ground squirrel, and mole, are important in soil movement, aeration, drainage, and the incorporation of organic matter. These animals are generally unwelcome in agricultural and lawn settings, however. Insects are numerous in the soil and important in many ways; their adverse effects will be discussed in Chap. 12.

Earthworms Long recognized as major contributors to the fertility and productivity of soils, earthworms pass tremendous quantities of soil through their digestive tract each year—as much as 40 metric tons per hectare. Certain elements are made more available by the digestive processes of the earthworm, particularly nitrogen. Organic matter is mixed and partially decomposed by worm activity. Their tunnels improve soil drainage and aeration, both vital characteristics of a productive soil. Earthworms are most numerous in soils which are moist but well drained and high in organic matter. In midsummer, when soil becomes dry and warm, earthworms burrow to deeper levels, returning to surface layers when moisture is replenished and temperatures decline.

Microorganisms

Protozoa and Nematodes Although more than 200 species of single-celled protozoa are present in extremely high numbers, little is known about them and their possible interactions with plants. Of major consequence are the usually microscopic nematodes, or eelworms. Certain species are plant-parasitic and do great economic damage to crops (Chap. 12); others are nonparasitic and live on organic matter; a third group is parasitic on animals.

Bacteria These single-celled organisms occur in vast numbers, often estimated to exceed 1 billion per gram of soil. Conservative estimates put the fresh weight of bacteria at 1100 kg per hectare-furrow-slice (a depth of 15 to 20 cm). Fantastic reproduction rates enable bacterial populations that have been drastically reduced by adverse climatic conditions, such as drought, to rebuild rapidly.

Soil bacteria are classified according to their source of energy as *heterotrophic*, bacteria which obtain energy from organic matter and function in its decay, and *autotrophic*, bacteria which obtain energy by oxidizing inorganic substances, e.g., ammonium ions to nitrite ions and nitrite ions to nitrate ions (Chap. 11). Soil bacteria are also classified by their need for oxygen:

Aerobes—require oxygen
Anaerobes—do not require oxygen
Facultative Anaerobes—can grow with or without oxygen

Without bacteria the soil could not support luxuriant plant growth.

Fungi Also of major consequence in soils, fungi are present in vast numbers (normally 10 to 20 million per gram of dry soil) and are active in breaking down organic matter. Fungi are multicellular, usually filamentous forms that commonly produce mushrooms as their fruiting bodies. The most important soil fungi, the *molds*, are adapted to a wide range of soil pH and are abundant in acid soils, where bacteria and actinomycetes are less numerous. Some fungi can enter into the symbiotic *mycorrhizal* relationship with roots of higher plants, which favors both members of the symbiosis, the higher plants benefiting from improved water and nutrient absorption. Other species of fungi cause tremendous damage to crop plants, particularly the damping-off fungi (*Pythium* spp.) and the stem-, crown-, and root-rot fungi (*Phytophthora* spp.), discussed in Chap. 12.

Actinomycetes Morphologically between the molds and bacteria and having some characteristics of each, actinomycetes are active in decomposition of organic matter and the subsequent release of nutrients. They are active under drier conditions than bacteria and fungi. Actinomycetes are seriously hindered in acid soils and decline drastically at a pH of 5.0 or below. This characteristic is used to control the actinomycete which causes potato scab by acidifying soils with the addition of sulfur to keep the pH at 5.3 or below.

Algae Usually present in large numbers, most algae are chlorophyll-containing cells that tend to be concentrated in the upper few centimeters of soil where sunlight penetrates. Algae populations are encouraged by high moisture, as in rice paddies. Under conditions of intense sun, blue-green algae may fix significant quantities of nitrogen.

TEXTURE

One of the major physical properties of a soil, from the standpoint both of crop production and classification, is its texture, i.e., the percentage of sand, silt, and clay-sized mineral particles. In a *mechanical analysis* the soil is separated into size fractions and the percentage of each determined. Then from the graph in Fig. 10-2 it is quite easy to classify the soil into one of the 12 categories, as follows. The three scales in the figure are divided into units from 1 to

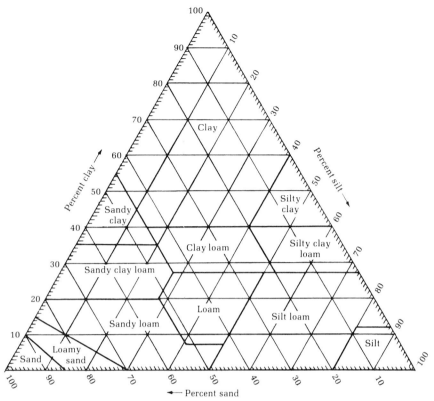

FIGURE 10-2
Guide for textural classification of soils. (*From L. M. Thompson and F. R. Troeh, Soils and Soil Fertility, 4th ed., McGraw-Hill, New York, 1978.*)

100; locate the intersection of the lines representing each of the three components, or *separates*; follow the line at the same angle as the number on the scale you are using. For example,

Sand, %	Silt, %	Clay, %	Type
40	40	20	Loam
60	10	30	Sandy clay loam
10	60	30	Silty clay loam

An experienced soil scientist can learn much about a soil by its feel. By assessing the plasticity and grittiness of a soil sample, an experienced person can classify it into one of the categories in Fig. 10-2 with considerable accuracy.

STRUCTURE

The texture of a soil is important but does not provide nearly enough information to assess the soil's potential productivity. Besides the particle-size distribution, the arrangement of the particles affects the value of a soil. How individual soil particles are grouped is called its *structure*. An example of a soil with no structure is sand, in which each particle is independent of all others. In soils containing some smaller particles and organic matter, at least limited amounts of clumping, or grouping, of individual particles occurs. Such *aggregation* has profound effects on the soil.

Soil aggregates may be from small to large, weakly to strongly persistent, and of several distinct shapes. Whatever its character, aggregation is highly desirable as it improves aeration, percolation, and root penetration. Since aggregates are larger than the individual particles, pore spaces between the aggregates are much larger. In a well-aggregated soil, the porosity is more like that of a coarser-textured soil, but within the aggregates it retains the small pore spaces of a fine-textured soil. The major cementing agents contributing to good soil aggregation include clay particles, partially decomposed organic matter, and certain microbes. Hydrogen and calcium ions promote aggregation, but sodium ions tend to break it up.

Good soil structure is slow to develop and can be rapidly destroyed by careless farming. A clay soil with good aggregation can be severely damaged by plowing, disking, or moving heavy equipment over it when it is wet. The aggregates break apart and the result is *puddled soil,* in which the pore spaces have been closed up. When dry, such a soil is crusted or in lumps, or clods, which are difficult to break up.

PORE SPACE

Total pore space, the volume of the soil not occupied by solid particles, ranges between about 40 and 60 percent of the total soil volume. The actual percent porosity varies with both soil texture and soil structure. Coarse soils, such as sands or sandy loams, have less total pore space than a silty clay. Important, however, is not merely total pore space but the relative size of individual pores. Small pores, which predominate in fine-textured soils, tend to fill with water and remain full, limiting aeration. Good aggregation in a fine-textured soil improves aeration by providing larger pores which drain readily. Although sandy soils have less total pore space, the individual pores are larger, providing excellent aeration but limited water-holding capacity.

SOIL PROFILE

If a cut is made down through it, the soil is usually seen to consist of layers of different color, particle size, and general appearance (Fig. 10-3). Good examples can be seen where a new highway has been cut through uneven terrain, leaving the soil exposed. The layers from the soil surface to the bedrock are

FIGURE 10-3
A soil profile showing A, B, and C horizons. The lines between horizons are indicated by white dots. Note the relative similarity between the A and B horizons but the distinct difference between B and C. (*Courtesy of William J. Edmonds.*)

called horizons, and the whole transect is called a soil *profile.* The A *horizon,* or upper layer (often called *topsoil*), is usually higher in organic matter and therefore darker than layers below. The middle layer of the profile, called the B *horizon,* usually contains less organic matter and smaller particles and is lighter in color than the A horizon. The B horizon is commonly referred to as the *subsoil.* The bottom layer, the C *horizon,* typically extends from the lower limit of the B horizon down to bedrock. This third layer is sometimes called the *substratum,* and bedrock is sometimes referred to as *parent material*, because it gives rise to the soil's mineral particles.

The A horizon is also referred to as the zone of leaching and biological activity. The level of nutrients is highest in this zone because of the breakdown of organic matter by microorganisms and the surface addition of fertilizer. As water moves down through the A horizon, it leaches soluble nutrients and tends to transport fine clay particles downward. The B horizon, where the fine soil particles tend to accumulate, is the zone of accumulation. The A and B horizons combined constitute the *rooting zone*. The C horizon, described as the zone of weathering, is a transition region between the subsoil and the parent material and has characteristics intermediate between the two.

The gradations between horizons may be fairly distinct, or the layers may merge gradually with little or no line of demarcation. The degree to which the horizons can be distinguished often depends on the age of the soil, older soils having more clearly defined horizons. Often distinct subhorizons allow a more complete description of the soil profile. The total depth represented by the profile is typically about 1 m for temperate-zone soils, although it may range from a very few centimeters to many meters. The depth and components of the profile are used in classifying soils into types and thus in predicting their value for crop production or other uses.

SOIL MOISTURE

Since almost all the water absorbed by plants is taken up by the roots, the soil serves as a reservoir upon which plants draw. Water uptake by roots takes place by diffusion and then osmosis, in which water moves from a region of high water potential in the soil to a region of lower water potential in the root.

The more water there is in a given soil the higher the soil's water potential. After a heavy rain or irrigation which saturates the soil, its water potential increases (becomes less negative) and approaches zero. As the soil dries owing to uptake of water by plants or surface evaporation or both, the soil's water potential decreases (becomes more negative). We shall see below how water in the soil is divided into categories to help explain the dynamics of soil moisture.

Classification

From a physical standpoint, soil moisture can be divided into four categories:

Chemically combined water is a component of the soil particles and as such is inactive and totally unavailable. It can be essentially ignored for our purposes.

Hygroscopic water exists as a very thin layer around soil particles which is held tightly and is unavailable to plants. At the upper limit of hygroscopic water, the soil water potential is -3.1 MPa.

Capillary water exists in films around and between soil particles at soil water potentials from about -0.03 to -3.1 MPa. Since capillary water moves and is held by the phenomenon of capillarity, this portion of the soil water is greatly influenced by the size and arrangement of soil particles. Essentially all the water taken up by plant roots is capillary water.

Gravitational water is that in excess of capillary water and therefore held at soil water potentials of less than -0.03 MPa. Gravitational water moves downward and is normally lost from the root zone 1 to 3 days after rain or irrigation.

Movement

Gravitational water moves downward because of the pull of gravity. Its downward rate of movement depends on the size of the pores. For example, excess water moves rapidly through a sandy soil, but a clay loam may still contain gravitational water 2 to 3 days after a rain.

In soils that no longer contain gravitational water, movement is largely by *capillarity*. When a fine glass tube is dipped in water, the liquid can be seen to rise in the tube. How high the water rises depends upon the diameter of the tube: in one the size of a drinking straw, the rise is quite small; in a very fine one, the water may rise several millimeters. Capillarity results from the attraction between water molecules and glass molecules. Water movement in soils is similar except that it takes place in the pores between the innumerable shapes and sizes of soil particles. The movement of water into soil pores is caused by the attractive forces between the water molecules and the soil particles (adhesion) and the attractive forces between water molecules (cohesion).

Although water movement by capillarity occurs in all directions, upward or horizontal movement is most obvious. The process of capillarity is utilized in subsurface irrigation, in trench irrigation between raised beds (Chap. 9), and in watering potted plants by adding water to the saucer underneath. Capillary movement is relatively rapid at high soil water potential but slows dramatically as soil water potential decreases. Although capillarity is rapid in sandy soils, distances involved are small because of large pores. In heavy clay soils, water may move great distances, but the minute pore spaces mean that its movement is very slow. For these reasons, capillarity is most important and useful in medium-textured soils.

Terminology

Additional terminology is often used in characterizing soil moisture. The *field capacity* of a soil is the amount of water it can hold against the pull of gravity. After a soil has been saturated by rain or irrigation, gravitational water continues to drain away for 24 to 72 h, after which the soil is at field capacity. At field capacity, all the water is held by capillary and hygroscopic forces. The soil water potential at field capacity is typically about −0.03 MPa. Water will have left the larger pore spaces but not the smaller ones. Field capacity is considered to be the upper limit of *available water*, i.e., water

available to plants. The amount of water present in a soil at field capacity is small in a coarse soil like sand and markedly higher in finer-textured soils (Fig. 10-4).

As a soil dries, either through surface evaporation or uptake by plants, the soil water potential becomes more negative. If a plant is grown in a limited volume of soil without periodic watering, it will eventually wilt during the day. The plant will recover turgidity at night unless soil moisture has reached the *wilting coefficient*, the level at which a plant cannot regain turgidity in a saturated atmosphere. The wilting coefficient is also called the *permanent wilting point*. At the wilting coefficient, the soil water potential is typically about −1.5 MPa. Since most plants can extract little moisture at soil water potentials below −1.5 MPa, the wilting coefficient is considered to be the lower limit of available soil moisture (Fig. 10-4). As is true at field capacity, the amount of water still in a soil at its wilting coefficient is a function of soil texture.

Figure 10-4 makes it clear that the amount of water a soil can hold is affected by texture. A clay loam soil holds more available and unavailable water than a sandy soil. In fact, a clay loam holds more water at the wilting coefficient than a sandy loam does at field capacity.

The *hygroscopic coefficient* is the moisture level when the soil water potential reaches −3.1 MPa,

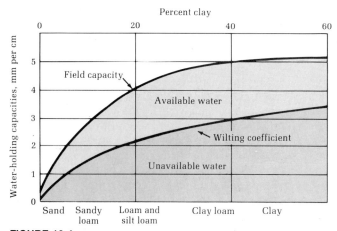

FIGURE 10-4
The water-holding capacity of soils varies with texture. A clay holds much more available and unavailable water than a sandy loam. (*From L. M. Thompson and F. R. Troeh, Soils and Soil Fertility, 4th ed., McGraw-Hill, New York, 1978.*)

and is the amount a soil can absorb and hold in equilibrium with a saturated atmosphere.

CHEMISTRY

Earlier portions of the chapter have dealt with the physical aspects of soils, but many chemical aspects must be considered as well, including pH, cation-exchange capacity, fertility, and fertilization.

Soil pH

The pH of a solution is a measure of the degree to which it is acidic or basic. With soils, the pH is also referred to as the *soil reaction*. The pH is the negative logarithm of the hydrogen ion (H^+) concentration. At neutrality, the pH is 7.0, which indicates an H^+ concentration of 10^{-7} mole per liter (mol L^{-1}). Since the product of the normalities† of H^+ and OH^- is always 10^{-14}, at a pH of 7 the concentration of OH^- is also 10^{-7} (mol L^{-1}). Because pH is based on a logarithmic relationship, at a pH of 6 there is 10 times more H^+ and 10 times less OH^- than at a pH of 7. Thus it should be apparent that soil pH is an expression of the relative proportion of H^+ and OH^- in the soil.

Most agricultural soils fall within a pH range of 4 to 9, although for nonagricultural soils the extremes may extend from 3 to 11. The pH that naturally develops in a soil depends on many factors, two of the most important being the amount of precipitation and the type of vegetation.

Humid areas, those with rather heavy rainfall, tend to have acid soils because basic ions (those that raise the pH of a solution) are removed from the soil by leaching. As such ions as Ca^{2+} and Na^+ are leached downward, they are replaced by H^+ ions, thus lowering soil pH. This natural effect can be accelerated by the addition of acid-forming fertilizers, e.g., some containing NH_4^+ or SO_4^{2-}. The type of vegetation can also affect soil pH over the years; grasses tend to cause soils to remain at somewhat higher pH than forests, especially coniferous types.

---

†Normality is one method of expressing concentration. A 1 N (normal) solution contains 1 gram equivalent weight per liter of solution. Thus a 1 N acid solution has 1 g of H^+ per liter and a 1 N base solution has 17 g of OH^- per liter.

Unless the pH is extremely low or high, direct effects on crop growth are minimal. In general, the most critical effects of soil pH are indirect. The availability of certain nutrients is strongly influenced by pH (Fig. 10-5). For example, copper, iron, manganese, and zinc become much less available in highly alkaline soils. For certain acid-loving horticultural crops, lowering the soil pH is the most effective method of curing iron deficiency. In other words, the plant is suffering from an iron deficiency not because of a lack of iron in the soil but because the iron is unavailable as a result of high pH. Since the nitrogen-fixing bacteria associated with legumes are seriously hindered in acid soils, maintenance of moderate pH levels is critical with such crops as bean and pea.

Cation-Exchange Capacity

Clay particles and humus are important in determining the chemical properties of soils. These materials are complex in structure and contain sites with negative charges which attract positively charged ions (*cations*). Cations at these sites are in contact with the soil solution and may be replaced by, or exchanged with, other cations. The *cation-exchange capacity* of the soil is a measure of such negatively charged sites. It is determined by refined laboratory procedures and often used as an index of the potential fertility of the soil. The cation-exchange capacity is typically expressed in milliequivalents† per 100 g of soil. Depending on the type and proportion of clay present, mineral soils have cation-exchange capacities ranging from 2.0 in sand to more than 50 in certain clays. Since humus has a very large cation-exchange capacity, its presence greatly improves this aspect of soil chemistry as well as soil structure and water-holding capacity.

To relate the cation-exchange capacity to plant nutrition, we must also ascertain the *percent base saturation*, defined as the proportion of the cation-exchange capacity occupied by basic nutrients such as Ca^{2+}, Mg^{2+}, K^+, and Na^+. Specifically excluded are H^+ and Al^{3+}, which are acidic in reaction.

The ions in the soil solution at any given time represent only a minute proportion of the total ions in the soil. It is estimated that in most soils more than 99 percent of the cations are adsorbed on the

---

†A milliequivalent is the amount of a substance that will react with or displace, 1 mg of hydrogen.

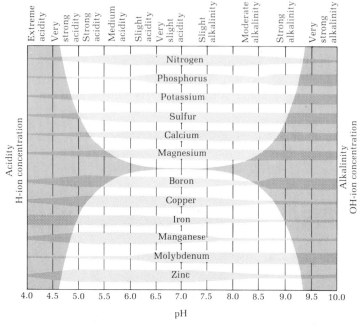

FIGURE 10-5

Effect of soil pH and associated factors on the availability of plant nutrient elements. The width of the band for each element indicates the relative favorability of this pH value and associated factors to the presence of the elements in readily available forms (the wider the band the more favorable the influence). It does not necessarily indicate the actual amount present since this depends on other factors, e.g., cropping and fertilization. (*Reproduced from Changing Patterns in Fertilizer Use, (1968). Fig. 7, p. 152, by permission of the Soil Science Society of America.*)

various particles exhibiting cation exchange. These charged colloidal† particles of clay or humus are called *micelles*. Various cations are adsorbed on the surface of the micelles but are in dynamic equilibrium with cations in the soil solution. Cations are exchanged between the micelles and soil solution on the basis of two phenomena: (1) The strength of adsorption indicates relative amounts held by a micelle if the cations are of equal concentration. From highest to lowest, the adsorption of cations is Al > Ca > Mg > K > Na. If H^+ is included, it falls between Al and Ca. (2) In certain circumstances the law of mass action is involved. When a relatively large concentration of one ion, say H^+, becomes available, other cations may be replaced on the micelles by H^+ because its larger numbers make it

more competitive for the available exchange sites. A similar phenomenon occurs when an application of lime is made, in that the Ca^{2+} replaces H^+ on the micelles. The replaced ions go into the soil solution and thus are subject to leaching.

Adjusting Soil pH

In the humid regions of the United States, most soils are naturally acid, primarily because the bases are leached downward, leaving an excess of H^+ ions. Crops vary considerably with regard to the soil pH to which they are best adapted. Plants requiring acid soils include azalea, blueberry, and rhododendron. Among those adapted to a mildly acid soil are wheat, corn, soybean, strawberry, apple, and tomato. Crops best suited to a neutral to slightly acid soil are alfalfa, asparagus, lettuce, and muskmelon. Another factor of considerable importance is the range of soil pH within which a particular crop can prosper. The blueberry will produce vigorous growth and full

---

†A colloid is defined as a particle small enough to remain suspended in water without agitation. Because of its small size, a colloid has a large surface area per unit weight.

crops only when growing in a soil pH of 4.3 to 4.8, whereas apple trees are often equally productive in soils with pHs ranging from 5.0 to 8.0.

Raising the pH If the pH is found by soil test to be lower than desired, the usual procedure is to add lime. Major factors to be considered before the lime is applied are soil type, how much the pH is to be raised, fineness of lime to be applied, and type of lime to be used. As shown in Table 10-1, the amount of lime needed to raise the pH of a sandy loam soil is much less than for a clay loam. This increasing lime requirement with decreasing soil particle size relates to the high cation-exchange capacity of clay particles. Obviously, the more one wishes to raise the pH the greater the lime requirement. These relative amounts are also given in Table 10-1. The smaller the particle size of ground limestone the sooner the lime reacts to neutralize soil acidity. The usual recommendation for ground limestone is that 95 percent pass through a U.S. standard 60-mesh screen and 30 percent through a 100-mesh screen. The range in particle size means that this lime will react over a few weeks to several years.

Lime is available as ground limestone, burned lime, and hydrated lime. Ground limestone consists of calcite, or calcium carbonate ($CaCO_3$) with varying amounts of dolomite, or calcium magnesium carbonate [$CaMg(CO_3)_2$]. If the limestone is mostly $CaCO_3$, it is called *calcitic*; when sizable quantities of dolomite are included, it is called *dolomitic limestone.* If limestone is heated in a kiln, CO_2 is driven off and calcium oxide and magnesium oxide remain.

$$CaCO_3 + heat \longrightarrow CaO + CO_2 \uparrow$$
$$CaMg(CO_3)_2 + heat \longrightarrow CaO + MgO + 2CO_2 \uparrow$$

This *burned lime*, or *quicklime*, can be used for raising soil pH but is unpleasant to handle. *Hydrated* or *slaked* lime, produced by reacting burned lime with water is even more disagreeable to handle than burned lime.

$$CaO + MgO + 2H_2O \longrightarrow Ca(OH)_2 + Mg(OH)_2$$

The primary advantage of burned or hydrated lime is that both react much more quickly than ground limestone, but usually they are more expensive on the basis of cost per hectare for an equivalent amount of neutralizing power. One exception would be where the lime must be shipped long distances because 1 kg of calcium oxide is equivalent to about 1.8 kg of ground limestone in acid neutralization.

The addition of lime has both direct and indirect effects on a soil. Direct chemical effects include a reduction in the concentration of H^+, an increase in the concentration of OH^-, and greater availability of Ca^{2+} and Mg^{2+}. These changes lead to an increase in the percent base saturation. Indirect effects of an elevated pH are decreased solubility of aluminum, iron, and manganese (Fig. 10-5), which in very acid soils may reduce toxic concentrations of these ions in the soil solution. This in turn results in increased availability of both phosphorus and molybdenum.

Lowering the pH To grow an acid-loving crop like blueberry, azalea, or rhododendron successfully, it is necessary to acidify the soil from time to time. In fields used for potato production, lowering soil pH may be desirable to reduce the problems with the scab organism. The substances most widely recommended for the purpose are sulfur, iron sulfate, or aluminum sulfate. If only a gradual reduction over a limited pH range is needed, it may be feasible to

TABLE 10-1
Recommended Approximate Metric Tons per Hectare of Limestone for Different Soil Types to Attain a pH of 6.5†

pH of unlimed soil	Sandy loam soils	Loam soils	Silt loam soils	Clay loam soils
4.8	6.72	8.96	10.08	11.20
5.0	5.60	6.72	7.84	8.96
5.5	3.92	4.48	5.60	6.72
6.0	2.24	2.80	3.36	4.48

† From A Handbook of Agronomy, *Va. Polytech. Inst. State Univ. Ext. Div. Publ.* 600, Blacksburg, Va., 1974, p. 60.

accomplish the change through selection of the nitrogen source. The nitrogen source most effective in lowering soil pH is ammonium sulfate [$(NH_4)_2SO_4$], but NH_4NO_3 and NH_3 are also acidifying. Since such nitrogen sources as $NaNO_3$, KNO_3, and $Ca(NO_3)_2$ are somewhat basic in reaction, they should be avoided where acidification is desired.

Alkaline Soils

In arid and semiarid regions, soils are frequently alkaline and often exhibit related problems. Arid soils are usually classified as saline, sodic, and saline-sodic. Despite slightly different characteristics, all three tend to occur in areas where there is inadequate precipitation to leach the soil. The problem is accentuated where additional bases are added to the soil by ground water or irrigation water.

Saline Soils These are soils which are high in soluble salts, usually so classified if the soluble-salt content is 2000 ppm or higher. Less than 15 percent of the cation-exchange capacity of saline soils is occupied by sodium ions. Soil pH is usually between 7.0 and 8.5 because sodium content is low and most of the salts present are neutral. The salts are mostly chlorides and sulfates of calcium, magnesium, and sodium. When they are allowed to accumulate on the soil surface, a white residue is formed, giving rise to the name *white alkali soils*. Leaching, an effective means of reducing the detrimental effects of such soils on crop growth, often requires improved drainage. With poor drainage, problems with excess salts may be accentuated by the application of large quantities of water, especially where irrigation water contains even moderate levels of salts. With good drainage, however, it is often possible to rid a soil of excess salts with one or two heavy irrigations. Excess soluble salts, which occasionally occur in heavily fertilized greenhouse and nursery soils regardless of the soil pH, can be removed by leaching.

Sodic Soils In sodic soils, more than 15 percent of the cation-exchange capacity is occupied by sodium (Na^+) ions, but these soils are low in total soluble salts. The detrimental effects of sodic soil are due to high sodium and high pH (8.5 to 10). Because of high sodium, such soils tend to be deflocculated† and have poor structure. In severe

cases, organic matter tends to be dissolved and deposited on the surface, giving rise to the name *black alkali soils*. Sodic soils are the most difficult of the alkali soils to reclaim, and often the cost cannot be justified. Not only are large quantities of gypsum ($CaSO_4$) required, but the high sodium makes soil structure so poor that incorporation of the *amendment*, as an added chemical is called, becomes difficult.

Saline-Sodic Soils Soils in this category not only have high concentrations of soluble salts (>2000 ppm) but are also sodic and usually between 8.0 and 8.5 in pH. Although they resemble saline soils in appearance, their successful reclamation requires a different treatment. A soil amendment, usually gypsum or sulfur, must be worked into the surface. It generally takes several metric tons of gypsum per hectare but only about 20 percent that amount of sulfur. The addition of $CaSO_4$ leads to the following reactions:

$$\begin{array}{l}Na^+ \\ Na^+\end{array}\boxed{\text{Micelle}} + CaSO_4 \longrightarrow Ca^{++}\boxed{\text{Micelle}} + Na_2SO_4$$

$$Na_2CO_3 + CaSO_4 \longrightarrow CaCO_3 + Na_2SO_4$$

Since the Na_2SO_4 is soluble, it can be leached, removing excess Na^+ and replacing it with Ca^{2+}. The second reaction shows that Na_2CO_3, which is soluble and a cause of high pH, can be reacted with $CaSO_4$, forming $CaCO_3$ as a precipitate; the Na_2SO_4 can be leached out.

Sulfur added to the soil is converted into sulfuric acid (H_2SO_4), which reacts as follows:

$$2S + 3O_2 + 2H_2O \longrightarrow 2H_2SO_4$$
$$H_2SO_4 + Na_2CO_3 \longrightarrow Na_2SO_4 + H_2O + CO_2\uparrow$$

The added H_2SO_4 can also react as follows:

$$H_2SO_4 + CaCO_3 \longrightarrow CaSO_4 + H_2O + CO_2\uparrow$$

If enough gypsum or sulfur is added to take these reactions well along to completion, a saline-sodic soil will have been converted into a saline soil reclaimable by leaching.

CLASSIFICATION

Soils differ in their potential value for producing crops and in how they should be managed for optimum yield. Soil scientists have described over

---

†Deflocculation is the breaking up, or dispersal, of aggregates.

10,000 soils in the United States, based on profile characteristics and such features as the texture of the surface soil. Trying to consider 10,000 different soils obviously would overwhelm even the most dedicated agriculturalist. Fortunately, soils, like other entities in nature, including plants and animals, can be grouped by consideration of specific features. This classification of soils provides a framework for considering a smaller number of related individual soils within a group, for comparing groups, and so on. The procedure of organizing and classifying individuals based on similarities and differences is most evident in plant and animal taxonomy.

Pre-1965 System

Before its replacement in 1965, a system adopted in the 1930s was used to identify soils in the United States. This system did not group related soils effectively. It was based strongly on descriptions of soils from the individual profiles, and names were arbitrary, usually reflecting the area in which the particular soil was first described. This older system recognized three major groupings of soils: zonal soils, azonal soils, and intrazonal soils. Within the zonal soils, soils common to large geographic areas were a great soil group (not the same kind as later defined by the 1965 system) and included soils such as the Podzols of the Great Lakes region and the highly fertile Chernozem soils of the prairies. Although this system of soil classification has not been used officially for more than 20 years, some of the terms are still in popular use. It is sometimes helpful to know them, although the newer system is now rapidly replacing virtually all reference to, and use of, the system of the 1930s.

Current System

In 1965, a new system of soil classification was adopted by the United States Department of Agriculture and soil scientists throughout the United States and much of the world. In this system, soils are classified into increasingly refined categories, starting with soil *order*. There are 10 orders; each soil is placed in only one order. Each order is subdivided into two or more *suborders*. Each suborder is divided into several *great groups*, and these are divided into *subgroups*. Each subgroup is composed of several *families*, and families comprise specific soils identified as a *series*. The individual series, which is determined by profile characteristics, provides the most explicit identification of a soil,

whereas the order provides only the most general description. The farther down the classification scheme one goes the more closely related the soils. Soils within a family, for example, are more similar than soils within a great group.

Besides the organized framework for the six classification levels, the new system names soils for each level of classification in such a way that the name reflects facts of importance to the agricultural value of the soil. Frequently this includes critical climatic features of areas where the soil is found. Names are structured to have the same meaning worldwide, specific roots being used for various categories in the classification. Thus, new soils can readily be fitted into the existing framework of classification.

The names of the 10 soil orders and information about them are given in Table 10-2. Note that each order name ends in *sol*. Thus, in soil classification, if a name ends in *sol*, it must be the name of one of the 10 orders. Some of the names will have a familiar meaning, e.g., Aridisol, associated with dry areas through the common term *arid*, or *Oxisol* for soils containing oxides. Although the characteristics of each order are general and those for Entisol and Inceptisol are quite similar, the name for each order is usually associated with a measurable property of a soil—base-exchange capacity, amount of organic matter, color, or the presence of certain chemical elements.

Of course, the order alone tells very little about the soil. More detail is developed as a soil is further classified. For example, consider a soil classified only to the next level, the suborder. Since, the suborder is the first subdivision of the order, to be meaningful, the name of the suborder must include the order's nomenclature root (Table 10-2). Consider, for example, a soil with the suborder name Xeroll. The order root *oll* is the root for the order Mollisol (Table 10-2). Hence the suborder Xeroll is a suborder of the order Mollisol. The suborder designation *xer* refers to dryness or an area with an annual dry season. A soil identified as being in the suborder Boralf would be in the order Alfisol; the suborder designation *bor* refers to boreal region, cool, northern areas with an average annual soil temperature of less than 8°C. Names can be built in a similar manner for great groups, subgroups, and families.

The series name is still that of the profile from which the soil was originally described. For example, some excellent agricultural soils in the Sacramento Valley of California are in the Yolo series. There are

TABLE 10-2
Soil Orders According to the Comprehensive Soil Survey System

Order name	Formative element	Derivation of formative element	General characteristics
Entisol	ent	Coined syllable	Young soil with little or no horizon differentiation
Vertisol	ert	Latin *verto*, "turn"	Cracking clay soils
Inceptisol	ept	Latin *inceptum*, "beginning"	Young soil with poorly developed horizons
Aridisol	id	Latin *aridus*, "dry"	Soils in dry areas; low in organic matter
Mollisol	oll	Latin *mollis*, "soft"	Rich in organic matter and exchange bases
Spodosol	od	Greek *spodos*, "wood ash"	Soils with an accumulation of amorphous material in the subhorizon
Alfisol	alf	Coined syllable	Gray to brown soils with an accumulation of clay in the subsurface soil
Ultisol	ult	Latin *ultimus*, "final"	Soils with a clay horizon low in exchangeable bases
Oxisol	ox	French *oxide*, "oxide"	Soils with a clay horizon high in hydrous oxides of iron and aluminum
Histosol	ist	Greek *histos*, "tissue"	Soils rich in organic matter

specific profile characteristics to describe this series. If the type is a loam or a sandy loam, the soil would be a Yolo loam or a Yolo sandy loam. Of course these soils can exist in places other than Yolo County, California, where they were first described. Unfortunately, unless a scientist has seen a Yolo profile, the name means little. Using the sequence to family in the 1965 system allows anyone to understand a great deal about a soil, including some idea of under what conditions it might be found, without ever having seen it. Obviously, to understand the classification fully, scientists agree upon key roots and meanings for all levels of classification. This has

TABLE 10-3
Comparison of the Botanical Classification of Corn and the Soil Classification of Clarion Soil

Plant classification	Soil classification
Division Spermatophyta	Order Entisol
Class Monocotyledoneae	Suborder Orthents
Order Graminales	Great group Xerorthents
Family Poaceae	Subgroup Typic xerothents
Genus *Zea*	Family Fine silty, mixed, nonacid, thermic
Species *mays*	Series Yolo

been done, and revisions or additions are made from time to time. The similarity between classification schemes for plants and soils is shown in Table 10-3.

TILLAGE

The techniques and machinery used to prepare land for crops vary tremendously depending on crop, topography, climate, and region. The real diversity of crop production systems throughout the world will become more apparent in Chaps. 13 and 17. The present brief discussion is merely a description of modern systems of land preparation in highly developed countries.

In most crop production schemes, land preparation is necessary before planting; in many cases additional tillage is done during the cropping cycle. The purposes of land preparation, though many, are all aimed at optimizing the soil environment for the particular crop. Important techniques include plowing, disking, cultivating and, in some situations, deep tillage.

Plowing

The first step in traditional land preparation is plowing to turn the top layer of soil to a depth of 15 to 20 cm. Two types of plows are in use.

Moldboard Plow When this type is pulled through the soil, turning it almost completely, it incorporates into the soil crop residues left on the surface (Fig. 10-6). In earlier days, a single-bottom plow was pulled by a horse or other draft animal and had long handles held by a worker following the plow. Modern moldboard plows are made not only in gangs, or multiples, of two or more bottoms but also as two-way models: at the end of a row the plow is inverted so that all the soil in a field is turned in the same way regardless of the direction of travel. Moldboard plows are not very efficient with very wet soil (at or close to field capacity), since the soil sticks to the face of the moldboard and tends to form large, compressed clods. Too dry soils are not turned properly either, and the amount of energy required to pull the plow through the soil is excessive.

Disk Plow In situations of excessively wet or dry soils, a disk plow is more suitable, although less efficient than the moldboard plow in turning under crop residues. The disk plow has three to ten or more 60- to 75-cm-diameter disks which cut into the soil and turn as they are pulled along.

Disking and Harrowing

Plowing leaves the soil surface ridged, rough, and often with large clods. Disks, or *disk harrows*, consist of gangs of smooth, concave disks which break up clods and smooth the surface. Designs vary, but usually there are two gangs of disks; the front gang throws soil outward and the second throws it inward (Fig. 10-7). Depth of penetration is varied by changing the angle of the disks in relation to the line of travel. The basic principle of soil preparation is the same in a 1.5-m-wide disk for a garden tractor and a 12-m-wide disk for large-scale field operations.

In a modification of the standard disk harrow called a *stubble disk* the disks are larger and their periphery is scalloped rather than smooth (Fig. 10-8). This type is used to cut residue on the surface before plowing or sometimes in its stead. Figure 10-8 shows that all front disks turn soil one way and all rear disks turn it the other.

Other types of harrows used after plowing to prepare a smooth seedbed, particularly for small-seeded crops, include the spike-tooth harrows and cultipackers. The *spike-tooth harrow* has vertical teeth which break up clods and loosen the soil as they are pulled through it (Fig. 10-9). The *spring-*

FIGURE 10-6
A moldboard plow turning under crop residue. (*Deere and Co.*)

tooth harrow has long curved teeth which lift, stir, and loosen the soil and which are made of spring steel to flex and bend around or over stones. A *cultipacker* rolls over the soil breaking up clods and forming a firm, smooth seedbed; unnecessary for large-seeded species, such as corn, cultipackers are widely used for small grains and grasses.

Cultivation

Tilling the soil between planting and harvest is called *cultivation*. Cultivators vary with the crop being grown and the particular production system in use. *Row-crop cultivators* are designed for tilling between rows of corn, soybeans, and many vegetable crops (Fig. 10-10). Teeth of many sizes and shapes can be used, all designed to break up the soil surface and kill weeds. The teeth should go deep enough to root

FIGURE 10-7
Disk harrow used to break up clods and smooth the surface layer of soil. (*Deere and Co.*)

FIGURE 10-8
Stubble disk has larger disks with scalloped edges to cut plant residues and thereby improve their incorporation into the soil. (*Deere and Co.*)

FIGURE 10-9
A spike-tooth harrow in which the vertical teeth break up clods, loosen the soil, and smooth the surface. (*Deere and Co.*)

FIGURE 10-10
A row crop cultivator being used to kill weeds. In many crops, the need for such cultivation has been largely replaced by herbicides. (*Deere and Co.*)

FIGURE 10-11
Deep chisel used to break up a hardpan to allow penetration of roots and also to improve drainage of water. (*Deere and Co.*)

out weeds without causing excessive damage to crop roots. Cultivation in most crops has largely been replaced by the use of herbicides (see Chap. 12). *Field cultivators*, used for weed control in fallow (uncropped) fields, have shovel-shaped teeth which penetrate deeply and lift the soil but do not turn crop residues under. The purpose is to kill weeds and loosen the soil to improve water penetration. Crop residues left on the surface are beneficial in controlling erosion and conserving moisture. *Rototillers*, which range from small garden tillers to large tractor-mounted models, are rotary plows consisting of a horizontal central shaft with curved teeth. As the shaft rotates, the teeth cut into the soil and break it up.

In earlier times, cultivation was recommended to control weeds, aerate the soil, improve water infiltration, conserve soil moisture, and loosen soil. More recent research, however, indicates that, aside from weed control, the cost of cultivation is not justified. In fact, cultivation may damage the roots of crops enough to counteract other benefits. Modern no-till production systems (Chap. 17) almost totally eliminate soil tillage, including cultivation.

Deep Tillage

In some agricultural soils, *hardpans* (Chap. 9) restrict the penetration of both water and roots. Equipment used to break up the hardpan includes the *deep chisel*, which penetrates to a depth of 60 to 120 cm, breaks up a hardpan, and loosens the soil profile (Fig. 10-11), and the *slip plow*, which penetrates even deeper and serves much the same purpose by means of a V-shaped blade which lifts the soil several centimeters.

CHAPTER ELEVEN

NUTRITION

The mineral nutrition of plants has held the interest of plant scientists since the early nineteenth century, when they learned that the soil provides certain elements required for plant growth. Although most of the dry matter produced by a plant comes not from the soil but from the air, the soil-borne nutrients are every bit as essential.

ESSENTIAL ELEMENTS

Until the early twentieth century, the elements considered absolutely necessary for normal plant growth were the original 10 listed in Table 11-1. Experiments testing the essentiality of elements for plants are conducted by withholding a specific nutrient while supplying all others for which a requirement has been proved. During this century seven more have been added to the list (Table 11-1). For maximum control of the elements available to the plant, such experiments are usually carried out by growing the plants in inert sand or a well-aerated nutrient solution.

The nutrients required in relatively large amounts, referred to as *macronutrients* or major elements (Table 11-1), are usually present (and measured) in a few parts per hundred (percent). *Micronutrients*, sometimes called *trace* or minor elements (Table 11-1), though equally essential, are needed in much smaller quantities, usually being present (and meas-

ured) in a few to several parts per million of the plant's dry matter. Although carbon, hydrogen, and oxygen are absolutely essential elements, they are not normally considered in nutritional studies because they are readily available from air and water. Vanadium, recently shown to be essential in some green algae, may be necessary in the growth of some higher plants. The uncertainty arises from the difficulty in growing plants in an absolutely vanadium-free medium. Cobalt and silicon are also considered by some plant scientists to be essential elements for at least certain of the higher plants.

Plant Analysis

We describe later in the chapter how analyses of plant tissues provide insight into their nutritional status. The concentrations of elements in a typical plant are presented in Table 11-2. Note that carbon, hydrogen, and oxygen account for more than 95 percent of the total dry weight. The other 4 to 5 percent of the plant dry weight includes all the macro- and micronutrients as well as other elements which have not been proved essential.

NITROGEN

Although there are at least 17 essential elements, nitrogen probably deserves the most attention. Except for carbon, hydrogen, and oxygen, nitrogen is present in most plants in greater concentrations than any other nutrient. Nitrogen is also the element to which plants are most likely to respond.

Atmospheric N_2

It is a paradox that nitrogen is the element most frequently deficient for crop production when about 78 percent of the atmosphere is nitrogen in the form of N_2. As an atmospheric component, nitrogen is an odorless, tasteless, inert gas; as such, it is *unavailable* and of no value to plants until it has been *fixed*, i.e., converted into a form available to, or usable by, plants. The major organisms capable of metabolizing nitrogen are a small group of bacteria and algae. Nitrogen fixation will be discussed later in this section. Atmospheric nitrogen can also be fixed by various industrial processes to make plant-available forms.

Function

Nitrogen is a vital component of protoplasm. It is found in chlorophyll, in the amino acids from which proteins are made, and in the nucleic acids. Without nitrogen in reasonably adequate quantities, crop growth is drastically reduced.

Deficiency Symptoms

The most common nitrogen deficiency symptoms include stunted growth and pale green to yellow (*chlorotic*) leaves, which are usually smaller than normal. The older leaves are most affected by chlorosis because nitrogen, a relatively mobile element, is withdrawn from the older leaves and translocated to the younger foliage. Older leaves may abscise prematurely; nitrogen-deficient perennials often exhibit early leaf drop in the fall.

Symptoms of Excess

Under certain conditions, e.g., application of too much nitrogen fertilizer, symptoms of excess nitrogen can occur. They vary with the plant but normally include dark green foliage, weak tissues, and succulent vegetative growth. Closely associated symptoms are delayed or reduced flowering and fruiting. In many perennial plants, high nitrogen levels in late summer delay the normal cessation of vegetative growth, predisposing plants to freezing injury in the fall and early winter (see Chap. 8). The optimum nitrogen level depends on the crop. High nitrogen levels in small grains lead to taller stems and increased lodging. Since with leafy vegetables, rapid succulent growth is desirable, nitrogen fertilization may often be heavy.

Sources

The gradual decomposition of organic matter in the soil (mineralization) is one source of nitrogen for plants. Although certain forms of organic nitrogen can be used directly by plants, most nitrogen is absorbed as inorganic ions, the two most readily available being nitrate (NO_3^-) and ammonium (NH_4^+). Although there may be sizable quantities of potentially available nitrogen in the soil, most of it is in organic form, either as a component of microorganisms or as a part of undecomposed organic matter.

TABLE 11-1
Essential Elements

Original ten	Later additions	Uncertain status	Macronutrients	Micronutrients
Carbon (C)	Manganese (Mn)	Vanadium (V)	N	Fe
Hydrogen (H)	Zinc (Zn)	Cobalt (Co)	P	Mn
Oxygen (O)	Boron (B)	Silicon (Si)	K	Zn
Nitrogen (N)	Copper (Cu)		Ca	B
Phosphorus (P)	Molybdenum (Mo)		Mg	Cu
Potassium (K)	Sodium (Na)		S	Mo
Calcium (Ca)	Chlorine (Cl)			Na
Magnesium (Mg)				Cl
Sulfur (S)				
Iron (Fe)				

TABLE 11-2
Elemental Composition of a Typical Plant†

Element	Percent of fresh weight	Percent of dry weight
O	81.0	45.0
C	6.8	44.5
H	11.5	6.0
N	0.2	1.5
K	0.15	1.0
Ca	0.05	0.35
P	0.03	0.20
Mg	0.03	0.20
S	0.02	0.15
Cl	0.015	0.10
Fe	0.015	0.10
Mo	0.008	0.05
Zn	0.003	0.02
B	0.003	0.02
Cu	0.001	0.01
Other‡	0.175	0.08
Total	100.000	100.00

† Expressed as a percent of the fresh weight, which is about 85 percent water, and of the dry weight (water removed).
‡ Trace and nonessential.

The Nitrogen Cycle

Nitrogen is in a constant state of turnover, a process referred to as the *nitrogen cycle* (Fig. 11-1). As organic matter decomposes, nitrogen is released as NH_4^+ in the process of mineralization. The rate of nitrogen mineralization depends largely on the carbon-nitrogen (C:N) ratio in the soil. The higher the C:N ratio, the more rapidly the organic matter is broken down. For example, if the C:N ratio is 20:1, organic matter will be broken down much more rapidly than if it is 10:1. The NH_4^+ released can be used in different ways—taken up by soil microorganisms, used by higher plants, held by the soil particles, or converted into NO_3^-. Conversion of NH_4^+ into NO_3^-, called *nitrification*, results from the activities of two different types of soil bacteria

$$NH_4^+ \xrightarrow{\text{Nitrosomonas}} NO_2^- \xrightarrow{\text{Nitrobacter}} NO_3^-$$

Ammonium Nitrite Nitrate

This process usually occurs rapidly enough for there to be much more NO_3^- than NH_4^+ in the soil.

Although plants can survive (and some thrive) on NH_4^+ as a nitrogen source, most nitrogen is absorbed in the form of NO_3^-. The uptake of available forms of nitrogen by microorganisms is called *immobilization* since the nitrogen is bound—at least temporarily. When a crop is harvested, nitrogen is removed from the cycle. If the crop is plowed under, the nitrogen in the plants eventually returns to the soil after the crop residues break down. The nitrogen absorbed by microorganisms is similarly returned to the soil when microbial cells die.

Fixation

Any discussion of soil fertility must include *nitrogen fixation*, the biological incorporation of gaseous N_2 from the atmosphere into a usable form. Three groups of microorganisms are capable of nitrogen fixation: (1) some free-living bacteria such as *Azotobacter* and *Clostridium*, (2) some blue-green algae, and (3) the bacteria *Rhizobium* in a mutually beneficial relationship with legumes. Some nonlegumes also fix nitrogen in a symbiotic relationship with certain

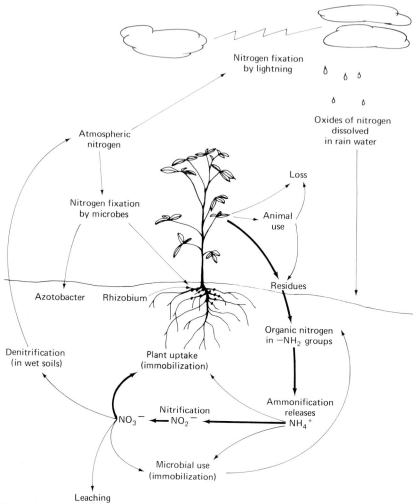

FIGURE 11-1

Nitrogen cycle. The darker lines indicate the main cycle of mineralization and immobilization. (*From L. M. Thompson and F. R. Troeh, Soils and Soil Fertility, 4th ed., McGraw-Hill, New York, 1978.*)

microorganisms. Most widely studied is the fixation of nitrogen by legumes whose roots are infected with *Rhizobium*. The specificity of the bacteria is such that a particular bacterial species will infect only certain legumes. To ensure good nitrogen fixation, legume seeds are usually inoculated with the appropriate bacterium before planting. Formation of the nodules (Fig. 11-2) is not only complex but influenced by environmental factors such as soil pH, calcium level, and the amount of nitrate in the soil. Legumes fix fairly wide ranges of nitrogen, but average amounts appear to be from 50 to 150 kg of nitrogen annually per hectare. Although nitrogen fixation by the free-living bacteria and the blue-green algae is probably of importance, their activities are not yet well understood. As the cost of nitrogen fertilizer continues at its high level, these alternative sources of nitrogen will doubtless be explored more thoroughly. The limited quantities of nitrogen fixed by lightning probably account for only 2 to 3 kg of nitrogen per hectare annually.

Losses

The loss of nitrogen through several processes can be sizeable. Leaching is a major problem in humid regions. Because of their negative charge, NO_3^- ions are not held by negatively charged soil particles and remain in the soil solution. Although NH_4^+ ions are held as cations, they are eventually converted into NO_3^-, which can be leached. Soil erosion can account for sizable nitrogen losses, but crop removal is probably the greatest cause of nitrogen loss where good soil-management practices are followed. Depending on the crop and what part of the plant is

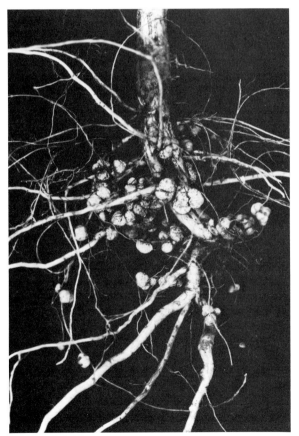

FIGURE 11-2
Nodulation of roots of a typical legume. (*U.S. Department of Agriculture.*)

harvested, this can represent 50 to 75 kg of nitrogen per hectare annually for crops such as potato and tomato. For hay crops, the annual figure can be more than 150 kg ha^{-1} since the entire top of the plant is removed repeatedly during the growing season.

The process by which nitrate nitrogen is reduced to gaseous forms is called *denitrification*. The causal microorganisms are thought to be facultative anaerobes, but the exact mechanisms by which they reduce nitrate is not well understood. The volatile forms of nitrogen are nitrous oxide (N_2O), elemental nitrogen (N_2), and nitric oxide (NO). Certain chemical reactions in the soil can also lead to the volatilization of nitrogen. Nitrogen losses by denitrification are difficult to measure but are estimated to be in the range of 10 percent of the nitrogen added per year. The losses are greatest where aeration is poor or where nitrogen levels are excessive.

Fertilizers

The many fertilizers which supply nitrogen are classified as mineral (some people prefer "chemical" to mineral), e.g., calcium nitrate, and organic. The oldest forms are obviously organic, such as animal manure, used for thousands of years. Other examples are bone meal, dried blood, and composted leaves. The release of nitrogen by organic sources tends to be slow and occurs by the same basic processes as for organic matter in the soil.

Chemical fertilizers are by far the major source of nitrogen. Major advantages of inorganic nitrogen fertilizers include rapid availability of applied nitrogen, ease of shipment and application, and lower cost per unit of applied nitrogen. Although under certain conditions slowly available nitrogen from organic sources is advantageous, the goal is usually to have high nitrogen availability early in the season, followed by declining nitrogen levels.

All major nitrogen carriers (Table 11-3) can now be manufactured from ammonia (NH_3), which is produced from N_2 and H_2 under high temperature and pressure. The NH_3 can be stored under pressure and injected into the soil as a liquid, mixed with water and sprayed on the soil surface, or used in the manufacture of dry nitrogen carriers.

Since the major factor in selecting a nitrogen source is cost, anhydrous ammonia is widely used. For many crops, however, dry fertilizers are preferable because they are better suited to band applications and other techniques. If there is no particular reason

TABLE 11-3
Common Mineral Nitrogen Sources

Source	Formula	N, %	Comment
Ammonia (anhydrous)	NH_3	82	Liquid under pressure; cheapest source
Ammonium nitrate	NH_4NO_3	33	Widely used, often cheapest dry source
Ammonium phosphate	$NH_4H_2PO_3$	11	Also source of P; used in mixed fertilizers
Ammonium sulfate	$(NH_4)_2SO_4$	21	Used on acid-loving plants
Calcium nitrate	$Ca(NO_3)_2$	17	Imported; source of Ca
Sodium nitrate	$NaNO_3$	16	Originally imported from Chile; now synthesized
Urea	$(NH_2)_2CO$	45	Used for both soil and foliar applications

for choosing NH_4^+ or NO_3^-, ammonium nitrate may be the best choice among the dry formulations as it is usually the lowest in cost per unit of nitrogen. For immediate availability, a NO_3^- source such as $NaNO_3$ or $Ca(NO_3)_2$ is probably best, the choice depending on cost and the desirability of either Ca^{2+} or Na^+. Urea is widely used for foliar fertilization because it is soluble in water and quickly absorbed by leaves, thus giving a rapid response. Another advantage of a urea spray is that, since the effect is quite temporary, fairly precise control of nitrogen levels is possible. Since the optimum number of nitrogen applications varies with the crop, soil management, soil type, level of organic matter, rainfall, and other factors, generalizations are meaningless.

In recent years, the use of *slow-release fertilizers* has become widespread for some crops. Particles of fertilizers are coated with waxes, resins, or other substances which by dissolving slowly release the nutrients over 3 to 6 months. Slow-release fertilizers, which usually contain phosphorus and potassium as well as nitrogen, are particularly useful with potted plants, greenhouse bench crops, and lawns, which thrive on a uniform supply of nutrients throughout the growing season.

PHOSPHORUS

The importance of phosphorus has long been known, and indeed phosphate fertilizers were the first to be used in large quantities.

Functions
Phosphorus is vitally important in energy storage and transfer and other aspects of plant growth. The formation of adenosine triphosphate (ATP), containing high-energy phosphate bonds, is essential in plant metabolism. Phosphorus is also a constituent of DNA, RNA, phospholipids, and the coenzymes NAD and NADP (Chap. 5).

Deficiency Symptoms
Because phosphorus plays a vital role in energy transformations in the plant, a deficiency shows up in altered metabolism and growth. Growth is stunted; older leaves tend to abscise because, like nitrogen, phosphorus is mobile and moves from the older to the young leaves; leaves are dark green and sometimes distorted. Carbohydrates tend to accumulate, encouraging the formation of anthocyanins and the associated red or purple coloration of leaves and stems.

Availability in Soil
Unlike nitrogen, of which there are tremendous quantities in the atmosphere, the supply of phosphorus is limited to that already in the soil or added as fertilizer. Soil phosphorus exists in both inorganic and organic forms. Inorganic phosphorus compounds are divided into those containing calcium and those containing iron or aluminum. The mineral apatite, a complex compound containing calcium phosphate, is the source of most of the native soil phosphorus and is also found in deposits from which phosphate

rock is mined. Phosphorus in soils is generally in unavailable forms, only a minute portion of the total amount being available at any time. Although this may seriously limit phosphorus availability for crop growth, it also minimizes losses due to leaching. Phosphorus availability is affected by soil pH because of associated changes in the concentration of various cations. Under alkaline conditions, phosphorus is tied up in various calcium containing compounds ranging from moderately soluble calcium phosphates to insoluble forms such as apatite. In very acid soils, phosphorus is tied up as iron and aluminum phosphates, which are insoluble, making the phosphorus unavailable for plant use. As a general rule, phosphorus availability is maximized by maintaining the soil pH between 6.0 and 7.0.

The uptake of phosphorus from the soil depends on the forms present in the soil, which vary with pH. Most of the phosphorus is taken up as dihydrogen phosphate ($H_2PO_4^-$), but the HPO_4^{2-} ion becomes more prevalent at a pH above 7.2.

Fertilizers

When phosphorus fertilizers are applied, the effect on the availability of soil phosphorus varies with the fertilizer. Most frequently used today are superphosphate or treble superphosphate, both of which provide sizable percentages of soluble phosphorus. Since bone meal provides mostly insoluble phosphorus, it has little immediate effect. Even the soluble forms become fixed, i.e., converted into unavailable forms in the soil. To minimize fixation, fertilizers containing phosphorus should not be mixed into the soil, but applied in bands below and on both sides of the seed. In other cases the fertilizer is broadcast and plowed or disked in. Regardless of the method of application, fixation limits recovery of the applied phosphorus to well under 50 percent.

The manufacture of superphosphate involves treatment of rock phosphate with sulfuric acid to yield phosphoric acid and gypsum

$$Ca_3(PO_4)_2 + 3H_2SO_4 + 6H_2O \longrightarrow 2H_3PO_4 + 3CaSO_4 \cdot 2H_2O$$

The content of phosphorus in superphosphate is about 9% which has been conventionally expressed as 20% P_2O_5. For use in high-analysis fertilizers, the rock phosphate can be treated with phosphoric acid

$$Ca_3(PO_4)_2 + 4H_3PO_4 \longrightarrow 3Ca(H_2PO_4)_2$$

This product, treble or triple superphosphate, contains about 20% P (equivalent to 46% P_2O_5). The higher phosphorus content is advantageous both in long-distance shipment and the formulation of high-analysis mixed fertilizers.

Although some phosphate is applied alone, most is used in *complete fertilizers*, containing nitrogen and potassium as well. The analysis always lists nitrogen, phosphorus, and potassium in that order. The nitrogen is expressed as N, the phosphorus on the basis of P_2O_5 and the potassium as K_2O equivalent. Thus a 10-10-10 fertilizer contains 10% N, 10% P_2O_5 equivalent, and 10% K_2O equivalent. The current trend is to provide the guaranteed analysis on an elemental basis for all three nutrients. To convert P_2O_5 to P, multiply by 0.44; and to convert K_2O to K, multiply by 0.83. Thus, the old designation of 10-10-10 becomes 10-4.4-8.3 on the elemental basis.

POTASSIUM

Potassium is the third of the three major nutrients supplied in a complete fertilizer. Widespread use of potassium fertilizers has lagged well behind that of both nitrogen and phosphorus because the native supply in most soils is high enough for deficiencies to develop only slowly.

Functions

Although there is no doubt that potassium is essential, its exact functions in plants have been elusive. No major organic compound in plants is known to contain potassium. Potassium is essential for photosynthesis, sugar translocation, and enzyme activation, yet its specific roles remain obscure. It is now known that potassium ions are pumped in and out of guard cells, thus regulating water potential and the resulting opening and closing of stomata (Chap. 9).

Deficiency Symptoms

Like those of both nitrogen and phosphorus deficiencies, the early symptoms of a potassium deficiency appear on older leaves first because of the element's mobility. These symptoms include leaf chlorosis, which is followed by *marginal scorch*, a term given to necrosis at the leaf tip and margins (Fig. 11-3). Metabolic changes induced by inadequate

FIGURE 11-3
Potassium deficiency of cabbage, showing chlorosis of leaf margin. As the deficiency worsens, the margins will become necrotic. (*From J. C. Walker, Plant Pathology, 3d ed., McGraw-Hill, New York, 1969.*)

potassium, namely, the accumulation of carbohydrates and soluble nitrogen compounds, are due to a lack of protein synthesis.

Luxury Consumption

Potassium is unusual in that excessive applications are relatively harmless, and for this reason, potassium is sometimes wasted. Many plants take up potassium in proportion to its availability. If a twofold application is made every other year to save application costs, the first crop uses excess potassium (*luxury consumption*), and crops in following years may not have enough. Excessive amounts may create an imbalance with other nutrients, e.g., magnesium, and lead to a deficiency of these elements.

Availability in Soil

The total potassium content of most soils is usually many times higher than that of phosphorus or nitrogen. Most of it, however, is in forms relatively unavailable to plants. Since potassium is not chemically bound into major organic molecules, it is readily leached from organic matter soon after organisms die: and, unlike nitrogen and phosphorus, is not held in slowly released organic forms. The inorganic potassium in soils includes forms which are available readily, slowly, or not at all. Most of the readily available potassium, constituting only 1

to 2 percent of the total, exists as exchangeable ions on soil particles; the remainder is in the soil solution. Slowly available potassium, ranging from 1 to 10 percent of the total, consists of potassium ions in nonexchangeable positions between layers of clay crystals. There is an equilibrium between the soil-solution potassium,, the exchangeable potassium, and the nonexchangeable potassium. When potassium is added by fertilizer or removed by plants or by leaching, the equilibrium shifts accordingly. For example, as K^+ ions are removed from the soil solution by a plant root, they are replaced from the exchangeable fraction. The nonexchangeable K^+ ions are gradually released until a new equilibrium is reached. With the addition of K^+ ions from fertilizer, the ions move from the soil solution to exchange sites and into the nonexchangeable fraction. The remaining 90 to 98 percent of the soil potassium, which is unavailable and mostly in such minerals as feldspars and micas, becomes available only as the minerals gradually weather.

Because of the fixation of applied potassium, application techniques resemble those for phosphorus more than those for nitrogen. In soils with a high potassium-fixation potential, it is common to band the application for row crops; surface broadcast applications are suitable for perennial plants with well-established surface root systems.

Fertilizers

The most common fertilizer source of potassium is potassium chloride (KCl), or *muriate of potash*, which occurs naturally in deposits in Europe, the United States, and Canada, laid down as briny lakes dried up. Potassium chloride contains about 50% K, or the equivalent of 60% K_2O. When the chloride ion is undesirable or sulfate is preferable, potassium sulfate (K_2SO_4) is used. This source, known as *sulfate of potash*, contains about 43% K, or 52% K_2O equivalent. A third source, potassium nitrate (KNO_3), provides nitrogen as well. The analysis of KNO_3 is about 13% N and 38% K (44% K_2O). The major limiting factor in its use is cost, since it is considerably more expensive than KCl or K_2SO_4.

CALCIUM

The concentration of calcium in plant tissue ranges from quite low up to several percent dry weight.

Because of its relative immobility, calcium must be continually available from external sources. Calcium used to be almost ignored in nutritional research because classic calcium deficiencies were rare; now, however, because we know of its involvement in several major disorders, research on calcium is in the forefront.

Functions

Much of the calcium in plant tissue is in the middle lamella, or cementing layer, between cells, immobilized as calcium pectate and unavailable for use elsewhere in the plant. Other functions of calcium, which are less well documented, include the normal growth of meristems, especially in the roots; the maintenance of membrane structure and function; and cell division. Calcium seems to be involved in regulating of the amount of organic acids present by neutralizing or precipitating excess acids, for example as calcium oxalate crystals.

Deficiency Symptoms

The classic symptoms of a calcium deficiency are curled and distorted leaves, which often show a hook at the leaf tip, and stunted growth. Roots are especially affected by a deficiency of calcium, as shown by a stubby, brown appearance.

During the past 20 to 25 years several serious disorders of horticultural crops have been proved to result from calcium deficiency including blossom-end rot of tomato (Fig. 11-4), blackheart of celery, and bitter pit of apple. Although each has been related to adverse water relations and other factors,

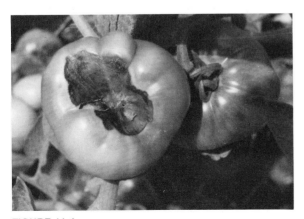

FIGURE 11-4
Blossom-end rot of tomato.

e.g., nitrogen level, it is now known that calcium deficiency is the primary cause. Calcium must be added, usually as gypsum ($CaSO_4$), to improve fruiting in peanuts. Although the classic symptoms of calcium deficiency in apples are unknown under orchard conditions, in many cases calcium levels are well below optimum, as reflected in the occurrence of bitter pit and other results of low calcium level, e.g., excessively soft fruit and high rates of respiration leading to a short storage life.

Availability in Soil

Although the total amount of calcium in soils is considerably less than that of potassium, usually a much higher proportion of it is available. Whereas only 1 to 2 percent of the total potassium is available at any one time, it is not uncommon for more than 20 percent of the total calcium to be in readily available forms. Well over 99 percent of the readily available calcium is held by soil particles as exchangeable calcium, and less than 1 percent is actually in solution. As calcium is removed from the soil solution by plant uptake or leaching, the equilibrium is reestablished by the release of exchangeable calcium, which in turn is replaced by hydrogen in humid areas. With this gradual replacement, soil acidity increases and can be reversed by the addition of lime (Chap. 10).

Fertilizers

The major source of calcium added to soils has been lime used to raise the soil pH. Superphosphate, which contains more calcium than phosphorus, is also a good calcium source. It used to be assumed that calcium would be sufficient if soil pH were adequately maintained, but the realization that such devastating problems as bitter pit of apples and blossom-end rot of tomatoes are calcium-related has led to extensive research on calcium nutrition. Perhaps the most effective treatment has been foliar sprays with calcium chloride or calcium nitrate since the problem with calcium is not so much its shortage in the soil as uneven uptake by the crop or inadequate distribution in the plant.

SULFUR

Sulfur is required by plants in much smaller quantities than nitrogen but in the same general range as phosphorus.

Functions

As a constituent of cysteine, cystine, and methionine, sulfur is necessary for the synthesis of proteins containing any of these three amino acids. It is also a component of the vitamins thiamine and biotin and coenzyme A, which is involved in the Krebs cycle (Chap. 5). Certain crops, e.g., onion, garlic, and mustard, owe their particular odor and flavor to sulfur-containing organic compounds.

Deficiency Symptoms

For many crops, the symptoms of a sulfur deficiency are similar to those of nitrogen deficiency since both result in a shortage of proteins. One difference is that the chlorosis from sulfur deficiency is generally worse on young foliage, due to sulfur's lack of mobility. Other less apparent results are the accumulation of soluble nitrogen materials, because of depressed protein synthesis and a shortage of carbohydrates, stemming from reduced photosynthesis.

Availability in Soil

Soils contain sulfur in both organic and inorganic forms. As a constituent of proteins, amino acids, and other complex molecules, the organic fraction is released slowly by decomposition. A small portion of the total sulfur in a soil exists as the sulfate ion (SO_4^{2-}) in solution and adsorbed on soil particles. A *sulfur cycle* exists which is similar to the nitrogen cycle described earlier. Small amounts of sulfur dioxide (SO_2) are released into the atmosphere by the burning of fossil fuels and the decomposition of organic matter. When SO_2 is present in very low concentrations in the air, it can be absorbed by plants, thus meeting their sulfur needs at least partially. However, since SO_2 is very toxic to plants at an atmospheric concentration as low as 1 ppm, it is normally considered an air pollutant. Because SO_2 is water soluble, sizable quantities are deposited in rainfall; in many places sulfur deposited in rainfall has been estimated to equal the sulfur lost by crop removal. It also contributes to lowered pH and is part of the "acid rain" problem. The increased strictness of air-pollution laws has lowered levels of SO_2 in the atmosphere, especially around industrial sites.

Under well-aerated conditions, the major inorganic form of sulfur is the sulfate ion (SO_4^{2-}), the form in which sulfur is absorbed by plants. The decomposition of organic molecules releases sulfur in various forms, including sulfides, but the final product is sulfate. When calcium is plentiful in the soil, gypsum ($CaSO_4 \cdot 2H_2O$) is formed as a slightly soluble precipitate. Conditions of poor aeration cause accumulation of reduced forms of sulfur such as iron sulfides.

Fertilizers

The application of sulfur-containing fertilizers is increasing. Reduced sulfur in rain due to more stringent air-pollution laws is partly responsible for this practice. Another reason is that with greater use of higher-analysis fertilizers, the amount of sulfur as an impurity in fertilizer has declined. For example, superphosphate contains considerable quantities of $CaSO_4$, but for high-analysis fertilizers, formulators use treble superphosphate, which contains no sulfur; and, as ammonium sulfate is replaced by ammonium phosphate, potassium nitrate, or anhydrous ammonia, sulfur is eliminated from the mix.

As sulfur is needed, it can be applied as superphosphate or ammonium sulfate or—perhaps more economically—as elemental sulfur or gypsum. Quantities necessary to meet the sulfur needs of crops are much smaller than those mentioned in Chap. 10 for the treatment of saline sodic soils.

MAGNESIUM

Because magnesium exists as a divalent cation (Mg^{2+}), it has many characteristics in common with calcium; there are, however, some important differences.

Functions

Probably the best-known fact about magnesium is its central position in the chlorophyll molecule (Chap. 5). Since magnesium is necessary for chlorophyll formation, it is obviously vital to photosynthesis. Nevertheless, much of the magnesium found in leaf tissues is not a part of the chlorophyll molecule but apparently functions in the chloroplasts as an enzyme activator which facilitates a wide diversity of reactions, especially in energy transfer.

Deficiency Symptoms

The most obvious symptom of a magnesium deficiency, related to its central position in the chlorophyll molecule, is exhibited as an interveinal leaf chlorosis. Since magnesium is a relatively mobile element in the plant, it behaves more like potassium than calcium. The deficiency appears first on the older leaves and advances upward on the plant. As the deficiency becomes severe, progressive defoliation may start at the shoot base.

With certain horticultural crops, symptoms of deficient magnesium affect the economic part of the plant directly. In the northeastern United States magnesium deficiencies have been common in apple orchards and frequently cause severe preharvest fruit drop. Citrus fruit, sweet potatoes, and the stone fruits are other crops likely to suffer from magnesium deficiencies.

Availability in Soil

The forms in which magnesium exists in the soil are much like those of calcium. The exchangeable fraction of magnesium, held by soil particles, is in equilibrium with the magnesium in the soil solution. The large percentage of the total soil magnesium in relatively unavailable forms is in minerals such as mica and dolomite, and is released only as they are acted upon by soil water containing carbonic and other acids. Since soil magnesium is depleted more quickly than either calcium or potassium, sandy soils are more likely to be deficient in magnesium than finer-textured soils with greater cation-exchange capacity.

Fertilizers

The material commonly used to add magnesium to soils is dolomitic limestone.† Where magnesium is likely to be deficient, it is advantageous to use dolomitic rather than calcitic limestone since not only is excess acidity neutralized but sizable quantities of magnesium are added at relatively low cost. Aside from dolomite, the major magnesium source is magnesium sulfate ($MgSO_4 \cdot 7H_2O$), epsom salt. For quick response, magnesium sulfate dissolved in water is applied as a foliar spray; repeated applications may be required to alleviate a severe deficiency. For a long-term effect, the magnesium sulfate is applied to the soil.

IRON

Although iron is considered a micronutrient, it is required in greater quantities than any of the others; it was the first to be proved essential.

---

†Calcite is mostly $Ca(CO_3)_2$, and dolomite is $CaMg(CO_3)_2$.

Functions

The involvement of iron in plant metabolism is very broad. Iron is necessary for the synthesis of chlorophyll although it does not form a part of the molecule; an iron deficiency leads to abnormal chloroplast structure. Iron is also involved in many of the enzymatic reactions in respiration and photosynthesis.

Deficiency Symptoms

The symptom of iron deficiency, which is quite specific, is termed *iron chlorosis* (Fig. 11-5). The young leaves show a yellowing or even a whitish coloration in the interveinal areas. Since the veins tend to stay dark green, they are in strong contrast to the light-colored areas between. Unless the deficiency is severe, plants show little leaf necrosis. As severity increases, plants become stunted, produce poorly, and may eventually die.

Availability in Soil

Although most soils contain large quantities of iron, much of it is in unavailable form; therefore the amount of iron in the soil is seldom the limiting factor in iron nutrition. Iron occurs in soils in both the ferrous (Fe^{2+}) form, which predominates in poorly drained soils, and the ferric (Fe^{3+}) form, which predominates in well-aerated soils. The availability of iron depends on soil pH. In alkaline soils, most of the iron is precipitated in compounds containing iron, hydroxyl, and phosphate ions. An iron deficiency resulting from high soil pH is often called *lime-induced chlorosis*. In more acid soils, iron

FIGURE 11-5
(*Left*) Normal leaves of rhododendron. (*Right*) Iron deficiency. The interveinal areas are chlorotic while the veins are still green.

availability is greatly increased and may actually reach toxic quantities.

Fertilizers

Foliar applications of iron sulfate can be quite effective in adding small quantities, but soil applications of iron sulfate frequently proved unsuccessful because the iron was rapidly converted into insoluble forms. The introduction of iron chelates has dramatically improved the success rate in curing iron deficiencies. *Chelates* are complex organic molecules which combine with cations but do not ionize. The iron is held in such a way that it can be absorbed by plants but it cannot react in the soil to form insoluble precipitates. The most widely used chelating agent is ethylenediaminetetraacetic acid (EDTA). In combination with iron to form the iron chelate it is called Fe-EDTA. Iron chelates, though rather expensive, are widely used and often are effective for up to 2 years. Foliar applications of iron chelates are also made.

BORON

Although the crop requirement for boron is very low, boron deficiencies in a wide variety of crops are common in some parts of the United States. An unusual feature of boron as a nutrient is the narrow range between deficient, sufficient, and toxic levels: although deficient boron can cause various maladies, excess boron causes necrosis and can be lethal.

Functions

The exact role of boron in the normal metabolism of plants has not been proved, but several hypotheses have been proposed, namely regulation of certain aspects of carbohydrate metabolism and involvement in carbohydrate translocation, cell-wall development, and RNA metabolism.

Deficiency Symptoms

The spectrum of boron-deficiency symptoms is broad, the most common characteristic being the stunting, discoloration, and eventual death of the apical meristems of both shoots and roots. Other symptoms include thick, brittle leaves, which may become dark and discolored as well as malformed. Flowering is suppressed by a lack of boron, and even if flowering does occur, fruit set is limited because pollen ger-

mination and growth of the pollen tube are inhibited by a boron deficiency.

Major disorders of crops shown to be the result of inadequate boron levels include internal cork of apples, hollow stem and browning of cauliflower, internal black spot of beets (Fig. 11-6), cracked stem of celery, hollow heart of peanut, and many others. Legumes are particularly susceptible to boron deficiency.

Availability in Soil

Total boron in most soils is quite low compared with that of many other nutrients. Boron availability is relatively high in acid soils, and some leaching may occur, especially in sandy soils. Although the mechanisms are not well understood, boron tends to be fixed at high soil pH, so that overliming can induce boron deficiencies. Boron uptake is affected by moisture in soils and tends to be depressed at very high or low levels of soil moisture.

Fertilizers

The common source of boron, sodium tetraborate (borax), contains about 11% boron and is mined from deposits in California. Boron trioxide is now available as fertilizer Borate 46 and fertilizer Borate 65, containing about 14 and 20% boron, respectively. Soil application of boron fertilizers is the rule, but for some situations foliar sprays are recommended. A spray on apple leaves provides quick response, danger of toxicity is lessened, and one spring treatment will satisfy the boron needs of the trees for the year. Many find this procedure preferable to a soil application every 3 years, formerly the standard practice.

ZINC

Zinc is often deficient, and the effects of a zinc deficiency can be devastating.

Functions

Zinc serves both as a component and an activator of enzymes and is believed to be involved in protein synthesis. Since zinc has a direct influence on the level of auxin in plants, it is hypothesized that zinc is required either for the synthesis of tryptophan or

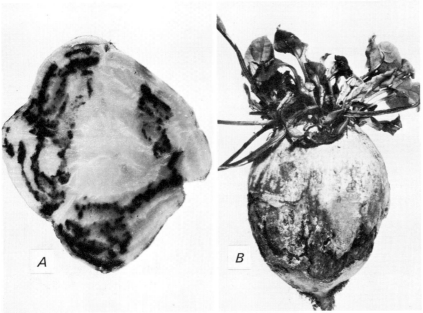

FIGURE 11-6
Boron deficiency of beet. (A) Internal necrosis of tissue in the secondary cambial rings; (B) leaves are stunted and dormant buds of the crown are stimulated, but form small leaves. Internal necrosis near the exterior of the root leads to collapse of the outer tissues to form cankers. (*From J. C. Walker, Plant Pathology, 3d ed., McGraw-Hill, New York, 1969.*)

in its conversion into auxin. Zinc deficiency is characterized by symptoms similar to those of an auxin shortage.

Deficiency Symptoms

Symptoms of inadequate supplies of zinc are related to a striking suppression of growth. *Little leaf* refers to severe stunting of leaves. *Rosette* describes the lack of internode elongation which causes several leaves to originate close to each other. Both effects are related to the suppression of auxin. Other symptoms include frenching, or mottle leaf, of citrus and white bud of corn. Fruit trees are particularly prone to zinc deficiency; vegetable and field crops are less widely affected.

Availability in Soil

As with many other nutrients, restricted availability of zinc in soils is often more important than low total levels. Since the solubility of zinc compounds

in soils is low, especially when soil pH is high, zinc deficiencies were first noted in the alkaline soils of the western United States.

Fertilizers

Following the discovery of zinc deficiency, the major source of zinc has been zinc sulfate. Applications to the soil are effective in acid soils which are not too sandy. In calcareous or sandy soils, it is more effective to spray the plant with a solution of zinc sulfate. When zinc deficiency was first diagnosed in fruit trees, it was cured by driving galvanized (zinc-coated) nails or glazing points into the trunk, enough zinc being absorbed to alleviate the problem.

MANGANESE

Manganese is essential in small quantities but is toxic when excessive amounts are taken up by plants.

Functions

Manganese is involved in many plant processes, including chlorophyll synthesis, respiration, photosynthesis, and nitrogen assimilation. Although precise roles remain to be defined, manganese seems to be associated with multiple enzyme systems, especially the Krebs cycle.

Deficiency Symptoms

Although symptoms vary with the crop, the most common sign of manganese deficiency is a mottled chlorosis of leaves with some necrotic spots developing as the deficiency worsens. Growth is poor, with both flowering and fruiting being suppressed. The foliar manifestations are worse on young leaves, and, like iron, manganese is not readily translocated.

Toxicity Symptoms

The usual symptoms of manganese toxicity are crinkling and cupping of the leaves, often accompanied by chlorosis of the leaf margins.

Availability in Soil

Most soil manganese is in relatively insoluble forms at high soil pH but becomes increasingly soluble as the pH is lowered. Manganese is sometimes toxic in soils at a pH of 5.5 or below. The logical cure for excess manganese is the addition of sufficient lime to raise the soil pH. Soils most likely to be deficient in manganese are those which are sandy, organic, or alkaline. Steam pasteurization of soils may cause manganese toxicity because the high temperature increases the solubility of soil minerals containing manganese.

Fertilizers

Manganese sulfate, applied to the soil or sprayed on the foliage, is effective in eliminating manganese deficiencies. Since manganese is bound in alkaline soils, application of manganese sulfate to such soils is ineffective. More expensive chelated manganese is often the best choice where soil pH is high.

COPPER

Copper deficiency is rarer than boron or zinc deficiencies, but may be fairly widespread under certain conditions. Crops are most prone to copper deficiency on soils that are sandy, organic, or have a high pH.

Functions

Copper is required for activation of several enzymes in plants, is needed for chlorophyll synthesis, and is involved in the metabolism of carbohydrates and proteins.

Deficiency Symptoms

Although the specific manifestations of deficient copper vary from species to species, leaf symptoms are the most common. Leaf growth is suppressed, the tip often becomes necrotic, and leaf color is darker than normal. If the deficiency is severe, dieback of shoots will occur.

Availability in Soil

Copper occurs in soil as both cupric (Cu^{2+}) and cuprous (Cu^+) ions, which are exchangeable cations. The solubility of copper is highest under acid soil conditions and declines as the pH is raised. Since most soil copper comes from weathering of mineral particles, deficiencies are quite common in organic soils. Calcareous soils are occasionally deficient due to the unavailability of the copper. Sandy soils lack copper because of leaching and lack of replacement by weathering.

Fertilizers

Copper sulfate is the most frequently used source of copper. If high soil pH renders soil applications ineffective, a foliar spray of copper sulfate or a copper chelate spray to soil or plant can be applied. Some fungicides with copper as the active element supply copper to potato and other crops.

MOLYBDENUM

Molybdenum, required in very small amounts, is present in most soils in minute quantities.

Functions

The two most clearly defined roles of molybdenum relate to nitrogen metabolism. Molybdenum is required by plants for nitrate reduction and assimilation and for the symbiotic nitrogen fixation by legumes and *Rhizobium* bacteria in their root nodules.

In both processes molybdenum apparently functions as an enzyme component.

Deficiency Symptoms

Since molybdenum plays a vital role in nitrogen metabolism, it is not surprising that early symptoms of deficiency resemble those of a lack of nitrogen. Older leaves become chlorotic in the interveinal areas, often with a mottled appearance. The margins may wilt, roll, and cup as the deficiency becomes more severe. In cauliflower and related species, the name *whiptail* has been given to molybdenum deficiency because the malformed leaves resemble a whip when the midrib develops predominantly.

Availability in Soil

Most soils contain some molybdenum, but deficiencies are not uncommon. Since unlike the other micronutrients, molybdenum is more available in alkaline soils, liming acid soils will often increase availability enough to eliminate the deficiency.

Fertilizers

Foliar sprays of sodium or ammonium molybdate are often effective. Thanks to the very small requirement, only 75 to 80 g ha^{-1} may be necessary. For some crops, dusting the seed with a molybdenum compound before planting may be sufficient.

CHLORINE

The general acceptance of chlorine as an essential element is a recent development, probably because chlorine deficiency is unknown under field conditions and since chlorine is present in soils, water, air, and many fertilizer materials, it is difficult to maintain an experimental chlorine-free environment.

Functions

Although the exact role is not well understood, chlorine is essential for that part of the photosynthetic process during which oxygen is liberated.

Deficiency Symptoms

Plants deficient in chlorine tend to wilt and show a buildup of free amino acids, but the mechanisms are not understood.

Toxicity Symptoms

Toxic levels of chlorine can be detrimental to a broad array of plants. Chlorides are of major concern in saline soils (Chap. 10), as well as in areas adjacent to highways salted for ice removal. Symptoms of excess chlorine include chlorosis and necrosis of leaves, growth suppression, lowered yields, and death if chlorine levels become very high.

Availability in Soil

Chlorine exists as the chloride (Cl^-) ion in soil, where it moves freely. It may come from fertilizers, e.g., potassium chloride, but perhaps the major sources are rainfall and irrigation water. Along coastal areas salt spray evaporates and the chloride is redissolved in rain water and deposited on the soil. In irrigated areas, the small amounts in the water accumulate in the soil because of inadequate leaching.

SODIUM

The essentiality of sodium, like chlorine, has been proven only recently. As discussed in Chap. 10, excessive sodium levels are detrimental in sodic soils.

Functions

Sodium is involved in plant-water relations, probably by affecting stomatal opening, but the exact mechanism is not clear. In C_4 and CAM plants, sodium has a role in the functioning of PEP carboxylase (Chap. 5); a lack of sodium can cause a shift from the C_4 to the C_3 photosynthetic pathway.

Availability in Soil

In the soil, sodium exists as Na^+. Although sodium is readily leached if adequate rainfall occurs, deficiencies are rare, problems with excess sodium being much more prevalent.

DETERMINING FERTILIZER NEEDS

In discussing the essential elements in this chapter, we briefly described the most frequently cited symptoms of a deficiency of each. For several reasons, however, agricultural producers do not usually depend on deficiency symptoms to guide them in applying fertilizers.

1. By the time deficiency symptoms are visible, a reduction in growth and perhaps yield and quality has already occurred.

2. The symptoms often are not clear-cut. For example, early symptoms of nitrogen and sulfur deficiencies are similar. In many cases, more than one element may be in suboptimal supply, and symptoms can be difficult to identify.

3. The time between application of fertilizers and response by the plant may delay recovery during a critical period in the life of the crop.

4. Crop productivity may suffer from a mild nutrient shortage without obvious deficiency symptoms.

For all of these reasons, researchers have sought better ways to determine the fertilizer needs of crops. The two most widely used are soil analysis and tissue analysis.

Soil Analysis

The most common method of estimating fertilizer needs for agricultural crops is analysis of the soil by agricultural experiment stations or private laboratories.

The soil sample submitted for analysis weighs only about 500 g. To receive a useful recommendation, the farmer must provide a sample which represents the field under examination. Soil laboratories provide directions for proper sampling procedure, which usually involves collecting samples from various areas of the field, mixing them together, and taking a relatively small but representative portion of the mix. For relatively shallow-rooted crops, a sample of the upper 15 to 30 cm of soil may be sufficient. For more deeply rooted crops, it is usually desirable to sample the upper 30 cm of soil and to take a second sample consisting of the next 30 cm as well. Soil-analysis results can be no better than the sampling procedure.

A record sheet submitted with the sample helps in the interpretation of the results. The information requested includes the type and amounts of fertilizer and lime added during the previous 2 to 3 years, recent cropping history, and the crop to be grown.

When the soil sample is received in the laboratory a small subsample is used for a pH determination, standard in all soil analyses. Another subsample is treated with an extractant, such as a weak sodium acetate solution, for a specific length of time. The soil extract is filtered to remove the soil particles and then subjected to various analytical procedures.

Nutrients which are routinely measured in soil analysis are phosphorus, potassium, calcium, and magnesium; iron and manganese may also be included. This list of nutrients includes only four of the macronutrients and two of the micronutrients. The others are equally important, but dependable, low-cost extraction techniques have eluded researchers.

The purpose of a soil analysis is to extract the available nutrients in the soil chemically and to measure their concentration. In other words, the extractant is selected to remove approximately the same amount of each nutrient as can be extracted by roots of a crop. On the basis of many experiments and much experience, a soil scientist can make recommendations for both fertilizer and lime application to meet the nutritional needs of the crop to be produced. Soil analyses and their interpretation are of tremendous value. Most fertilization and liming done today is based on such tests. Nevertheless there are inherent weaknesses in soil analyses as the sole source of guidance in fertilization. One has already been mentioned: soil analyses are not routinely available for nitrogen, the most frequently applied nutrient, or for several of the micronutrients. The other major problem is that soil analyses do not always reflect the uptake by the crop. Because plants absorb nutrients selectively, the availability of a particular nutrient in the soil may or may not reflect actual uptake.

Tissue Analysis

During the recent past, increasing emphasis has been placed on refining the use of tissue analysis to assess the nutritional status of crops. The tissue most frequently analyzed is the foliage. Since most photosynthesis takes place in leaves, it is logical to assume that nutritional levels in leaves are most important. Regardless of the amount of a specific nutrient available in the soil, it is the amount present in the leaf or other region of use that is important. Perennial plants, such as fruit trees, for example, are able to thrive on soils very low in certain nutrients such as phosphorus.

Sampling procedures have been refined for many crops, and tissue analyses have become quite routine. The tissue sample must be representative and usually consists of leaves, which are dried. The laboratory grinds a subsample into a powder and ashes it to burn off the organic matter. This ash is dissolved and the resulting solution analyzed for the various essential elements. A major advantage of tissue analyses is that essentially all nutrient elements can be determined.

Interpretation of soil and tissue analyses requires considerable training and experience. The background data used in interpreting tissue analyses have been generated through controlled sand and solution culture experiments, field studies, and nutritional surveys. For each element a general response curve has been developed. A generalized example is shown in Fig. 11-7. Nutrient concentrations below optimum are said to be in the *deficient range*; the more severe the deficiency the greater the suppression of growth and yield. At the lower end of the optimum range is a region known as *hidden hunger*, in which growth

is reduced but no visible deficiency symptoms are present. The *critical concentration* can be defined as the level of a nutrient at or below which yield will be reduced. The *optimum*, or adequate, range is where growth and yield will not be limited by this particular element. With some nutrients, e.g., potassium, the upper end of the optimum range may be designated as *luxury consumption* in that supraoptimal levels are present but are neither beneficial nor detrimental. With others, e.g., boron, excessive tissue levels are toxic and lead to decreased growth and even death.

Using this type of information for the particular crop in question, makes it possible to evaluate the status of each element. The data requested on the informational sheet include species, cultivar, previous fertilization and liming practices, rainfall, and notes on symptoms and overall health of the crop. For micronutrient analyses, a history of the spray program must also be included because many pesticides contain significant quantities of iron, zinc, and copper.

To be in the best position to make recommendations, the interpreter should have results from both

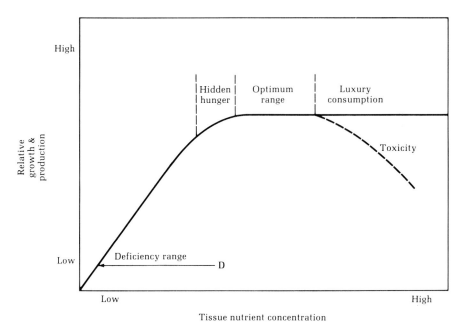

FIGURE 11-7
Schematic relationship between tissue nutrient level and plant growth or production.

soil and tissue analyses, because the most practical way to adjust tissue levels may be to adjust soil pH. For example, excessive tissue manganese levels may be due to low soil pH. An application of lime will probably be the most effective treatment. When the results and recommendations are received by the grower, they must be evaluated in the context of field observations, experience, and economic considerations.

FERTILIZERS

With the tremendous diversity of crops, soils, and production systems in use, it is not surprising that methods of fertilizer application vary widely. The choice of the elemental composition of fertilizer to meet the nutritional needs of the crop in the most economical way is of major importance. The method and timing of the applications is also vital.

Types

Dry Most fertilizers used today are prepared, sold, and applied in dry form, but modern dry fertilizers differ from those used in the past, chiefly in terms of their physical condition. Since many chemicals used in fertilizers absorb moisture, early formulations tended to harden or get sticky. Modern fertilizers are made in granular or pelleted form instead of powder, an improvement that not only reduces their tendency to absorb moisture but also makes them easier to handle and apply. Using plastic liners in fertilizer bags has also reduced moisture uptake.

With increasing labor costs, other major changes have been made in the fertilizer business. For many years, fertilizers were bagged at the fertilizer plant, shipped, and applied by the farmer. Today, however, more and more fertilizers are blended and handled in bulk. The farmer orders a particular blend of fertilizer from a local distributor, who prepares the desired mix and delivers it to the farm in bulk. In fact, large amounts of such bulk fertilizers are spread on the fields by trucks owned by the distributor (Fig. 11-8).

Liquid Increasing amounts of fertilizers are formulated, delivered to the farm, and applied to the soil in liquid form. Labor savings are significant because the fertilizer can be mixed and transported in tanks and readily pumped from one container to another. The extra weight of the water in liquid

FIGURE 11-8
Dry fertilizer spreader. Local fertilizer distributors blend and deliver the desired fertilizer mixes and often spread the fertilizer. (*Tennessee Valley Authority.*)

FIGURE 11-9
Liquid fertilizer application. Note large flotation tires. (*Tennessee Valley Authority.*)

formulations may make them uneconomic if transport over any distance is required.

Application of Dry Formulations

The method of applying dry fertilizers varies with the crop. Vegetables, which are heavily fertilized, usually receive a broadcast application before planting, which is disked into the soil. Several weeks later, an additional application, called *side-dressing*, is made on the soil surface along each side of the row. This division into two applications serves to keep the nutritional level relatively high and stable during the life of the crop and reduces losses due to leaching, which may occur if the entire amount is applied early in the season. With many row crops, e.g., corn, potato, and cotton, the fertilizer is applied during seeding. As the seed are planted, a band of fertilizer is applied several centimeters below the seed and on both sides. By the time the roots of the seedling plants reach the fertilizer, it will have been at least partially dissolved, diluted, and spread into the adjacent soil region. If the fertilizer were put in direct contact with the seed, toxicity and burning of the roots would occur.

Small grains are also usually fertilized as they are seeded. If very small amounts of fertilizer are used, it can be applied very close to the seed, but most modern small-grain drills (planters) put the fertilizer a small distance away.

Pastures and lawns are usually fertilized before planting, by fertilizer broadcast on the surface and then disked into the upper few centimeters of soil. Often pastures are fertilized annually with a broadcast application. Lawns, golf courses, and other areas maintained for aesthetic reasons are usually fertilized repeatedly to keep them lush and green during as much of the year as possible. When fertilizer is applied to the surface, rain or irrigation water dissolves the nutrients and carries them into the root zone.

Orchards are normally fertilized once a year, usually in late winter or early spring. The goal with fruit trees is generally to have high fertility early in the season during flowering and fruit set. As the season progresses, decreasing fertility is desired to encourage good fruit color and quality, to have the plant stop vegetative growth, and to develop maximum cold hardiness.

With container-grown plants in a greenhouse or nursery, dry fertilizers are often mixed into the potting medium before the crop is planted. In many cases, these are slow-release formulations. If necessary, additional fertilizer can be added through the irrigation system.

Ornamental trees may be fertilized in an unusual way if they are surrounded by a lawn. Since a fertilizer application on the surface around the tree would cause excessive grass growth, holes are bored into the subsoil around the tree and the fertilizer placed in the holes, below the roots of the grass, thus meeting the fertilizer needs of the tree but avoiding uneven fertilization of the grass.

Some dry fertilizers are also used to make fertilizer solutions for foliar applications, a procedure widely used in specific situations. With a micronutrient such as boron, where only a few grams or kilograms are needed per hectare, uniform distribution of small amounts is difficult unless it is included in a bulk-blended fertilizer broadcast over the field. A common alternative is to spray a dilute boron solution onto the foliage of the crop. Foliar sprays provide even coverage and a rapid response, since the nutrient is absorbed directly by the leaves, and they avoid fixation in the soil, a problem associated with many micronutrients.

Application of Liquid Formulations

Liquid formulations may also be applied directly to the soil. Materials that are in liquid form at atmospheric pressure can be distributed on the soil surface. Others, such as anhydrous ammonia, must be injected into the soil. Since ammonia is a gas at atmospheric pressure, it is shipped and applied under pressure to keep it in a liquid form. The liquid is injected into the soil, where it is quickly adsorbed by the soil particles. Liquid fertilizers are frequently applied during seeding, either on or below the soil surface (Fig. 11-9).

In irrigated regions, an increasingly popular practice is to inject fertilizers into irrigation water. Precise metering of the amount of liquid fertilizer ensures even distribution and minimizes cost of application. In some cases, dry fertilizers are dissolved in irrigation water, but liquid formulations are easier to work with. The addition of fertilizers to irrigation water, called *fertigation*, can be used with various irrigation systems but is most efficient with sprinkler and trickle systems because of the more even distribution of water.

CHAPTER TWELVE

PEST MANAGEMENT

This chapter discusses the pests of crops (insects, nematodes, diseases, and weeds) and their control. World population trends make it imperative to avoid the disastrous crop failures that have occurred from time to time in the past. One of the most notable is the Irish famine of the mid-1840s, when late blight destroyed several successive crops of potato, upon which the people were almost totally dependent for food. Tens of thousands died of starvation, and even greater numbers emigrated to North America. More subtle than the insect predators or disease but no less important for crops is their battle for supremacy against a horde of weed competitors. The damage from weeds ranges from loss in crop yield and quality to total failure.

HISTORICAL BACKGROUND

Primitive agriculture involved gathering wild plants for food. Both the dispersal of the harvested plants and their genetic diversity meant that extreme outbreaks of a particular pest were probably rare. Obvious exceptions, such as plagues of grasshoppers were probably uncommon. As long as a plant species was not concentrated physically or restricted genetically, it was less susceptible to diseases and other pests.

Agriculture has gone through an evolutionary process, starting with the first farmers who began to plant seeds of particular plants. Modern agriculture has little resemblance to primitive practices. Associated with the changes in production practices has been a dramatic surge of crop pests. A modern example is the outbreak of southern corn blight in the United States in 1970, when one disease organism caused a yield loss in 1 year of 1 billion dollars. What are the reasons for such events?

Intensive culture of a single crop represents a severe disruption of the natural ecology of an area. Modern agriculture is largely a monoculture, in which the aim is to grow one plant species to the exclusion of most other species in the same field. Although generally a very efficient production system, monoculture may encourage the buildup of a host of pests that afflict particular plant species.

Before the development of chemical pesticides, pests were controlled by selection of resistant types, crop rotation, such cultural practices as plowing and sanitation (destruction of crop refuse, removal of infested plants, etc.), elimination of alternate hosts, and related techniques. Nevertheless, some pests persisted and even increased, so that better control techniques were sought. The many natural materials tried included ashes, urine, soap, alcohol, and various plant extracts.

One of the earliest success stories in the control of a major plant disease involves downy mildew. In the mid-1800s, European grapevines became heavily infested with an aphid that attacked the roots. American grape rootstocks were imported because of their resistance to this pest, but unfortunately some of the imported rootstocks brought with them downy mildew, a fungal disease previously unknown in Europe. Since the popular French cultivars were highly susceptible, an epidemic of downy mildew threatened the entire grape industry in France and Germany. In 1882, while walking among the vineyards near Bordeaux, a French professor of botany, Alexis Millardet, observed that some vines along the road had a strange blue color but were free of mildew. The vineyard owner explained that a mixture of copper sulfate and lime had been applied to the vines along the road to discourage passersby from stealing the grapes. In the laboratory Millardet determined the optimum concentrations of copper sulfate and lime needed to control downy mildew. *Bordeaux mixture*, named after the city in France, became important throughout much of the world and proved effective not only against downy mildew of grapes but also against late blight of potato, apple scab, and many other devastating fungal diseases.

The last two decades of the nineteenth century ushered in the beginning of the modern era of chemical pest control, but successes with insecticides date from the 1940s and the widespread use of dichlorodiphenyltrichloroethane (DDT). First synthesized in 1874, DDT had insecticidal properties that were not discovered until 1939. During World War II, it was widely used by the Allied forces for the control of insect carriers of malaria, typhus, and cholera. After the war, when DDT and many other organic pesticides were introduced commercially, their effectiveness, low cost, and availability meant that chemical pest control all but replaced other less effective or more time-consuming methods.

Despite the spectacular success of modern pesticides, there have been problems and concerns for many decades. As early as the 1920s, questions were raised about pesticide residues on fruit from orchards treated with lead arsenate. Eventually, legal residue tolerances for arsenic were established and were the forerunner of the tolerances set for many pesticides on food crops.

The resistance of pests to specific pesticides is another problem. As early as 1908, growers began to notice that an insect, San Jose scale, was becoming resistant to the lime-sulfur sprays used in their apple orchards. Many other examples could be cited for both insects and diseases.

In spite of the warnings of some scientists, the use of pesticides continued unabated for many years. Unfortunately, as the new materials proliferated, many older but safer techniques were largely ignored. The recently initiated thorough reevaluation of pest-control strategies has many causes, perhaps the most significant being public health concerns and the pressure to protect our environment. A contributing factor is the rising cost of pesticides and the realization that as production systems become more intensive, strictly chemical approaches to pest control have not been fully successful.

INTEGRATED PEST MANAGEMENT

Integrated pest management (IPM) is now the subject of intensive research. As defined by the Department of Agriculture, "IPM is the selection, integration, and implementation of pest control based on the predicted economic, ecological, and sociological consequences." A basic hypothesis of IPM is that no single control procedure will control a pest successfully. As a consequence, IPM seeks to integrate a variety of physical, biological, and chemical methods into an overall scheme aimed at providing long-term protection against economic yield losses. Major emphasis is put on such naturally occurring features of the pest environment as weather, predators, diseases, and parasites which may keep the pest below damaging levels. Artificial control measures, such as chemicals, are used only as required to reduce the pests to acceptable levels and keep them there. These treatments are selected to pose minimal risks to people, beneficial nontarget organisms, and the environment.

Five principles form the basis for IPM programs:

1. *Potentially harmful species will continue to exist at tolerable levels of abundance.* Before the advent of IPM, the goal in pest control was often eradication of the pest species (a goal seldom, if ever, achieved). With IPM, low levels of a pest are tolerated as long as they remain below the *economic injury level*, defined as the pest density (or amount of crop damage) at which control costs equal losses in crop value. In other words, if it takes $40 per hectare to control a pest that is lowering yields by the equivalent of $40 per hectare, there is no economic advantage in controlling it.

2. *The ecosystem is the management unit.* The ecosystem is a complex system of biotic and abiotic factors which exist together in a community. Agricultural cropping systems, e.g., a soybean field, an orange grove, or a cranberry bog, are called agroecosystems. Woodlands or forests are often more natural ecosystems. Any modification of an ecosystem can dramatically alter the pest populations. The intensive production systems characteristic of modern agriculture have produced ecosystems markedly different from those of earlier days. For example, the monoculture of corn in vast areas has built up pests of corn, while pests of other crop plants have decreased. The vast numbers of interacting factors in a given ecosystem make it no wonder that progress analyzing the interactions involved is slow. Research in defining and understanding particular ecosystems is in its infancy, but with the help of modern computers progress is being made.

3. *Use of natural control agents is maximized.* IPM aims at taking maximum advantage of factors in the ecosystems that regulate pest numbers, e.g., natural enemies, competition between species, weather hazards (heat, cold, drought, or rain), and limitations of food or shelter. Particularly important in the suppression of many species of insects and mites are natural predators. Considerable progress has been made in careful evaluation of the effects of pesticides on predator populations. The aim is to identify pesticides with minimal effects on insect predators while providing at least partial control of the target pest. Even these will be used only when pests exceed the economic injury level. Besides trying to enhance control by naturally occurring enemies, there is increased interest in identifying and introducing new enemies, as discussed later in the chapter.

4. *Any control procedure may produce unexpected and undesirable effects.* Before and after implementation of a new pest management procedure, potential side effects must be evaluated. For example, before a new predator is released, it must be shown that it is not detrimental to other crops, the environment, people, or animals. Before releasing for planting, a new cultivar bred for resistance to a particular disease must be tested for susceptibility to other diseases, insects, and mites. These potential problems are not limited to IPM strategies. Examples of unexpected and deleterious effects of chemical pesticides are common and have involved the decimation of pred-

ator populations as well as detrimental effects on the environment. Indeed some of the side effects of chemical pesticides have provided impetus to the IPM approach.

5. *An interdisciplinary systems approach is essential.* A successful IPM program for a farm or a forest must be an integral part of the overall management scheme. By its very nature, IPM cannot be compartmentalized into entomology, plant pathology, plant physiology, agronomy, horticulture, and forestry but must pool information from all related disciplines. Much effort is aimed at developing computer models to simulate the entire ecosystem under study. As more information is gained about each of the many interacting factors, predictions will become more precise. For example, just a list of a few of the factors affecting pest management in a soybean field (light, temperature, rainfall, cultivar, time of planting, fertilization, and plant spacing) makes it obvious why the use of computers is a vital part of modern IPM.

The ultimate effectiveness and economic advantage of IPM programs remain to be proved in most cases, but the basic premise is widely accepted.

INSECTS

Insects are found in most parts of the world and appear from fossil remains to have been on earth for at least 250 million years. The number of insect species in existence is not known but certainly exceeds 1 million. By current estimates, insects account for approximately 70 to 80 percent of the total number of animal species. In the United States and Canada alone, entomologists estimate that there are in excess of 80,000 species, but their distribution varies with locality and climate. Although the effects of most insect species are either beneficial or neutral, insect damage combined with control costs runs into billions of dollars annually.

In the animal kingdom, the phylum Arthropoda includes insects, mites, spiders, and similar animals. Insects typically reproduce sexually, the fertilized eggs being laid in a diversity of places, depending on the type of insect. The number of eggs laid varies with the species from one to thousands. In certain species, the female is capable of *parthenogenesis*, producing offspring without males. In certain aphids (not true insects) although normal sexual reproduction occurs in the fall, the young are produced parthenogenetically during the summer.

A newly hatched insect may or may not resemble the adult of the species. Some species go through *metamorphosis*, a series of distinct changes of form from the young to the mature insect, whereas others merely enlarge, changing little in appearance or form. In types showing a *complete metamorphosis*, the distinct stages are the egg, larva, pupa, and adult. The Japanese beetle (*Popillia japonica*) hatches from the egg to become a small white grub (larva) in the soil. During its development, the larva molts, i.e., sheds its outer skin, 3 times and then is full-grown. The larva pupates, or changes to the pupa, an inactive stage that neither feeds nor moves about while its larval structure is transformed into that of an adult. The adult beetle emerges from the soil and feeds for 4 to 6 weeks (Fig. 12-1). It then burrows into the soil and lays eggs from which the next generation of larvae develops. Although the Japanese beetle has only one generation per year, other species may have several. Other insects exhibiting complete metamorphosis are butterflies, beetles, and flies. With some, like the Japanese beetle, both the larva and adult

FIGURE 12-2
Cabbage plant damaged by feeding of cabbage looper.

cause damage; with others, e.g., the European corn borer (*Ostrinia nubilalis*) and cabbage looper (*Trichoplusia ni*), only the larval stage causes damage (Fig. 12-2).

Insects that undergo a gradual or slight metamorphosis, e.g., grasshoppers and leafhoppers, resemble adults when they hatch. The major changes are in size and the growth of wings in winged species. A limited number of species, e.g., the silverfish, undergo no metamorphosis, the young being essentially minute forms of the adult. No agriculturally important insects are included in this group.

Benefits

Many insects assist our efforts to feed the expanding world population. Commercial producers of such crops as cherries, apples, cucumbers, melons, and alfalfa seed are almost totally dependent on honeybees (*Apis mellifera*) for pollination, without which there are neither fruit nor seed. Although wild bees are effective pollinators, they are usually too few for the task, especially for tree fruits, where the entire crop may need to be pollinated during a few hours of good weather. The ladybird beetle (*Hippodamia sinuata*), commonly referred to as the ladybug, is another beneficial insect feeding on aphids, and there are many other predator insects and mites. As emphasis on biological rather than chemical control of insects grows, the importance of insect predators is likely to increase.

FIGURE 12-1
Japanese beetles on a peach branch. Note the large mass of beetles as well as the damage to the leaves. (*U.S. Department of Agriculture.*)

Damage

Although insects are classified in several ways, we use a system based on mouth parts, the two major groups being those with *chewing mouth parts* and those with *sucking mouth parts*. This distinction is important not only for the type of damage resulting but also in selecting an effective chemical control, as discussed later in the chapter.

Chewing Insects The damage caused by chewing insects is readily apparent. The total devastation that can result from large populations of voracious grass-hoppers is legendary. They denude the land and leave the soil barren, exposing it to erosion by wind and water. Normally the damage chewing insects cause in most crops is much less striking but still may be of major economic proportions. Important crop pests include the Mexican bean beetle (*Epilachna varivestis*), Colorado potato beetle (*Leptinotarsa decemlineata*), and Japanese beetle, all causing damage as adults. Important insects destructive in the larval stage include the cotton boll weevil (*Anthonomus grandis*), the cabbage looper, corn ear worm (*Heliothis zea*) (Fig. 12-3), codling moth (*Cydia pomonella*), elm leaf beetle (*Pyrrhalta luteola*), leaf miners, and the gypsy moth (*Lymantria disper*).

Major belowground damage can also result from chewing insects, usually from larval stages of a variety of insects, including weevils, maggots, borers, and cutworms. Root damage is often severe before the top of the plant shows symptoms clear enough

FIGURE 12-3
Corn earworm damage to sweet corn.

FIGURE 12-4
Apples infested with San Jose scale; the adult females are covered by a waxy scalelike covering. The injury caused by each insect causes a red ring on the skin.

to lead the grower to check on the cause. Control may therefore be especially difficult because the problem is underground and difficult to treat and the insect population may be large before symptoms become apparent.

Certain chewing insects are particularly damaging and hard to control because they feed on internal parts of the plant. Eggs are laid on or near the surface, but the young larva quickly enter the plant, where they are difficult to control. Examples are leaf miners, borers, and bark beetles, of which there are many species. Damage to stored seed or grain by chewing insects may be severe enough to render the material useless as seed, food, or feed.

Sucking Insects Although the injury resulting from the feeding of sucking insects and mites is not so immediately apparent, the damage is still tremendous. Groups of major importance include the aphids, scale insects (Fig. 12-4), mites (considered here for convenience although they are not true insects), leafhoppers, mealybugs, and whiteflies. Some, e.g., aphids and mites, are troublesome both in greenhouses and in the field; others, e.g., whiteflies are primarily troublesome in greenhouses.

Sucking insects have mouth parts that resemble a hypodermic and cause injury by extracting cell sap or phloem juices from the plant. The damage usually shows up as discoloration, curling, and malformation of leaves and stem tips. When an infestation becomes severe, growth is suppressed and the overall productivity of the plant drops. Many of these insects excrete a sugary substance called *honeydew*, which supports an unattractive fungal growth.

Other Insect Damage The damage discussed above results largely from feeding activities. Some insects cause commercially significant damage by egg laying and the spread of disease. The periodical cicada (*Magicicada septendecim*) spends most of its life in the soil as a larva (*nymph*) and emerges as a adult only once every 13 or 17 years, depending on the race. (The fact that there are several broods explains why they appear more often than every 13 or 17 years.) The aboveground damage occurs because in laying groups of eggs in twigs of hardwood trees, each female makes a slit deep enough to cause the twig to break. With heavy populations of cicadas, injury can damage trees seriously, especially young ones.

Many plant diseases are spread by insects, particularly those with piercing or sucking mouth parts. During the late 1950s and 1960s a disease called *pear decline* killed thousands of pear trees in the western United States. After intensive research to rescue the industry, pear decline was found to be caused by a viruslike organism, or mycoplasma, transmitted by the pear psylla (*Psylla pyricola*). Except for using rootstocks with resistance to pear decline, the best method of control is to minimize the population of pear psylla. Other important insects which disseminate plant diseases are leafhoppers, aphids, thrips, and whiteflies. Mites can also transmit diseases. Important diseases and associated insect vectors include Dutch elm disease (*Ceratocystis ulmi*), a fungal disease transmitted by elm bark beetles, or fireblight of apples and pears (*Erwinia amylovora*), a bacterial disease transmitted by bees and other insects, and tobacco mosaic virus, which can be transferred from one plant to another by aphids and other sucking insects.

NEMATODES

Nematodes, also known as threadworms, roundworms, and eelworms, cause widespread injury to crop plants. Plant-parasitic nematodes are generally about 1 mm long, although they vary in both size and shape. The eggs laid by females hatch into nymphs, or juveniles, which go through four molts,

after which they are adults. Plant nematodes have a *stylet*, or spear, used to puncture cell walls, after which a secretion is injected. It presumably contains enzymes that liquify cell contents, making them easier to ingest. The secretion may also alter cellular metabolism and cause suppressed cell division, cell proliferation, or necrosis. The visible effects of these changes are often used in classifying nematodes into such categories as root-knot (Fig. 12-5), cyst, lesion, and stubby-root nematodes.

Although certain symptoms are peculiar to nematodes, the general effects may be similar to those caused by a number of other plant pests. The typical symptoms of nematode damage to roots are lowered plant vigor, lessened resistance to adverse environmental factors, and low productivity. Of less impor-

FIGURE 12-5
Tomato plant infested with root-knot nematodes. (*U.S. Department of Agriculture.*)

tance than the root feeders, nematodes that feed on leaf, stem, or bud may cause malformed stems and leaves, necrotic spots on leaves, and leaf galls.

Nematodes also serve as vectors of some virus diseases. The dagger nematode has recently been indicted as the vector for tomato ring-spot virus which causes serious diseases in both apples and peaches.

DISEASES

The yield and quality of crops are reduced by a broad spectrum of plant diseases, but the definition of a diseased plant often depends on who is giving the definition. To a plant pathologist, a diseased plant may be a plant whose anatomy, morphology, or physiology is sufficiently abnormal to cause visible symptoms or to result in lowered growth, yield, or quality. With such a broad definition, adverse environmental conditions, nutrient deficiencies, and air pollutants often lead to "diseased plants." Other crop specialists generally prefer a more restrictive definition and limit the term *diseased plant* to one which is abnormal as a result of an infectious *pathogen*, i.e., a microorganism that causes a plant disease. The three major groups of plant pathogens are the fungi, bacteria, and viruses. Other pathogens (discussed only briefly here) are classified as actinomycetes, mycoplasmas, and viroids.

Fungi

The diseases causing the most widespread and serious damage to crops are those caused by fungi. The financial loss from fungal diseases is heaviest and the number of different diseases is greatest. Fungi are usually considered members of the plant kingdom.

Not all fungi are harmful. Beneficial types include those used in cheesemaking and for producing penicillin and other antibiotics; wine and bread yeasts; edible mushrooms; and the fungi that decompose organic matter. Of special importance are the *mycorrhiza*, which enter into a symbiotic relationship with the roots of some plant species and help them in the uptake of water and nutrients. In turn, organic substrates, such as carbohydrates, move from the roots to the fungus. The fungi are divided into four classes, largely on the basis of how spores are produced and the appearance of the fruiting body,

FIGURE 12-6
Tomato leaf showing the dark water-soaked spots characteristic of late blight. (*U.S. Department of Agriculture.*)

namely the Phycomycetes, the Ascomycetes, the Basidiomycetes, and the Deuteromycetes (Fungi Imperfecti).

Phycomycetes Characterized by having no cross walls in the *mycelium* (fungal filaments), these are considered to be rather primitive. Diseases caused by members of the Phycomycetes are late blight (*Phytophthora infestans*) of potato and tomato (Fig. 12-6), some downy mildews (Fig. 12-7), and some soft rots of fruits.

Ascomycetes These are somewhat more advanced on an evolutionary scale, have cross walls in their *hyphae* (the mycelium mass), and also form *asci*, saclike structures in which the ascospores (sexual spores) develop. This class includes the largest num-

FIGURE 12-7
Early stages of downy mildew on cucumber leaf. The yellow-green spots eventually kill the older leaves. (*U.S. Department of Agriculture.*)

ber of disease-causing fungi, e.g., anthracnose of soybean (*Glomerella glycines*), common leaf spot of alfalfa (*Pseudopeziza medicaginis*), Dutch elm disease (*Ceratocystis ulmi*), apple scab (*Venturia inaequalis*) (Fig. 12-8), and black spots of rose (*Diplocarpon rosae*).

Basidiomycetes These, the most highly developed group of fungi, are characterized by sexual spores found in club-shaped structures called *basidia*. This class contains the very damaging diseases called *rusts* and *smuts*, e.g., white pine blister rust (*Cronartium ribicola*), cedar-apple rust (*Gymnosporangium juniperi-virginianae*), and corn smut (*Ustilago zeae*).

Deuteromycetes Also called Fungi Imperfecti, the fourth class is not clear-cut, as there is partial overlap with the other three; they are in this class because no sexual production of spores is known. Representatives of this class cause anthracnose of beans (*Colletotrichum lindemuthianum*), *Verticillium* wilt, *Botrytis* rots of fruits and vegetables, and early blight of potato.

The spread of fungi is largely passive since only in certain types and parts of life cycles are fungal cells motile. Some Phycomycetes form spores with

flagella which can propel themselves. In rare cases, spores are discharged forcibly. Thus the spread of fungi whether by spores or fragments of mycelium relies on such agents as wind, water, animals, and insects. The numbers of spores produced are tremendous and since they are small enough and light enough to be easily carried on wind and water, they spread rapidly.

A Case Study A widely studied fungal disease, stem rust of wheat, caused by *Puccinia graminis*, has been a major problem in wheat areas of the central United States from Texas to the Dakotas and into Canada. The fungus attacks both the stems and leaves. A heavily infected plant may mature early and lodge (fall over); with lighter infections, kernels are few and small.

The disease cycle of stem rust is complex, involving the production of several different types of spores. As infected wheat plants mature, the fungus produces spores which will infect only the *alternate host*, American barberry. Following fertilization, a second type of spore is produced, which is windblown to the wheat plant, the only plant it can infect. A third type of spore, produced by the fungus on the wheat plant during the growing season, is also windblown and infects other wheat plants.

The survival of the different types of spores varies with the climate. In the early 1900s, an effort was begun to eliminate the alternate host, barberry. Al-

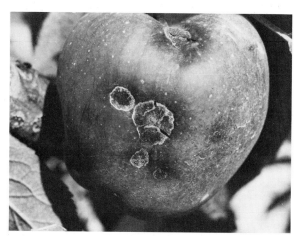

FIGURE 12-8
Apple scab on 'Delicious' apple.

though partially effective in northern areas, this approach was not a total success. In Texas, one type of spore can overwinter in the absence of the alternate host, and with strong southerly winds the disease spreads northward for hundreds of miles. Therefore, for adequate control of stem rust of wheat, resistant cultivars were developed. Fungicides can be used, but the vast expanse of wheat fields makes application impractical or the cost prohibitive.

Bacteria

Bacteria are one-celled, microscopic plants varying in shape from spherical to spiral and tubular. Although each bacterium cell is independent, bacteria usually grow in colonies. Essentially all plant-pathogenic bacteria are rod-shaped, usually 0.4 to 0.7 μm in diameter and 1 to 3 μm in length. Bacteria differ from fungi in that they lack a membrane-bound nucleus, but like the fungi they contain no chlorophyll; many have flagella. Reproduction is usually by cell division and under optimum conditions can be rapid. The diseases induced by bacteria are blights, rots, galls, leaf spots, and vascular wilts.

Some bacteria are beneficial to people, especially the nitrogen-fixing bacteria, which infect legumes and convert atmospheric nitrogen into forms usable by plants. Other bacteria are valuable as decomposers of plant litter.

Bacteria enter plants in many ways, the most common sites for entry being natural openings and wounds. The fireblight bacterium (*Erwinia amylovora*) enters the apple or pear tree through the blossom, presumably through the nectaries or other natural openings. The bacterium causing blackleg of cabbage (*Phoma lingam*) is thought to enter through the hydathodes, natural openings in the leaves through which liquid water may be exuded at night. Many of the rot-causing bacteria enter the host through wounds. Once inside the plant, bacteria are carried by the sap stream; motile cells may be propelled by flagella over short distances.

A Case Study Fireblight, an important disease of apple and pear, serves as an example of a bacterial disease. The symptoms of fireblight occur on blossoms, leaves, shoots, and spurs and in severe cases kill the whole tree. When a flower becomes infected, it takes on a water-soaked appearance, and within a few days, the whole spur is often affected and turns brown. Infected terminals wilt; the leaves and shoot

dry and turn brown to black. A characteristic crook in the terminal shoot and the persistence of dead leaves are common symptoms. Depending on the susceptibility of the tree, the bacterial infection may be confined to small shoots and spurs or may kill scaffold limbs or entire trees. When fireblight bacteria are multiplying during wet weather, a liquid *ooze*, which can often be seen coming from infected tissue, contains bacterial spores.

The life cycle of fireblight starts in the overwintering cankers on branches. Ooze formed in early spring is spread by rain and insects, the primary infection occurring in the open blossoms. Secondary infections occur as the bacteria are spread by bees, other insects, or rain. A major influence on the severity of fireblight is temperature, as warmth encourages rapid spread. Further infections occur through injuries caused by insects feeding or hail. The susceptibility to infection declines after bloom because the flowers are gone and the vegetative tissues gradually harden and become less easily infected. Unless they are removed by pruning, the cankers provide inoculum during the following spring. Care must be used in pruning to avoid spreading the disease with the tools used. Disinfectants are sometimes used to clean the tools, but if cuts are made several centimeters below the visible infection, the tools should not be contaminated.

Viruses

Viruses are generally considered to be infectious nucleoproteins. Isolated virus particles are not capable of metabolism; since they do not reproduce by growth and division, they are not considered to be living. Viruses reproduce only within living host cells, whose metabolism the virus redirects to produce new virus particles. Plant viruses typically contain nucleic acid (DNA or RNA) surrounded by a protein sheath. Since virus particles are not visible under a light microscope, their structure was not known until the electron microscope was developed.

Although various systems of nomenclature for viruses have been suggested, the one most widely used is based on the common name of the disease and the host in which it was first described, for example, tobacco mosaic, tomato ring spot, cucumber mosaic, and citrus tristeza. The plant symptoms caused by the *mosaics* include a mottled appearance of leaves and sometimes flowers (Fig. 12-9). Portions of the tissue become chlorotic and may be small or

FIGURE 12-9
Leaves of cucumber (*top*) and tobacco (*bottom*) infected by cucumber mosaic virus. (*From J. C. Walker, Plant Pathology, 3d ed., McGraw-Hill, New York, 1969.*)

involve the whole leaf. Some plants may also exhibit dead (necrotic) spots on leaves, fruits, and stems.

The transmission of viruses occurs in several ways. Tobacco mosaic virus is readily transmitted by mechanical means. Since the sap from an infected plant will infect a healthy plant if rubbed on the surface or injected, it is quickly spread by workers suckering tobacco or tomato, i.e., removing unwanted shoots from the lower stem. Most viruses are much less readily transmitted by mechanical means. For viruses affecting horticultural plants, transmission by vegetative propagation is particularly troublesome. Except for some seeds, essentially all cells of an infected plant contain the virus particles, so that propagation by cuttings, grafting, or budding gives new plants infected with the same virus or viruses as the mother plant. The mature seeds of most plants do not contain the virus; there are exceptions, but fortunately seed propagation often avoids virus problems.

The most important insect vectors of viruses are the aphids (instrumental in transmitting the mosaic group), thrips, and leafhoppers. Some virus transmission also occurs from mites and nematodes.

Mycoplasmas

The mycoplasmas are very small parasitic organisms known to cause a range of plant diseases. Between viruses and bacteria in size, mycoplasmas have a structure consisting of protoplasm surrounded by a membrane but no cell wall. Unlike viruses, mycoplasmas are living entities, which reproduce and have enzyme-catalyzed energy systems.

Mycoplasmas occur largely in the phloem of the host plant, where they interrupt normal translocation of photosynthates. Many diseases formerly attributed to viruses are now known to be caused by mycoplasmas. Common symptoms include yellowing, stunting, leaf curling, and excessive branching. Diseases include peach yellows (Fig. 12-10), aster yellows, corn stunt, and pear decline. The name of mycoplasmas usually reflects the symptoms exhibited by the host plant from which it was first isolated. The spread of mycoplasmas, like viruses, is largely by sucking-insect vectors such as leafhoppers, aphids, and psylla.

Viroids

These disease-causing molecules are only about one-tenth the size of the smallest known virus particle. They consist of low-molecular-weight RNA but lack the protein sheath characteristic of viruses. Another name for these molecules is *infectious RNA*. Diseases attributed to viroids include citrus exocortis and potato spindle tuber.

WEEDS

Weeds are not merely a nuisance: they suppress crop yields and cost millions of dollars to control. A precise definition of a weed is hard to find, but one of the most useful is "a plant growing where it is not wanted." It is common for a plant to be considered a weed in some situations but a desirable plant in others (Fig. 12-11). Certain plants, e.g., pigweed, are always weeds, but others, e.g., bermudagrass, can be a turf and pasture crop but also an undesirable weed in a vegetable field. Weeds can be detrimental in many different ways, but the damage can be grouped into five main categories.

Reduction in Crop Yields

Because of their great numbers and rapid growth rate, weeds effectively compete with crop plants for moisture, nutrients, and light (Fig. 12-12). The effect of weeds on the crop can range from slightly reduced

FIGURE 12-10
Peach yellows. Symptoms include deformed leaves, multiple upright shoots bearing small chlorotic leaves, and small insipid fruits. In 1 to 5 years, the tree dies. (*U.S. Department of Agriculture.*)

FIGURE 12-11
In a field of soybeans, a corn plant is considered a weed.

yields to total crop failure. The degree of damage varies but is often most devastating where moisture or nutrients are barely sufficient for the crop and thus the competition is most serious. Weeds are often particularly competitive when they have the C_4 pathway of carbon fixation (Chap. 5).

Contamination
Weeds harvested with food or feed crops can markedly lower the quality of the crop. They can be both difficult and expensive to remove. Since contamination of crop seed with weed seed can be troublesome, reputable seed growers make every effort to produce seed as free as possible of weed seed. A wise expenditure of money is the extra percentage paid for certified seed, which has minimal contamination by weed seed.

Hosts for Insects and Disease
Weeds serve not only as hosts for insects and diseases but in some cases as alternate hosts. For example, cedar-apple rust must have both apple and red cedar plants to complete its life cycle. Before modern fungicides became available, the main control for

cedar-apple rust was the mass removal of the red cedar; now, however, effective fungicides have largely eliminated that necessity. The elimination of barberry, the alternate host for stem rust of wheat, provides at least partial control of this important disease. Certain weeds may be hosts for insect pests of crop plants, and weedy areas often harbor troublesome insects and encourage their reproduction.

Toxicity
The presence of poisonous plants such as larkspur and lupine in forage can make the crop toxic to livestock. Under certain conditions, other usually desirable forages such as sudangrass can be toxic. In horticultural crops, the problem is usually most serious with poison ivy, which can cause discomfort or severe allergic reactions in many people who touch it. A perennial climbing vine, it is particularly troublesome in orchards, around fence rows, or any area not cultivated, tilled, or mowed closely. Adequate control of poison ivy is also important in pick-your-own fruit and vegetable operations.

Appearance
Weeds can be very undesirable from an aesthetic standpoint. Millions of dollars are spent annually in controlling weeds in lawns, gardens, golf courses, and other area where weeds detract from the appearance.

FIGURE 12-12
Compare the rows of potatoes on the left, which were treated with a herbicide, with the untreated rows on the right.

PEST-MANAGEMENT STRATEGIES

Considering the staggering array of plant pests, it is not surprising that there are also many means of pest management, including legal, genetic, cultural, physical, chemical, and biological strategies. Modern IPM programs tailor the mix of these options to the particular situation.

Legal Controls

Complete exclusion of a specific pest from the entire country is the ultimate means of control. Most governments have passed laws and set up agencies to attempt to prevent the entry of new insects, diseases, nematodes, and weeds. In 1912, Congress passed the Plant Quarantine Act, which gave the Department of Agriculture the authority to regulate the entry of plant material. More recently the legislation has been expanded to include pests of animals, and the agency is called the Animal and Plant Health Inspection Service (APHIS).

Both commercial and private entry of plants is regulated by APHIS, from whom a permit must be obtained for the commercial importation of potential pest-carrying materials. The shipper must have the material certified as being free of damaging pests. APHIS inspectors are often stationed at major points of origin, such as the Netherlands, which ship large quantities of plant material to the United States. Some imports must be fumigated to minimize the risk of contamination.

Individual travelers are also potential importers of plant pests. A person bringing back fresh fruit, vegetables, or plants from foreign countries may unknowingly bring in pests. As citizens or tourists enter the United States, they are questioned and their luggage subjected to search for potential pest-bearing materials. Such items are surrendered or confiscated and destroyed. Plant quarantine inspectors monitor airports, seaports, and other points of entry. A major part of this total program is educational. Most people who are made aware of the potential damage from foreign pests willingly cooperate.

Foreign plants are sometimes imported for the introduction of new and potentially useful plant material. They may be introduced or utilized in breeding programs. The plants are imported in cooperation with APHIS and grown in isolation at a plant quarantine facility until they are known to be free from pests.

Interstate movement of plant materials is regulated by both federal and state statutes. These laws are aimed at preventing or at least slowing the movement of pests into areas where they are not already established. Such regulations affect not only commercial movement by truck, rail, and air but also mail shipments and hand-carried plants.

The U.S. Department of Agriculture also directs programs to eradicate pests which appear in spite of efforts to exclude them. Several outbreaks of the Mediterranean fruit fly have occurred over the past few decades. The first step in complete elimination is to quarantine shipment of any plant material which may carry the insect out of the affected area. An intensive program is immediately initiated to eradicate the fly while it is still localized. Several attempts have been successful in both Florida and California since the late 1920s. After the eradication of the pest has been confirmed, quarantines on shipments are lifted.

Although legal methods of pest exclusion or containment are not totally successful, they have prevented millions of dollars in potential losses.

Genetic Controls

Major strides have been made in breeding crop plants with increased resistance to pests. A prime asset of genetic resistance is that it is usually the most economical of the pest-control strategies. Much of the progress in developing pest resistance has been with annual crops, because new cultivars can be developed in a few years, whereas similar progress with perennials may take decades. Even with tree fruits, however, genetic resistance to some pests is a reality.

In breeding for pest resistance, the greatest progress has been made in the development of cultivars able to resist disease attack. Major diseases for which resistant cultivars are available include

Fungal—wheat rust, southern corn blight, corn smut, verticillium and fusarium wilts of tomato and related crops, downy mildew of soybean, red stele of strawberry, apple scab
Bacterial—bacterial wilt of corn; bacterial canker of almond, cherry, and peach; fireblight of pear
Viral—curly top of sugarbeet, tobacco mosaic

Although work by plant breeders on insect resistance has lagged behind that on disease resistance, significant progress has been made. Alfalfa cultivars are available with resistance to the spotted alfalfa aphid, as are wheat cultivars with resistance to the wheat stem sawfly and the Hessian fly. Apple rootstocks developed in England have genetic resistance to woolly apple aphids.

Nematode resistance is of growing importance in plant breeding and selection. 'Nemaguard' is a peach rootstock resistant to attack by most species of the root-knot nematode. Resistance to nematodes is also being bred into such crops as tomato, soybean, tobacco, pepper, and sweet potato.

Even genetic resistance to pests has its shortcomings. Since diseases, insects, and nematodes are living entities, they are characterized by genetic diversity and change. New strains, or races, of a pest organism may arise and be able to overcome the resistance of a particular cultivar, forcing plant breeders to develop new cultivars with resistance to the new races. (Similar genetic shifts have markedly reduced the effectiveness of many pesticides.)

With the growing emphasis on IPM, breeding for pest resistance will play an increasingly important role in disease and insect control in the future.

Cultural Controls

Among cultural methods for pest control, crop rotation is the one most frequently cited. If a crop such as corn, cotton, or tobacco is grown in the same field each year (monoculture), pests of that specific crop tend to build up and may become progressively harder to control. Changing to other crops which are not as susceptible to the same insects, diseases, or nematodes will cause the pest population to decline. In actual practice, however, crop rotation may not be a viable alternative, either because of economics or because it is not effective.

From an economic standpoint, many modern farms are designed to produce large quantities of one crop or, at most, a very limited number. A farmer in the corn belt has the farm, the equipment, and the management skill to grow corn. Even if it is desirable from a pest-management point of view to rotate many crops, it is not practical for the farmer.

The effectiveness of crop rotation in suppressing pest populations varies greatly with the specific organism. Pest species that survive in soil for many years without the presence of a susceptible host are called *soil inhabitants*. Those which cannot survive for long in a soil without susceptible hosts are called *soil invaders*. Simple logic tells us that crop rotation will be effective only against soil invaders. Since many insects, diseases, and nematodes attack a wide range of host species, selection of a suitable crop to use in a rotation is often difficult.

Traditional production methods of many field crops such as corn involve plowing crop residues under. This frequently gives improved insect and disease control. In minimun and no-tillage systems, however, the crop refuse often remains uncovered and both insect and disease problems increase to some degree. This is but one of many examples showing how one change in a production system may affect total crop management. It also underlines the need for cooperation between many disciplines in designing modern IPM strategies.

Some of the major crop diseases require the presence of an alternate host for the completion of their life cycles. We have already described the relationship between wheat rust and barberry on the one hand and apple and cedar on the other. For many years, efforts were made to eradicate barberry in major wheat-producing states and cedars in apple-growing regions. Cedars were not eliminated, and other ways to control cedar-apple rust were developed.

Roguing is the removal of infected plants from the field, orchard, or other crop area. Although impractical in a wheat or soybean field, with hundreds of thousands of individual plants, roguing is practiced in orchards. Peach yellows is a disease caused by a mycoplasma and is readily spread by insects. As soon as a peach tree exhibits symptoms of peach yellows, it should be quickly removed from the orchard and burned. This procedure removes a large source of inoculum and greatly reduces spread of the disease.

Another technique of cultural control, *trap cropping*, means planting a small plot of the crop about 2 weeks earlier than the main crop. It is effective in controlling damage to soybeans from bean leaf beetle and Mexican bean beetle in Virginia. The insects overwinter as adults and choose the earliest-planted soybean field upon which to feed and deposit their eggs. The small plot (2 to 3 percent of the land area to be planted to soybeans) concentrates the insects where they can be controlled with pesticides. Late-season beetle populations are reduced because so many of the overwintering adults are eliminated before they reproduce. Major benefits include reduced pesticide costs, less environmental contamination, and minimal damage to beneficial insects.

The timing of both planting and harvest can be used effectively to minimize damage from specific pests. Delayed sowing of winter wheat has been shown to avoid attack by the Hessian fly. Early planting and harvest of potato in California have avoided tuber attack by nematodes, which make late-planted potatoes unmarketable.

Other cultural methods such as flooding and burning are occasionally used. Cranberry bogs are readily flooded, and some insect control is achieved. Burning of crop residues has been recommended to destroy both insects and diseases. One has to balance pest control against the destruction of beneficial insects and the loss of organic matter, which would otherwise benefit the soil.

Cultural methods for weed control include pulling by hand, hoeing, cultivation, mowing, burning (flaming), and smothering. The oldest technique is hand pulling; although slow, it is still practical in small areas, e.g., flower beds. Hoeing was standard procedure for many crops, but it has generally been replaced by cultivation. Cultivators were first pulled by draft animals, and more recently by tractors. Many types of cultivators can be used with crops grown in rows. The aim of any cultivation system is to destroy weeds with minimum damage to the roots of the crop.

The tremendous number of weed seed present in most soils makes hoeing or cultivation almost endless. Burning has application in certain situations. With lowbush blueberries, for example, competing vegetation is kept in check by periodic burning of the area; the blueberries send up new shoots from the roots and reestablish themselves faster than competing vegetation. For certain crops, a special implement flames the rows to destroy the young weeds by directing the flame to avoid harming the crop. Of major importance in horticulture are many techniques which smother weeds. Organic mulches, e.g., bark, sawdust, ground corn cobs, and straw, not only provide excellent weed control but also stabilize soil temperature and conserve moisture. Black plastic (polyethylene) mulch, very effective in weed control, is used in vegetable production. Both the organic and black plastic mulches smother young weed seedlings by preventing them from getting light.

Where the control of weeds by competitive means is possible, it is both effective and probably the least expensive alternative. A good example of this technique is turf; where a good stand of grass is maintained, weed problems are minimal because germinating weeds tend to succumb to the competition of the grass or not germinate at all.

Physical Controls

With some plant pests, physical means of control are the most practical. For example, planting only disease-free seed or plants is one of the best means of controlling certain diseases. Crop seed is often produced with irrigation in arid regions, where fungal diseases are not a serious problem. Fruit-tree nursery stock is frequently grown in fumigated soil for control of both diseases and nematodes. Screen houses (structures like greenhouses covered with screen wire instead of glass) are used to produce disease-free foundation plants of strawberry. Fumigated soil and the screens exclude insect transmitters of virus disease.

Wire or plastic guards around tree trunks keep off mice and rabbits. Tall fences prevent deer from entering orchards and damaging the trees. Bands of a sticky substance placed around tree trunks trap crawling insects. Various devices capture and destroy specific insects, attracted to the trap by a particular wavelength of light, a specific color, or an odor.

Artificial temperature control can be used. Some grain elevators control insects with heat of 60 to 65°C. Hot-water treatment is used on bulbs, corms, tubers, and fleshy roots. Temperature control must be precise, because the lethal temperature for the insect or nematode may be quite close to that for the plant material. Soil pasteurization by heat, a common practice in greenhouses, gives good control of insects, diseases, and nematodes, but is obviously limited to a relatively small scale.

Chemical Controls

Insects Insecticides are classified as stomach poisons, contact poisons, or systemics. Some materials may fit into more than one category.

Since *stomach poisons* act when ingested by the organism, they work only on chewing insects. At present one of the major stomach poisons is carbaryl (Sevin). Because of its low toxicity for mammals and the broad spectrum of insects it controls, its use has been widespread since its introduction in 1956. From 1930 through the mid-1950s, the arsenicals, e.g., lead arsenate, were widely used stomach poisons.

Contact poisons need not be ingested because they are absorbed by the insect through its body surfaces. Although also effective against chewing insects, contact insecticides are most widely used for controlling sucking insects, e.g., mites, aphids, leafhoppers, and scale insects, not affected by stomach poisons. Common contact poisons include the organophosphates, such as malathion, which has relatively low mammalian toxicity, and parathion, which is highly toxic to human beings, animals, and insects.

Systemics are poisons absorbed and readily translocated throughout the plant, thereby giving effective control against sucking insects feeding on it anywhere. Some systemics are applied to the soil and taken up through the roots; others are applied to the aboveground parts for absorption and translocation. Systemic insecticides are not as effective against chewing insects, which may not ingest sufficient toxicant to receive a lethal dose.

Diseases The chemical control of diseases is based on two approaches: *protective* materials prevent the entry and infection by the organism; *therapeutic* chemicals control the pathogen after entry into the host plant. A recent development in disease control is the introduction of systemic fungicides, which have longer-term effectiveness because once they have been absorbed, they cannot be washed off by rain and they will be translocated to new leaves.

Many fungicides are available. Ionic forms of sulfur and copper were the first, but although effective against many diseases, both tend to be toxic to the crop plant. In the 1930s, the first organic fungicide was patented, and today most of the widely used materials are organic. They include several metal salts of dithiocarbamic acid, e.g., ferbam, ziram, and maneb. Another important type is the heterocyclic nitrogen compounds exemplified by captan. The newer systemic fungicides are also organic materials.

The protectant fungicides are applied in many ways, all directed toward having the fungicide present before appearance of the inoculum (Fig. 12-13). Treatment of seeds is widely practiced to protect young seedlings from organisms on the seedcoat as well as soil-borne pathogens. Most protectant fungicides are applied to the foliage and fruits of crops. The control of apple scab and peanut leaf diseases

in humid regions requires several protective fungicide applications. Apple growers are much less concerned with apple scab in the semiarid northwest than in other areas. Brown rot of stone fruits (peach, cherry, apricot, plum) also requires multiple fungicide treatments, especially as harvest approaches. Producers of many vegetable and agronomic crops apply fungicides to control late blight on potato and tomato, downy mildew of many of the cucurbits (cantaloupe, cucumber, and squash), leaf diseases in wheat, and others.

Therapeutic fungicides are less common because the first line of control is protection. When it fails, however, eradication of an existing infection is called for. With apple scab, infection sometimes results from excessive rainfall and inadequate protection. Infections in their early stages can be eradicated with a limited number of fungicides which serve as both eradicant and protectant.

Postharvest disease control is mandatory if fresh fruits, vegetables, and other plant products are to reach consumers in prime condition. Even with lettuce, spinach, and strawberries, which are marketed quickly after harvest, diseases can be troublesome. With apples, pears, potatoes, etc., which are stored for up to several months, good disease control may require both pre- and postharvest treatment with fungicides. As the product is washed after harvest, small quantities of a fungicide are often added to the rinse water. As with all pesticide treatments, only materials and application rates approved by the Food and Drug Administration (FDA) can be used on commodities destined for human consumption.

Nematodes Chemical control is widely adopted as the most effective method for nematode control despite relatively high costs. Soil fumigants are especially effective. Small quantities are injected into the soil and gradually vaporize. Chloropicrin (tear gas) provides good soil sterilization by killing fungi and weed seeds as well as nematodes. Methyl bromide is also effective against a broad range of soil pests. Since it is a gas at atmospheric pressure, methyl bromide is injected under a plastic cover to prevent escape. After a few days, the cover is removed and the fumigant allowed to dissipate. The field must be thoroughly disked and allowed to ventilate before the crop is planted. Such expensive procedures are economically feasible only for high-cash-value crops,

FIGURE 12-13
A fungicide application by helicopter to an orange grove near Lake Wales, Florida. (*U.S. Department of Agriculture.*)

such as strawberries, tobacco, or young fruit trees in a nursery. Some of the newer nematicides can be applied to the soil surface and provide good control.

Weeds By a wide margin, the most widely used techniques of modern weed control involve *herbicides*. Salt and other chemicals have been used for centuries to kill all vegetation, but only in the mid-1890s were the first *selective herbicides* discovered. These chemicals applied to a mixed stand will kill only certain plants, leaving others unharmed. Early work was with a variety of salts which would eliminate broadleaf weeds from grain fields.

The modern age of herbicides began in 1944 with the introduction of 2,4-dichlorophenoxy acetic acid (2,4-D). The first of the hormonelike herbicides, it is effective at low dosages, is somewhat selective, and is readily translocated in the plant. The tremendous potential of phenoxy herbicides was quickly realized, and many similar compounds followed. This group of herbicides is highly toxic to broad-leaved species but does not appreciably harm grasses. Use on a worldwide scale has been tremendous for weed control in grasses such as turf, cereal grains, and corn. Other organic herbicides include the amides, aliphatics, benzoic acid derivatives, carbamates, and a host of other groups.

Contact herbicides, which kill only the tissues they touch, are largely the inorganics, petroleum oils, and certain organics such as paraquat and diquat. A major use of paraquat is to provide rapid knockdown of existing vegetation, especially annuals. For minimum-tillage corn production (Chap. 17), paraquat is often used to kill the cover crop. Since paraquat is deactivated when it comes in contact with the soil, it has no effect on the germination or growth of the corn.

Translocatable herbicides are applied to the foliage or to the soil but are absorbed and translocated so that they affect the entire plant. Major herbicides applied to the foliage include 2,4-D, dalapon, dicamba, and glyphosate. Soil-applied herbicides, which are absorbed by the roots, include a very large number of commercial compounds.

Problems After the introduction of highly effective, relatively inexpensive chemicals for pest control, agriculture became increasingly dependent upon these chemicals and tended to look upon them as the ultimate weapon. It has become increasingly apparent, however, that chemical control is seldom, if ever, the total answer. Many pest species have developed resistance to specific pesticides and even to whole groups of related materials. Resistance has developed most rapidly when the same or similar chemicals have been applied repeatedly. A common recommendation is to alternate pesticides to slow the buildup of resistance to a particular chemical or group of chemicals.

The widespread use of broad-spectrum insecticides has controlled some insects but worsened the damage by others. For example, before the advent of modern insecticides, European red mites were a minor problem in apple orchards. Since new insecticides not only control the target pests but kill beneficial insects which had kept mites at low levels, European red mites have become a major problem in apple orchards.

Similar cases can be cited in chemical weed control. When orchards were mowed to minimize competition between the ground cover and the trees, such perennials as wild raspberry and blackberry were not a serious problem. The wide use of herbicides under the trees provides excellent control of grasses and annual weeds, but the perennials, with less competition, have become a major pest.

Other concerns with chemicals include consumer safety, environmental contamination, health hazards to applicators, and increasing costs. The first three are regulated by the Environmental Protection Agency. Before a new pesticide is approved for use, it must be proved to be safe to consumers of the ultimate product, the environment, and people applying the chemical and to be effective against a specific pest. Estimates of the cost to develop and receive label approval for a new pesticide range from 6 to 12 million dollars. Huge compilations of data on efficacy, residues, breakdown products, toxicology, etc., must be provided by the chemical company before receiving clearance.

As we move in the direction of a truly integrated program in IPM, chemicals will be more selectively used but will remain a vital component of pest management.

Biological Controls

Biological control is the manipulation of natural organisms or phenomena to suppress damaging levels of insects, nematodes, diseases, or weeds.† The greatest successes to date in biological control of plant pests have been with insects. Specific examples offer insight into the range of techniques which have been effective.

The first introduction and release of a natural insect enemy was successfully carried out in 1888, when the vedalia lady beetle (*Rodolia cardinalis*) was released in California to control cottony cushion scale (*Icerya purchasi*) on citrus. It was an immediate success and became the forerunner of hundreds of attempts all over the world, many of which have been successful and have greatly reduced the need for pesticide application. This approach is currently receiving widespread attention not only for predators but also for parasites and diseases of insects.

For many native insects, predators are already part of the ecosystem. When new crops and their pests are introduced, however, as early settlers did in North America, natural enemies may not accompany the crop and its major pests. In such cases, entomologists often travel to the native range of the host plant in search of potential control mechanisms for its pests. As mentioned under legal methods of pest control, such organisms must be proved to be host-specific and not harmful to other crops or beneficial organisms before they are released.

Less dramatic than the importation of a predator, manipulation of natural enemies already present is under intensive study. The development of biological control programs for specific crop pests is a goal in many parts of the United States. The specific natural enemies of major insects on a particular crop often vary with the region and climate. For example, mites are a major problem on apple throughout the United States. In many states, the predatory mite *Amblyseius fallacis* is an effective control whereas in Pennsylvania and Virginia, a beetle, *Stethorus punctum*, is a prime control species. The management of a long list of insect and mite predators is a major part of apple IPM programs.

Other biological control techniques being tested and sometimes used against insect pests are diseases and even nematodes. A bacterium, *Bacillus thuringiensis*, provides excellent control of many larvae of the order Lepidoptera, including cutworms, army

---

†Although breeding of disease-resistant cultivars could fit this definition, we chose to put such techniques in a different category.

worms, tent caterpillars, and cabbage loopers (Fig. 12-2). Another bacterium, *Bacillus popilliae*, is an effective control for Japanese beetle. Other disease organisms which infect insects, including viruses and fungi, are being studied for possible use in insect control. Some nematodes attack insects and may also have potential.

An even newer area is control of fungal and bacterial diseases through the use of antagonistic members of the same genus. Crown gall is a disease of many crops, including stone fruits and roses, caused by *Agrobacterium tumefaciens*. Researchers in Australia isolated agrobacteria, called Strain 84, which do not cause crown gall. When seeds and roots are inoculated with this isolate, crown gall is essentially eliminated. Strain 84 produces an antibiotic that selectively inhibits most pathogenic agrobacteria.

The release of sterile males has been effective in controlling screwworm, an insect that burrows under the skin of cattle and can kill them if infestations are severe. Male screwworms reared in captivity and sterilized by exposure to gamma radiation, when released, compete with the wild males for mating. Since the female screwworm mates only once in her lifetime, this technique has proved remarkably effective. The approach is being tested for other insects, e.g., the pink bollworm in cotton-growing areas of California.

Biological control of weeds has also been successful in some areas. The prickly pear cactus was introduced into Australia and became a serious weed. The Argentine moth borer was imported and released and has cleared large areas of this cactus. Thistles, a major weed problem in Virginia, are being partially controlled by weevils imported from France and Italy. Another success is the effective control of alligatorweed in parts of Florida, Alabama, Louisiana, and Texas through the introduction of a flea beetle (*Agasicles hygrophila*). This immersed or floating plant, which clogged waterways for years, has been reduced to a minor problem.

These examples of biological control are but an introduction to a rapidly developing field. As IPM programs are refined, it seems likely that biological control will play an increasingly important role.

ANOTHER LOOK AT IPM

Since an IPM program is a composite of pest-control strategies, a great deal of research is necessary to clarify the complex interactions in the specific agroecosystem before a new IPM program can be implemented. The economic injury level has to be established for the major pests, and the effects of each control method on both pest and beneficial organisms has to be understood.

Scouts are a vital link in the implementation of an IPM program. These highly trained specialists periodically determine the populations of pests as well as predators in the field or orchard. On the basis of data collected by the scouts, recommendations for specific actions are made. In the initial stages of an IPM program, funding is usually by federal and state agencies. As the benefits become apparent, however, the scouting costs are often paid by the producers as a per hectare fee. The number of self-employed crop consultants who advise growers on pest management is increasing.

CHAPTER THIRTEEN

CROP ECOLOGY

Ecology is the science dealing with the relationships between living things and their environment. The widespread interest in preserving the natural environment may lead some people to think that ecology and ecological principles apply just to plants and animals in the wild. Although we recognize that human activities affect particular ecological systems, as our influence becomes greater, we tend to assume that ecological laws are somehow suspended. This chapter should dispel that notion. Ecological principles† that apply in virgin forests, wild prairies, or untouched tundra operate just as fully and powerfully in Kansas wheat fields, Korean rice paddies, or Canadian greenhouses.

ECOLOGICAL PRINCIPLES AND CONCEPTS

Competition

Life is popularly viewed as a struggle in which each living thing survives by depriving other organisms of their life-sustaining needs. Such competition for various factors certainly exists, but in ecological systems, or *ecosystems*, the intensity of interaction between organisms is not always great, even when they are living side by side.

---

†Our discussion of key ecological principles and concepts is too brief to do them full justice. To learn more about them, refer to any good ecology text. The emphasis here is on the operation of those principles in agricultural settings.

Consider two different plants growing next to each other in a forest. Each requires certain things from its environment: light for energy to drive photosynthesis, CO_2 for use in photosynthesis, water to carry out all its important functions, minerals for nutrition, and space. Two adjacent plants seem inevitably to be in competition, and the closer they grow to each other the more intense that competition seems. For example, if one forest plant is quite tall, it may get enough light but at the same time taking light away from its neighbor. The lower-growing species appears to be at a distinct disadvantage for light, to be outcompeted and in danger.

Niche

Another ecological principle at work in the forest ecosystem means that the lower-growing species may not be at a disadvantage at all. Some plants cannot tolerate full sun and would be at a distinct disadvantage if they were *not* being shaded. Every plant, in the forest and elsewhere, has a genetically determined physiological need for, or tolerance of, certain conditions. The conditions peculiar to a plant that permit it to survive in a specific environment represent its *niche*.

In general usage, "niche" means a place, a physical or geographical setting, but in the strict ecological sense it refers also to the physiological, genetic requirement for such a place. Thus its meaning is more that of life-style or a way of fitting into an environment. In theory, each species expresses slightly different demands on its environment and therefore can fill a slightly different niche—one that is wetter or drier, cooler or warmer, shadier or sunnier, more acid or more alkaline, and so forth. It is because they have different niches to fill that two species can grow side by side without competition. One plant does well in the shade created by its neighbor because its niche, or life-style, is shady.

Very slight differences in the environment, sometimes created by the plants themselves, may make it possible for two different plants to exist side by side without competing to the detriment of either. Slight variations in *microenvironment*, the conditions found at a particular spot, favor one species here and another species there. Microenvironmental variation creates niches that permit a sort of peaceful coexistence between neighbors. Each species can capitalize on the microenvironment that favors its existence until a *population* of that species essentially fills the niche.

Adaptation

Both cacti and rice require water to grow, but cacti can grow at much lower levels of water availability; i.e., they are *adapted* to a drier niche. Paddy rice, on the other hand, grows in soil wet enough to kill most species (Fig. 13-1). The differences in the morphology and physiology of cacti and rice are so great that the two would no more be found growing side by side than fire and ice could coexist.

Apparently slighter differences in environment or microenvironment can have almost as dramatic an effect on plant distribution. Soil differences of only a few tenths of a pH unit may allow two species to grow side by side with no tendency to invade each other's territory. Peaceful coexistence brought about by diversity of niches is more the rule than the exception in natural ecosystems. The tremendous diversity of microenvironments available within a small area creates many different niches to be filled by various adapted plants, whose presence tends to diversify the microenvironment even more. A *community* of many different species develops throughout the various niches.

Succession

As plants modify the microenvironment, other plants can appear and occupy the new niches. This *succession* of species filling changing niches is another general ecological rule or principle. For a particular

FIGURE 13-1
Lowland or paddy rice, a type of plant adapted to continuously saturated soil conditions. (*Courtesy of E. Pehu.*)

climate and locale, succession is marked by a fairly predictable sequence of appearance and disappearance of particular plant (and animal) species. The particular species involved in a succession will vary with the geographical location and the initial state of the environment.

When a new piece of land appears, e.g., by volcanism, or when an existing community is essentially destroyed, e.g., by forest fire or human activity, the first plants to appear are called the *pioneer species*—lichens or mosses, whose spores were blown in from great distances, or hardy seed-bearing plants, whose seeds were delivered by wind or bird or survived in the soil. The niches occupied by such species often seem rather harsh, but the plants are well adapted to such microenvironments. In fact, some species flourish *only* in recently disturbed or burned-over soils.

Pioneer species alter their microenvironment. The changes may occur only very slowly, as with the weathering of rock by lichens and mosses; but the changes create new niches. New plants appear and fill those niches, and immediately begin to alter the microenvironment further. Eventually shaded microenvironments develop where once there was only glaring sun. Organic matter begins to accumulate and to have its highly desirable effects on the soil. Changes in pH, soil temperatures, soil moisture, available nutrients, and the like are occurring throughout the disturbed or revegetating area; and thousands of individual, peculiar microenvironments are produced.

At some point the rate at which additional species appear and microenvironments change is slowed. In theory at least, some set of species will come to occupy a more or less permanent or stable place in the flora of an area. The endpoint in plant succession is called the *climax vegetation* or *climax community*. The climax vegetation of a particular locale is the endpoint of the natural succession of species at that locale.

The number of different kinds of climax communities is relatively small. Large geographical areas were once characterized by a rather uniform and stable vegetation. The extensive prairies of the Great Plains, the coniferous forests of the northeast, and the deciduous forests of the southeast (Fig. 13-2) were large climax communities in ecosystems that covered millions of hectares.

Stable Communities

A climax community is characterized by a balance of components and efficiency of resource utilization, both working to ensure the stability and perpetuation of the whole community: plants, animals, and microorganisms. The balance is seen in the compatibility and diversity of niches. All available niches tend to be occupied and to blend into a harmonious whole. No species (plant, animal, or microorganism) increases beyond the stable climax level. In this balanced state, the climax ecosystem uses its life-

FIGURE 13-2
Undisturbed areas in the Great Smoky Mountains (and elsewhere) contain virgin forests—climax communities that are populated by a diverse group of organisms occupying a variety of niches. This gigantic specimen of a tulip poplar might be more than 300 years old. (*Courtesy of W. L. Daniels.*)

supporting resources (light, water, nutrients, etc.) in a most efficient way, and many are used to their maximum. For example, although the light levels on the floor of the forest are not sufficient to support the growth of many species, some green plants are found there and are capturing essentially the last usable fraction of light filtering down.

In a climax community, certain resources may be in especially short supply and likely to become a *limiting factor* for growth of all members of the community. If the resource is nonrenewable, it is especially susceptible to depletion. Natural ecosystems often avoid this threat by recycling the resource. A prime example of such conservation and reinvestment is the *nutrient cycling* that occurs in tropical rain forests. Although the soils in these environments are typically unfertile and the mineral portions of the soil may be deficient in many nutrients, some of the world's lushest, most productive, most stable climax vegetations grow there. This is possible largely because the plants are able to recapture nutrients from decaying plant material before they have a chance to enter the mineral layer of the soil. The roots of many trees, vines, and herbs proliferate throughout the rapidly decaying litter on the floor of the rain forest. Decomposition and mineralization take place through the activity of various microorganisms, other important members of the ecosystem (Chap. 10). As the litter decomposes, the living plants absorb the newly mineralized nutrients immediately, before they can be leached away. This tight cycling of key minerals reallocates the limited supply and keeps them from becoming depleted (Fig. 13-3).

Cycling may keep the ecosystem's nutrient supply continually available, but the cycle can be broken if the ecosystem is disturbed. The destruction of tropical rain forests is a prime example. Recently cleared tropical land may be productive for a year or two, but the soils are rapidly depleted of the nutrients released by decomposition (or burning) of the native vegetation. The land can quickly become sterile and nonproductive, a phenomenon of great concern in tropical agricultural systems. The nutrient cycles in temperate forests are not generally as "tight" as in tropical rain forests, and usually the soils under temperate forests are inherently more fertile; so the concern of nutrient depletion is not so great when clearing such land for cropping.

Stable ecosystems like a climax rain forest are in ecological *balance*. They do not tend to exhaust or

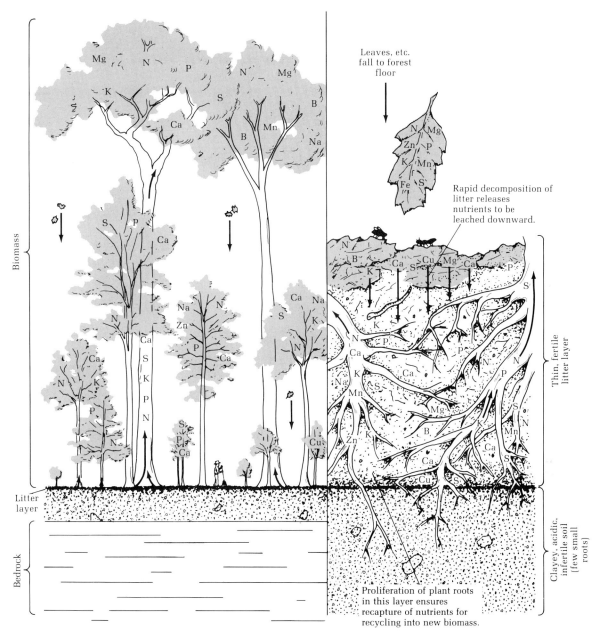

FIGURE 13-3
A rain forest ecosystem and nutrient cycling.

deplete the environment. Limited resources are especially well conserved. Plants and other organisms in a stable ecosystem return as much to their environment as they take from it. The only exception is sunlight, a constantly renewed resource that is provided from a more or less unlimited supply. Otherwise, natural ecosystems approach or reach a state of equilibrium, where outputs never exceed inputs and resources are continuously recycled.

Energy and Matter Flow

One more important ecological concept involves the *flow* of energy and matter through ecosystems. The energy can take various forms, but the most important to us and plants includes light energy and the chemical energy associated with complex organic compounds.

Light is the primary source of energy for biological systems. In photosynthesis, the energy of photons is converted into chemical energy. The plant can consume its photosynthates for energy to perform various functions, incorporate them as building materials into new plant matter, or simply store them directly or in chemically altered forms. Thus the photosynthates represent a source of energy to the plant or to any organism that obtains them from the plant.

Energy flows from sun to plant to plant consumer, which may be consumed in turn by other organisms. This idea of a *food web* or *food chain* has to do with the flow of energy and matter through an ecosystem. Food is important both for its energy content (calories) and for complex materials (proteins, vitamins, etc.). As the *primary producers*, plants stand at the beginning of food chains and provide both energy and complex materials directly or indirectly to all the consumers.

The photosynthates and other matter a plant incorporates or stores in its tissues constitute the plant *biomass*. More generally, biomass is the net biological dry matter (plants, animals, and microorganisms without their water content) accumulated in the system. The amount of biomass accumulated in a community in a given time reflects how efficiently the plants capture and transform the sun's energy and other organisms use that photosynthate; thus biomass is a measure of ecological productivity. In general, ecosystems that have progressed to a stable climax community produce the maximum biomass sustainable in a balanced, nonexhaustive way. To boost the biomass output the system must have additional inputs of resources that are in limited supply. The full significance of this idea will become more apparent as we discuss agroecosystems.

AGROECOLOGICAL PRINCIPLES AND CONCEPTS

Agroecosystems are populations or communities of harvestable crops, together with all other organisms and factors influencing them, that are being *managed* or manipulated to some degree by people. The "crop" may be plants or animals, but we concentrate on agroecosystems that are essentially plant-based. The distinction between natural ecosystems and agroecosystems becomes blurred in situations where management is limited, as in some timber stands. In other cases, the agroecosystem is so intensively manipulated that it might seem that all ecological rules would be suspended. Although the greenhouse environment is so artificial that it bears little resemblance to natural settings, the same principles of competition, limiting factors, niche, adaptation, resource utilization, and energy flow that apply in a climax deciduous forest also apply in a greenhouse.

While ecological principles certainly apply in agricultural settings, the very nature of human agricultural activity often tends to disrupt the continuity and balance that would otherwise develop. Succession cannot occur where fields are plowed yearly; the succession process is reset to its starting point each spring; however, an old field left untended will quickly show signs of succession, as various species appear and then are replaced by others.

The disturbances imposed on ecosystems in the name of agriculture are, of course, designed to be of human benefit. Removing natural vegetation, turning the soil, cultivating certain plants while excluding others, controlling insect populations, etc., can provide the landholder with food, fuel, and fiber.

Niches and Competition

The concept of niche suggests that the various organisms in a community are able to reduce or avoid competition by occupying different microenvironments. Each species is genetically and physiologically suited to particular conditions and tends to predominate where such conditions exist. In natural ecosystems, diversity of niches reduces the competition for potentially limiting factors.

But think now of a soybean field, where the farmer tries as nearly as possible to create a plant community consisting of a single species. If the farmer were perfectly successful, there would be no weeds on the land, the soil would be fertilized to uniform levels of the desired nutrients throughout, soil pH would be at the preferred level across the entire field, and all other growth factors would be optimal. In such an ideal situation, how many different niches would be there to be filled? Obviously, only one. Every spot in the field would be theoretically just like every other. Presumably, there would be no areas of greater sunlight or less water or areas where soil pH would be more favorable. Although this is a very unlikely situation, it is the grower's theoretical goal to produce a uniform, favorable environment and then to plant the species or cultivar that is best suited to that environment (Fig. 13-4). The corn belt of the central United States and the cotton belt across the southern United States are examples of such large-scale matches between crop and environment.

Continuing with the soybean field, even though a uniform and ideal environment for soybean is never achieved, the farmer plants only the single species. Therefore, even though there are microenvironmental differences, there is no diversity of species to take advantage of the differences. There are only soybeans, and that species will typically be represented by only a single cultivar in the field. Every soybean plant theoretically will have exactly the same genotype.

The idealized soybean field provides a single niche, something never actually realized; but the plants of a genetically pure cultivar all "seek" to fill a single niche, the one for which their genotype suits them. If a single niche is imposed genetically, what can be predicted about competition between the plants? It should be very keen. Two identical plants growing side by side cannot avoid competition, since each needs everything the other does. The more alike the plants the greater their competition.

Niche remains an important concept in many agricultural situations because it is almost "missing." To be sure, some complex agroecosystems use diverse plants and multiple niches (Chap. 17), but in the more familiar single-crop situation, only one crop species is grown and individually managed on a particular piece of ground. Niches are conspicuous by their absence, and competition is keen. Under this competitive system, limiting factors become all

FIGURE 13-4
An agroecosystem, a corn field, with essentially only one niche and a plant population consisting of only one species to fill that niche. A high degree of manipulation of the environment is inherent in agroecosystems. (*Courtesy of E. W. Carson.*)

the more crucial. If some factor (water, light, nitrogen, etc.) is needed in relatively large amounts by one plant, it is needed by all. There is no favorable accommodation between plants, all of which have the same demands. The grower must understand this fact.

When a species requiring abundant nitrogen to produce its grain is planted, special efforts must be made to keep that nutrient available throughout the critical nitrogen-requiring periods. During times of most active uptake and utilization, soil nitrogen can be depleted rapidly because every plant in the field simultaneously draws on the available supply of that potentially limiting factor. Nitrogen management must take into account the plants' demands and other factors, e.g., leaching that depletes nutrient supplies. With some grain crops, it may be desirable and economical to make several applications of nitrogen fertilizer to keep the nutrient at optimum levels throughout the plants' life cycle (Fig. 13-5). The need for split applications of fertilizer (Chap. 11) is occasioned in part by the single-crop, single-niche agroecosystem, a pattern that maximizes competition and accentuates limiting factors.

Creating and Matching Niches

We have already mentioned several things the grower does to improve the match between environment and the plant's genetically determined niche. In some

FIGURE 13-5
Spring fertilization of wheat. To achieve maximum production, some growers apply nitrogen fertilizers 2, 3, or even 4 times during development of the crop. (*Courtesy of M. M. Alley.*)

cases, the management is more or less passive; almost by default, for example, the grower must choose a crop as nearly suited to the climate as possible. One does not try to grow sugarcane in Ohio or sugarbeet in Hawaii. Major climatic factors (especially temperature in the case of the sugar crops) must be considered and accepted as fixed. The corn and cotton belts are areas of especially good match between environment and species niche. Growers naturally gravitate to producing certain crops in certain areas because of the good match between crop and climate, as well as for some economic reasons.

In other cases, the grower works actively to help create a more favorable environment: seedbed preparation (Chaps. 10 and 15), liming (Chap. 11), irrigation (Chaps. 9 and 17), fertilization (Chap. 11), and pest control (Chap. 12) are all efforts in that direction. Sometimes the management practices cause changes that seem subtle but have dramatic effects on plant growth. The management of the residues from a previous crop is one example. Leaving litter on the soil surface instead of turning it under can reduce soil temperature fluctuations in the spring by 50 percent and raise soil moisture in the critical seed zone two- or threefold, both usually desirable effects. The residues of some crops can also have adverse effects on the following crop. Litter from some plants can inhibit the growth of following crops; e.g., the straw and stubble from a wheat crop reduces the growth of a succeeding sunflower crop. Although there are some difficulties in trying to match and create the proper environment for a crop, the intelligent manager is constantly seeking ways to improve the fit.

Succession

The progressive appearance and disappearance of different species with eventual establishment of a climax ecosystem is missing in agricultural situations. (Naturally occurring and intelligently managed rangelands and forests may be the only types of climax community that have perpetual economic value.) Succession is generally undesirable in agriculture since the grower wants only particular crop species to grow during each cropping interval. Since succession has no regard for human economics, only for available niches, it will occur unless the manager takes definite steps to prevent it. The weeds that appear in a field or garden or flower bed are merely plants that have found a suitable niche. They will thrive and be succeeded by other weeds unless the grower alters the niche and blocks the succession. Similarly, many insect pests will fill available niches.

Succession is normally blocked or interrupted by the periodic destruction of vegetation on a piece of land. Plowing, disking, and herbicide applications kill many plants. Plants with underground portions that survive tillage or herbicides will reappear, and the seeds of many species will also survive; but turning the soil or killing the vegetation resets the ecological stage. Only species that can sprout and survive in the newly disturbed or bare soil will appear quickly. Others must wait until microenvironmental conditions again favor their germination and growth.

Cultivation, a shallow stirring of the soil to dislodge and destroy weeds, is an old practice. Although it has been replaced to a large degree by herbicides (Chap. 12), it remains an effective method of weed control. The practice not only turns up and dries out the roots of the current weed crop but since many weed seeds must be brought to the surface before they will germinate, it can be used in the fall to promote germination of weeds that will then be killed by frosts. The difference in numbers of weed seeds between tilled and untilled conditions is striking. Weed-seed populations in the soil may be reduced by 30 to 60 percent (up to 90 percent, with some species) per year with tillage. That is about twice the rate of decline in weed-seed populations under chemical weed control with no tillage. (Of course, with no weed control, one would expect weed-seed numbers to increase.)

The grower may never visualize cultivation or tillage as a succession-interrupting practice, but it is just that. Planting does not require tillage. Successful planting of many crops is possible without time-consuming and expensive plowing, disking, harrowing, and cultipacking to prepare a seedbed. Many agronomic crops are planted using a *no-till* method. Special planters open a narrow slit in the soil, drop in the seeds, and then close the slit, without *any* general preparation of the soil.†

The description of weeds as plants that appear where they are not wanted says nothing about them ecologically. Weeds are just as subject to the laws of ecology as other plants; they survive only in a suitable microenvironment in a suitable climate. In other words, they will occupy a niche. It is no coincidence that the most noxious weeds for any crop are generally those species whose niche closely resembles that of the crop. When growers create conditions to favor soybean, they are also creating a favorable environment for other ecologically related species. When those species appear and fill their niche, they are simply obeying the laws of succession and adaptation.

Many food crops are, in some sense, pioneer or early succession species. They can survive, produce, and be harvested successfully only under the conditions created by severe disturbances of the natural vegetation. Their typical niche requires full sunlight, so no taller species should grow with them. Their seeds typically must be *planted,* i.e., buried as opposed to *sowing* or scattering on the soil surface, so that at least some soil must be disturbed. Their generally high productivity usually requires high levels of water and nutrients, so that competition from other species often must be minimized and additional amounts of water and nutrients supplied. The farmer or gardener must go to considerable trouble and expense to create and maintain a productive environment for crops; in other words, the niche that many crop species fill must be artificially maintained to ensure high productivity. Good crop management includes practices that are designed to create in an otherwise favorable environment an early succession niche to match the crop's niche. Management for pests often means trying to create an unfavorable environment for the pest and depriving it of its niche (Chap. 12).

---

†No-till plantings are successful only if herbicides are used to control the weeds that would otherwise appear with the crop species.

Most weeds are weeds because they are also early successional species. In fact, they unfortunately tend to be hardier pioneers. Where ground is cleared, soil is disturbed, and nutrient levels are artificially boosted, the environmental stage is set for many actors, not just a particular crop. Dandelions, crabgrass, and pigweed are but three examples of weed species that need disturbed, open areas to become established (Fig. 13-6).

Succession, an important ecological force in agriculture, has both positive and negative aspects in cropping situations. It works to the grower's advantage when the successional stage is reset to favor crop species, but it is sometimes a nuisance in that many noncrop species also have early successional niches.

Energy Flow

In the final analysis, agriculture is a system for capturing energy from sunlight and transforming it into other forms which find use as food, fuel, fiber, or any of a hundred other plant-derived products that enrich and vary our lives.

The flow of energy and matter in agroecosystems can be described in a simplified sequence.

1. Energy flows into the system as light.

2. The light energy is used to raise various chemicals (ADP, NADP, etc.) to higher energy levels.

FIGURE 13-6
Weeds are typically plants that thrive under the disturbed conditions associated with agriculture.

3. The chemical energy is used to assimilate simple matter found in the system (H_2O, CO_2, minerals) into biomass.

4. A portion of the biomass is removed (harvested).

We have already discussed steps 1 to 3 in some detail. In photosynthesis, light energy is used to build CO_2 and H_2O into energy-rich matter. Part of the energy and part of the matter in the photosynthate are combined with simple mineral and elemental substances to produce the complex stuff of biomass. *Assimilation* is the biological conversion of simpler matter into more complex matter, and the assimilated materials, or *assimilates*, are the biomass.

Only a portion of the biomass is typically harvested from an agroecosystem. The harvested fraction may be just the grain, fruit, or seed. In other cases flowers (broccoli), leaves (lettuce and tobacco), or stems (sugarcane) may be harvested. With some crops essentially the entire aboveground biomass is taken for its economic value, e.g., alfalfa hay and corn silage (Fig. 13-7).

The biomass produced by an agroecosystem consists of both harvested (crop or economic) and unharvested (residue) portions. The weight or volume of the crop portion is the *yield*. While biomass is a measure of ecological productivity, yield is the measure of economic productivity. (With alfalfa hay and corn silage, aboveground biomass and yield are essentially synonymous.)

The interception of light (energy) and its conversion is the primary yield-producing process of the crop; growers devote most of their management efforts to ensuring conditions that will favor maximum interception and conversion of the available light. Skillful growers are successful largely to the degree they can increase the crop's efficiency in utilizing sunlight.

FIGURE 13-7
In cutting corn for silage, essentially all the biomass is removed from the field. (*From D. A. Miller, Forage Crops, McGraw-Hill, New York, 1984.*)

Managing Sunlight

Competition for sunlight between the plants of a crop can be intense. There is no genetic tendency for one soybean plant in a field to grow any taller than another or for some corn plants to tower over other corn plants. When crop plants are crowded in the field, they will tend to shade each other's leaves, leading to reduced productivity simply because each plant receives a smaller portion of the available light.

Up to a given point, the grower will accept a lower yield *per plant* if planting more plants increases the yield *per hectare*. If 50,000 corn plants per hectare produce 0.2 kg of grain per plant, the area's yield is greater than with 30,000 plants producing 0.3 kg per plant. However, many crops naturally decline in yield as the density of planting increases beyond a certain point. When that optimum point has been reached, competition for light and other factors becomes so severe that individual plants may fall off sharply in production of their harvestable portions, and yields per hectare may decline as well. Careful management of plant spacing or planting rates to produce the desired density permits the skillful grower to convert the maximum amount of sunlight into harvestable form.

The skillful manager selects crops carefully and schedules planting to take full advantage of the light available during the *growing season,* the segment of the year bounded by unsuitable growing conditions.† Crops vary in the time required to reach maturity. Radish may be ready for harvest in 3 weeks, while sweet corn may take 3 months or more. Such variation can be planned for and used to the grower's advantage. Crops also vary in their response to temperatures. Some do not germinate well in cool soils and are sensitive to light frosts or temperatures slightly above freezing. Others, like peas or wheat, are able to tolerate mild freezes but do not grow well in hot weather.

Ideally, a crop will capture all available sunlight throughout the growing season, but the ideal is never realized. Some compromise is necessary. This may mean choosing a cultivar that can be planted near the last frost of spring and then be expected to reach maturity just before the first frost of fall. In other

---

†Freezing weather is the usual limitation to growth in temperate climates, whereas drought is the most common constraint on growing season in the tropics.

FIGURE 13-8
Canopy closure is about to occur in these soybeans. (*Courtesy of G. R. Buss.*)

cases, the grower may plant two successive crops to take fuller advantage of the growing season (Chap. 17). The grower can improve productivity per unit of land by keeping crops growing on the land for the entire growing season, a simple concept not always simple to achieve.

Part of the grower's problem in achieving maximum interception of light (and therefore yield) during the growing season is the inherently slow early growth of crops. Much of the sunlight that falls on a soybean or corn field during the first month or two falls unintercepted on the ground. The few, small leaves of young plants do not begin to cover the field with even a single layer of photosynthetic surface. The arrangement of leaves on individual plants and the spacing between plants mean that a crop does not intercept most of the daily input of light energy until the total leaf area of the plants is about 3 or 4 times greater than the soil surface area. This generally happens about when leaves from plants in adjacent rows begin to touch and is called *canopy closure* (Fig. 13-8). At canopy closure, there are generally enough leaves to intercept the light before it falls uselessly to the soil between the plants.

Anything the grower can do to hasten canopy closure will potentially improve yields by increasing the capture of light energy for driving photosynthesis. Although some things can be done to improve leaf development, the major problem is inherent: the plant cannot produce leaf area rapidly until it has developed enough leaf area to support rapid leaf growth. Plant breeders have been able to help by providing cultivars with leaves that appear and expand more rapidly in seedlings.

Careful management of soil nutrients and water by the grower can sometimes hasten canopy closure. Anything else the grower can do to reduce early stress while planting as early as possible may also result in earlier canopy closure. To some degree, planting at higher densities hastens canopy closure, but later crowding may work against improved yields. Soybean cultivars planted late in the season after a grain crop has been harvested are often planted in closer rows. Since these more rapidly maturing cultivars normally stay smaller when planted late, closure might not occur at all in widely spaced rows.

Many crops are planted in a protected or favorable environment and later transplanted into their final growing place. Most such crops are horticultural, but some agronomic crops, e.g., tobacco and rice, may also be started in beds or greenhouses and then moved into the field. Although the process is both time-consuming and labor-intensive, the advantage to the plant more than offsets the effort. In the protected environment, the seedling achieves sufficient leaf area and root development for rapid growth when it is transplanted. The transplants are moved to the field as soon as conditions permit and grow considerably faster than new seedings would. Thus transplants develop leaf area and canopy closure occurs well ahead of seeds planted at the time of transplanting. This often means earlier harvests and bigger yields. In some cases, the head start spells the difference between a good yield and no yield at all. In some longer-season areas, it may permit two or more crops to be produced.

Other things that the grower does to or for a crop—fertilize, control pH, control pests, or irrigate—can also increase the plant's ability to capture sunlight and convert it into photosynthates.

Balance and Resource Conservation

In succession, natural ecosystems move toward a climax community characterized by stability and balance. The resources of the environment are recycled within the community or otherwise used in a nonexhaustive way. No growth requirements ever become depleted although they may become limiting. The ecosystem is in balance and can remain productive year after year, even century after century.

Agroecosystems are not naturally self-sustaining. Some of the resources may be depleted by agricultural activity. The land may become so unproductive that it provides insufficient return for the investment of labor and capital, often the case in colonial and frontier America.

In early days, when land was plentiful, farmland was often simply "farmed out" and then abandoned. When there was plenty of new, free land a few weeks to the west, there was little incentive for achieving balance and conserving resources in agroecosystems, and the land was used up or depleted agriculturally. Probably the major reason was a lack of awareness. The pioneers did not know about the limitations that low nitrogen availability and adverse soil pH levels could place on plant growth, and they certainly had no clear understanding of the depletion and loss of soil nutrients.

What can we say about ecological balance in agriculture today? Is American agriculture still "mining" the land and other resources, or are our agroecosystems balanced, with no tendency toward depletion of resources? The answers to these questions are of tremendous importance for now and for the future. If agriculture is not a balanced system, if it is depleting nonrenewable sources to produce food and fuel, production of food and fuel will inevitably decline as resources are exhausted. If the data suggest that agriculture is acting as a drain on nonrenewable resources, what are the implications for agriculture? The rest of this chapter deals with these issues.

ECOLOGICAL IMBALANCE IN AGRICULTURE

Agriculture as practiced today in industrialized nations can indeed act as a drain on nonrenewable resources. Critical materials being depleted more rapidly than they are being replaced by natural processes include nutrients, water, soil organisms, other beneficial organisms, solid organic matter, and even soil itself. Before documenting a few of these cases, we emphasize that the problems are neither universal nor inevitable. Many agroecosystems have approached ecological balance. In many places the cropping systems have been used for centuries and harvests have been stable for generation upon generation (Chap. 17). The discussions that follow therefore describe worst-case situations where agriculture is ecologically out of balance.

Depletion of Soil

One of the most important resources for agriculture is soil. Highly productive soils vary considerably from place to place, but essentially all are characterized by an A horizon, or topsoil layer, of 15 to 30 cm or more. The surface horizon is especially critical for crop growth because it generally contains the highest concentrations of organic matter, responsible for many desirable physical and chemical properties. Nutrients, water, aeration, and biological factors are generally optimal for root growth in a rich topsoil. Nevertheless, topsoil is a fragile, potentially exhaustible resource, which agriculture tends to reduce quantitatively and qualitatively. It is a cruel paradox that agriculture, which so greatly increases the productivity of soil for human purposes, simultaneously poses a threat to the soil's continued productivity.

Several processes that deplete topsoil are intensified by our efforts to grow crops. Most obvious is *physical erosion* of the topsoil by water and wind, but in some agricultural environments, this is not so damaging to soil productivity as *chemical erosion,* the loss of organic matter and nutrients by means other than wind or water. The turning, stirring, and baring of soil that accompanies most crop production accelerates erosion of both kinds.

Physical Erosion Physical erosion of topsoil is a natural process that was occurring long before agriculturalists appeared on the scene; but in a natural situation topsoil-forming processes are offsetting losses. Forming topsoil takes time. By some estimates, the *maximum* rate of topsoil development is about 1 mm (or 9 metric tons per hectare) annually, and in most situations, the accumulation of inorganic and organic matter to build and enrich topsoil is much slower. In nature, the rate of development of soil generally equals or exceeds the rate of erosion, so *topsoil is potentially a renewable resource* but a very slowly renewable one when viewed in human lifetimes. It would take a minimum of 150 to 300 years to form a topsoil 15 to 30 cm deep, the depth needed to support good plant growth. In many environments, it would take much longer.

Soil erosion by water and wind removes the valuable topsoil from the place of its formation and deposits it in stream beds, ditches, roadways, and

even in your lungs and on your furniture. Tremendous amounts of soil are moved in this way. The *average* loss of soil from cropland by erosion is estimated to be about 1 mm, or 9 metric tons per hectare annually throughout the continental United States; an alarming figure since soil formation cannot proceed at that pace in most places. On the bright side, less than one-third of United States cropland loses as much as 9 metric tons per hectare annually. In certain regions, however, including some of our most productive agricultural areas, the average yearly physical erosion of soil approaches 50 metric tons (5 mm) per hectare and occasionally over 150 metric tons (16 mm) per hectare is washed or blown away from extensive areas. That largest, most alarming figure is the equivalent of 160 m^3 of soil, enough to fill a convoy of dump trucks.

The slow regenerative processes of soil formation obviously cannot keep up with drastic erosion losses. In areas of hilly terrain and high rainfall or open expanses and high winds, good topsoil is disappearing down the streams or into the sky much faster than it is being renewed. In fact, some farmers are no longer working with topsoil at all. The A horizon is long gone, and the usually much less desirable B horizon must be primed with expensive fertilizers to provide acceptable yields. In other cases, the land has simply been abandoned agriculturally.

Studies in the late 1970s showed that erosion is the most pressing conservation problem on more than half of United States cropland. Almost one-quarter of the total cropland was judged to be losing soil faster than it was being replaced. Such soils are being "mined" and their long-term productivity is in danger.

The threat to productivity caused by erosion frequently is not apparent because, despite soil losses, yields are often maintained on the eroding lands. This productivity, however, has been maintained at an increasing cost to the producer. Fertilizers, lime, special tillage practices, and irrigation must be used more and more to coax higher (or simply equivalent) yields from the physically eroded soils. By some estimates, the value of nutrients in soil carried away by wind and water approaches $10 billion each year. Since farmers must replace those nutrients or face a drop in yields, eroded land produces crops at a considerably higher cost per kilogram of grain or other commodity produced. One study showed that moderately eroded soils exacted a 20 percent increase in production costs to produce yields equivalent to noneroded soils; severely eroded soils required 50 percent more.

Wind Erosion Major factors influencing the severity of soil loss by wind include soil-particle size and density, moisture of the soil, wind speed, smoothness of the soil surface, and the presence or absence of a covering on the soil. Although large soil particles (sand) are readily moved by wind, they are seldom truly air-borne and move relatively short distances, as in the gradual shifting of sand dunes. Smaller soil particles (clay) are usually not dislodged as easily, but once suspended in the air, they may travel hundreds or even thousands of kilometers. Moist soil is not particularly subject to wind erosion, but it takes only a few hours of a drying wind for the surface to become dry enough for wind erosion to begin. Since the severity of wind erosion naturally increases with wind speed, rows of trees as windbreaks (Fig. 13-9) have long been recommended in areas susceptible to wind erosion. Unfortunately, many established windbreaks have been destroyed in recent years because they make operation of big equipment difficult. One subtle but important factor in wind erosion is the smoothness of the soil surface. A rough surface slows surface wind speed enough to reduce erosion and trap moving particles. In some

FIGURE 13-10
Soil losses in this severely eroded and gullied agricultural land have made it much less suitable for farming, and the sediments carried away pose a threat to downslope and downstream ecology.

cases, therefore, it may be advantageous to plow the soil to raise clods.

A vegetative cover on soil is tremendously effective in reducing wind erosion. The dust bowl in the 1930s had several causes, a major one being the farming of large areas of the Great Plains formerly protected by grasses. In many erodable areas where wheat or other small grains are grown today, the stubble is left in the field to protect against erosion. This *stubble mulch* is vital when land is left fallow for a year or more to allow soil moisture to accumulate.

Water Erosion Physical erosion by water is a serious concern in much of the world (Fig. 13-10). How severe the problem is depends on the force with which raindrops strike the soil, the amount and speed of runoff water, the kind and amount of soil cover, kind of soil, steepness and length of slope, and other factors.

Raindrops striking a bare soil surface can detach soil particles, break up aggregates, and transport soil particles through splashing. The particles loosened by raindrops are readily carried downhill in water flowing across the surface. As aggregates are broken up, the smaller particles are much more readily carried in water. With a moderate rain on an absorbent soil, there may be no runoff at all. A torrential rain on any soil or even a moderate rain on a soil

FIGURE 13-9
Windbreaks. These rows of trees help break the force of wind sweeping over the fields. (*U.S. Department of Agriculture.*)

with a slow infiltration rate can cause considerable runoff. As runoff increases, soil erosion increases. Runoff is greatly affected by soil type, soil cover, and, obviously, slope of the land.

A vegetative cover, whether forest or grass sod, is tremendously valuable in minimizing soil erosion. The foliage reduces the impact of raindrops as well as the rate of runoff. When a soil and its cover are disturbed by deforestation or cultivation, erosion becomes a major threat (Fig. 13-11). A major goal of modern soil management is to minimize disturbances of the soil and the time during which a soil-protecting crop is not being grown on tilled land. No-till and minimum-tillage techniques markedly decrease erosion since crop residues and stubble act as a partial soil cover and reduce soil splash and runoff.

In other potentially erodable situations, a slightly different approach is sometimes effective. For example, in establishing grasses for a lawn or a road-bank, a mulch is often added to the bare soil. In small areas, a few bales of straw are spread by hand, but for larger areas, a mixture of water, seed, and mulch is applied with a machine that sprays them all on at the same time. The mulch slows erosion by reducing raindrop impact and increases seed germination by conserving moisture.

Physical characteristics of soils also affect erosion by water. A soil into which water infiltrates rapidly is naturally less subject to erosion than one which absorbs water slowly, because water that infiltrates cannot also run off. Infiltration is affected by soil particle size, soil structure, amount of organic matter, and the presence of subsurface layers which restrict water movement. Coarser textured (sandy) soils allow water to penetrate more rapidly than clayey soils.

The steepness of the slope affects both the speed at which runoff water flows and the amount of runoff. Length of the slope is important because the total volume of water running off naturally increases with the surface area involved. Several techniques, besides those already mentioned, can reduce erosion on sloping land. *Strip cropping* involves strips of close-growing vegetation running at a right angle to the flow of water. Conventional planting of such row crops as corn or soybeans increases erosion because of the bare soil between the rows. If strips of rye, wheat, or other closely planted species alternate with the row crop, considerable control of erosion can be achieved. As water flows through the row crops, it picks up soil, but the runoff is slowed by the small grain strip, where the soil may be redeposited.

The flow of water tends to be heavy in swales which receive runoff from adjoining slopes. Deep gullies often develop that make further cropping impossible. It is better to leave these low areas in sod to minimize water erosion. The percentage of a field occupied by such *sod waterways* is usually minimal, and the savings in both soil erosion and resulting inconvenience can be great.

Another widely practiced method of mechanical control of water erosion, *contour tillage*, means cultivating soil and planting rows around hills rather than up and down the slope (Fig. 13-12). This practice is easiest on smooth and uniform slopes; if the land is rolling and the topography not uniform, the contours are more difficult to follow. With contour tillage, each plow ridge catches water, thereby increasing infiltration and reducing runoff. On steep slopes, however, runoff and erosion can still be excessive in spite of the ridges running along the slope.

On slopes too steep for contouring to be effective, an alternative is *terracing*. Water runs off the slopes but stops and deposits its soil load on the terrace. Terrace building was practiced by the Chinese, Romans, Incas, and other ancient civilizations. The earliest terraces, built by hand, often included vertical stone walls (Fig. 13-13). Modern terracing is done by machine. A major drawback to the use of terraces is the expense of their establishment, but sometimes terraces are the only alternative if rela-

FIGURE 13-11
A clear-cut site. The exposure of soils on such steep slopes (without the benefit of a vegetative cover) often sets the stage for severe erosion. (*Courtesy of D. W. Smith.*)

FIGURE 13-12
Contour tillage. Plowing along the slope, rather than up and down, reduces soil erosion. (*U.S. Department of Agriculture.*)

tively steep sloping land is to be farmed by conventional methods.

Chemical Erosion Physical erosion creates ecological imbalance and reduces productivity because it carries away nutrients and exposes subsoils inherently less suitable for plant growth because of poor texture, adverse pH, or other characteristics. A process that can deplete soil nutrients even if physical erosion is minimal is *chemical erosion*, the loss of vital soil components through the harvest and removal of crops. When corn or soybeans or tomatoes are harvested, the grower is taking from the land nutrients that were assimilated into the fruits of the plants. The rate of chemical erosion of macronutrients like phosphorus and potassium can be considerably higher than that from losses by wind and water. A good harvest of soybeans will remove from the soil about 35 kg of phosphorous and 60 kg of potassium per hectare, a level 5 to 10 times greater than would occur with physical erosion of 35 metric tons of soil per hectare. Table 13-1 shows the chem-

ical losses associated with harvests of corn. If those nutrients are not returned in a good fertilization program, they will become depleted. To avoid repeating the ecological imbalance over many cropping cycles a well-managed fertilization program will maintain a satisfactory, economically feasible level of fertility from season to season and crop to crop.

Loss of Humus Humus formed from decaying biomass decomposes naturally at a slow rate. The carbon-containing matter can be oxidized (biologically and chemically) to produce CO_2. In many ecosystems, e.g., a forest, new humus forms at least as fast as the old decomposes, but in agricultural situations, the oxidation and depletion of organic matter often exceeds replacement. The loss of organic matter from soil can adversely affect soil structure, water-holding capacity, and other important properties.

One cause of the imbalance between humus buildup and breakdown in agricultural settings is the aeration of soil that accompanies tillage, which raises O_2

FIGURE 13-13
Ancient terraces still in use in this mountainous area create level land for rice production. (*Courtesy of E. Pehu.*)

concentrations in the soil and hastens oxidation. The humus almost literally evaporates into CO_2 in repeatedly plowed and disked soils. This problem is most dramatic in highly organic soils, such as drained areas in the Florida Everglades, where soil scientists estimate 135 tons of CO_2 per hectare are released into the atmosphere each year by oxidation of the organic matter (Fig. 13-14). Another cause is that not enough biomass stays in the field to replace the decomposed humus. The problem is most serious when most of the aboveground biomass is harvested. Crops harvested for silage or hay are often entirely removed except for the roots and cannot make a major contribution to soil organic matter. Efforts by the grower to leave or return plant residues and other organic materials can help restore the balance between deposition and decomposition of humus. Still another cause of increased loss of organic matter in agricultural soils is liming to raise soil pH.

TABLE 13-1
Some Nutrients in Harvested Portion of Maize Producing 6275 kg of Grain ha^{-1} (Equivalent to 100 bu acre^{-1})

Nutrient	Grain, kg ha^{-1}	Grain and stems (for silage) kg ha^{-1}
N	105	235
P_2O_5	20	60
K_2O	20	260
Mg	20	75
S	10	30

Depletion of Water

Water is another critical resource being used faster than it is replenished in some agricultural situations. It is not destroyed by being used in agriculture, but its quality and availability at a specific location may be reduced. Although not the chief culprit in deterioration of water quality, agriculture is clearly the largest consumer of water. In some places the depletion of water has reached the point where agriculture has been curtailed. We are not talking about periodic imbalances between rainfall and crop demand that occur during drought years but areas like the deserts of New Mexico, Arizona, and southern California, where the rainfall is never sufficient to raise the crops being planted. Growers in such places routinely and perennially resort to irrigation to provide the necessary water by delivering surface water (streams and rivers) to crops or pumping water from underground. In either situation, the amount of water used to support crop growth can exceed the amount of water being returned to the streams and groundwater, an imbalance with serious, if not catastrophic, long-range implications for agriculture in those regions.

FIGURE 13-14
Oxidation and depletion of a drained, highly organic soil. The top of this concrete post was at soil level 40 years ago. (*Courtesy of W. Kroontje.*)

Water is a renewable and essentially indestructable resource. *Surface water* in streams, rivers, lakes, and ponds is often quite rapidly renewable at the point where it is withdrawn for irrigation, since water flowing downstream from rain or snow-fed sources can provide a continuous supply; but the water withdrawn is lost to potential users downstream, a matter of great concern in many places. Diversion of water from the Colorado River, for example, turns the stream into a mere trickle before it reaches the Gulf of California.

In addition to the more or less rapidly renewable surface source, irrigation water may also be drawn from underground supplies. Such *groundwater* may accumulate in natural reservoirs in soil and bedrock crevices. Water that seeps vertically into the soil often moves laterally as an underground river, or *aquifer*. The volume of water in aquifers and underground reservoirs greatly exceeds that in all surface streams and reservoirs. Almost 85 percent of the water resources of the 48 adjoining states is groundwater (Fig. 13-15).

Over one-third of the water currently used for irrigation is drawn from wells, i.e., groundwater, and the amount has been increasing. (Over two-thirds of all the water drawn from wells is used for irrigation.) Aquifers and underground reservoirs are replenished or *recharged* primarily by vertical movement of water. The lateral movement of water in aquifers is generally so slow that there is little tendency for recharge similar to that occurring by flow in a surface stream. A problem arises when withdrawals (primarily for irrigation) exceed the rate of recharge. The level of the groundwater, the *water table*, falls as the aquifer is *overdrawn*.

To be sure, some of the water drawn from surface streams and aquifers returns in runoff or percolation to the streams and aquifers, but most does not. It is

FIGURE 13-15
Some major aquifers (shaded areas) in the United States. (*After G. T. Trewartha et al., Elements of Geography, 5th ed., McGraw-Hill, New York, 1967.*)

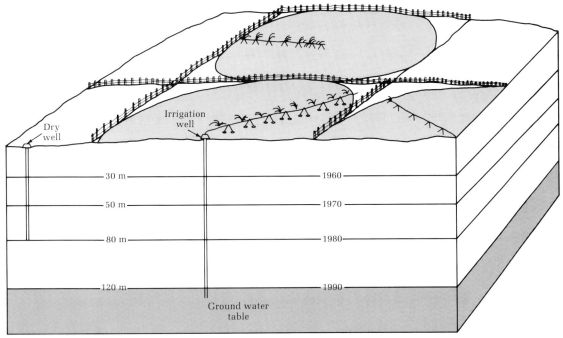

FIGURE 13-16
The lowering of the groundwater supply by irrigation makes pumping water to the surface more difficult and expensive. At some point, it will no longer be profitable.

lost into the atmosphere through evapotranspiration. This loss of liquid water into the vapor state constitutes *consumptive water use*, and agriculture is a major consumptive water user. Almost 80 percent of United States consumptive water use is directly attributable to irrigation. Although agriculture's percentage is expected to drop, the actual amounts consumptively used in irrigation are expected to increase for several years.

How serious or widespread is the water-depletion problem? Many areas have no problem at all; rainfall regularly recharges streams, reservoirs, and aquifers as they are drawn off. Occasional shortfalls in annual precipitation put an area into a temporary overdraft situation, but eventually the balance is restored. Crops are irrigated during the growing season when local rainfall is insufficient to support crop growth. Rain or snow later replaces the surface or groundwater lost to the temporary imbalance between evapotranspiration and recharge.

Perpetual overdraft occurs in regions where, year in and year out, annual rainfall does not equal crop consumptive water usage. Areas where such an imbalance exists are classified as *arid* or *semiarid*.†

It is a natural but cruel paradox that evapotranspiration potential is greatest where rainfall is lowest. The sunny, nonhumid, usually hot and often windy conditions that accompany low rainfall patterns combine to cause high rates of evaporation from soil surfaces and transpiration from plants. Crops that normally fill humid climate niches can be grown in arid or semiarid areas only with irrigation. Therein lies the problem. Crops adapted to high rainfall are being grown in areas where evapotranspiration demands the equivalent of 150 cm of rain but only 5 cm or less may fall. In the United States, the arid southwest is the region where potential evapotranspiration most exceeds precipitation.

About 85 percent of the irrigated land in the United States is in the 17 western states. In several of those

---

†Areas where potential evapotranspiration from a full vegetation cover is less than or equal to precipitation are considered *humid* or *subhumid*.

states, more than 90 percent of the market value of all crops is derived from crops grown on irrigated farms. In 12 of the 17 states, more than 60 percent of the state's farm income from crops and forest products is due to commodities sold from irrigated farms. It is obvious that irrigation is crucial to productivity (and profit) in this region. It is also crucial in maintaining stable levels of productivity on a national scale; about one-quarter of the national income from marketed crops and livestock is earned on irrigated land, which represents only about one-eighth of the total hectarage harvested.

Irrigation has served a great need in agriculture and continues to do so. Land that otherwise would not be highly productive can produce crops like corn, grapefruit, or potato. The United States enjoys the standard of living and diversity of diet that it does in part because of irrigation. If United States agriculture had to depend strictly on *rain-fed* and *dryland* farming, production would be lower, prices would be higher, and less would be available for export.†

In short, irrigation is absolutely essential for production of many valuable crops on soils that are fertile but arid. Where groundwater is drawn on to

---

†The commodities that constitute the bulk of international trade (corn, soybean, and wheat) are not generally grown under irrigation, but some of our groundwater is being pumped out to produce crops that help balance the trade deficit.

provide for irrigation, recharge is insufficient. The water table will subside until it is no longer possible (or profitable) to pump water to the surface for irrigation (Fig. 13-16). In the western states where irrigation from wells is important, groundwater levels are declining at annual rates of 15 to 180 cm. In the high plains of western Texas and Oklahoma, the yearly overdraft of water is equivalent to a 30-cm-deep lake with an area of 5.7 million hectares. Many farmers in this part of the country have already been forced back into dryland farming. Perhaps half of the presently irrigated land in the Texas high plains will no longer be irrigable by the year 2020. Significant decreases in other areas presently under irrigation from groundwater sources are expected to occur by the middle of the next century.

Although solutions to this problem are not appearing as rapidly as the problem is growing, there are hopeful signs. Greater efficiency in use of water through trickle irrigation and other techniques (Chap. 10) will allow growers to extend their dwindling supply. Breeding for improved water-use efficiency by the plant itself may reduce evapotranspirational demand. Regulation of water use is already a major factor in western irrigation programs, but even stricter or more inclusive laws may be needed to enforce conservation of the groundwater resource. Ultimately, though, unless there is a drastic change in rainfall patterns or in groundwater consumption, the imbalance between the two will inevitably lead to a reversion to much less productive dryland agriculture in many areas.

PART THREE

CROP
MANIPULATIONS

CHAPTER FOURTEEN

CROP IMPROVEMENT

Nobody knows when, where, or how people first brought plants under cultivation—or even why. One theory holds that the food value of certain seed-bearing plants led to their intentional propagation after seeds gathered from the wild and accidentally dropped around nomadic camps produced more of their own kind. After thus discovering the relationship between seed and food, some prehistoric person, probably a woman, began to scatter seed intentionally. Another plausible theory holds that the first cultivated plants were propagated purposely and vegetatively rather than from seeds. Roots or tops of edible plants buried to appease the spirits responded by growing more of the edible portion, thus turning the ritual into a more widespread practice.†

Whatever the prehistoric source of plant cultivation, the plants that now form crops arose from wild species selected because they were found to be of value. The people who first gathered seeds of wild plants and preserved them for later planting began practicing *crop improvement*. In its simplest form, it involves *selection* of desirable plants and then using seeds or vegetative parts to produce more plants of the same type. The crops responded dramatically to selection. Cultivated lima beans found in the ruins of ancient Peruvian villages are 100 times larger than the wild limas growing in the area. The great increase in size presumably resulted from generation after generation of selection of plants with bigger seeds. Although the plants were crossing naturally in fields and garden plots, human intervention selected larger and larger seeds. Ultimate *domestication* of wild plants resulted when accumulated changes in genotypes produced phenotypes able to survive only as cultivated species. Corn is perhaps the extreme example of a domesticated crop; cabbage is another that apparently bears no resemblance to any wild plants now living. Domestication of our major crops by selection occurred thousands of years before Mendel discovered the genetic basis for crop improvement, i.e., the prehistoric processes leading to crop improvement were accidental, or at least unplanned. We now have the scientific basis for manipulating crops genetically, and crop improvement has moved ahead by a quantum leap as a consequence.

OBJECTIVES OF PLANT BREEDING

Before discussing how plant breeders and biotechnologists work, let us see what they are trying to accomplish. The ultimate goal is to produce a superior type of plant; this may be accomplished by any of three general approaches: increasing yield, improving quality, or raising production efficiency.

---

†In humid tropical and semitropical regions branches stuck in the ground for whatever purpose will often sprout and grow.

The last goal is especially important in these days of dwindling resources and spiraling production costs, because plants that can use fertilizers and water more efficiently with no loss in yield or quality enable growers to improve their profit margin. Reducing the need for inputs such as fertilizer and water means that those resources and the energy consumed in their production are also conserved—significant and worthwhile objectives in plant breeding.

Some of the breeding objectives discussed below are specifically directed toward one of these general goals—increasing yield, improving quality, or raising efficiency; others serve two or more of these ends. For example, eliminating certain diseases can lead to more fruit of better quality commanding a higher price. On the other hand, some specific breeding problems do not seem to have a fully satisfactory solution; improvement in quality may reduce yield, and vice versa. For example, increasing disease resistance in tobacco through breeding has typically reduced yield under disease-free conditions; breeding for higher yields of wheat seems to involve an inescapable reduction in protein content; and breeding for higher lysine content in grain sorghum has resulted in yield reductions for that important food crop in many subarid areas.

Disease Resistance

Many plant breeders work to develop plants that are resistant to particular diseases, which may be caused by bacteria, fungi, viruses, or other microbes. Each crop has its own set of more or less unique diseases, and quite different diseases are important in different geographic areas. For example, southern wilt, a fungal disease of sugarbeet (Fig. 14-1), is so serious in a warm, humid environment that the crop usually is not planted in such areas. Some sugarbeet breeders at the southern edge of sugarbeet production areas are trying to breed for resistance to southern wilt, but breeders in the major sugarbeet-producing areas do not breed for resistance to this disease; it poses no problem for their growers and other diseases keep them busy.

Eliminating a disease by breeding a resistant cultivar may increase yields (because the disease no longer disrupts the plant's metabolism or destroys its tissues) or it may improve the quality of a crop. (The food and feed value of many crops is lost if the plant is diseased.) Some organisms that infect food

FIGURE 14-1
Southern wilt of sugarbeet, a fungal disease that can devastate a crop, especially in warm, humid areas.

crops produce toxins. Ergot, a notorious fungal disease of cereal grains, has caused hallucinations and even mass deaths from only a small amount of infested wheat. Breeders of small grains have improved quality by selecting for types less susceptible or totally resistant to such diseases.

Breeding for disease resistance has another great benefit. By eliminating the need for fungicides or other chemicals, natural or genetic resistance saves money for the grower and reduces the threat that some substances pose to both applicator and environment.

Unfortunately, plant breeders who develop disease-resistant plants cannot rest on their laurels but often face a continuing challenge. The organisms that cause diseases mutate or reproduce sexually (or both) to produce new genotypes often capable of reinfesting the temporarily resistant plant. Successful breeding programs can become unsuccessful overnight when a new strain of an old ("conquered") disease organism appears. This is not unlike the great number of different types of human influenza that have spread across the globe from time to time, each caused by a slightly different, presumably mutated form of a common human virus. Resistance or immunity to one does not ensure resistance to the next.

Broadening the Genetic Base Between 1846 and 1851, Ireland was devastated by a famine in which millions died or were forced to emigrate. The potato was essentially the sole staple food for the general populace, and all the cultivars of potato being grown in Ireland at that time were genetically similar and therefore equally susceptible to a disease. When a new fungal disease organism found this niche, the entire Irish crop was blighted and lost. Unusually wet, cool springs aggravated the situation, but the potato's narrow genetic base had already set the stage for disaster.

The potato originated as a crop in the Andes of present-day Peru. The plants that apparently served as progenitors of our modern cultivated potato varieties can still be found in this *center of origin*. In fact, many widely different potato genotypes are still present in the Peruvian Andes, because human activity and natural selection tended to make the region a *center of diversification* also. During eons of natural selection, many different genotypes evolved. Crop improvement (by human selection for desirable cultural or eating characteristics) then had its dramatic effects on phenotypic (and, of course, genotypic) diversity (Fig. 14-2). Some phenotypes produce tubers with better keeping quality for longer storage. Some are better suited for particular food uses. Some grow better in a given soil situation or on a slope that faces south. The early potato cultivators and

their descendants, who are still living in the center of origin and diversification, took advantage of the genetic diversity available to them. They probably never experienced a famine as severe as Ireland's, because the broad genetic base provided insurance against catastrophic crop failure. A new disease, a drought, or an outbreak of some insect pest would not automatically pose a threat to the whole crop because the crop was so diverse genetically.

The Europeans *introduced* the potato into Europe in only a very few rather similar genotypes. Since the potato is propagated vegetatively, the crops of the Irish countryside were all of the same genotype or very similar. When the new disease organism appeared in one Irish field, it met no resistance in the next field or the next county.

The story is of more than historical and sociological significance, because it illustrates a genetic concern that is just as important today. If the genetic base of a crop is narrow and the crop is being grown on a large scale, the situation is potentially disastrous. A disease organism (newly mutated or introduced from another geographic region) can cause an epidemic.

This is not a remote theoretical possibility. In 1970, a fungal disease of corn, southern corn leaf

FIGURE 14-2
Diversity of sorghum phenotypes from around the world. Each of the genotypes is a product of centuries of selection for certain locally important characteristics. (*Courtesy of S. VanScoyoc.*)

blight, caused an estimated 10 percent reduction in the United States crop, which meant a 1 billion dollar loss to the economy. Although there are many genotypes of corn resistant to southern corn leaf blight, certain positive factors associated with a particular susceptible genotype meant that it was being grown on millions of hectares throughout the United States. A rapid conversion to blight-resistant genotypes prevented further losses, so that the consequences were not so drastic, in terms of human misery, as the Irish potato famine; but 1970 was a dark year for many United States farmers.

Leaf rust, a serious fungal disease of wheat, has broken out at least three separate times in this century in the major wheat-producing regions of this country. Again the seriousness of the problem was aggravated by the lack of diversity and genetic resistance in the cultivars being planted.

These examples highlight the need for genetic diversity within a crop species. Yet the very nature of crop introduction and crop improvement tends to work against it. The broadest genetic diversity generally exists at the center of origin and center of diversification of a crop (or both), but only a few of our crops, e.g., tobacco and sunflower, originated or diversified in this country; instead they typically have been introduced as a relatively restricted number of genotypes. Subsequent crop improvement tends to involve rearrangement of genes within that small genetic base, which may lack diversity in resistance to diseases, insects, drought, etc.

Recognizing a problem is a first step in its solution. Plant breeders are quite aware of the potential for disease in a crop with a narrow genetic base. An objective of many breeding programs is to increase disease resistance by reintroducing genetic diversity into the population, but it is not easy. One cannot simply find the wild, weedy progenitors of potato, corn, or wheat and breed them to currently popular cultivars. The offspring of such crosses are almost inevitably lower-yielding and otherwise unsuitable even though they may have improved disease resistance. We examine below some successful approaches to broadening the genetic base.

Resistance to Other Pests

Diseases are, of course, not the only pests of crops. Insects, nematodes, and higher animals can seriously reduce the yield and quality of crops. Many plant breeders work to develop plants that are less vul-

nerable to attack by various plant predators. In some cases, the solution may be as simple as selecting for plants that are pubescent, i.e., produce hairs on their leaves and stems. For example, aphids are less likely to feed on pubescent varieties of sorghum, and leafhoppers seem to be deterred from attacking soybeans bred for greater pubescence. Or the crop's resistance to attack by predators may be based on plant-produced chemicals that render it less palatable to the pest. Bird-resistant grain sorghums, for example, are less susceptible to depredation by birds because the grain contains bitter or astringent tannins.

Natural or genetic resistance to insects, nematodes, and other plant predators can improve crops quantitatively and qualitatively and reduces the need for chemical control. Once again, however, there is the problem of mutation of the pest organisms to types for which the plant is nonresistant. This genetic shift has occurred more than once in nematodes that attack soybeans. No sooner have new resistant soybean lines been developed than new strains of the nematode appear.

Environmental Adaptation and Stress Tolerance

Many environmental factors cannot be altered by the grower at all or only at prohibitive expense; these include day length, length of the growing season, temperature extremes, high salinity of the soil, and toxic concentrations of substances such as aluminum. Although such factors may seriously limit plant growth or even prevent proper development of the crop, plant breeders can sometimes manipulate a crop genetically so that it fits into an otherwise unproductive situation. *Breeding for adaptation* to a particular environment is often a major aim of plant breeders.

Corn and most other annual crops show considerable variation in the length of time required to reach *maturity*. Some types of corn take 90 days to develop from a seedling to a harvestable plant; others may take 150 days or longer. The *early* types are needed in northern latitudes, where the frost-free period may last only from mid-June to early September. In southern regions, however, longer growing seasons are available, and later-maturing corn types benefit from the additional frost-free days. Corn breeders have developed an extremely wide range of maturity times within corn types. *Long-season,* or *late-maturing*, plants, which take up to about 180 days to mature, are generally higher-yielding than shorter-season, or earlier-maturing, ones. Thus the breeder provides an invaluable service to the grower by developing a corn plant that can take advantage of the available growing season.

Breeders are developing plants for adaptation to other local, unchangeable conditions. Efforts are being made in California to increase the heat tolerance of lima beans, production of which now extends inland from the southern coastal area to the San Joaquin Valley. Plant breeders are also working to develop more salt-tolerant varieties of many horticultural and agronomic crops. The need is especially great in areas like the San Joaquin Valley, where evaporation of slightly saline irrigation water tends to increase the salinity of the soil with time. Greater plant tolerance to the accumulating salts would enable production to continue on some soils that are becoming unproductive, though other longer-term solutions to the problem of salinization are obviously needed.

Adaptation of a plant to a particular locale can occur by natural selection as well as the conscious selection by plant breeders. For years, breeders have known that productive midwestern varieties of wheat have much lower yields when grown in the east. The reasons for yield reductions were not apparent but did not seem to be related to disease, insects, or other pests. Only recently has the answer been found. The levels of aluminum in eastern soils are naturally much higher than in the midwest. As eastern wheat varieties were selected consciously over the years for resistance to certain diseases or insects, they were also being selected unconsciously (or naturally) for tolerance to aluminum. Only cultivars that could thrive in the high-aluminum soils were selected, although the breeders were generally unaware of it. Midwestern varieties, which were bred for resistance to many of the same diseases and insects, were not developed in the high-aluminum environment and therefore showed little tolerance for aluminum when brought east. This example points out the importance of doing the breeding work for a plant in the environment for which it is being developed, where there can also be the additional benefit of natural selection.

Cultural Adaptation

Besides breeding for characteristics that make crops more suited to particular climates and local condi-

tions, the plant breeder may also select for traits that make it easier for growers to produce a crop efficiently and economically. Many examples of such *cultural adaptations* are available.

One of the first traits to be selected for in many grain-producing crops was that of grain retention, i.e., *nonshattering* seed heads. This selection certainly began long before people knew anything about genes and probably before they even knew anything about agriculture. The progenitors of wheat and rice and other small grains probably had seed heads with very loosely held grains. Wild oats (*Avena fatua*), a close relative of the important crop *Avena sativa*, or oats, has an inflorescence that shatters readily when the seeds are mature. The wind, an animal, or a combine passing through the field at harvest time can dislodge the seeds and scatter them on the soil.† Wild oats actually has some desirable traits, higher protein content and a longer grain-fill period than common oats, but its shattering makes it totally unacceptable as a grain crop.

In the wild state, there would presumably be no advantage to retaining seeds in a cluster at the top of the plant. On the contrary, there would be great advantage in producing seeds that could be released easily at maturity. This natural tendency to *shatter* would permit seeds to be scattered by wind, animals, or other agents. Shattering obviously would not be desirable for grain harvested by people, because seeds that fall off too readily at ripeness would not be available for efficient gathering.

Apparently, however, some plants of some grain-producing species occasionally carried (or developed them by mutation) genes causing seeds to be retained by the grain head. It was not a desirable trait, perhaps, for a wild plant, but for a crop it had a great cultural advantage. Such plants would be the most likely to be harvested by some prehistoric gatherer and eventually produce more nonshattering plants. As with aluminum tolerance in wheat, the line between natural selection and crop improvement is not distinct. The unconscious or accidental selection for nonshattering seed was a powerful force in early crop improvement and domestication.

Recent breeding programs for improving cultural characteristics are numerous. Norman Borlaug (Fig.

---

†The grains have a peculiarly twisted awn that acts to burrow the seeds into the soil as the awns coil and uncoil in response to relative humidity changes.

14-3) won the Nobel Peace Prize in 1970 for his efforts in producing varieties of wheat suitable for culture in some of the world's poorest regions, the tropics. One of his most dramatic successes was the development of day-length-insensitive wheats with shortened, sturdy stems. Such traits were critical for successful culture of this important cereal grain in the tropics. The tropical varieties of wheat grown before Borlaug's work were fairly well adapted to tropical conditions but not highly productive. Part of the difficulty related to their photoperiodic requirement for producing a grain head (Chap. 4). The day length in many tropical areas was such that the

FIGURE 14-3
Norman Borlaug, winner of Nobel Peace Prize and father of the Green Revolution. (*U.S. Department of Agriculture.*)

plants were genetically unable to flower in a timely fashion in that environment. This obviously relates to environmental adaptation mentioned in the previous section, but there was another problem that relates to cultural adaptation. When fertilizers were supplied, the native wheat varieties produced much larger grain heads, but the tall, slender stems could not support the weight. The top-heavy plants fell over, and the grain spoiled when it came in contact with the soil. Borlaug's day-neutral, short-stemmed cultivars retained all the traits that made them environmentally adapted to the tropical climate and could also support large grain heads when fertilized. The higher-yielding varieties produced much-needed food for countries like India and Mexico. Borlaug has been called the father of the Green Revolution for his efforts in modernizing and greatly increasing agricultural output in many third world countries.

Breeding for cultural adaptation takes different turns in different crops. Tomato breeders have worked in recent years to develop varieties with several improved cultural characteristics. Because mechanical harvesters expose the fruits to knocks and drops, less easily bruised types were needed. Equally important for mechanical harvest is uniform maturity of the fruits. Tomatoes are gathered with vine-destroying harvesters that cut the plant off at the ground in a single pass (Fig. 4-14). Breeders have developed tomatoes that mature simultaneously and are firm enough to withstand mechanical handling. As discussed in Chap. 4, tomato breeders have also joined with plant physiologists to develop tomatoes with delayed ripening, so that green fruits can be harvested and chemically ripened later. Another achievement has been to develop "square" tomatoes, which although not truly cubic do have a cuboidal shape that packs into shipping containers with greater ease and less wasted space.

Efficiency of Resource Utilization

Much of the input into an agroecosystem must come from limited and nonrenewable resources, e.g., fertilizer nutrients used to replace those lost by physical and chemical erosion must be taken from elsewhere on the earth. Those nutrient-rich sources are dwindling, and eventually there will be no more in a readily available form. Phosphate is of special concern. Water is another limited resource in many areas. Anything that can be done to reduce the dependency of crops on some of these critical resources will be a welcome achievement. (Breeding for disease and pest resistance can also conserve the energy and other resources that go into production and application of pesticides.) Thus, one of the goals of crop breeders is to increase the efficiency of crop utilization of limited resources.

Plant breeders and microbiologists are cooperating to develop improved cultivars of legumes that can fix more nitrogen. Although this will require changes in the genes of the legume and the nodule-forming bacteria, some successes have been reported. Scientists are even trying to develop associations with nitrogen-fixing bacteria for plants other than legumes. It is not beyond the realm of possibility that corn or wheat might enter into a symbiotic relationship with some nitrogen-fixing organism, making nitrogen fertilization unnecessary.

Genetic differences in the efficiency of nutrient utilization, which have been noted in many species, are related to more extensive root systems or to altered metabolism of roots and shoots. In either case, the cause is genetically controlled and can therefore be bred into plants. The ability to breed for nutritional efficiency will become increasingly important as some nutrients become scarce.

Specific Crop Quality Characteristics

Since the specific quality factors being bred for vary tremendously between crops and even within a crop, several different programs for improving quality of a species may be going on simultaneously.

In food and feed grains, protein content is an important quality consideration. Grains like wheat, corn, rice, and sorghum are normally high in starch and relatively low in protein. People or livestock subsisting on such grains are generally getting enough calories but not enough of all of the amino acids needed to make muscle and cellular proteins. Lysine is the essential amino acid in shortest supply in cereal grains. High-lysine lines of corn and sorghum have now been developed. Swine and other nonruminant animals fed high-lysine corn make much faster growth than those fed normal corn and no protein supplement (Fig. 14-4). High-lysine sorghum is of special importance to millions of people in regions of Africa where sorghum is a staple.

In peaches, fruit size, color, and texture are quality traits important to the canning industry. Genetic

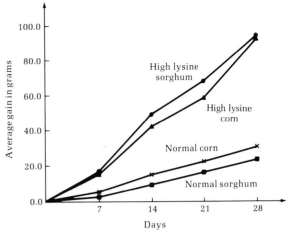

FIGURE 14-4
Animals fed on high-lysine corn or sorghum gain weight faster than those fed on normal corn or sorghum with no protein supplement. (*From Crop Science, Vol. 13, 1973, pp. 535–539, by R. Singh and J. Axtell.*)

differences between peaches for these factors provide the opportunity for breeding for them. As with breeding for harvestability of tomatoes, the distinction between breeding for quality characteristics and for cultural adaptation is not a sharp one.

In work with forages, the *nutritional factors* sometimes bred for, e.g., protein and vitamin content, are so strongly influenced by stage or time of harvest that breeding for positive quality factors is difficult. Many forage breeders concern themselves instead with eliminating *antiquality factors* from the forage crops. For example, tall fescue and ryegrasses contain alkaloids that can be harmful to livestock. Forage breeders have had some success in breeding the offending alkaloids out of improved cultivars. The reduction in alkaloid content frequently results in greater *palatability,* and since the improved varieties taste better, the animals eat more and gain more weight. At least they gain more weight as long as the *digestibility* of the forage is high. Some breeders have increased forage quality by eliminating substances that interfere with digestion or by breeding for higher content of the readily digested fractions.

Plant breeders working with vegetable crops look for many different characteristics to satisfy a diverse market and to develop new markets. The pea experienced something of a revival as a garden crop when types with edible pods were developed. "Burpless" cucumbers have certain advantages over the normal types. Sweeter, low-acid tomatoes have become popular (although they are not as safe for home canning). Potato breeders have developed different varieties for baking, chipping, and frying. Breeding work with various types of dry beans has reduced their tendency to produce flatulence.

Aside from wheat, probably no crop receives as much breeding work on obvious quality improvement as flowers and other ornamentals. Plant breeders working with ornamentals look for unusual flower colors and shapes and unusual foliage. Although most breeders of ornamental crops are professionals, developing new types is also a rewarding hobby for amateurs.

METHODS OF PLANT BREEDERS

The usual product of a successful breeding program is an improved plant, with a new and unique combination of desirable characteristics. Breeders look for new genes, new alleles, or new combinations of genes and alleles, i.e., new genotypes, that can produce superior phenotypes. That superiority may be expressed as higher yield, better quality, or greater production economy.

When breeders develop a new genotype, usually they do not start from scratch but insert superior genes into genotypes otherwise well suited to a particular location or production system. The altered genotype represents a "new, improved" crop, and the genotype can then be marketed or otherwise made available as a named variety or hybrid. Each marketable variety (or *cultivar*, as *cultivated varieties* are called), genotype, or hybrid of a species is genotypically unique and can be propagated to maintain its uniqueness. Propagation may be by seeds or vegetative means (Chap. 15), but the genotype must be preserved to produce the same phenotype year after year. A cultivar or hybrid is a salable product only if it is reliable—always performing as advertised.

The new, improved product from a breeding program will be genetically packaged in one of three different ways: variety (or cultivar), hybrid, or synthetic variety. The type is determined largely by the mating system of the species. Generally speaking, self-pollinating row-crop species are bred and sold as genetically pure cultivars; open-pollinating row-crop species are bred as hybrids; and open-pollinating (and sometimes self-pollinating) turf and forage species are handled as synthetic varieties. We shall discuss each approach in more detail and provide specific examples as we proceed.

Plant breeders also work with species that are not normally propagated by seeds. Chapters 6 and 15 consider the vegetative propagation of various crops, where one is dealing with a *clone*, individuals derived asexually from a common genetic source and alike in their genotype.

Apomixis represents a category intermediate between sexual (seed) reproduction and asexual (vegetative) propagation. In apomictic plants, a normal-looking seed is formed, but its embryo is not the result of meiosis, gamete production, pollination, and fertilization. Instead, cells from the ovular (maternal) tissue divide mitotically and undergo embryonic development to produce a seed that is genotypically identical to the seed-bearing plant. Kentucky bluegrass and citrus species are commonly (but not exclusively) apomictic. Apomixis causes these species, which are otherwise open-pollinating and therefore heterozygous, to appear more like self-pollinating species in the consistency of genotype from generation to generation. Because viruses commonly are excluded from seeds, apomixis is a good way to eliminate viral diseases from an apomictic crop while maintaining its genotype, which is crucial to consistent performance.

The methods of breeders vary tremendously with the crop, whether it is self- or cross-pollinating, and the outlook of the breeder. Several different sources or methods for obtaining "new" genes or alleles or combinations of genes are available to breeders, and there are various ways of getting such genes or alleles into genotypes that are otherwise suitable.

We pause for a moment to make a distinction between plant genetics and plant breeding. Plant geneticists may study inheritance in plants with no other goal than a greater understanding of genes and their operation. Although this information may ultimately be of great practical value, it need not have been the original intent of the work. By contrast, the plant breeder has the practical goal of developing plants that will be superior to existing cultivars or hybrids in economically important traits. Breeding may involve basic genetic research to reach the goal or rely on basic research findings of other scientists, but frequently the plant breeder is a highly skilled plant geneticist and vice versa.

The plant breeder, seeking a new genotype superior to existing genotypes, usually *crosses* to obtain new genotypes, *selects* the best offspring, and *field tests* to determine the success of the program.

Beginning with a well-formulated goal, the plant breeder identifies genetic differences for the phenotypic traits that are to be improved. Genetic effects on phenotype are distinguished from environmental effects, and the nature of the inherited traits is determined. When these steps have been accomplished, the plant breeder determines how best to cross the plants with the desired traits in order to obtain the desired new genotype. If this seems straightforward, we have misled you because, while seeking to improve one or two traits, the plant breeder must also work to retain at least existing levels of many other traits. For example, little is gained from developing a new disease-resistant, non-lodging, high-yielding, high-protein cultivar of bread wheat if the grain is inferior for milling and baking.

Screening

As one of the early steps in many breeding programs, the breeder looks for genes or alleles that produce the desired phenotypic character. Because breeders cannot see a genotype, they look for phenotypes that reflect the desirable genes. In other words, a plant breeder selects desirable phenotypes for breeding work on the assumption that their desirable traits are genetically determined. A few generations of breeding will clearly establish whether the trait is *heritable* and whether it is inherited by simple Mendelian inheritance. If the desirable trait is under simple genetic control, it may be a good candidate for further breeding work. This process of finding desirable heritable traits is called *genetic screening*.

The many different approaches to genetic screening all have the same objective—to find genes and genotypes that will produce a superior phenotype. In some screening programs, the breeders look for natural mutations. Mutations of a particular gene occur naturally at a fairly low but consistent rate of around once per 500,000 individuals. On that basis, one could expect to find a mutation for any specific gene about once in every 8 ha of corn or every 42 ha of commercially grown tomato, but the odds are that the mutation found would not be a desirable one. This type of screening obviously would be time-consuming and require tremendous space.

The rate of mutation in a species can be greatly increased, however, to raise the likelihood that a desirable mutant will occur. Some breeders and geneticists expose plants to *mutagens* such as ionizing radiation or various chemicals in order to speed up mutation. Although this greatly increases the potential for finding new (mutant) traits, the breeder cannot create a specific desirable mutation. In fact, in both natural and mutagen-caused mutations, the vast majority of changes have no positive value. On the contrary, most mutations are harmful. For these reasons screening for mutations is not a particularly popular or productive approach for finding useful new alleles.

A screening method that has been both more popular and more productive looks for desirable genes or alleles not in the present population but in members of the species growing elsewhere and derived from different lineages. For example, wheat that has grown for many centuries in a particular tropical locale will naturally evolve a distinctive germplasm. The alleles for many of the wheat genes will be different from those of wheat that has grown for centuries in a temperate region. Different genotypes that develop in this way, called *landraces*, develop where populations of a crop have been isolated from others of its species for hundreds or thousands of years. Through time they come to bear the stamp of natural and artificial selection in their germplasm. Each landrace of wheat is unmistakably wheat, but each has many genes or alleles that are unique and especially well suited to its particular environment. Given the hundreds or thousands of years during which such landraces developed, it is not surprising that many desirable mutations accumulate. The greater number of undesirable mutations occurring over the centuries would tend to be eliminated by natural selection. By screening germplasm from different regions of the world, a plant breeder hopes to find genes for resistance to a troublesome disease, for higher protein content, for earlier or later maturity, or for any number of other traits.

Borlaug, the breeder of the day-neutral, short-stemmed, productive wheats for the tropics, and others screened wheat from all over the world to find a type with shorter, sturdier stems and found it in the germplasm of a Japanese landrace, where it presumably developed by mutation of a long-stemmed plant long ago. The short-stem trait was apparently successful enough in Japanese agriculture to be selected for, naturally or artificially, over the centuries. Of course, the Japanese landrace was not suitable for growth in the tropics. Aside from its shortness, it was rather unfit for Borlaug's work since it was not adapted to the climate, diseases, and other conditions of the tropics. By crossing wheats with tropical and temperate germplasms, selecting for day-neutral, short-stemmed phenotypes, Borlaug eventually produced cultivars having all the genes and alleles necessary for survival and productivity in the tropics plus stems to support the productive heads. Because these cultivars were day-neutral, they could grow into many tropical areas with widely differing photoperiods.

A Different Approach

A commercially successful cultivar is stable genetically. When a breeder produces a superior genotype, and hence a superior phenotype, every effort is made to maintain the identity of the genotype for generation after generation. With an open-pollinating species like corn, it is not possible to produce the same genotype over many generations of open pollination. The mixing of alleles that occurs in cross-pollination produces heterozygous offspring, each with a different genotype, obviously not conducive to generation of a consistently productive genotype. In some open-pollinating species, breeders have abandoned propagation by seed altogether, propagating the desirable genotypes instead by strictly vegetative means. We mentioned several such cases in Chap. 5. In corn, another approach has been successful, but to understand it we need to know more about plants and inheritance.

Inbreeding Depression When a crop is self-pollinated, there is a tendency for each gene present to become homozygous, whether the plants are naturally or artificially self-pollinated. Corn can be made to fertilize itself by manually delivering pollen from the tassels to the silks of the same plant (Fig. 14-5).

FIGURE 14-5
Artificially self-pollinating corn, an important step in the development of hybrid corn. Pollen captured in the bag in the breeder's left hand is being sprinkled onto the stigmas (silks) of the corn plant. The envelope in his right hand will prevent any other pollen from reaching the silks. (*Courtesy of Jacques Seed Company.*)

The resulting seeds can be planted and a second round of self-fertilization performed. Several successive generations of this artificial *inbreeding* eventually produce plants that are highly homozygous; each gene locus occurs twice (once on each homologous chromosome), but the same allele is present on each homolog. Such highly inbred plants are said to be *true-breeding* because they produce more of the same genotype when self-pollinated or when crossed with plants of the same homozygous genotype. Although true-breeding plants appear to be fine candidates as corn cultivars because they can consistently and repeatedly be used to produce the same genotype, such inbred corn plants are generally poor specimens—small and quite low-yielding (Fig. 14-6).

Corn is not the only plant faced with this problem. *Inbreeding depression*, a reduction of vigor and productivity, is generally associated with self-fertilization of all normally open-pollinating species, including apple, pine, squash, sorghum, and sugarbeet. The more generations of inbreeding the greater the inbreeding depression. As their genotypes become more homozygous, the offspring become more depressed.

Two theories have been offered to explain how inbreeding depression occurs. According to the *dominance hypothesis*, the problem with open-pollinating plants apparently lies in recessive, undesirable alleles in the genotype. Normally, deleterious alleles are not expressed because they are *masked* in the heterozygous genotype by the dominant allele for that same gene. (Remember Mendel's work with the short (recessive-allele) peas.) When a heterozygous plant is self-fertilized, such recessive alleles will be expressed as they are matched homozygously. As more and more homozygous, deleterious recessive pairings occur in the germplasm with succeeding generations of self-fertilization, the offspring will become less vigorous.

The *overdominance hypothesis* assumes that the heterozygous condition per se is superior to homozygosity in that *Aa* produces a more vigorous phenotype than either *AA* or *aa*. Perhaps genotype *AA* produces too much and genotype *aa* too little of the enzyme controlled by the gene; or perhaps the two slightly different forms of the enzyme produced by *A* and *a* can do a better job than either alone.

The dominance and overdominance hypotheses are not mutually exclusive; each mechanism could be operating on different genes. In any event, the heterozygous genotype of normally open-pollinating species produces generally more vigorous phenotypes than inbred, homozygous genotypes of the species.

Heterosis As we have seen, inbreeding corn causes phenotypic depression in the offspring. Thus inbred, true-breeding corn is an unlikely candidate for a cultivar. The homozygous genotypes can be produced time after time, but their phenotypes are not productive. Heterozygosity generally improves yield but not uniformly, and those heterozygous genotypes that are productive cannot be propagated via seeds.

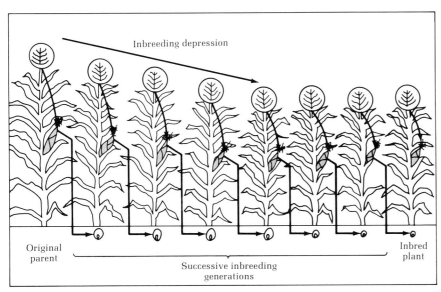

FIGURE 14-6
Inbreeding depression in corn; the reduction in plant vigor results from self-pollination. (*Courtesy of DeKalb-Pfizer Genetics.*)

In other words, heterozygous genotypes do not lend themselves to consistent production of a productive phenotype, something highly desirable in agriculture.

In the first quarter of this century, corn breeders solved this dilemma by producing different inbred, homozygous lines and then crossing them with each other. The startling results of this *single-cross* hybridization were that the offspring were much more vigorous and productive than either parent. The *hybrids*, as they are called, were often 25 or 30 percent taller and bore much larger ears (Fig. 14-7). The phenomenon of *hybrid vigor*, or *heterosis*, is not really startling when one considers what is happening genetically. When two *different* homozygous genotypes are crossed, the odds are that some of the

FIGURE 14-7
Hybrid vigor or heterosis results in offspring (center) that are much larger and more productive than the parents (left and right). (*U.S. Department of Agriculture.*)

harmful recessive alleles from one parent will be masked by dominant alleles from the other (or that the heterozygous, overdominant condition will be created). Furthermore, with some luck and breeding skill, *most* of the harmful recessive alleles will be masked. The result is a heterozygous genotype far superior to either of its homozygous parents.

Because the parental lines for hybrid corn are true-breeding, they can be maintained in open-pollinating situations like a cornfield without laborious artificial self-pollination. As long as the genetic line is homozygous, it makes no difference genetically whether a plant fertilizes itself or is fertilized by another plant of the same line. Since every gamete, whether sperm or egg, will be carrying an identical haploid genotype, the offspring will develop the same homozygous genotype. As long as the inbred line is kept isolated from pollen of other genotypes, it will remain pure.

Because the parental lines for hybrid corn are true-breeding, they can be used year after year in the hybrid cross to produce the same hybrid genotype. The two different inbred parental lines only have to be grown together to permit the two homozygous genotypes to mix. But since it may make a difference which line is the male parent and which the female, only one plant line is allowed to produce pollen and the hybrid seeds are harvested only from the other. To keep pollen from the maternal line from also supplying pollen to the open-pollinating environment, its tassels must be removed or made infertile. Genetically male-sterile maternal lines have been developed to eliminate the need for costly tassel removal. In such cases, the paternal and maternal lines are simply planted in adjacent rows, and only the maternal rows are harvested for hybrid seed. The hybrid seed can then be planted where their vigorous phenotypes will produce yields that so dramatically exceed those of their parents, and the sequence of hybrid seed production from two inbred parents can be repeated time and again to produce the same productive hybrid genotype.

It is *not* possible to replant seed from the hybrid generation and hope to get the same uniform, heterozygous genotype. In all probability, the mixture of alleles when the F_1, or hybrid, generation crosses will produce as many different genotypes as there are kernels of corn. It obviously is not a good practice to save the grain produced from the hybrid plants for replanting. They are not true-breeding and will not consistently produce high-yielding, vigorous plants.

Hybrid corn is one of the most dramatic successes of agricultural research in the United States. Following its introduction in the early 1930s, acreages grown with hybrids jumped from about 15 percent of the total field corn grown in 1938 to over 75 percent in 1950 and nearly 100 percent currently. Yields in the same period have more than tripled—but only partially as a result of growing hybrid corn. With the adoption of hybrid corn and its much greater yield potential, farmers were encouraged to practice better overall management, including weed control and increased applications of fertilizers. Seeding rates were also greatly increased. In order to realize its *yield potential*, hybrid corn must be intensively managed. This is the case with many agricultural improvements and demonstrates again that crop production must be viewed as a complex picture, not separate steps.

Working with Other Cross-Pollinating Species

Other cross-pollinating species† also present special problems and special opportunities. Because they are open-pollinating, it is not possible to produce specific, desirable genotypes via seed consistently, but artificial inbreeding makes it possible ultimately to produce homozygous, true-breeding genotypes. These lines can be maintained by open pollination as long as each is kept isolated from other pollen sources; but, as with corn, such inbred lines tend to be low in vigor and productivity. The breeder of open-pollinating species therefore usually resorts either to hybrid development, already discussed, or to development of *synthetic varieties*.

Plant breeders can sometimes create a satisfactory level of vigor through open pollination between many parents upon which inbreeding and selection have been practiced. Bringing together several inbred, true-breeding lines for open pollination allows mixing of genes and expression of heterosis in numerous genotypes. Each of the many different genotypes created will be heterozygous and therefore vigorous. Breeders of cross-pollinating forage and turf species often use this method. They develop several inbred lines of a species and then plant them together, allowing them to mate randomly. The multiple parents of such synthetic varieties ensure many different heterozygous genotypes in the product.† Since the parents are inbred and true-breeding, the same general set of alleles can be brought together to create the synthetic variety time after time. Reproducibility combined with heterosis makes for a reliable and marketable product.

In many other open-pollinating species, the breeder practices a compromise; some inbreeding and selection produce a population of plants somewhat heterozygous genotypically but largely homozygous for many important phenotypic characteristics. The balance avoids inbreeding depression in the population as a whole and provides fair uniformity within the population for characteristics of particular concern such as height, productivity, or disease resistance. One other consequence of the heterogeneity of the population is that its seed can be used for replanting. The open pollination of the population tends to keep the genotypes at the proper level of heterozygosity and the offspring at the proper level of vigor.

It is at least theoretically possible to develop highly productive inbred lines of normally open-pollinating species. To be successful, though, the breeder would have to find a genotype that is homozygous essentially for good alleles only. With 100,000 or more different genes (each with perhaps several different alleles), there is very little chance of producing a homozygous genotype without at least a few of the harmful recessive alleles. Such efforts have not resulted in any homozygous cultivars competitive with hybrid corn, but carefully selected inbred lines can be fairly productive. If a highly productive homozygous corn cultivar ever is developed, it could revolutionize the corn industry as much as hybrid corn once did.

Working with Self-Pollinating Species

Plants that are self-pollinating‡ tend naturally to be highly homozygous. Many generations of self-fertilization will bring essentially every gene to the homozygous state. The only exceptions would be for newly

---

†Defined as species in which cross-pollination normally exceeds 40 percent.

†In a typical single-cross hybrid, there is only one genotype because there are only two parents, each contributing the same alleles to the hybrid generation each time the hybrid cross is made.

‡By definition, these are plants that are naturally cross-pollinated less than 5 percent of the time.

occurring mutations, occasional cross-pollinations, and a vanishingly small percent of heterozygotes theoretically always present. For all practical purposes, however, plants that are strongly self-pollinating, e.g., wheat, barley, soybean, pea, and tomato will be highly homozygous.

How do self-pollinating plants cope with natural inbreeding? If *artificial* inbreeding of some plants leads to a serious loss of vigor in their offspring, how can a self-pollinating plant be a successful species when it continually exposes itself to the effects of inbreeding? The answer must be that natural selection over the eons has selected *against* harmful recessive alleles whenever they appeared in a homozygous genotype. The selection is especially effective if the allele is lethal and kills the plant before it can produce seed, effectively eliminating those alleles from the future lineage. Conversely, recessive beneficial alleles can be selected *for* when they appear homozygously. If a recessive allele provides some natural advantage to the plant, survival of the fittest will operate to increase its occurrence. Natural selection cooperates with the naturally self-pollinating species to accentuate the positive and eliminate the negative alleles.

Although we see how self-pollinating species can be successful as a species and how they come to be so highly homozygous, how can a plant breeder operate within the uniformity of genotypes imposed by self-pollination? Plant breeding depends upon diversity, and self-pollinating plants seem to be creating uniformity. Self-pollination is not actually eliminating diversity. Although individuals tend to be homozygous, a population of a self-pollinating species will contain many different genotypes. On a global scale, the genotype variation within a species can be immense.

Crossing followed by *backcrossing* has been used extensively by plant breeders working with self-pollinated (and inbred open-pollinated) crops to introduce variation into homozygous genotypes. Sometimes a cultivar is excellent except for a single trait. All its genes are in the homozygous state and produce a desirable phenotype except, say, for resistance to a new disease. If a gene for resistance to the disease is found in otherwise inferior germplasm (a *donor parent*), the plant breeder can transfer the gene for disease resistance into the otherwise productive cultivar. The resistance trait can be introduced by artificially cross-pollinating, but as a result

of sexual reproduction, the cross made to bring the allele for disease resistance into the cultivar also brings in undesired traits from the donor parent. To eliminate them, the hybrid offspring are crossed back to plants of the desired cultivar (the *recurrent parent*), and disease-resistant plants most like plants of the original cultivar are selected for repeated backcrossing to the recurrent parent (Fig. 14-8). After backcrossing has reestablished a great preponderance of the superior germplasm in a disease-resistant plant (usually three to six *recurrent selection* cycles), the offspring may be selected to move on to homogeneity by their inherent self-pollination. Ultimately a new, improved cultivar is developed that is like the original cultivar except for the addition of disease resistance. Borlaug practiced backcrossing to tropical recurrent parents while selecting for the short-stemmed, day-neutral characteristics from the donors.

Complications

Plant-breeding efforts become complicated for many reasons. For example, plant-breeding programs commonly strive to improve two or more traits simultaneously. This greatly affects the selection phase following introduction of the desired traits. The number of plants with both of the desired new characteristics during a backcrossing sequence will be reduced.

Unfavorable *linkages* between genes can be a significant problem. For example, a gene for disease resistance may be located on the same chromosome as a gene that causes poor crop quality. Because they are linked, they will be transmitted as a pair through meiosis and gamete formation and introduction of disease resistance into good germplasm simultaneously reduces quality. This situation requires far more backcrossing to obtain resistance and good quality together, but crossing-over can eventually separate the linked traits and they can be selected for (or against) individually.

Another common difficulty in breeding programs is distinguishing genetic effects from environmental effects. (Chap. 6 discussed the relationships between genotype, phenotype, and environment.) Any trait that is greatly affected by environment may be difficult to improve through breeding and selection. Increasing kernel weight in cereal crops is a good example. Selection for kernel weight is difficult because the trait is greatly affected by rainfall, tem-

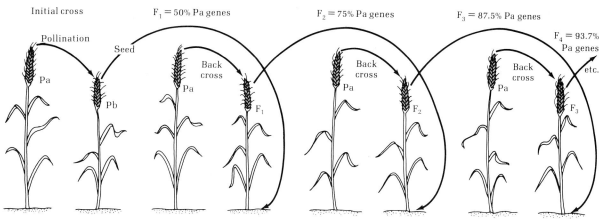

Initial cross $F_1 = 50\%$ Pa genes $F_2 = 75\%$ Pa genes $F_3 = 87.5\%$ Pa genes $F_4 = 93.7\%$ Pa genes

Pollination

Seed

Pa

Pb

Back cross

Pa

F_1

Back cross

Pa

F_2

Back cross

Pa

F_3

etc.

FIGURE 14-8
Backcrossing is a means of reintroducing traits into offspring. Pa is the recurrent parent and
Pb is the donor parent.

perature, and other nongenetic factors. Even plants with genetic potential for larger kernels have small kernels under conditions of moisture stress. Differences in reaction to diseases also affect the expression of the inheritance of kernel weight. Diseased plants commonly have shrunken, small kernels. Because of the frequent but not easily distinguished effects of environment on phenotype, the breeder finds it difficult to select for improved genotypes. Of course, many traits, e.g., petal color, are relatively unaffected by environment. A red rose will be a red rose in a rainy or a dry year.

Pure-line (homozygous) cultivars of self-pollinated annual crops, e.g., soybean and tomato, seem to offer the greatest potential for breeding efforts because matings are most readily managed. (When the goal is a hybrid, as in corn or sorghum, inbred lines must be developed and maintained and mechanisms implemented to foster production of hybrid seed.) In some instances, plant breeders work to produce *multiline varieties* from self-pollinating crops. This method of packaging genetic material involves blending true-breeding lines that differ in only one or two genes. A typical application of this technique would be in providing wider disease resistance to a constantly changing pathogen. Two or more genotypes differing in their mode of resistance to the pathogen mixed and planted as a blend offer potential improvement of the population's resistance to the disease. Some will be affected by the pathogen, but

others will survive and may compensate for those lost to the disease. All genotypes in the multiline blend are alike except for the mode of resistance to the disease.

GENETIC ENGINEERING

In all the crop-improvement methods considered thus far, the breeders have worked within the system of sexual reproduction. Although there were several examples of altering natural pollination patterns to produce hybrids, backcrosses, etc., the new genotypes were all produced by fertilization of an egg by a sperm. In some cases we have not discussed, sperm and egg from two different species are brought together to form an interspecific hybrid, but the methodology still involves sexual reproduction. Triticale, a hybrid of wheat and rye, is a prime example of mixing germplasms from two different species to produce a new crop (Fig. 14-9).

Now there is an entirely different approach to crop improvement. *Genetic engineering,* or *biotechnology,* is genetic in that it relies on the manipulation of genes, but it no longer depends on sexual reproduction for transferring genes and generating new genotypes.

Genetic engineering is actually a multiple technology whose diversity rivals (and in some ways parallels) that of computer technology. The economic

FIGURE 14-9
Triticale, an interspecific hybrid of wheat and rye. (*U.S. Department of Agriculture.*)

successes registered by genetic engineers have been rather modest as of this writing, but by the time you read this, significant agronomic and horticultural breakthroughs may have occurred.

Genetic engineers, or biotechnologists, manipulate genes and genotypes without resorting to meiosis, flowering, and fertilization. Their techniques are often rather straightforward and can be accomplished with ease, some apparently sophisticated ones being technically no more difficult than a conventional backcross. The genetic engineer's approach may be more direct than that of conventional plant breeders, since theoretically specific desirable traits can simply be inserted into an already existing productive genotype. Biotechnology offers opportunities to work directly on the germplasm and avoid some of the difficulties inherent in plant breeding.

The Revolutionary New Technologies
The dramatic rise in productivity resulting from the application of Mendelian genetics and plant-breeding techniques to agricultural problems during the middle half of the twentieth century has been called the first agricultural revolution. Between 1930 and 1980, farm production in the United States more than tripled, while farm acreage declined by almost

10 percent. Hybrid corn was a major factor, but plant breeding for many other crops contributed to the productivity boom; many other factors like fertilizers, pesticide, and improved management practices also played a part. The new wheat and rice cultivars introduced in the 1950s and 1960s became the basis of the Green Revolution, during which world production of these two staples increased fivefold by some estimates.

The technology associated with conventional methods of plant breeding has limitations, some of which we have alluded to, and it appears that future gains in agricultural productivity through conventional plant breeding may come more slowly. This prediction, based on recent reductions in the growth rate of productivity, comes when world population is predicted to double in 40 years. The stage now seems to be set either for a global food catastrophe or a second biological revolution. The new technology that may revolutionize crop improvement involves both somatic-cell genetics and gene transfer.

Somatic-Cell Genetics
In Chap. 6, when we discussed techniques of cell and tissue culture whereby somatic cells are caused

to divide and sometimes produce whole plants, we stressed the theoretical and practical value of cell and tissue culture in demonstrating totipotency and taking advantage of it. The cultured cells can be made to produce millions of new plants with a genotype identical to the original cell. The asexual reproductive technique relies on the faithful transmission of germplasm during mitosis.

In somatic-cell genetics, biotechnologists rely not on the inherent, almost monotonous faithfulness of genotype transmission but on a degree of unfaithfulness—of mutation and diversity. Attempts to increase the chances of mutations occurring in mitosis and then screening for the altered types represent a much more efficient screening method than working with the phenotype.

A single petri dish or culture flask holds 4 or 5 million individual cells. It would take about 1 ha to produce 4 or 5 million individual plants of wheat, over 50 ha for that many corn plants, and almost 250 ha for tomato. If one could screen for mutations in a cell culture, the savings in time and space would be enormous. To use this approach, for example, in developing cultivars of crops with greater resistance to herbicides, one might need only add the offending herbicide to the culture medium to screen for resistance (Fig. 14-10). Cells with resistance to the herbicide would survive and could be grown into whole plants; the plants could then be screened for stability and inheritance of the resistance trait. One flask or petri dish with 5 million soybean cells would provide for as much screening of germplasm as 15 ha of conventionally grown soybeans, and the initial cell-culture screening could be accomplished in a few days.

A similar approach might work in screening for disease resistance, heat or salt tolerance, or more efficient use of nutrients. Of course, many desirable crop traits cannot be selected for in a cell culture. Plant height, for example, is not manifested in a single cell; nor can it be selected for by introduction of some chemical agent into the culture medium. Despite such obvious limitations, somatic-cell genetics represents a great new advance in the technology of crop improvement.

We have failed to mention one critical requirement for somatic-cell genetic screening. Where does the genetic diversity in a population of cloned cells come from? Since mitosis will faithfully pass on the original donor or cloned genotype from cell generation to cell generation within the culture (or will it?), one might not expect to find much genetic diversity to screen.

In somatic-cell genetics, variation in genotypes that create phenotypic differences with the cultured cell population arise from at least two sources. Protoplast fusion relies upon sophisticated tampering with germplasm, but somaclonal variation is spontaneous and still unexplained.

Protoplast Fusion One form of cell culture involves isolating and regenerating plant protoplasts, the living portion of the cell without the cell wall. Although generally more difficult to maintain in a somatic-cell culture, the isolated protoplasts offer opportunities that whole cells do not because the cell wall can prevent chemicals or other biotechnological agents from reaching a cell's genome. Working with protoplasts in culture, it is sometimes possible to induce two diploid cells to fuse (Fig. 14-11). Their genomes may also fuse to produce a unique genotype much as happens in fertilization. When the two cells that fuse are of the same species and genotype, the resulting polyploid genotype may impart important new phenotypic characteristics to a plant, *if* it can be regenerated from the fused cells.

When the two protoplasts that fuse are from different genotypes or even different species, the protoplasts containing the hybrid genomes can potentially be induced to form a cell wall, proliferate, and then regenerate as new hybrid species. In this way hybrids might be produced between plants that could never be crossed sexually. The possibilities may seem like science fiction but they are close to being realized.

To date, success with protoplast fusion and somatic-cell genetics has been limited. The protoplast-fusion technique has been used to fuse nuclei of two different clover species into a unique hybrid. Fusions of protoplasts from widely unrelated species have been achieved, but the *recombinant genomes*, i.e., genome in which genes have been mixed or recombined from two different sources, have not been successfully reconstituted into a plant. Recombining genomes from such widely divergent germplasm sources may be like trying to fuse the blueprints for a skyscraper and a football stadium—the structure cannot be built.

Genetic engineers have done some wholesale recombining, mixing genotypes of different species together and waiting to see what grows. By fusing somatic cells from potato and tomato and then

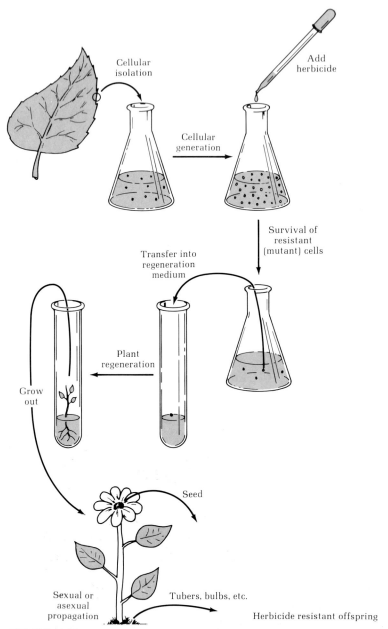

FIGURE 14-10
Screening for herbicide resistance in somatic-cell culture.

conductors with incredible speed and microcircuit complexity. With the help of those very computers, plant scientists are now working to solve some of the problems encountered in protoplast fusion. It is tempting to predict a rapid solution.

Somaclonal Variation The second means by which genetic diversity arises in somatic-cell culture, although unanticipated and as yet unexplained, has more immediate promise. When plant scientists first began to grow somatic-cell cultures, it was assumed that the genomes of cells in a culture derived from a clump of cells would be identical to each other and to the donor from which the clump was taken. But plants regenerated from cloned cells often look very different from each other and from the parent plant (Fig. 14-12). This variation commonly seen following the *cloning* of *somatic* cells was dubbed *somaclonal variation*. It appears to be widespread, having been observed in several of our most important crops—rice, wheat, corn, potato, alfalfa, and sugarcane.

The mutations that produce somaclonal variation may result from mitotic accidents that break chromosomes, rearrange DNA within a chromosome, or substitute a single nucleotide. Some mutations are

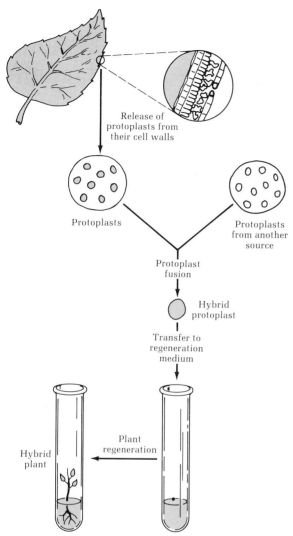

FIGURE 14-11
Protoplast fusion in somatic-cell culture.

FIGURE 14-12
Somoclonal variation in wheat. The eight plants in the center were produced by regeneration from cells in a somatic cell culture that started with cells from the genotype shown on the right and left. (*Courtesy of P. Larkin, S. Ryan, R. Brettell, and W. Scowcroft.*)

culturing the recombinant cells, scientists have produced cells (but no whole plants) with some characteristics of each parent. Perhaps someone will develop a "pomato" that has both tubers and edible fruits. Other possibilities are almost endless.

Protoplast fusion to produce genetic diversity in recombinant genotypes has enjoyed only limited success, but it is an infant technology. Think how rapidly computer technology moved from a slow, cumbersome vacuum-tube stage into the era of semi-

not stable in the regenerated plants, but others reappear in successive generations, which is good news. From somatic-cell cultures of sugarcane cultivars with no resistance to a disease have come somaclonal variants that are resistant. From potato somaclonal mutants have come genes for resistance to the blight that caused the Irish famine. From wheat cell cultures have come plants with shorter stems. Perhaps Borlaug's painstaking, multiple-generation efforts to produce tropical wheats could be duplicated without resorting to the first cross or any of the long series of backcrosses.

Somatic-cell culture offers an opportunity for a rapid, highly efficient screening for many traits, and the genetic diversity introduced into the clones by protoplast fusion and natural somaclonal variation holds out the hope of a renewable, almost limitless source of genetic traits. As we learn to solve the technological problems, crop improvement may well embark on a second biological revolution.

Gene Transfer

Although the biotechnology of somatic-cell culture has attracted well-deserved attention, the greatest excitement in crop improvement research centers on techniques for isolating genes for a specific trait and inserting them into a species. The present state of the art in gene transfer does not permit such manipulation of specific genes; there is still an element of chance or plain luck in getting a functioning gene safely into the genome of a diploid nucleus. The gene-transfer successes scored so far have been with relatively simple bacterial systems, but, again, the technology is still in its infancy. *Gene splicing*, or *recombinant DNA technology*, as gene transfer is also known, is probably in its "first generation" or will be so viewed in a few years.

Nevertheless, genetic engineering has given a glimmer of what can be accomplished. In a historic genetic transplant, the genes for nitrogen-fixing enzymes in *Rhizobium* were transferred to a yeast cell. Although the transformed yeast was not able to fix nitrogen, the work showed that the genes could be transferred. Eventually someone may transfer functional nitrogen-fixing enzymes from bacteria into corn or wheat. Or if the approach is to recombine nodule-forming genes from legumes into other crops, the bacteria will still be the nitrogen fixers, but the nodulated organism may be corn, rice, or wheat. The

ultimate direct savings in fertilizer and indirect savings in fossil fuel would be tremendous.

The first transfer of genes from one higher plant into another was a scientific curiosity, not a potential new crop, but it clearly demonstrated that the genetic barrier between species can be breached. No longer does the developer of an improved cultivar have to look for superior genes within the species. Potentially, any gene from any species is available for use, and the pollination and fertilization barriers between species can be sidestepped.

The Techniques

Gene transfer is simple enough in theory. The basic four-step procedure involves identification, isolation, duplication, and then insertion of the desired gene. Only the last two steps have been achieved with any consistency, and even they are not yet widely applicable.

Identification To identify a specific gene in the immense genome of a plant is less like looking for a needle in a haystack than looking for a specific piece of hay in a haystack. The figure of 100,000 genes to regulate the growth and development of a plant, suggested earlier, is conservative. There is enough DNA in a typical plant cell's genome to constitute 5 million genes. Whether all this DNA is organized into actual protein-coding sequences is unknown. Many molecular biologists believe that much of the DNA is regulatory, controlling not so much what proteins are to be made but when. Much of the DNA is known to be repetitive; some genes that code for a specific protein may be represented hundreds or even thousands of times in the genome of a single cell. The tremendous volume of DNA in a genome poses a great problem in locating a gene that controls a specific desired trait. Techniques are being refined, and others are theoretically possible to help identify the portion of a chromosome that contains the gene; but location or identification of the desired gene remains a big hurdle.

Isolation If or when a gene can be identified on its precise locus, there is still the very formidable problem of isolating it, separating it cleanly from the rest of the genome. Biotechnologists have already developed some powerful tools for this task. When they learn how to use them more precisely, it may be possible to lift a gene from a chromosome as neatly as a magnet will draw a needle from a haystack. The tools that can work on DNA so precisely are enzymes. A large group of similar-acting enzymes known as *restriction endonucleases* can cut DNA molecules at specific points, i.e., specific nucleotide sequences (Fig. 14-13). The first such enzymes were discovered in the early 1970s, and hundreds of different specific endonucleases are now available commercially. If the nucleotide sequence surrounding a gene is known, it may be possible to snip the gene out of the genome using restriction endonucleases specific for the points to be cleaved; but that's a big "if." A protein-coding gene has three essential regions:

Promoter Region—signals where to begin transcription
Code Region—determines amino acid sequences
Terminator Region—stops the transcription properly

The genes isolated by the restriction enzymes must be cut off in such a way that the promoter, code, and terminator are included. Anything less or more may be nonfunctional.

Duplication and Insertion Once a gene has been properly isolated, it must be cloned, or duplicated, to give enough copies to increase the likelihood of success in transferring it into cells. Duplication is currently accomplished with a naturally occurring genetic structure called a *plasmid,* a tiny, wayward, ring-shaped piece of DNA existing in the cytoplasm of certain bacteria (Fig. 14-14). The circular DNA of the plasmid can be cleaved by restriction endonucleases and a gene from some other source inserted into the opening. Special enzymes called *ligases* reclose the ring with the new gene inserted. When the bacterium containing the transformed, or recombined, plasmid divides, it makes copies of the plasmid including the new gene.

A plasmid, whether transformed or not, can be biotechnologically transferred to another bacterium.

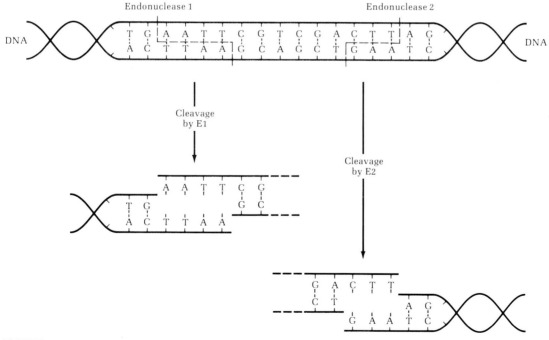

FIGURE 14-13
Restriction endonucleases cleave DNA at specific points.

If it has been transformed, it can cause the production of medically important substances such as insulin or antibiotics. Commercial successes are being scored in this area now.

Even more exciting from the perspective of crop improvement is the ability of some plasmids to enter into and transform higher plant cells. One such plasmid is found in the soil-borne bacterium *Agrobacterium tumefaciens*, which can infect a number of different plants, causing tumors or galls to form in the plant tissue. In fact, the plasmids in the bacteria are the tumor-inducing agent, hence the name *Ti plasmid* (Fig. 14-15). The Ti plasmid induces gall formation by inserting a portion of its DNA into the genome of an infected cell. The transferred DNA causes the tumorous growth. The biotechnological significance of the Ti plasmid lies not in the crown gall disease it causes but in its natural ability to clone DNA and then insert it into a plant's genome.

It can potentially act as a replicator and vector for genes that have been identified, isolated, and then spliced into the plasmid's DNA.

This description of the Ti plasmid's function in the transfer of DNA is oversimplified; the plasmid is not ready-made for gene-transfer work. Many problems had to be solved before taking advantage of its vector function. In 1983, two laboratories continents apart working independently finally solved enough of the remaining problems to accomplish a specific gene transfer into a higher plant. Each group isolated a functional bacterial gene for antibiotic resistance and spliced it into the appropriate portion of the plasmid's DNA. Petunia cells in a cell culture were then infected with *A. tumefaciens* containing the transformed plasmid. The plasmid carried the resistance gene into the genome of the plant cells, which then became resistant to antibiotic added to their medium. Moreover, when a whole petunia plant

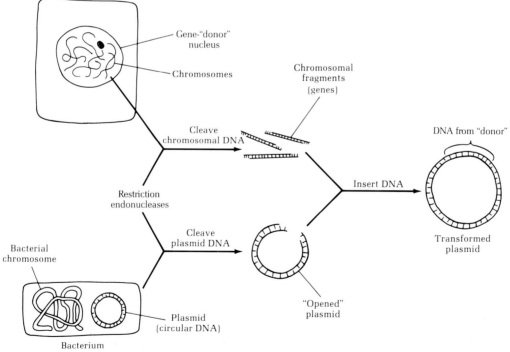

FIGURE 14-14
Plasmid transformation involves adding DNA to its genome.

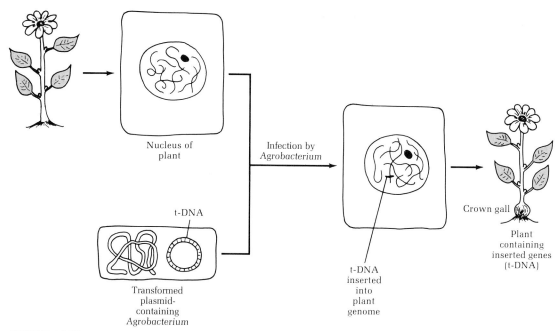

FIGURE 14-15
The Ti plasmid and *Agrobacterium tumefaciens.* When the Ti plasmid's t-DNA is added to the higher plant's germplasm, it causes the plant to produce a gall.

was regenerated from the cells, it exhibited the resistance as well, showing that the bacterial gene had actually become a functioning part of the petunia's genome.

The success of this gene transfer shows the value of the Ti plasmid vector, but much more work is needed to make the plasmid a precision tool for genetic engineering. At present, for example, Ti plasmid can be used as a vector only in plants that can be infected by *A. tumefaciens.* Many important crops are susceptible to the bacterium, but the entire grass family is not, so that rice, wheat, corn, and many more species are excluded from consideration. Either another bacterial-plasmid vector must be found, or the grasses' resistance to *A. tumefaciens* must be broken down.

Unanswered Questions

Many problems remain to be solved before gene transfer becomes a standard technique for crop improvement. We have just alluded to the lack of suitable vectors for the grasses; other problems may be more fundamental. One has to do with control of gene expression. For example, it does no good to insert a gene for senescence into a plant if the plant cannot turn the gene on or off at the appropriate times. A transferred gene that would direct the synthesis of a more efficient Calvin cycle enzyme ought to function only in chloroplast-containing cells. If root and stem cells are not able to prevent the gene's expression, its benefits may be lost overall. Molecular biologists do not yet know enough about the regulation of gene expression in higher plants to provide gene splicers and plasmid workers with specific guidance.

Another difficulty will be the need to transfer many different genes into a genome to assemble all the gene products (enzymes) causing a particular trait or property. Nitrogen fixation, for example, requires at least 17 different genes. Such *multigene* traits can pose monumental problems. When the nitrogen-fixing *Nif* genes were transferred from a bacterium to a yeast cell, the job was relatively easy because the Nif genes are clustered as a unit in the bacterial genome. But transfer of the Nif genes to yeast cells did not make the yeast cells competent in nitrogen fixation, because still other genes are needed to create the conditions in which the nitrogen-fixing enzymes can work.

Another major obstacle to crop improvement via gene transfer lies in the frequent inability to regenerate plants from the cells that may have been transformed. We have mentioned the necessity for working with isolated protoplasts during transformation efforts, but in many cases protoplasts are difficult to regenerate. For reasons unknown, some species (carrot, tomato, alfalfa, and petunia) are relatively easy to manipulate, but others (corn is a major recalcitrant crop) often defy efforts toward regeneration. But great progress is being made in many areas of biotechnology. For example, while this book was in preparation, a group of workers found how to regenerate soybean plants from cell cultures. The prospects truly seem great for a hybridization of crop breeding and biotechnology.

CHAPTER FIFTEEN

PLANT PROPAGATION

To survive as a species, all living things must reproduce and perpetuate their kind. In nature there are a number of ways in which plants make copies of themselves. Early agriculturalists learned how to take advantage of natural reproduction in plants, and gradually artificial methods were developed to perpetuate, increase, and control plant growth. Seeds, bulbs, and tubers are examples of natural propagules; vegetative cuttings are an example of a method of propagation that does not occur in nature. Such cuttings are a successful method of reproducing some of our most valuable plant materials. Some plants, e.g., corn, now require human intervention for species survival since the early progenitors have long been extinct, and they are unable to survive naturally. In nature most plants reproduce sexually, but we propagate plants by whatever method best serves our interest. The procedure used commercially depends on the time required, the cost, the botany of the plant, and genetic considerations.

SEXUAL PROPAGATION

Most higher plants naturally reproduce themselves by seeds—many exclusively so. Most agronomic, vegetable, and forest crops, many annual and perennial flower crops, and some woody ornamentals are commonly seed-propagated (Fig. 15-1). In cultural situations, propagation by seed is usually cheaper and easier than other methods and is used whenever possible. Some plants, e.g., potato, sweet potato, hydrangea, and pineapple, do not regularly produce viable seeds, and many plants, e.g., apple, chrysanthemum, and raspberry, display such genetic variability that their seed usually produces inferior plants or fruits and they are not commonly propagated sexually.

Pollination

Pollination, the transfer of the pollen grain from the stamen to the stigmatic surface of the pistil, was discussed in Chap. 6. *Self-pollination*, in which the pollen comes from a flower of the same plant (or technically, the same cultivar) as the stigma, is the norm in many species, e.g., pea, snap bean, tomato, lettuce, and oats. Pollen shed from the stamens falls directly upon the stigma, or the stamens may be in such close contact with the pistil (as in the pea) that pollination occurs as soon as the anthers release pollen. An example of *cross-pollination*, in which the stigma is pollinated by pollen from a different plant or cultivar, would be the transfer of 'Golden Delicious' pollen to the stigma of a 'Stayman' apple flower. Some species and cultivars of crop plants set fruit equally well with either type of pollination; others require cross-pollination. The fact that field crops such as corn, soybean, and wheat and vegetable crops such as cucumber, tomato, and sweet corn set fruit readily after self-pollination makes it possible

FIGURE 15-1
Geranium plant production from seed. (*Ball Seed Company.*)

to plant large acreages of a single cultivar. Many tree fruits, however, require cross-pollination, so that at least two different cultivars must be planted relatively close to each other.

 Pollen Transfer Besides the transfer of pollen by gravity and direct contact between the anther and pistil, the major agents of pollen transfer are wind and insects. In wind-pollinated species, tremendous numbers of very light pollen grains are carried by air currents. The flowers of wind-pollinated plants are typically not very showy, presumably because they have no need to attract insects. Wind-pollinated crops include many of the grasses, e.g., corn, the nut trees, e.g., pecan and walnut, and many forest species. Hay fever sufferers are all too aware of when certain species of weeds, e.g., ragweed, simultaneously shed their pollen.

Many crop species depend on insects for the transfer of pollen, whether for self- or cross-pollination. The flowers of such species are usually colorful, and many produce nectar, which attracts insects. The most common insect pollinator is the honeybee *(Apis mellifera)*. Growers of such diverse crops as cucumber and apple routinely keep or rent hives of honeybees to ensure adequate pollination and fruit set. Although fruit set plays no part in growing onions, carrots, or celery as vegetable crops, producers of commercial seed for these crops are also dependent upon insect pollination.

The flowers of some species are so constructed that only one kind of insect can reach their stamens and stigmas and effect pollen transfer. Red clover depends upon certain species of the bumblebee; yucca, or Spanish bayonet, depends on a specific moth, *Pronuba*, for pollination; and another striking example, the fig, will be discussed in Chap. 21.

When we are concerned with the production of edible fruits such as cucumber, muskmelon, or apple, pollination is necessary for fruit set but we do not care whether self- or cross-pollination occurs because we plan to use the fruit and not the seeds. In most plants, the pollen source (male parent) does not affect the size, color, or flavor of the fruit, which develops from maternal tissue. An exception, called *metaxenia,* in which the pollen parent affects both fruit size and date of ripening, occurs in the date palm.

Since seeds develop from the union of male and female gametes, they are directly influenced by both parents. With most plants, the effects are of concern only if the seeds are to be used for propagation. With some crops grown for the seed, however, the appearance or quality of the seed can be affected by the pollen parent: if sweet almonds are pollinated with bitter almond pollen, the resulting nuts (seeds) are bitter and inedible; if a white sweet corn cultivar such as 'Silver Queen' is grown beside a yellow cultivar which sheds pollen at the same time, the ears of 'Silver Queen' will have a mixture of white and yellow kernels. Self-pollinated stigmas (silks) of the ears of 'Silver Queen' produce white kernels, and cross-pollinated ones produce yellow kernels. The direct effect of the pollen source on embryonic or endosperm tissue is called *xenia.* The multicolored kernels of Indian corn are the result of xenia.

Plants propagated sexually can be divided into four groups:

Naturally Self-Pollinated—As a rule, plants in this group have less than 4 percent of cross-pollination.

They include wheat, barley, oats, tobacco, tomato, pea, bean, soybean, flax, rice, and lima bean.

Often Cross-Pollinated—Self-pollination is more common than cross-pollination, but cross-pollination occurs frequently enough to require some method of preventing it between cultivars and strains of different genotypic constitution to keep the cultivar homozygous for seed production. Crops belonging to this group are sorghum and cotton.

Naturally Cross-Pollinated—This group includes corn, clover, rye, sunflower, sugarbeet, many fruits, cucumber, squash, watermelon, carrot, and cabbage. For commercial seed production, cultivars must be widely separated so that there will not be mixing of genotypes.

Dioecious—Dioecious plants are unisexual, having the male and female flowers on different individuals. They are obviously, then, cross-pollinating. Asparagus, hemp, and date palm belong to this group. Again, if seed true to type are to be maintained, cultivars must be separated by long distances for seed production.

SEEDS

From a morphological standpoint, a seed is an entire, embryonic plant surrounded by protective tissues from the parent plant and (in some cases) embedded in a nutritive tissue derived from both parent and offspring. Physiologically speaking, a seed represents a resting stage in the life cycle of an organism, following intense developmental activity. This period of *quiescence* ends with the onset of germination, when growth resumes and the embryonic plant attempts to establish itself as a fully functioning autotroph. It can no longer depend on seed reserves or the parent plant but must rely on its own ability to harvest sunlight, collect raw materials, and respond to the environment.

For most agronomic and horticultural crops, the period of quiescence is accompanied by the almost complete drying out of the seed at maturity. Metabolism depends on the presence of water: when it is lacking, normal metabolism ceases. Obviously, seeds which dry down at maturity must be able to prepare for and survive such dehydration. This critical ability of seeds is uncommon in other living matter. In some important weed plants and valuable fruit and forest species, a condition called *dormancy* or *rest* is imposed or regulated by physiological means. Such seeds often do not dry out, but will not germinate prematurely even under optimum external conditions of moisture, oxygen, and temperature.

The unique morphological and physiological properties of the seed enable the plant to survive adverse periods (too cold, too hot, too dry, too wet, etc.) and to disperse easily. Growers can take advantage of these properties, especially if the seed is only quiescent (not in rest) and ready to grow, but only if they are aware of the restrictions and requirements imposed by the seed's quiescence and germination.

Propagation by Seeds

The success of propagation by seeds depends upon the production of flowers, normal meiotic behavior in the formation of the microspores and megaspores, fertilization, subsequent development of the embryo and endosperm to form the seed, and finally germination of the seed and establishment of the seedling.

Germination

Germination (Fig. 15-2) is the sequence of events leading to reinitiation of active growth by the embryo, rupture of seed coverings, and emergence through the soil of a new seedling. Germination is a dramatic and potentially disastrous phase in the organism's life cycle. Such a drastic change in its metabolism and growth exposes the young seedling to conditions which, though not really adverse for an established plant, may be catastrophic to an immature one.

Germination of dry, quiescent seeds begins with the absorption of water. Water makes possible the production or release of key enzymes and hormones, hydrolysis of insoluble stored foods into soluble forms, and the translocation of these soluble foods to the growing points. Then growth of the axes can begin. Availability of water and oxygen are obviously critical and can be potentially limiting factors during this early period. Usually temperature and other environmental conditions such as light also affect germination. A frequently overlooked factor that figures heavily in failure or success is the original vigor of the seedling. We shall take a closer look at each of these factors in turn to see how they help or hinder the seed in its attempt to become an established plant.

FIGURE 15-2
Germinating corn (*Zea*) seedling with young primary root nearly covered with root hairs. Note that no hairs are on the root tip. As the corn seedling develops, its primary root grows for a time and then produces secondary roots. (*Carolina Biological Supply Company.*)

Water Water, the biological solvent, must rehydrate the dry seed before it can resume activity. Dry seeds are normally planted in soil or a planting medium, from which they must obtain the necessary water. The tremendously low water potential of many dry seeds enables them to absorb enough water from even a fairly dry soil to come to life and begin to grow. A rain or thorough watering, of course, hastens the process, as can carefully soaking some seeds before planting. The moisture content required to trigger germination varies from about 40 percent of the seed's dry weight for corn to nearly 100 percent for sugarbeet and soybean.

Hard seeds have impervious seed coats and cannot absorb moisture or oxygen without special treatment. *Scarification* is any process of breaking, scratching, or otherwise altering hard seed coats to make them permeable to water and gases. Although some scarification often occurs during harvesting and cleaning of the seed, germination of most seeds with hard seed coats is improved by additional treatment such as rubbing the seed over sandpaper or similar surface, soaking for a short time in concentrated sulfuric acid, or leaving seeds outdoors through the winter, thus keeping them moist and subjecting them to alternate freezing and thawing. A natural degradation of the hard seed coat by organic acids or microorganisms present in the soil is common among weed and forest species.

Imbibition of water increases the volume of the seed and sometimes softens and breaks the seed coat. Water is also required to activate the enzymes which transform stored starch, protein, and oil into mobile, usable forms. Adequate levels of water must be continuously available once the germination process begins because even temporary drying can kill the awakened seed or young plant.

Oxygen During germination, oxygen must be present to permit respiration of carbohydrates, fats, and other energy reserves. To assure a free flow of oxygen, the germination medium should be well aerated. Seeds in heavy soils show poor germination, especially during a wet season, because the soil often lacks sufficient oxygen. Careful, deep preparation of the soil to allow it to drain properly and to improve aeration is important in benches or greenhouses. The germination medium there is often a special mix; finely milled sphagnum moss, which has natural fungistatic properties, is generally used, alone or with other components.

Temperature Temperature affects both percentage of germination and rate of seedling growth. It influences the rate of absorption of water, of translocation of foods and hormones, of respiration, of cell division and elongation, and of many other physiological reactions. For example, lettuce seed may remain dormant at a temperature of 30°C or higher. Low temperature may not prevent germination of this cool-season crop but reduces growth rate enough to prevent emergence of the seedling from the soil. In some species, moderately low temperatures (5 to 10°C) have an adverse effect on the metabolism and survival of seeds. Soybeans and lima beans are notoriously sensitive to such low temperatures. Most seeds germinate well around 21°C (see Table 15-1).

Light Some seeds germinate equally well in dark or light; others, e.g., lettuce and certain grasses, are stimulated to germinate by exposure to particular wavelengths of light. Still others, e.g., garlic and

TABLE 15-1
Cardinal Temperatures (°C) for Germination of Vegetable Seeds†

Crop	Minimum	Optimum	Maximum
Warm-season:			
Bean	16	16–29	35
Lima bean	16	18–29	29
Corn	10	16–35	40
Cucumber	16	16–35	40
Squash	16	21–35	38
Tomato	10	16–29	35
Watermelon	16	21–35	40
Cool-season:			
Beet	4	10–29	35
Cabbage	4	7–35	38
Carrot	4	7–29	35
Lettuce	2	4–27	29
Onion	2	10–35	35
Pea	4	4–23	29
Radish	4	7–32	35

† From E. P. Christopher, *Introductory Horticulture*, McGraw-Hill, New York, 1958.

onion, require continuous darkness. The germination of many weed seeds is triggered by light, and seeds have been known to lie dormant in the soil (literally for centuries in some documented cases) only to germinate when exposed to light by excavation or other soil-turning action. That is one reason why fall plowing and/or disking is often a good practice; weed seeds are exposed and germinate and then are killed by the freezing temperatures of winter before they can reproduce.

Vigor The quality of seeds undergoing germination, though of great significance, is not easy to assess. Of course, when seeds deteriorate too much, they become incapable of germinating, but long before that, they may show signs of decreased vigor. Vigor, which is the ability of a seed to produce a robust seedling under a variety of conditions, decreases with age or improper storage. Checking the germinability and vigor of the seedling can avoid expensive crop failures. *Certified seed* come with an indication of germinability on the package, and many seed producers are now experimenting with putting an index of vigor on the seed package as well.

Dormancy

Viable seeds that fail to germinate when environmental conditions of moisture, aeration, and temperature are favorable are said to be *dormant*. Dormancy is due to such physical causes as a hard seed coat or internal physiological causes such as natural chemical inhibitors.

An afterripening period is sometimes required before germination will take place. The amount of time required varies from a few days to several months and is sometimes influenced by postharvest treatment. During this period, physiological and mechanical changes occur: germination-promoting hormones are produced, key enzymes are released, and the seed moves toward germination. Often the after-ripening period is satisfied or dormancy is broken by stratification, in which seed are placed in a moist medium at about 4 to 7°C for 1 to 3 months. Many seeds in the wild are exposed to such conditions naturally, and this stratification requirement is similar to the chilling requirement for bud break or fruit trees and flowering of small grains.

In physical dormancy a physical barrier must be broken before germination can occur, e.g., hard seeds, whose moisture-impermeable seed coats must be scarified before sowing; hard shells or pits, which must be mechanically cracked, as in the stone fruits.

The juices and fruit flesh of several species contain chemical inhibitors which prevent or delay germination. For example, tomato seeds will not germinate until removed from the fruit and separated from the adhering tissue. In other cases, applying the plant hormone cytokinin is fairly successful or leaching the seed with water. In many cases, however, the simple act of planting and watering is enough to break chemical dormancy.

Most important agronomic and horticultural crop species do not have dormant seeds; they will germinate whenever and wherever planted as long as environmental conditions are permissive. Most ancestors of our crops had strong dormancy, which helped them survive. In the course of selection and domestication, types with the culturally desirable characteristics of nondormancy were retained and dormant types discarded, resulting in a seed which is no longer dormant and cannot survive in the wild.

Food Reserves

Food reserves nourish the seedling until photosynthesis occurs rapidly enough for the seedling to be fully autotrophic. Plump seeds usually have more food reserves than small, shriveled seed. This is why many cultivars of field corn germinate better than sweet corn cultivars and immature wheat or soybeans do less well than plump, fully matured grains or seeds. Thus seed appearance can often be an indicator of seedling vigor or even germinability.

Emergence

After the radicle emerges from the seed coat and begins to grow downward, producing root hairs and new roots to acquire water and minerals and to anchor the seedling for its upward push, the shoot begins to grow. As germination proceeds, two different patterns of shoot elongation become apparent. In *hypogeal* germination, the hypocotyl does not elongate and the cotyledons therefore remain underground. The epicotyl grows and only the shoot apex emerges from the soil. With *epigeal* germination, the hypocotyl grows and the cotyledons are pulled above the soil surface. Among dicots, epigeal germination is much more common. Among the legumes, pea, vetch, and scarlet runner bean have hypogeal germination, while most others are epigeal. Peanut is peculiar in that both hypocotyl and epicotyl are active during emergence. The cotyledons are pushed upward until they lie near the soil surface but the epicotyl usually is the only emerging part. Grasses, including small grains, turf species, and corn, have a peculiar seedling morphology which includes a hypogeal growth pattern.

The upward push of the stem of dicots poses a threat to the fragile plumular leaves in hypogeal plants or to the cotyledons in epigeal plants, particularly if the soil is heavy and compacted or crusted. The possibility that cotyledons or plumules will be torn away during emergence is reduced when they are pulled up through the soil rather than pushed through it. This is accomplished when the hypocotyl or epicotyl bends or hooks near the point of attachment of the leaves and elongates behind or below the hook, so that the tender leaves are drawn up through the passage created by the hook. Once the stem emerges and is exposed to light, the hook straightens out and the plumules or cotyledons spread out. A stem grown in the dark or underground will continue to elongate and remain hooked until the energy reserves are used up. When the plant emerges into the light, elongation is slowed and unhooking occurs. In grasses, a special organ, the *coleoptile*, protects the leaves during emergence.

Emergence of a growing seedling obviously depends on many factors. With enough water and oxygen and suitable soil temperature, the plant may still fail to emerge if it is planted too deep or the soil is restrictive. The depth of planting must be no greater than the capacity of the hypocotyl or epicotyl to elongate. If a seed's food reserves are exhausted before it reaches the sunlight to make new photosynthate, it will die. If the expanding seedling is unable to penetrate a heavy, compacted, or crusted soil, it will likewise die when it runs out of energy. Proper preparation of the seedbed and attention to planting depth can do much to improve success in achieving desirable stands. Preparation includes working the soil to ensure adequate contact between seed and soil particles to increase water transfer but not overworking until it is powdery and much more likely to crust. Planting at the appropriate depth—determined by the characteristics of both seed and soil—and some tamping to improve contact are key practices. Deeper planting may be permissible or even advisable in lighter (sandier) and drier soils, but the seed should not be buried so deep that its ability to emerge is lost.

Once established, the seedling usually passes through a morphological progression of increasing size and complexity (Fig. 15-3). The first true leaves, i.e., those produced above the cotyledonary node, often differ from later leaves. In legumes, where leaves of mature plants are usually compound, the earlier leaves may be simple or unifoliolate and the earliest compound leaves may have leaflets that differ in shape from later ones.

Branching patterns and rooting patterns begin to manifest themselves in the young seedling. In legumes like clover and soybean, which are normally taprooted plants, the radicle develops into the primary root and often becomes the dominant root. The stem of a legume often shows very early whether it will be erect or prostrate. The epicotyls of white clover, for example, begin to grow horizontally soon after emergence.

The pattern of axis elongation in emerging seedlings, whether hypogeal or epigeal, determines whether there are nodes and therefore axillary buds belowground. The elongating epicotyl of pea and other hypogeal species may produce two or three nodes before emerging; axillary buds on these nodes are potential growing points if the aboveground portion of the plant is lost to frost or predation. Peas killed or eaten back to soil level can still put up new shoots. On the other hand, epigeal plants such as soybean and tomato have no such growing points beneath the soil and when frozen or eaten back to soil level are unable to produce new growth. The growing

(a)   (b)   (c)

FIGURE 15-3
The development of a bean plant: (*a*) bean seed, (*b*) germinating bean seed, and (*c*) bean plant. (*Courtesy of F. D. Cochran.*)

point of grasses like corn and sorghum remains below soil level for some time after leaves emerge and is therefore relatively safe from light frost and predation.

Seed Production

The growth of plants specifically for seed production is widespread and ranges in scale from the home gardener who saves enough for next year to large commercial enterprises that grow many hectares of a crop for seed. Seed collecting from wild plants in their natural habitat is also practiced on both the amateur and commercial levels. Seeds from ornamental shrubs and forest trees are collected in large commercial operations.

Commercial growers in this highly specialized industry devote large acreages to the production of particular cultivars or hybrids and may produce seeds of field crops, grasses, vegetables, or flowers for both commercial growers and home gardeners. Many seed companies conduct active plant breeding programs for the major seed crops they produce. This has become especially important with the extension of plant patent laws to cover seed-propagated plants. For field crops and certain other species, the production of certified seed is gaining in importance.

Environmental conditions necessary for economical production of high-quality seed largely determine the location of commercial seed operations, which are often located in areas different from those in which the crops will later be grown. Seasonal temperatures, light levels, photoperiod, rainfall, and relative humidity are especially important. For example, seeds free of fungus and bacterial diseases are much more easily produced in an arid than a humid climate. For many species, low humidity and minimum rainfall during maturation are desirable conditions for drying seed prior to harvest. The Pacific Northwest and some irrigated regions of the Southwest provide such conditions. For some species, a low humidity causes premature shattering of the seed pod before or during harvesting. Many flower seeds are produced in Puerto Rico, Florida, or the Lompoc Valley of California, where moist winds from the ocean and frequent fogs prevent early dehiscence of the seed pods.

The methods used in seed production depend upon the genetic makeup of the species and cultivar and its pollination characteristics. If a cultivar is homozygous and self-pollinated, true-to-name seed can be produced quite readily. Plants which will cross-pollinate with other cultivars or even species are more difficult. For seed production, such plants are grown in strict isolation to ensure self-pollination.

With some crops such as corn, essentially all cultivars grown are F_1 hybrids. To produce F_1 hybrid seeds, two purebred (homozygous) cultivars must be crossed (Chap. 14). Although expensive, one method of producing such hybrid seed was to interplant the two parents. Before the release of pollen, the female parent plants were detasseled so that the only source of pollen was the desired cultivar, leading to hybrid seed. When plant breeders learned that a genetically transmitted male sterility factor could render pollen from the female parent sterile, essentially all hybrid corn was grown using male sterility in place of detasseling. In 1970 an epidemic of southern corn-leaf blight dealt a devastating blow to corn throughout the United States. Since this particular strain of the fungus attacked only plants containing the male sterility factor, corn seed producers had to temporarily abandon the male-sterility method of controlling pollination and return to detasseling.

Harvesting Because they are dry and incapable of carrying on metabolism for cellular repair or maintenance, seeds must be protected from mechanical, environmental, and microbial damage. In theory, this means harvesting seeds as gently as possible as soon as they reach maturity and then storing them under conditions which prevent damage or decomposition from environmental influence. In practice, one must wait until the seeds are dry enough to escape being easily damaged in the mechanical thresher, which removes the seeds or grains from other unwanted plant parts. They must not be too dry, however, or they will be fractured or broken. Of course, drying depends on the weather conditions, and the grower can only wait until the seed reaches the right stage. The various threshing devices in a combine, which harvests and separates seeds, must be adjusted to minimize damage to seeds while separating them from unwanted plant parts and weed seeds.

Tedious and careful hand labor is necessary in the harvest of many flower and vegetable seeds, especially when the seed head or pod shatters easily or the seeds continuously mature over a long period of time so that flower buds, flowers, and immature and mature seeds may be on the same plant at one time. The chief advantage of hand picking is that cleaning is greatly reduced. In such crops as carnation, hollyhock, and sweet pea, the entire plant is cut, placed on canvas and dried so that the seed can be threshed out. Many crops, especially those with fleshy fruits, require the seed to be milled from the fruit and then cleaned. Tomato seeds are removed after maceration of the ripe fruit and fermentation of the pulp.

Cleaning After harvesting and field threshing, seed often need cleaning because of contamination with plant residues, soil, weed seeds, or other foreign material. The centuries-old method of winnowing the chaff from grain is still practiced in many places, and the same principle has been adapted in some types of modern seed-cleaning equipment. Differences in size, density, and shape of seed compared with plant debris and other undesirable objects generally determine the method. Sieves with different mesh sizes are used to separate the larger particles from smaller seed or larger seed from smaller particles. Smaller, lightweight particles can be removed by blowing an air current through the seed as they pass from one screen to another, across a porous bench, or against an inclined plane. The heavier seed remain at the base while the lighter materials are blown into another plane.

By using an *indent machine*, or disk separator, desirable seed can be separated from other seed or particles of the same density but of different shape. A disk with indentations, the size and shape of which are determined by the particular crop being cleaned, is passed through a batch of seed, and each indent picks up a seed of the appropriate size. Some seeds can be separated on the basis of color (bean or pea). Single seeds are picked up by suction through perforations on a hollow wheel and are then passed through a photoelectric device; when an object of the wrong color is detected, the vacuum is released and the seed ejected.

Seed separation and cleaning are delicate operations. The machinery must be adjusted carefully so that the seed are not damaged or chipped. Damaged seeds may fail to germinate, be more susceptible to storage diseases and insects, or produce weak seedlings.

Storage It is usually necessary to store seed for some time between harvest and the next planting season. Viability of seed after the storage period depends on the initial vigor of the seed at harvest, the rate of physiological deterioration inherent in the particular species, and the environmental conditions under which the seed are stored.

Seeds which dry down at maturity and are quiescent until rehydrated must be stored under conditions which will keep them dry and inactive. For such seeds, which include most of our field, flower, and vegetable crops, storage in a cool, dry environment is most desirable. Up to a certain point, the drier and cooler the better.

The commercial seed company, which is capable of drying and storing seeds in large quantities, often stores them at optimum moisture and temperature levels. Temperatures of -18 to $0°C$ appear optimal for most species. A common rule of thumb for dry, quiescent seeds is that storage life is doubled for each 1 percent decrease in seed moisture or each $5°C$ decrease in storage temperature. Seeds dried to a low moisture content are often stored in airtight bags or other containers to exclude moisture. Small packets of seeds sold to the home gardener now frequently come foil-wrapped or polyethylene-lined to keep the seeds dry and viable.

In some cases, it is possible to modify the storage atmosphere with beneficial results. A vacuum can be created, carbon dioxide content raised, or oxygen reduced. Controlled-atmosphere storage greatly extends the seed storage life of most dry, quiescent seeds.

On the other hand, seeds of many plants, some with great economic value, cannot dry down or they will die. Seeds of maple, oak, apple, pear, citrus, and others must remain moist and hydrated from maturity until germination. Water content of 40 percent or greater is necessary to keep the seeds viable. Obviously such seeds must be stored under different conditions from seeds which must be kept dry. High humidity and low temperature promote storage life and often satisfy stratification requirements common to many of these species.

Longevity Desiccation plays a role in reducing the longevity of seeds which must remain moist, while soybean, wheat, peanut, corn, and others have the reverse problem: high moisture can drastically reduce their lifespan to a year or less. Seeds of medium life are those which normally retain their viability for 2 to 3 years and may remain viable as long as 15 years, depending on storage conditions. Some crops, especially the small-seeded legumes, may remain viable for a century or more. Some weed seeds have even germinated after hundreds of years.

Treatment Seeds can be treated to protect the seed and/or the seedling from fungi, bacteria, insect, and animal pests. Three different types of seed treatments are used.

Disinfestants such as calcium hypochlorite eliminate organisms present on the surface of the seed, whereas *disinfectants* such as formaldehyde and hot-water treatment (50 to 56°C for 15 to 30 min) eliminate most organisms within the seed itself. Disinfestants also help protect against the fungi which cause damping off in seedlings. These fungi, which are often found in seedbed soil, destroy a wide range of host plants by attacking the stem at ground level, causing the seedling to collapse. *Protectants* are insecticides and fungicides which prevent insects or fungi from attacking the seed after treatment. As yet, there is no treatment for seed-carried virus.

Testing Careful protection of the embryo and retention of viability is worth little unless the seed has certain other desirable characteristics. Good seed is clean and free from soil, debris, and weed seed. This can be accomplished by good milling and cleaning operations. Good seeds have high germinability (Fig. 15-4). Although germinability can be tested by planting seeds in sand or soil or starting

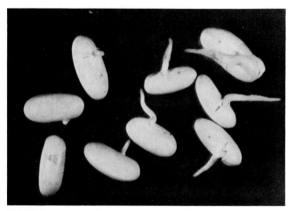

FIGURE 15-4
Closeup of a bean-germination test showing normal seedlings and ungerminated seeds. Some seeds sprout but make abnormal plants that will not survive in the field. (*U.S. Department of Agriculture.*)

them between folds of blotting paper, cotton flannel, or burlap, vigor is much more difficult to measure.

Seed certification programs with production standards set by the certifying state's crop improvement association maintain and make available crop seed of good seedling vigor that are *true to name* (belonging to the hybrid or cultivar listed on the certifying label). Seeds are classified as follows:

Breeder's seed is directly controlled by the originating or sponsoring plant breeder.

Foundation seed is seed stock handled carefully to maintain specific genetic identity and purity under supervised or approved production methods certified by the agency.

Certified seed is the progeny of *registered* seed stock; it is certified with a seal and tag.

Seed Planting

Usually small areas, flats, greenhouse beds, and home gardens are planted by hand, while larger areas and fields are planted by machine. Seeders, which are implements used for planting, may be hand-operated or be accessory equipment for a tractor. Some grain drills and other planters may be used both to plant seed and place the fertilizer at the time of planting. (The name "drill" is generally used for planters which sow close-seeded crops such as cereal grain.) A correctly adjusted planter places the seed at a uniform depth, a predetermined distance apart, and firms the soil around the seed. Desired depth of planting depends upon the size of seed, soil texture, structure, and moisture and oxygen content and the time of year. Since small seeds have less stored food, they must be planted nearer the soil surface to decrease the time and energy needed for the seedling to reach the surface and begin to produce its own food by photosynthesis.

If water is the limiting factor, as it often is during the germination period, seeds need to be planted at a greater depth to take advantage of the higher soil moisture. If the soil is nearly saturated, as it often is during late fall, winter, and early spring, seeds should be planted nearer the surface, since the oxygen necessary for respiration may be very limited in wet soils.

Temperature requirements of the crop and temperatures of the soil at planting depth determine the best time for planting. Warm-season crops are usually sown after the last killing freeze of spring; cool-season crops may be planted earlier. Soil temperatures also affect seeding depth. Clay soils are usually colder and stay cool longer; sandy soils are warmed earlier by the sun.

Tapes and Pelleting When seed are small or irregular, an even distribution is often a problem. Planting in narrow rows improves distribution patterns in some field-grown crops. To make seeds of garden and flower-bed species round and uniform in size, they are often pelleted, i.e., coated with montmorillonite, a fine clay. Pelleting permits more uniform planting, eliminates a great deal of thinning, and permits the incorporation of fungicides onto the coat. This procedure is used on a field scale with sugarbeet and some small-seeded legumes. Seed tapes or ribbons are available for many flower and vegetable cultivars. Seeds are attached at the proper intervals on a plastic tape. When the tapes are buried in a shallow furrow, the tape itself disintegrates from moisture as the seeds germinate. This convenience practice works well for the house gardener but is very expensive when one considers the cost per seed. Fluid drilling is a newly developed seeding technique in which hydrated seeds or barely germinated seedlings are injected into the soil with a special fluid drill. The planter is more expensive, but the technique shows great promise for many small-seeded, high-value crops, e.g., peppers, tomato, and eggplant.

ASEXUAL PROPAGATION

Asexual propagation, the duplication of a whole plant from any living cell, tissue, or organ of that plant, is possible because normal cell division (mitosis) and cell differentiation occur during growth and regeneration. Mitotic cell division is involved whether it is root and shoot initiation or callus-tissue formation in the process of budding and grafting. A single totipotent cell can generate a new plant because each cell contains all the genetic information necessary to reproduce the entire organism (Chap. 6). A group of plants originating from a single individual and propagated by vegetative means is referred to as a clone. Some clones have been maintained by people for hundreds of years; others have existed in nature by vegetative reproduction from bulbs, rhizomes, runners, and tip layers. Since clonal

plants are genetically identical to the parent, the desirable characteristics of the "parent" are transferred to each new plant. The 'Bartlett' pear, which originated in England in 1770, and the 'Delicious' apple, discovered over 100 years ago, are well-known asexually propagated clones. Other plants propagated by asexual means include strawberry, blueberry, sweet potato, potato, bermudagrass, sugarcane, chrysanthemum, rose, and most nursery and indoor plants.

Because of their heterozygous genetic makeup, cultivars of these plants do not pass their desirable genotype via seeds. Seeds from a 'Bartlett' pear will produce pear trees, but the fruit is inferior. The same is true for many other fruit and ornamental species and some vegetables.

Many valuable plants must be propagated asexually because they produce little or no seed, due to season of flowering, genetic incompatability, or sterility. Hybrids may be sterile if lack of homology between chromosomes results in irregular cytological behavior and reduced seed production. Double-flowered species in which flower petals replace reproductive parts, e.g., double petunia, carnation, and African violet, must also be vegetatively propagated.

Dwarfing effects, resistance to certain insects and diseases, better adaptability to a given soil, and hardiness of certain parts of a plant can also be obtained by combining two different clones of a species vegetatively. Disease-resistant rootstocks make it possible to grow cultivars susceptible to soil-borne pathogens. The European grape grown on American grape rootstocks is not attacked by root aphids, nor are peaches attacked by specific nematodes when grown on certain nematode-resistant rootstocks. Growth-regulating rootstocks are sometimes used in fruit trees and landscape plants to control size or produce other desirable characteristics.

CUTTINGS

Cuttings are detached vegetative plant parts which will develop into a complete plant with characteristics identical to the parent plant when placed under conditions favorable for regeneration (Fig. 15-5). Any vegetative plant part capable of regenerating the missing part or parts when detached from the parent can be a cutting, i.e., roots, stems, leaves, or modified stems such as tubers and rhizomes can all be used

FIGURE 15-5
Rooted stem cuttings. (*Courtesy of E. V. Jones and J. P. Fulmer.*)

photosynthetically. Moisture will be lost from the cutting and rot-producing organisms can invade the cutting through the unhealed wound. Normally, after the wound is made, the intercellular spaces and parenchyma cells near the wound become filled with unsaturated fatty acids, which combine with oxygen to form *suberin*. The thin, varnishlike layer of suberin seals the moisture of the cutting inside the tissue and keeps rot-producing organisms out. Suberin is inelastic and therefore effective only for a short time. If the cutting heals properly, a permanent layer of tissue is soon formed. These new cells become meristematic and produce other cells, which become filled with suberin, tannin, and similar materials to produce a corky layer. This layer is constantly renewed and elastic enough to withstand the pressure changes due to water loss and intake. The root system can arise from the pericycle, from callus, or from epidermis of young stems and from the cambium of older stems.

as cuttings. The type used to propagate a given plant is determined by ease of root or shoot formation, presence or absence and type of leaves, facilities for propagation, and season of the year.

Stem Cuttings

Segments of shoots containing lateral or terminal buds are obtained for stem cuttings with the expectation that, under the proper conditions, adventitious roots will form and the buds will develop, thus producing independent plants (Fig. 15-5). The type of wood, the stage of growth used in making the cuttings, and the time of year when the cuttings are taken can be important to rooting success. The four types of stem cuttings are classified according to the nature of the wood used in making the cuttings.

Softwood These cuttings are taken from soft, succulent, new spring growth of deciduous or evergreen species of woody plants. Typical examples are magnolia, weigela, spirea, and maple. The cuttings, composed of a stem portion and a growing tip, are usually 8 to 12 cm long, and the lower leaves are removed. Care must be exercised to aid wound healing and root development. Since the tissue is immature and few reserve carbohydrates will be available to provide energy for new growth, it is necessary to see that the leaves are turgid and active

Herbaceous These cuttings are made from succulent herbaceous plants such as chrysanthemum, coleus, geranium, and many indoor plants. Most florists' crops are propagated in this way. Herbaceous cuttings 7 to 10 cm long are rooted under conditions similar to those for softwood cuttings. Under proper conditions, rooting is rapid and in high percentages. Herbaceous cuttings of plants that exude sticky sap, e.g., the geranium, cactus, or pineapple, do better if the basal ends are allowed to dry for a few hours before insertion into the rooting medium, permitting a partial sealing to occur and preventing the entrance of decay organisms.

Hardwood Hardwood cuttings can be made from deciduous species and narrow-leaved evergreen species. Deciduous hardwood cuttings from the current season's growth do not require leaves for rooting; in fact, they wilt when softwood, i.e., recently produced, leaf-bearing stem is used. Examples are forsythia, wisteria, fig, grape, currant, and mulberry. Hardwood cuttings made during the dormant season should be from vigorous stock plants. They are taken after the chilling requirement for breaking bud dormancy has been satisfied but before midwinter or spring. Roots that develop from the cut derive from the cork cambium, but roots that develop at nodes derive from the vascular cambium. Cuttings vary from 10 to 75 cm in length. During the winter the wound heals and roots form; in the spring, the rooted cuttings can be planted in the field or in pots, where they remain a year or so before being transplanted. Hardwood cuttings of grape and some other species can be made, stored over winter, and then planted out in the spring when they will root. Ideally, roots form before bud break so that the transpirational needs of the new leaves can be met.

Narrow-leaved evergreen cuttings are taken between late fall and late winter. Cuttings of this type have leaves and must be rooted under moist conditions that will prevent excessive drying, since they are slow to root, taking several months to a year. In general, *Chamaecyparis*, low-growing *Juniperus* species, and the yews root easily; the upright junipers, the spruces, and hemlocks are more difficult. The cuttings are 10 to 20 cm long and all leaves are removed from the lower half of the cutting. Mature terminal shoots of the previous season's growth are generally used. In certain narrow-leaved evergreen species, an additional basal wound is often beneficial in inducing rooting.

Semihardwood Cuttings of this type are usually preferred for woody, broad-leaved, evergreen species. They are taken during the summer from new shoots just after a flush of growth has taken place and the wood is partially matured. Ornamental shrubs such as camellia, evergreen azalea, and holly (Fig. 15-5), and fruit species such as citrus and olive can be propagated by semihardwood cuttings.

The cuttings are made 7 to 14 cm long with leaves retained only at the upper end. If the leaves are very large, they should be cut smaller to lower the water loss and allow closer spacing in the cutting bed. The shoot terminals are often used in making the cuttings, but the basal parts of the stem will usually root also. The basal cut should be made just below a node.

Leaf Cuttings

Leaf cuttings produce more plants per parent plant and per unit of greenhouse space than stem cuttings. Entire leaves with or without the petioles are normally used. Adventitious shoots and roots develop from the leaf blade or petiole (or both); the blade normally rots away, rarely becoming part of the new plant. *Begonia rex*, *Sansevieria*, *Peperomia* (Fig. 15-6), and African violet are propagated by this method. Plants with thick, fleshy leaves such as *Begonia rex* are propagated by severing the large veins on the underneath side or cutting the leaves into sections and laying the leaf flat on the propagating medium. With high humidity a new plant will form where each vein is cut.

Leaf-Bud Cuttings

To make a leaf-bud cutting, a leaf blade, petiole, and a short piece of the stem with the attached axillary bud is placed with the bud end in the rooting medium and covered enough to support the leaf. Care must be taken because, if the bud is injured, the plant will produce roots but not a new plant. This method is used when plants are able to initiate roots but not shoots from detached leaves. It is most valuable when propagating material is scarce because more new plants can be produced by this method than with stem cuttings. Blackberry, lemon, camellia, rhododendron, and black raspberry can all be propagated by leaf-bud cuttings.

Root Cuttings

Plants such as blackberry and raspberry which develop shoots from the root system are easily propagated from root cuttings. In one method, root cuttings 5 to 15 cm long are taken in the fall or winter and are stored in moist sawdust or sand in a cool place to allow new shoots to develop from adventitious buds. (Roots do not have true buds, but many are capable of forming adventitious buds.) These plants are transplanted in the spring. Cuttings can also be started in hotbeds or greenhouse beds in the winter

FIGURE 15-6
Rooted leaf cutting of *Peperomia caperata.*

to be transplanted in the spring. The cuttings are placed in the medium either vertically or horizontally. Care must be taken that the original root-end orientation is maintained when the vertical position is used or roots will not form. In other plants, e.g., sweet potato, the whole fleshy root is planted in a warm medium, where shoots develop with their own root system. Each rooted shoot (slip) can be transplanted. In some situations, root cuttings are used as rootstocks to reproduce a cultivar which is grafted to a special rootstock. For example, lilac can be dwarfed by grafting to root cuttings of privet.

Environmental Factors

Environmental factors play an important role in the ability of cuttings to heal and regenerate parts. Since stem and leaf cuttings lack roots, shoot growth must be slowed until a root system can be developed to balance and support it. A balance between humidity, water, temperature, and sunlight should exist for optimum growth without wilting. The removal of leaves should be minimal since as large a turgid-leaf area as possible should be maintained to produce hormones and carbohydrates for the support and development of the plant. With a high relative humidity, less water is required by the plant, and stomata may remain open in the light. Thus, carbon dioxide can enter the leaves, photosynthesis can occur, and carbohydrates will be manufactured. Low air temperature will slow respiration and the growth of the top. If the medium in which the cuttings are placed receives bottom heat, in which the base of the cutting is heated to 24 to 27°C by electric resistance wire or steam or hot water in pipes, the rate of root-cell division will be increased (Fig. 15-7). The high root-zone temperature encourages rapid oxidation of the fatty acids to suberin, which heals the wound and helps develop a new root system, while the lower air temperature maintains top growth at a slow pace. Oxygen required for the oxidation of the fatty acids to suberin and for the activities of the meristem is provided by using an adequately aerated rooting medium.

A relatively high light level must be maintained to promote photosynthesis, the products of which are used for root initiation and growth. Since a high light level and consequent high temperature can also cause rapid transpiration, wilting is likely to occur, resulting in decreased photosynthesis and possibly desiccation of the cutting. To combat this, mist propagation (Fig. 15-8) is commonly used. Mist maintains a thin layer of water on the leaf, and the evaporative effect means that leaves are often 5 to 8°C cooler than the surrounding air temperature. The intermittent mist system syringes the cuttings with water and gives best results if applied only during daylight. Continuous mist can result in severe leaching of the foliage nutrients and potential disease

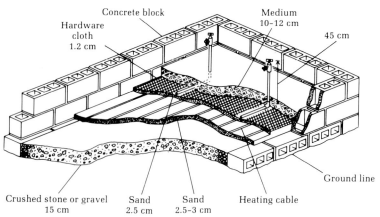

FIGURE 15-7
Cutaway view showing construction of a plant bed for outside conditions. Width of bed shown is 152 cm; nozzles spaced 76 cm apart and 38 cm from side and end of bed. The first run of concrete blocks is laid horizontally for drainage. (*Redrawn from North Carolina Agricultural Extension Bulletin 506.*)

FIGURE 15-8
Closeup of mist nozzle and propagation beds.

problems since many serious fungal diseases are water-borne.

The optimum length of a mist period varies with the plant material and environmental conditions during rooting. Except during extremely hot, dry weather, mist for 6 s min^{-1} from the time the cuttings are stuck until rooting begins is recommended, with gradual reduction in the frequency of misting as the roots elongate. Therefore, by the time the roots are 2 to 3 cm long the cutting is acclimated to a drier atmosphere. This hardening off reduces the severity of shock and minimizes transplant losses. In a continuous succession of propagation where a misting regime is not possible, the newly dug cuttings should be placed in a shady area and syringed frequently until they harden off.

Propagation can be done in outdoor areas (Fig. 15-9), cold frames (Fig. 15-7), and greenhouses (Fig. 15-10). The rooting medium should give the necessary support and environmental conditions for the development of roots. It should be lightweight and moisture-retentive while allowing good aeration and drainage. Galvanized mesh can be placed beneath the medium to enhance both drainage and aeration. Clean sand, finely milled peat moss, vermiculite, perlite, or a combination of these materials is fre-

quently used as a rooting medium. Generally a combination of several components gives the best results. For most plants, a wide range of materials may be used successfully; but, with plants that are more difficult to root, both the percentage of plants rooting and the quality of the root system that develops are influenced by the medium.

Growth Regulators

With many species, the formation of roots on stem cuttings can be enhanced by the application of PGRs to accelerate root initiation and development, to increase the rooting percentage, to increase the number and quality of roots produced per cutting, and to increase uniformity of rooting. The roots produced following PGR treatments generally have the same origin as natural roots. The common rooting hormones are synthetic auxins, applied singly or in combination. Also effective in stimulating root formation in some species are ethylene, acetylene, and propylene.

To be most effective, PGRs must be used in a specific concentration range for each individual species. Excessive concentrations may injure or kill the base of the cutting, while low concentrations may be ineffective. Within given limits, high concentra-

FIGURE 15-9
Outdoor intermittent-mist propagation beds.

FIGURE 15-10
Propagation of plants in miniature greenhouses. Note the ventilation holes in the plastic.

tions for a short time are comparable to a lower concentration for a longer time.

Auxins such as indolebutyric acid are applied in a variety of ways. The commercial powder preparations are used simply by dipping freshly cut stem tips in powder spread on wax paper or aluminum foil. Cuttings can also be treated by soaking the basal end in a dilute solution for 24 hours before placing them in the propagation medium. In a third method, a concentrated solution of the chemical in 50% alcohol is prepared and the basal ends dipped in it for a short time, generally 5 s. This last method takes less time and equipment and results in more uniformity.

LAYERING

Layering causes portions of plants to produce adventitious roots, after which the new plant is severed from the parent. The wood should be young so that it can form adventitious roots more easily. Conditions for layering are quite similar to those necessary for rooting cuttings. Species that are hard to root lend themselves to layering since the parent plant continues to provide the food and water for as long as it takes for roots to develop. The method has two basic disadvantages, however, in that only a few individuals can be started from a given stock plant at any one time and the hand labor involved makes the cost prohibitive for most large-scale use.

In *tip layering*, rooting takes place near the tip of the current season's shoot, which naturally falls to the ground. The shoot tip begins to grow down into the soil but eventually curves upward, with roots developing from meristematic tissue at the curve. Trailing blackberry, dewberry, and black and purple raspberry frequently propagate themselves naturally by tip layering. New plants are limited by the number of canes available. *Simple layering* (Fig. 15-11), which is used for forsythia, grape, jasmine, and various broadleaved evergreens, is the same as tip layering except that the stem behind the end of the branch is covered with soil and the tip remains above ground.

In *trench layering* several new plants can be obtained from a single stock plant. Rose, muscadine grape, spirea, and various deciduous shrubs lend themselves to this method. The basal and middle portions of a young stem are injured by notching the nodes to increase likelihood of root formation. The stems are then placed in a shallow trench and covered with 5 to 10 cm of moist soil. The tip is left exposed to ensure continued growth and food production.

After being cut back to the ground during the dormant season, some plants develop many new shoots from the base (apple rootstocks, quince, and currant). In *mound* or *stool layering* (Fig. 15-12), soil is mounded around the bases of the new shoots as they develop in the spring, enhancing root formation. This mounding process is repeated 3 or 4 times as

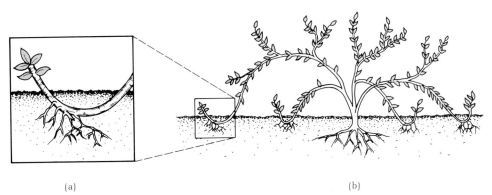

(a)   (b)

FIGURE 15-11
Propagating plants by simple layering. (*Adapted from E. L. Denisen, Principles of Horticulture, Macmillan, New York, 1958.*)

FIGURE 15-12
Mound layering, showing new plants still attached just after soil has been removed.

the shoots elongate, each time leaving the upper portion uncovered. The soil is removed during the dormant season and the new rooted plants are detached; the parent plant can be used year after year. This technique is widely used in producing clonal apple rootstocks.

Air layering can be used for foliage plants such as rubber plants, dumb cane (Fig. 15-13), and *Dracaena*. A ring of tissue about 2 cm wide is removed from the stem and the wound is surrounded with moist peat or sphagnum moss. Plastic wrap such as polyethylene, which is permeable to carbon dioxide and other gases but impermeable to water, is used to cover the rooting medium. When the roots have developed, the stem is severed from the parent plant below the root ball.

(a)

(b)

FIGURE 15-13
Air layer of *Dieffenbachia*: (a) polyethylene used to cover the rooting medium; (b) polyethylene and rooting medium removed, showing roots.

GRAFTING

Grafting consists of joining two separate plant structures, e.g., a root and stem or two stems, so that by tissue regeneration they form a union and grow as one plant. The upper part of the union is the *scion* (cion), and the lower part is the *stock* (rootstock). Some graft combinations contain an *interstock* or intermediate stem piece grafted between the scion and stock. This process is called *double-working*.

Grafting is used to reproduce clonal cultivars, e.g., almond, apple, walnut, and eucalyptus, which are not easily or successfully reproduced by other methods. The ability of certain dwarfing clones, used as either rootstocks or interstocks, to produce a dwarfed and early-bearing fruit tree has been known for centuries. Plants with special growth habits such as a rose tree or weeping cherry can be obtained only by grafting. *Topworking* is grafting a new cultivar into a tree, so that a tree found to be an undesirable cultivar can be changed to the one preferred by the grower. Several different cultivars of the same species can be grown on a single root system; pink and white dogwood branches can be grown on the same plant. Similarly, one could graft an apple tree to produce 10 or more different cultivars on the same tree.

Whip Grafting

The whip graft, often used to propagate apple and pear trees as well as ornamentals such as *Hibiscus syringa*, is also referred to as the bench graft since the work is commonly done at a bench or table (Fig. 15-14). The graft is made in the winter or spring before bud break. With scion and rootstock of approximately 0.6 to 1.2 cm in diameter, similar cuts should be made at the top of the stock and at the bottom of the scion to ensure as perfect a cambial fit as possible. A long, smooth, sloping cut 2.5 to 5 cm long is made in the scion. About one-third the distance from the top of this cut, a second cut is made about half its length and roughly parallel to it. The stock is prepared in a similar manner. The scion and stock are tightly fitted together, wrapped, and covered with wax or other waterproof coating to prevent the graft union from drying. Whip grafts heal quickly, forming a strong union.

Cleft Grafting

Cleft grafting, performed late in the dormant season, is used to topwork fruit trees and some large orna-

mentals, such as camellia. The scion should be fully dormant 1-year-old wood cut in the fall and stored or cut early in the spring. The stock is usually 5 to 7 cm in diameter and is split smoothly down the center for several centimeters (Fig. 15-15). Usually, two scions with the basal end cut in a long, gradually tapering wedge are inserted into the stock which has been split, and, as always, the cambium of the scion must be in contact with the cambium of the stock. The use of two scions allows selection for vigor and position, promotes speed in the healing process, and generally ensures a more successful graft union. The second scion also prevents dieback of one side of the stock, which is common if only one scion is used. When the poorer scion is removed, it is usually done gradually. The objective is to have the scion reach the diameter of the stock in a minimum number of years. Both scions should never be left because they form a weak crotch (Fig. 15-16).

Bark Grafting

Bark grafting can be adapted to a stock of any size but requires the stock bark to separate readily from the wood. Therefore, the procedure is always performed after active growth has started in the stock. The dormant scion, shaped like a wedge, is inserted between the bark and wood of the stock. Several scions can be inserted to allow selection. One of the most vigorous is selected and the others are pruned to keep them subordinate until they are eventually removed. Bark grafting is rapid and easy to perform and gives a high percentage of success.

Notch Grafting

Notch, or saw-kerf, grafting can be performed after the stock initiates growth and is generally used for topworking trees with large branches or for curly-grained wood which does not split evenly for cleft grafting. Usually three scions 10 to 12 cm long with two or three dormant buds are inserted into the stock where 3- to 4-cm cuts have been made on the perimeter into the center and 10 cm down the side. The wedge-shaped scions, with the outer side slightly wider than the inner edge, are inserted where the cambium matches. The exposed surfaces are then waxed.

Approach Grafting

Approach grafting, i.e., joining two plants on their own roots, is necessary for plants difficult to graft

(a)   (b)

(c)   (d)

FIGURE 15-14
Whip grafting: (*a*) scion and stock cut into one-sided wedges; (*b*) scion and stock cut one-third of that distance down for about 1.5 cm; (*c*) ends inserted so that cambium matches; (*d*) graft wrapped with tape. (*Courtesy of J. D. Ridley.*)

by other methods, e.g., conifers. A length of bark 2.5 to 5 cm long is removed from both plants and the wounded surfaces pressed and held tightly together. Formation of the union is slower than in other methods but can be speeded up if the grafting is done during the growing season. Upon successful union, the top of the stock is removed, and the root system of the stock is retained along with the new scions.

Formation of the Graft Union

The formation of the graft union is an intricate process. The first prerequisite for a successful graft is that the scion and stock align and fit tightly. The cambial layers of the scion and the stock must be in contact so that callus cells from the cambium tissues can unite (Fig. 15-17). The cambium of the stock and scion produce parenchyma cells, which soon interlock, forming the callus tissue. Cells of the callus

(a)

(b)

(c)

FIGURE 15-15
Procedure for making a cleft graft: (a) scion after wedge has
been cut; (b) inserting scions into stock with thick side of
scion wedge toward the outside; (c) graft after waxing cut
surfaces. (*Courtesy of J. D. Ridley.*)

(a)

(b)

FIGURE 15-16
Cleft graft on apple tree: (a) both scions left on stock, showing development of weak crotch;
(b) one scion suppressed and then eliminated, stock and scion the same size.

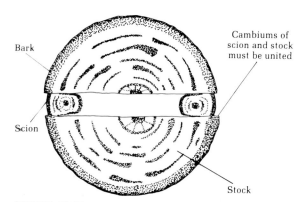

FIGURE 15-17
Cross-section of cleft graft, showing matching scion and stock cambiums. (*From N.Y. State Agric. Exp. Statn. Bull. 817.*)

tissue begin to differentiate into new cambium cells in line with the intact cambia of the stock and the scion. These cambium cells then produce vascular tissues, xylem to the inside and phloem to the outside, which establish the essential continuity between the vascular systems of the scion and the stock. Food materials and plant sap can then be translocated through the graft union.

Incompatibility The most important factor determining the success of the graft is incompatibility, the inability of the two members of the graft to form a lifelong union. This condition is frequently caused by latent viruses, which are present but produce no symptoms until introduced into a susceptible host. A rootstock infected with a latent virus shows no symptoms; when it is grafted to a susceptible scion cultivar, however, the scion will exhibit symptoms and the union may be poor. Incompatibility can result in weak unions which eventually die, nutritional disorders, stunted plants, physiological abnormalities, and even the failure to form a union at all. Symptoms of incompatibility occasionally develop many years later to an extent that varies with cultivars and taxa.

Generally, the closer the plants are botanically related the better the chance of a successful graft, provided no virus or other incompatibility problems exist. A graft between distantly related species generally will not be successful in the long run.

The Plant and Kind of Graft Even though the stock and scion may be compatible, the union may be extremely difficult to initiate. Once the union is successful, however, the plant grows very well. Hickory, oak, and beech are examples. In addition, some plants respond to specific techniques and kinds of grafts better than others. For example, topworking *Juglans hindsii* with *J. regia* (English walnut) is more successful with a bark graft than with a cleft graft. Some species are so recalcitrant that approach grafting is used, allowing both plants to remain on their own roots until a graft union has been formed. Muscadine grape and *Camellia reticulata* are propagated by the approach-graft method.

Environmental Conditions Temperatures that stimulate cell activity generally range from 13 to 21°C. Grafting is usually done when the stock tissues, especially the cambium, are in an active state but the scions dormant. Since the callus tissues of thin-walled, tender parenchyma cells are easily desiccated by drying air and soon die, a high moisture content is necessary to prevent dehydration. For most plants, a wax or other waterproof layer over the graft is all that is necessary to maintain the natural moisture of the tissue. A sufficient supply of oxygen is also necessary at the graft union, where cellular activity and respiration reach high levels requiring considerable available oxygen.

Stage of Growth The success of the graft also depends on the stage of growth of the stock plant. Most successful unions in many species take place when the bark is *slipping*. This is the state of rapid cell division when the vascular cambium is producing thin-walled cells on each side. The activity is initiated in the spring and may be the result of a stimulus from auxin from the expanding buds. Plants which exhibit extensive bleeding or abundant sap flow when cut for grafting will not form a graft union.

Poor Techniques Many failures are the result of poor technique. Delayed waxing, uneven cuts, lack of contact between cambium layers, and grafting under unfavorable conditions may lead to failure of the graft union.

Grafting of deciduous species is generally performed in late winter or early spring using dormant scion wood from shoots of the previous summer's growth. Scion wood with strong leaf buds is taken from 1-year-old shoots of a size suitable for grafting. Older wood can be used if buds are present. Some species, e.g., fig, can be grafted more successfully with 2-year-old wood. Care must be taken to see that scion wood is properly stored in cold storage or a refrigerator at slightly above 0°C. Freezing temperatures may cause injury. A cool cellar or similar location is satisfactory only for short periods. To keep the wood slightly moist, bundles may be packed with moist sawdust, sand, or peat and wrapped in heavy waterproof paper or polyethylene. Avoid soaking the wood as more damage results from the wood's being too wet than too dry. Keep scion wood away from fruits because the ethylene produced as they ripen is toxic to propagation materials. Grafting of broadleaf evergreens is done in the spring before active growth starts; the scion wood may be taken as needed from basal wood with dormant buds.

BUDDING

Budding involves placing a vegetative bud in a stock plant. Since only one bud is used, the use of scion tissue is economical.

T Budding

The stock bark is cut to form a T, which allows the bud piece to be slipped behind it (Fig. 15-18). The bud piece is attached by placing the bud against the freshly exposed cambial tissue of the stock. Raffia, rubber strips, soft string, or tape is used to hold the scion and stock tightly together. Xylem and cambium

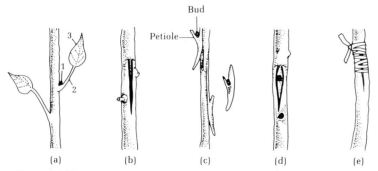

FIGURE 15-18
T budding: (a) scion, 1, bud; 2, petiole; 3, leaf blade; (b) T-shaped cut through the bark of the stock with bark raised to admit the bud; (c) bud stick; (d) bud in place; (e) bud wrapped with raffia. (*Redrawn from N.C. Agric. Bull. 326.*)

tissues of the stock and scion produce the callus tissue. For T budding, the bark of the scion must slip or separate readily from the xylem. In most areas of the country, T budding is done between mid-August and early September. In the southeast, T budding is done in June. With late-summer budding, the stock is left intact until the following spring, when it is cut off just above the bud. With June-budded trees, the stock is cut off as soon as the union has formed, i.e., in about 2 weeks. Using June budding, a nursery can produce a salable tree in 1 year compared with 2 years for August-budded trees. T budding is the usual method of propagation for fruit trees because it is much quicker than whip grafting.

Patch Budding

In patch budding, which is similar to T budding, a rectangular bark patch is taken from the stock plant, exposing the cambium. A bud is removed from the desirable cultivar with a bark patch identical in size to the patch removed from the stock plant and held in place with waxed cloth or budding tape. Thick-barked trees, e.g., pecan and walnut, adapt well to patch budding although it is slower and more difficult than T budding. It is usually done in late summer and early fall, when both stock and bud bark slip readily. In *flute budding*, a variation of patch-budding, a bark strip is removed from the circumference of the stock, leaving a thin vertical strip approximately 3 mm wide. A piece of the same size is removed with an attached bud of the scion cultivar and placed on the stock. The connecting strip serves as a supply route until a union is formed.

Chip Budding

Because it does not require that the bark be slipping, chip budding can be performed in early spring before growth starts or in late summer after growth has ceased, but it takes longer and is more complex than T budding. On the scion cultivar, a downward cut at 45° is made below the bud. A second cut is made about 12 mm above the bud inward and downward in a transverse manner, meeting the first cut. The bud chip is removed and placed near the base of the stock between two nodes, where an identical cut has been made to receive it. Wax, string, or nursery adhesive tape must be used to secure the graft. Late-summer grafts are topped the following spring and spring grafts approximately 10 days after budding.

SPECIALIZED STEMS AND ROOTS

Although the chief purpose of bulbs, corms, tubers, rhizomes, and fleshy roots is food storage, they are also valuable for propagation. Using bulbs and corms in propagation takes advantage of the naturally detachable parts in a process known as *separation*. *Tunicate bulbs*, e.g., onion and daffodil, have a series of scales arranged concentrically around the base and protected by a papery sheath, or tunic (Fig. 15-19). *Scaly bulbs*, e.g., lilies, have overlapping scales unprotected by a sheath (Fig. 15-20). Meristems develop in the axil to produce miniature bulbs, known as *bulblets*, which are grown to full size as *offsets*. In various species of lilies, bulblets form in the leaf axils on either the underground or aerial portion of the stem. Aerial bulblets are called *bulbils*

FIGURE 15-19
Tunicate bulbs for producing new plants. (*Courtesy of F. D. Cochran.*)

and underground parts *bulblets*. Other monocots, e.g., gladioli and crocus, produce corms, enlarged stems surrounded by dry, scaly leaves with roots extending from the bottom of the corm and a single stem developing from the top (Fig. 15-21). From axillary buds, the mother corm may also produce *cormels*, which can be planted to produce new plants. Some dicots produce *tubers*, which are enlarged tips of underground stems serving as storage organs of starch or related materials (inulin in Jerusalem artichoke). The buds of tubers are called *eyes* (potato, caladium). Tubers can be cut into sections and placed in a rooting medium to produce plants. Both monocots and dicots have rhizomes, i.e., stems, nodes, and internodes growing horizontally and separated by division, each segment containing a node to produce a new plant. Although nodes and internodes are not produced on fleshy roots, adventitious buds can develop to produce new shoots and plants, as in the sweet potato.

APOMIXIS

In plants which routinely form embryos without meiosis and fertilization, *apomixis* is a type of asexual propagation. *Obligate apomicts* produce only

FIGURE 15-20
Bulb scales of lilies producing new plants. (*Courtesy of J. P. Fulmer and E. V. Jones.*)

FIGURE 15-21
Gladiolus corm with cormels. (*Courtesy of W. P. Judkins.*)

FIGURE 15-22
Culture of explants on agar medium. (*Courtesy of R. C. Cooke.*)

apomictic embryos; *facultative apomicts* can produce both sexual and apomictic embryos. Apomixis ensures genetic uniformity through seed propagation since the apomictic cultivar produces genetically identical progeny seedling, e.g., Kentucky bluegrass. Since viral diseases are not usually transmitted by seeds, apomictic seedlings can be used to avoid some viruses in a clone which has become infected.

Sometimes additional embryos develop outside the embryo sac, as in *Citrus*. Two or more embryos developing within a single seed is referred to as *polyembryony*. One type is that of a true fusion embryo, as in *Citrus*, with up to 16 embryos from one seed. Only one embryo results from fertilization; the others result from sporophytic tissues such as the integuments or nucellus. Most plants that reproduce through apomixis require that pollination take place even though fertilization of the egg does not occur.

TISSUE CULTURE

Plant tissue culture techniques are the ultimate expression of totipotency and have become vital in pursuing a wide range of fundamental and applied problems in production, research, and development. Basic techniques are callus culture, organ culture, and cell suspension culture. The source of the plant tissue (shoot tip, leaf, or stem) is disinfected and then cut up. The explant is put in a container prepared with liquid or semisolid medium and kept in a controlled environment (Fig. 15-22). The growth and culture of the tissues are observed until plantlets reach desirable size, when they are subcultured and transferred to larger containers. Plantlets must be properly acclimated before being transferred to soil containers.

Growing a mass of unorganized cells (callus) on a semisolid medium or in liquid suspension is wide-

FIGURE 15-23
Photoperiod exposure of tissue cultures on nutrient agar medium in test tubes. (*Courtesy of R. C. Cooke.*)

spread in biochemical and growth studies. The culture of segments of stems, roots, leaves or of callus provides systems for the study of differentiation, morphogenesis, and plant regeneration. Shoot apex culture methods leading to plant regeneration have been adopted for plant propagation and production of virus-free stock. The culture of anthers and pollen provides new approaches to haploid plant formation. The technology has recently been extended to include the isolation and culture of plant protoplasts, which are employed in fusion and somatic-cell hybridization.

Plant tissue culture is useful for rapid asexual multiplication of carnation, orchids, and various ferns. Herbaceous plants have been used almost exclusively in this procedure, but woody types such as *Nandina* and *Ficus* are now also being propagated by tissue culture. Another important outcome of plant tissue culture is the mass production of disease-free material. Through research this method can also be used to improve plants genetically.

Special laboratory equipment and trained personnel are necessary for plant tissue culture. The laboratory should include an area for preparing nutrient formulations, autoclave, chemicals, glassware, and balances for weighing chemicals; an area equipped with a dissecting microscope and sterile bench for isolating the plant part and putting it in culture; and an area with controlled temperature and light for growing the cultures (Fig. 15-23).

Plant tissue culture holds tremendous potential for the production of a given clone and the propagation of plants that are slow or difficult to propagate. In some cases, tissue culture can produce a million genetically uniform plants from one plant part in a year's time, enabling plant breeders to develop a large volume of plants quickly. Pathogen-free plants, such as geraniums free of xanthomonas, chrysanthemums without stunt and other viruses, and carnations free of ringspot mottle, are all advantages of tissue culture.

CHAPTER SIXTEEN

DIRECTING PLANT GROWTH

Earliest recorded history indicates that our predecessors were actively involved in pruning and grafting as methods of plant improvement and growth control. The early Egyptians practiced surprisingly advanced horticultural techniques, as did the ancient Greeks. It is likely that many of the early treatments were an attempt to imitate nature. Many plants exhibit their own methods of growth control as part of their inherent makeup. Observant gardeners noticed instances of self-grafting, blossom drop, increased production on a branch bent toward the ground, and many other natural and accidental occurrences. Prompted by curiosity, those early agriculturists learned that copying nature could lead to dramatic results.

Today, genetic, physical, chemical, and environmental growth control is an advancing science. Our management of the plant kingdom is steadfastly growing with increased productivity, improved plants, year-round production, disease- and pest-resistant material, and a wide range of other results of growth control.

GENETIC CONTROL

Plant scientists have made remarkable progress in the control of plant growth by breeding and selection of new cultivars. For many crops, smaller-statured plants have been developed. Many of these dwarfs are probably the result of hormonal changes. We shall discuss only a few of many possible examples.

Cultivars of tomato have been bred for determinate growth to simplify mechanical harvest. A determinate tomato plant produces a limited number of nodes, leaves, and fruit clusters and then ceases to elongate. Photosynthates are used in maturing a limited number of fruit rather than producing more growth and additional fruit. Most of the fruits mature at about the same time, making these cultivars well suited to mechanical harvest. Large machines cut the plants just above the ground and remove the fruit (see Fig. 4-14). If most of the fruit did not mature simultaneously, such harvest techniques would be impractical.

Conventional indeterminate tomato cultivars produce new leaves and fruit clusters continuously, so that at any given time, mature fruit, immature fruit, and flowers are present. Ripe fruit on such plants can be harvested over a period of weeks or even months. Such cultivars are used where a concentrated harvest is not desired, in greenhouse tomato production and home gardens. Indeterminate cultivars are also used in growing fresh market tomatoes in the field. Because of their growth habit, the vines are usually held upright by stakes, strings tied to a wire support, or wire cages.

Bush cultivars of snap bean, lima bean, and pea have also been developed. Climbing or pole cultivars of snap bean not only required some type of support but had to be hand-harvested repeatedly. The bush cultivars are adapted to once-over mechanical harvest because a large percentage of the crop matures at one time. Although many home gardeners still prefer the pole bean or climbing pea, they are no longer grown commercially.

Members of the Cucurbitaceae family (squashes, cucumbers, melons) produce long trailing vines, which fruit continuously. Plant breeders have bred new cultivars better adapted to mechanical harvest. A major improvement has been the introduction of gynoecious cucumber cultivars, which produce only female flowers, so that a given length of vine produces more fruit than in conventional monoecious cultivars, which also produce male flowers.

Some new cultivars of small grains are shorter and therefore less subject to lodging. (Recall Borlaug's work with short-statured, tropical wheat.) These types often mature in a shorter growing season, an additional benefit in some situations.

Breeding and selection of dwarf cultivars of ornamental species have received great attention. New poinsettia cultivars are more compact and both easier to grow and more attractive. Dwarf marigolds grow only 15 to 20 cm tall compared to others which are 3 or more times taller. Among the woody ornamentals many types have been bred with different growth rates and habits. Instead of the large upright growth of many junipers, new selections are not only much smaller but some are prostrate and suitable as ground covers (Fig. 16-1). Dwarf forms of many other evergreen ornamentals have been introduced (Fig. 16-2). Many of the native rhododendrons are tall and leggy; compact types much better suited to landscape use have been developed and selected. Since they must be vegetatively propagated (Chap. 15), they are more expensive but well worth the cost.

Woody ornamentals with a weeping growth habit, i.e., with branches that droop or grow outward and then downward, are also popular not only because they provide diversity in the landscape but also because they are naturally dwarfed and require minimal pruning. Weeping forms of several evergreens and ornamental cherry are available.

Genetic control of growth is being used in tree fruit production. There are several genetic approaches to regulating size of apple trees.

1. Varying vigor in the scion cultivar is one method. 'Stayman,' 'Delicious,' and 'McIntosh' produce large trees; 'Rome' and 'Golden Delicious' produce smaller trees. There is also a choice within a given cultivar of either standard or spur-type growth habit. Spur-type trees have shorter internodes but produce similar numbers of leaves and are often called *compact mutants*. About 15 to 20 percent size control can be gained by using spur-type scion cultivars; they also tend to bear fruit a year earlier than standards.

FIGURE 16-1
Dwarf forms of juniper are effective in landscaping; note the use of *Juniperus conferta* as a part of the foundation planting.

FIGURE 16-2
Dwarf forms of azaleas are used as a form in design compositions. Note 'Gumpo' azaleas in center of photograph.

2. Using size-controlling rootstocks gives even greater control of apple tree size (Fig. 16-3). When the same cultivar of apple is budded on different rootstocks, mature tree heights can range from 7 m on seedling rootstocks down to 1 to 2 m on the most dwarfing rootstock. Dwarfing rootstocks are being more widely used because they also induce earlier production in the life of the orchard.

3. Fruit trees are also dwarfed by using an inter-stock, or intermediate stempiece (see Chap. 15). If a 10- to 15-cm-long piece of dwarfing rootstock is grafted between a vigorous rootstock and the scion cultivar, dwarfing of the tree will result.

Efforts are underway to develop dwarfing root-stocks for other nut and fruit trees, but the perennial nature of fruit trees makes genetic improvements much slower than with annual crops such as tomato, corn, or wheat. After the many years it takes to develop a new rootstock for tree fruits, adequate testing requires still more years. New rootstocks must be shown to induce early and heavy fruiting and to be well anchored, adaptable to a wide range of soils and climates, and resistant to major diseases and insects. A cooperative research project initiated in the late 1970s to test new fruit tree rootstocks throughout the United States and Canada will allow researchers to evaluate the potential of new rootstock candidates faster and better.

On the other end of the spectrum are the cultivars of crop plants which have been developed for increased vigor. Essentially all corn cultivars (both sweet and field corn) are F_1 hybrids. Hybrid vigor has markedly increased corn production.

There is new emphasis on development of more vigorous forest trees. With the increasing demand for forest products for a wide range of uses, foresters are searching for both genetic and cultural ways to accelerate tree growth. Millions of trees of major forest species are intensively grown in nurseries and then transplanted into forest settings. The superior selections grow faster, and the use of transplants increases the uniformity of the stand.

PHYSICAL CONTROL

The active, physical control of plant material causes specific material to look the way we want it to and regulates plant size. Such control can also improve

FIGURE 16-3
Apple tree height is affected by using size-controlling rootstocks. Trees are the same age; tree on left is on a vigorous rootstock while tree on right is on a dwarf.

the size and quality of flowers, fruit, or foliage; increase the structural strength and enhance the physical beauty of plant material; invigorate weak or old plants; or simply maintain good plant health.

Pruning and Training

Pruning involves the removal of plant parts such as buds, shoots, and roots (Fig. 16-4). If the natural form of the plant is undesirable, growth can be directed, at least to a degree, to the preferred form through pruning.

Pruning is both invigorating and dwarfing. In general, a pruned plant will produce longer shoots (frequently used as a measure of vigor) than an unpruned one. If we remove 5 or 10 percent of the shoots and growing points (buds) and do not prune the roots, each remaining bud (and shoot) has a greater supply of water and nutrients. Therefore, shoot growth is invigorated. Pruning can also be a dwarfing process. In spite of the increased vigor of the pruned plant, by the end of the growing season, it will be smaller and lighter than one left unpruned because the extra growth in longer shoots is not sufficient both to equal that of the unpruned plant and to replace what was pruned away.

In some fruit plants, pruning also offers partial regulation of fruiting. Highbush blueberries are pruned to regulate plant size but pruning also removes some of the flower buds. The remaining buds develop into berries that are larger, have better quality, and mature more uniformly. Pruning peach trees also removes many flower buds, reducing the number of fruits which must be thinned during the early summer.

Knowledge of the plant's fruiting habit is a prerequisite to proper pruning. On a peach tree, for example, all fruit are borne laterally on 1-year-old growth. As the current season's shoots develop, flower buds form in the leaf axils for next year's crop. If one pruned away all of last year's shoots, the entire crop would be lost. In apple trees, however, most flower buds are formed terminally on short spurs. Few flower buds form laterally on shoots, and all of last year's shoots can be removed with little effect on this year's crop.

Both the practice of pruning and the condition of the plant itself govern the results. A mature apple tree unpruned for several years will usually be in low vigor and quite unproductive. If such a tree is pruned rather severely, it will be invigorated and will become more productive. With young, vigorous

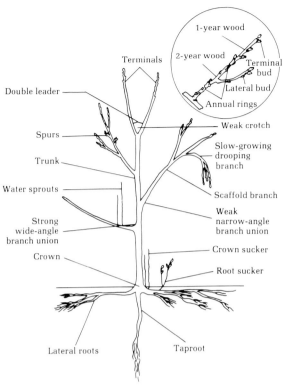

FIGURE 16-4
The major gross structure of a tree. (*Redrawn by permission from E. P. Christopher, The Pruning Manual. Macmillan, New York, 1954. Copyright © 1954 by Macmillan Publishing Co., Inc.*)

apple trees, however, severe pruning causes excessive vigor and suppresses fruit production.

Pruning helps develop a strong and durable framework which is critical for fruit- and nut-bearing trees and desirable with ornamentals (Fig. 16-5). Shaping and training should begin when the tree is very young. For example, the 'Delicious' apple tree naturally develops an upright growth habit and has narrow-angled branches with weak crotches. The branches with narrow angles between the trunk and branch are cut out, leaving the stronger ones with wider crotch angles. A *scaffold limb* is a branch originating from the trunk and forming a permanent part of the major tree framework; branches selected to form such limbs should be about 15 cm apart to avoid later crowding.

Narrow-angled branches are weak because bark (phloem) is enclosed between the xylem of the trunk and branch (Fig. 16-6). An angle of 45 to 60° between the trunk and branch is the strongest because the xylem layers will be continuous between the trunk and the branch. In many orchards, spreaders are used to increase branch angles and prevent formation of bark inclusions. Round wooden toothpicks can be used to spread young shoots; wire or wooden sticks are used in 1- to 2-year-old branches (see Fig. 16-10d). Spreading the 1- to 2-year-old limbs also suppresses vegetative growth and encourages flower-bud formation.

Limb orientation has a major effect on growth and on flower bud initiation. In many plants, the growth

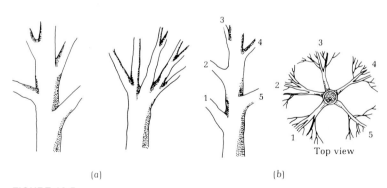

(a)   (b)

FIGURE 16-5
(a) Well-spaced branches have stronger attachments than those growing close together or in a cluster. (b) Branches with good scaffolding require proper vertical and radial spacing on the trunk. (*Adapted from Ohio State Univ. Bull. 543.*)

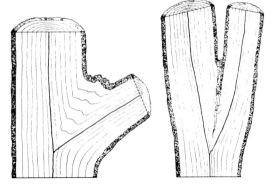

FIGURE 16-6
A wide-angle crotch makes scaffold branches stronger. Note the discontinuous cambium and phloem inclusion in the narrow angle on the right. (*Adapted from New York State Agricultural Experiment Station.*)

of a branch is roughly proportional to its orientation. In a vertical position the growth is strong; if the branch is put in a horizontal position, growth of the shoot apex is slowed dramatically; if taken below horizontal, growth abruptly ceases. In training young apple trees, branches are often spread to an angle of 45 to 60° from horizontal. Vegetative growth continues at a reduced rate, but this spreading encourages the initiation of flower buds on the lateral spurs on the branch. When the young tree comes into production, vegetative growth of the whole tree slows. Spreading thus serves three major functions: it strengthens the structural framework of the tree; it induces earlier fruiting and return on investment; it helps control tree size by accelerating the transition from a vegetative to a reproductive stage.

Root Pruning Root pruning is the removal of part of the root system, including the growing points, to produce a denser root system with more small roots. Since this technique temporarily decreases absorption of nitrogen, other essential elements, and water, cell division and enlargement slow down and carbohydrates are stored rather than used. Root pruning promotes the reproduction phase of growth; if initial vigor is high, root pruning may stimulate flowering and fruiting. Root pruning is often done several months before transplanting evergreens to encourage a more compact root system, a larger portion of which will be contained in the root ball when the plant is dug and transplanted.

Disbudding The removal of flower buds is widely used in commercial production of cut flowers, e.g., carnation, standard chrysanthemum, and rose. As soon as practical, the lateral flower buds are removed, leaving only the terminal flower bud to develop. Although the number of flowers is reduced, disbudding is essential in producing large, high-quality flowers. With some cultivars of carnation, when the goal is not one large flower per stem but a cluster of three or four smaller flowers, the larger terminal flower bud is removed, permitting the side flowers to develop equally and open at the same time.

Pinching When a terminal bud or shoot apex is removed, the process is called *pinching* because it is usually done by hand, using the thumb and a finger. The main purpose is to cause branching by the removal of apical dominance. For example, in poinsettia and chrysanthemum, a young plant can be pinched to induce two to four more stems, each of which will flower. After pinching, it is often necessary to select the desired number of branches and remove additional ones. For example, if we wanted to grow poinsettias with six inflorescences per pot, we could plant six cuttings and leave them unpinched or plant three cuttings, pinch them, and select the two strongest branches on each cutting.

Pruning Techniques

Choice of pruning procedures should be based on the life-span, structure, and growth and fruiting habits of the plant. Pruning ranges from pinching or removing a few buds in order to produce a more compact and bushy plant to removal of large branches in order to rejuvenate a tree. Herbaceous perennials are generally pruned only to remove dead or diseased parts or to thin out the plants. Chrysanthemums are disbudded to control flower number and thereby increase flower size. Usually it is the woody species, both deciduous and evergreen, that are pruned on a regular basis. The woody species can be grouped into shrubs, which have several main shoots, and trees, which have one to a few distinct trunks.

The two major ways of pruning, heading and thinning, evoke different responses. Both kinds of cuts are normally used in almost every type of pruning. In *heading*, or *heading back*, only the terminal portion of the shoot is removed (Fig. 16-7a). (Heading of ornamental flowering and foliage plants to increase branching is functionally the same as pinching.) Heading is used to contain a plant within particular bounds or to rejuvenate vegetative growth in older trees and shrubs.

When heading back a large branch, the cut should be made near a lateral branch, leaving no stub to create dead wood and delay healing. Dead wood arises when a portion of a branch beyond the last bud or branch is deprived of the hormonal influences that keep the cambium active. When unbranched shoots are headed, it is desirable to cut just above a node, leaving the uppermost bud or buds to develop. To control the direction of growth, a bud that points in the desired direction should be chosen. For example, in training young fruit trees, a bud facing the direction of prevailing wind is preferred. *Thinning*, in which entire shoots or branches are removed, is an extension or complement of heading back (Fig. 16-7*b*).

Growth response varies with the type of pruning cut. Extensive heading of young shoots, as in trimming a hedge, removes large numbers of shoot apices, thereby eliminating apical dominance. The result is that a large number of axillary buds grow into shoots. In effect, for each shoot apex removed, three or four side shoots develop. If the goal is to produce a thick, bushy hedge, heading is ideal. For a fruit grower, however, as we shall see below, mass heading of shoot tips is undesirable because the new shoot growth on the periphery severely shades the inner part of the canopy.

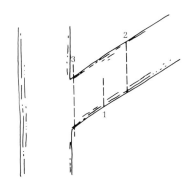

FIGURE 16-8
Three cuts are required to remove a large limb from a tree.

Thinning, on the other hand, has quite a different effect. Since only a limited number of shoot apices are removed, much less new shoot growth occurs. Removing fewer but larger branches creates openings in the canopy which allow the entry of light. Unless adequate light reaches the inner part of the canopy of fruit trees, size and color of the fruits formed there are inferior and flower-bud formation is inhibited.

To remove a sizable limb, the first cut is made 25 to 40 cm from the trunk one-third through the limb from the bottom up (Fig. 16-8). The second cut is made a little farther out than the first but from the top all the way down. Cutting in this manner reduces the chance of wood splitting or bark peeling below the branch being removed, which would invite insect or disease infestation. The last cut is made clean and flush with the trunk. Sometimes chisels or knives are used to ensure a smooth finish.

After pruning cuts have been made, rapid healing is essential. Wounds less than 2 cm in diameter heal rapidly and usually do not need protection; if the cut is over 2 cm in diameter, a wound dressing may be desirable, if only to prevent desiccation of the healing tissues. A good wound dressing should be waterproof, durable, and an antiseptic or a good barrier to disease organisms. Special wound dressings are available, but common latex paints perform many of the same functions.

Training Methods

Systems for training plants vary throughout the world. In Europe and Japan, trellis systems are sometimes used with apples and other tree fruits to make maximum use of limited space, but in the

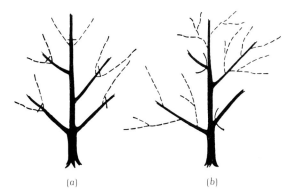

(a)   (b)

FIGURE 16-7
Removing one-half the growth by (a) heading back and (b) thinning gives different results in pruning. (*Adapted from E. P. Christopher, The Pruning Manual. Macmillan, New York, 1954.*)

United States the central-leader system (discussed below) is most popular. It provides good light penetration and maximum bearing surfaces. In landscaping, training methods can be used to develop plants for screens, accent, space savings, and architectural features. Trellis, espalier, and bonsai systems all involve sophisticated pruning and training techniques. The critical time for training is when the tree is young. Many of these training systems require a great deal of technical knowledge and constant maintenance.

Central-Leader Training In this technique, a single central leader is selected early and subordinate side branches are developed along it (Fig. 16-9a). Attention is paid to selecting well-spaced lateral branches and maintaining the proper degree of subordination among them. The result is a tall, upright tree whose upper branches are progressively shorter. Some fruit trees, e.g., apple, naturally lend themselves to this system, but others, e.g., plum, are difficult to train to a central leader.

The tree should be headed back to 70 to 90 cm when it is planted (Fig. 16-10a). During the next summer, any competing branches as well as new shoots on the bottom 0.5 m of the trunk should be removed. During the first dormant season, lateral branches are selected to form the lower scaffold limbs (Fig. 16-10b, c). From above, the tree can be visualized as the hub of a wheel with the laterals forming the spokes. After scaffolds are selected, all other shoots should be removed. The leader is headed back to induce branching and to strengthen it. Scaffold limbs may be headed for similar reasons.

Open-Center Training In this system, the main stem is headed and growth is encouraged from the upper end of the trunk, where a number of branches originate (Fig. 16-9c). Three to five primary lateral branches are trained to be of equal size by pruning each year. The selected branches should not originate too close together. This method affords maximum light penetration, resulting in a more uniform distribution and better color of fruit. The low, widespreading tree that develops is easier to prune, thin, spray, and harvest. The main disadvantage is that the arrangement and equal size of the scaffold system often produces weak, crowded crotches, and limbs must sometimes be propped or braced to prevent breaking. This system is widely used with peaches produced for the fresh market and is good for shake-and-catch mechanical harvesting (see Fig. 21-3), because the fruit tend to hit fewer branches in falling from the tree to the catching frame.

Modified-Leader Training The modified-leader system combines the best qualities of the central-leader and open-center systems (Fig. 16-9b). A leader is allowed to develop on the young tree until it reaches the height of 3 or 4 m. It remains dominant until the tree's architecture is formed and then is restricted by pruning. Laterals are selected to ascend in a spiral fashion up the central trunk. The lowest lateral should be 60 cm or more above the ground. When the central leader is headed, the tree develops a top that is rounded and open. The limbs are well-spaced and the fruiting wood is well distributed, not drastically different from the central-leader system. Many apple growers are now using the modified-leader system.

Espalier Training Espalier training is a time- and labor-intensive practice being used more in the United States today as a result of limited space in urban living. Espalier plants are trained to grow in a preplanned pattern against a flat fence or wall. With

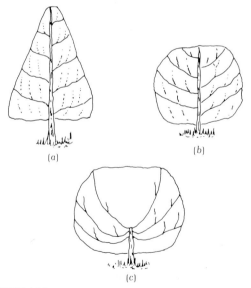

FIGURE 16-9
Plants are pruned to three basic forms: (*a*) central leader; (*b*) modified leader; (*c*) open center.

FIGURE 16-10
Training a young apple tree. (a) 1-year-old apple tree being pruned immediately after planting; (b) well-grown 'Delicious' tree following its first year's growth; (c) the same tree after the first dormant pruning; note the early selection of framework branches and maintenance of the central leader; (d) 4-year-old tree after pruning. (*From U. S. Department of Agriculture Bulletin 1897.*)

proper care, plants can be trained to almost any shape but need constant maintenance to maintain the design. Many kinds of trees already trained are available in nurseries or from specialists. Ornamental plants can be espaliered to serve as the focal point on a wall or more prominently displayed around borders of the private and public spaces in the landscape. Restrictive training systems of this sort usually require that dwarf cultivars or other means of genetic growth control be used.

Trellis Training Many vines and annual herbaceous plants can be successfully trained to a trellis, another rigid form of training, demanding planning, supervision, and care. Grapevines can be trained on a trellis with canes (previous season's growth from the trunk), tied to wires (Fig. 16-11). Fruit trees are occasionally trained to a trellis or some other framework, and some bramble fruits (raspberry, etc.) are grown commercially on a type of trellis system. Crops that can be grown on a trellis also include hops, pole bean, tomato, and cucumber. The home gardener can make much more economical use of space by trellising some vegetable crops.

Bonsai Training Bonsai is a sophisticated and intricate system of growing plants, usually trees, in small containers. The Chinese and Japanese have used bonsai as an art form for centuries and bonsai specimens may be passed from generation to generation. Some trees hundreds of years old are only 40 to 50 cm tall. Dwarfing mechanisms include bending, twisting, root restriction, and precise control of soil composition and water and nutrient supplies. Growth is redirected and restricted rather than being removed by pruning. Bonsai plants are manipulated to take on the characteristics of age in a miniature form (Fig. 16-12). To grow bonsai plants, one must have not only an appreciation of plant growth but also great patience.

Pruning Equipment

Hand pruners are of two different kinds. The *scissor type* has sharpened blades which overlap, producing a cutting mechanism similar to that of household scissors. *Anvil pruners* have a straight-edged top blade which comes down on a soft metal anvil (Fig. 16-13c). Although both are acceptable, a closer cut can be made with the scissor type. *Lopping shears* are hand shears on a larger scale; heavier blades on

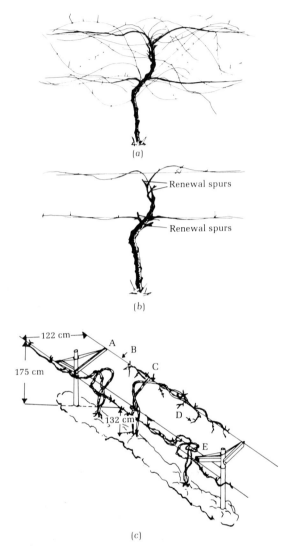

FIGURE 16-11
Pruning of bunch grapes to four-arm Kniffin system: (a) unpruned; (b) pruned. (*From Clemson Univ. Agric. Bull. 123.*) (c) Geneva double-curtain system for training American bunch grapes. A, cordon wire support; B, cordon wire; C, cordon; D, fruiting cane; E, renewal spurs. Posts are spaced 7.3 m apart. (*From U. S. Department of Agriculture Bulletin 2123.*)

longer handles give more leverage so that cuts too large for hand shears can be made (Fig. 16-13b). *Pole pruners* are used to make cuts in hard-to-reach areas or in trees without using a ladder. *Pruning saws* (Fig. 16-13d), used to make cuts too large for pruning

hard work out of the job. Power loppers and saws typically operate with hoses and compressed air.

Mechanical Pruners A recent innovation, mechanized pruners vary widely in design and operation but are all made to prune a plant mechanically. They include large circular saws which rotate on arms, small overlapping circular saws, and cutter-bar mechanisms (Fig. 16-14). The large saws will cut branches up to 5 cm in diameter, whereas the cutter-bar type is designed to cut branches up to 1 to 2 cm in diameter. Some are used as toppers for cutting back the top of the trees; others, called *hedgers*, cut the sides of the tree, either vertically or at an angle. It is possible to prune a hedge-row of citrus or apple trees to the shape of a box or even a Christmas tree.

FIGURE 16-12
Bonsai culture involves redirecting and restricting growth rather than conventional pruning.

shears or loppers, have coarse, wide-set teeth to prevent sticking.

Power Pruners Electric hedge shears are in common use. Most large-scale pruning in orchards is done with power pruners which take much of the

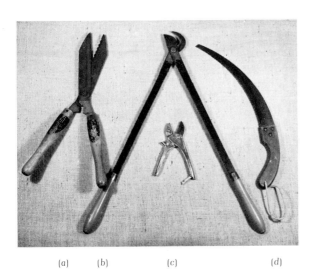

(a) (b) (c) (d)

FIGURE 16-13
Pruning equipment: (a) hedge shears; (b) lopping shears; (c) hand pruning shears; (d) pruning saw. (*Courtesy of W. P. Judkins.*)

FIGURE 16-14
Mechanical pruning of cherry trees with a cutter-bar-type pruner.

Much of this type of pruning is done on a contract basis because of the cost of the equipment.

Although the speed and relatively low cost of mechanical pruning mean that it has begun to be widely used, serious problems develop. It is quite possible to cut the tree back too severely, throwing it into a vigorous vegetative state. A partial solution is to anticipate the pruning by a year and omit nitrogen applications to reduce the rank vegetative-growth response. It is wise to bring trees back down to the desired size over 3 to 5 years rather than all at once. Another problem is that if the tree is hedged and topped only moderately, a vast number of new shoots will develop owing to the mass removal of apical dominance. The result can be thick growth, heavy shade, and poor fruit quality. The only solution is to use hand pruning to thin out the periphery of the tree. Although mechanical pruning is a great help, for many plant species it only partially does the job.

Positioners It has become increasingly difficult to hire labor willing and able to prune during the winter. One partial solution is the use of *positioners*, which actually put the worker up in the tree without climbing or using ladders (Fig. 16-15). Although expensive, these so-called bucket trucks or ''cherry pickers'' both increase the efficiency of workers and make the job more attractive. Many positioners are attached to a tractor and equipped with a power

FIGURE 16-15
Positioners are helpful to persons pruning fruit and ornamental trees.

source so that pneumatic pruners can be used. Similar equipment is used in pruning shade trees under power lines.

Mowers In addition to trees, shrubs, and flowering plants, a major type of growth control is cutting grass in yards, golf courses, and median strips of interstate highways. Millions of dollars are spent on equipment and labor for maintenance of grass areas; it ranges from large rotary mowers pulled by tractors to the much smaller models used by homeowners. The mowers used on golf greens are specially designed to cut the grass very close. Grass plants are unique in that the intercalary meristems at the nodes enable them to continue growing even after repeated mowings. Mowing not only controls the height of the desired grass but also helps control weeds by killing many and preventing others from going to seed.

CHEMICAL CONTROL

A variety of natural and synthetic chemicals is used in regulating plant growth. The chemical regulation of plants has a beneficial impact on a wide range of crops. The judicious application of appropriate growth-regulation measures enables growers to increase plant productivity, enhance general resistance and hardiness, and grow better fruit, larger flowers, and stronger stems and foliage. We can slow down or speed up specific plant processes according to our needs. Plant-growth regulators (PGRs) were described in Chap. 4. Here we briefly describe examples of their use in controlling plant growth and development.

Chemical Pinching

In many crops grown for their flowers, apical dominance minimizes the number of flowers per plant. Although hand pinching of the shoot apices is effective, the cost is high, and using chemical pinching agents is much more economical. These chemicals, which inhibit or kill apical buds, increase the number of new shoots per plant, control the shape of the plant, increase flower number, and regulate time of flowering.

Fatty acid methyl esters and alcohols of chain lengths C_8 to C_{12} (8 to 12 carbon atoms) are often effective in killing or inhibiting apical bud growth.

The effectiveness of the chemical varies with the chain length, C_{10} being the most effective. These *chemical pruning agents* are not translocated to other parts of the plant and kill only the meristematic cells. Lower alkyl esters of fatty acids destroy terminal buds, in effect pinching the plant, but the efficiency varies with the concentration of the chemical and the stage of bud development.

Chemical pruning is used on azalea (Fig. 16-16) to stimulate branching and increases the number of flowers per plant. Treated plants stop growing temporarily, but in a few weeks the growth may overtake that of an untreated plant. Many tedious hours of disbudding are required to produce chrysanthemums commercially. For each large flower that develops, growers must remove 10 to 25 lateral flower buds. Alkyl naphthalenes can kill most of the chrysanthemum floral meristems, to give fewer but larger flowers.

Growth Retardants

Widespread use is now made of synthetic *growth retardants*, chemicals which suppress the elongation of vegetative growth, largely by affecting internode length. Their precise mode of action is unclear, but most appear to be antagonistic to gibberellins or auxin. Some growth retardants affect cell division in the apical meristem; others inhibit cell enlargement below the shoot apex.

Although growth retardants are used on a range of plants, the widest application has been on floricultural crops, particularly potted flowering plants, partly because these crops are particularly responsive and partly because approval for use on food crops requires a great deal more research. Material cannot be used on any food crop until it has been shown to leave no toxic residue in or on the edible portion.

Growth retardants make potted flowering plants more compact and attractive and easier to handle in both production and marketing. The growth retardant Amo-1618, developed in 1950, though effective for dwarfing plants, was too expensive for commercial use. The related material Phosfon is widely used for restricting the height of chrysanthemum and lily. Chlormequat, a more recent introduction, can be used as a foliar spray or a soil drench to control the height of poinsettia and azalea. Used as a foliar spray, daminozide effectively dwarfs chrysanthemum, bedding plants, poinsettia, azalea, and hydrangea and has been used to control vine growth of peanuts.

(a)

(b)

FIGURE 16-16
Chemical pinching of azalea. Note new shoots in (*b*) which have developed because of the death of the apical bud. (*Courtesy of F. D. Cochran.*)

One of the most recently developed growth retardants, ancymidol, controls plant height at a low concentration (Fig. 16-17).

Chlormequat (CCC) is also used on agronomic crops in some parts of the world. Some European growers of wheat and other small grains use CCC to control stem length and reduce lodging. In general, however, the vast acreages of field crops seriously hinders the use of growth retardants because of the high cost of the chemicals and their application.

(a)   (b)

FIGURE 16-17
Growth of Easter lily controlled with A-Rest (ancymidol): (a) treated; (b) untreated. The major effect is the suppression of internode elongation. (*Courtesy of R. A. Larson.*)

In tobacco production, the greatest yield of high-quality leaves is obtained by removing the terminal flowers from each plant, but as soon as the flowering-shoot apex develops (and especially if it is removed), axillary shoots begin to develop and must also be removed or suppressed for optimum yield. Wide use of maleic hydrazide by tobacco growers to inhibit sucker growth has resulted in sizable savings in labor costs. Other and perhaps better chemicals are now becoming available to provide for the control of the "suckers," as the unwanted branches are called.

Ethephon is a chemical that releases ethylene, an extremely active PGR (Chap. 4). When applied in the spring to young apple trees, ethephon inhibits vegetative growth and induces the formation of flower buds. Ethephon is also used instead of or in combination with CCC to reduce stem height in barley and wheat.

Other Uses of PGRs

Growth regulators are used on fruit trees for a variety of other purposes, e.g., thinning fruit, controlling preharvest drop, timing fruit ripening, accelerating color development, and modifying fruit shape. These PGRs include synthetic auxins, auxin antagonists, ethylene-releasing agents, and cytokinins.

An intriguing element of PGR use is that a particular plant may have diverse responses to the same PGR applied at different times. For example, naphthaleneacetic acid (NAA) at 5 to 15 ppm is an effective chemical thinner for apples when applied 2 to 3 weeks after bloom. By causing the abscission of a large percentage of the fruit, NAA allows the remaining fruit to grow larger and be of better quality. The same concentration of NAA applied to the same tree a few days before normal harvest will dramatically delay formation of the abscission layer and thereby prevent preharvest drop.

In several crop species PGRs alter the ratio of female to male flowers. In monoecious cucumbers, for example, gibberellin treatment increases the number of staminate and ethephon the number of pistillate flowers. Gibberellic acid (GA) also causes the production of more pistillate flowers on hops. Auxin and ethephon have all been reported to increase femaleness in muskmelon.

Sprouting of potatoes, sweet potatoes, and onions during storage decreases product weight and eventually makes the product unmarketable. Several PGRs have been tested to inhibit such sprouting; the greatest success has been in using maleic hydrazide on potato.

A combination of GA and benzyladenine (a cytokinin) alters the shape of apple fruit. 'Delicious' apples are considered to be most desirable when they are elongated with very prominent calyx lobes. A spray of GA and benzyladenine applied during full bloom increases the length to diameter ratio. This treatment is used by some apple growers, particularly in the eastern United States, where 'Delicious' apple shape tends to be less elongated than in the northwest.

Another side of the PGR story involves herbicides used to kill weeds, discussed in Chap. 12. Of interest here is the fact that a particular compound may be used at one concentration as a herbicide and at a much lower concentration as a growth regulator. The use of 2,4-D as a herbicide is one of the early success stories in chemical weed control; at low concentrations, however, 2,4-D has been shown to be effective in delaying preharvest drop of apples with no sign of phytotoxicity.

ENVIRONMENTAL MANIPULATIONS

To complete the discussion of growth control, it is appropriate to review and put into perspective some of the important environmental factors described in earlier chapters.

Light
We regulate or modify plant growth and development by level, duration, and quality of light. The most striking effect of light is doubtless the photoperiodic flowering response of many plants. No synthetic PGR can mimic the effect of short days in inducing

synchronous flowering of an entire greenhouse of chrysanthemum or poinsettia.

Temperature
Temperature also has dramatic effects on plant growth and development. Until the rest period of many deciduous trees and some seeds has been met by an extended cool period, they remain dormant regardless of environment. Another example is the vernalization of small grains by cold winter temperatures which hastens their flowering in the spring.

With actively growing plants, considerable growth control is possible by varying temperatures. Temperature control is particularly critical in greenhouse crop production. In general, warm temperatures cause rapid succulent growth. Cooler temperatures slow growth and cause plants to develop thicker, stronger stems. Temperatures in field situations are not normally under the control of the grower, but time of planting can be varied to avoid or take advantage of normal seasonal fluctuations. For example, in warm areas of the country, cool-season crops are grown in spring or fall. Mulches are widely used to modify soil temperature, which has a major impact on the growth of many species.

During the summer, certain woody plants do not grow continuously but in one or more *flushes* of growth. Nurseries growing young forest trees often find that the desired species produces only one flush of growth per year, so that it takes 3 to 4 years to grow a tree to transplant size. Recent research has shown that when young trees are given a cold treatment at the end of each flush of growth, multiple flushes may occur during one growing season. If the mechanics can be worked out, the equivalent of 2 or 3 years of growth may be induced in 1 year.

Water
Plant growth is strongly influenced by the availability of water. Suboptimal soil moisture can suppress growth and production of crops. If the water shortage is severe, some crops die; others, e.g., some turfgrass species, merely go dormant and start growing again when water becomes available.

Available water can be manipulated for various purposes relating to crop growth, an obvious example being the use of irrigation. A less striking but important example is the hardening of bedding plants by withholding water. Recent research with peach trees raises an interesting possibility. It has been shown that by withholding water early in the season, when vegetative growth is most active, considerable control of tree size can be obtained without serious reduction in fruit size. Later in the summer, when fruit growth is most rapid, plenty of water is applied. By this time, vegetative growth is much less active. Such a procedure is obviously limited to irrigated land, but in such areas it may not only provide control of tree size but save water as well.

Nutrition

For optimum growth and production, an adequate supply of the essential elements must be available. In many cropping situations we regulate plant growth by careful control of nutrients, especially nitrogen.

The effects of nitrogen fertilization on growth, flowering, and fruiting are of major importance with crop plants. Excessive nitrogen stimulates vegetative growth at the expense of reproductive growth in most species. Before plants will flower, they must make some minimal amount of vegetative growth. In most cases, we want early rapid growth to establish the leaf surface and structural framework to bear the crop. After the basic vegetative stage, a lower nitrogen availability may favor the reproductive stage. Control of nitrogen availability is by time and method of application as well as the amount and source used, factors that vary with the crop, soil, and rainfall.

There are exceptions to the generalization just made. For pasture and forage crops, a continuous high nitrogen level is needed to maintain a high growth rate of vegetative plant parts. To achieve moderate but continuous growth of turf, multiple applications of fertilizer or in some cases slow-release fertilizers are used. For many grain crops that produce large volumes of seeds and then senesce, e.g., wheat, rice, and corn, high rates of nitrogen may be needed during the critical seed-fill period.

Interactions

The factors which regulate plant growth do not operate independently but interact in highly complex ways. This can be exemplified by the production of apples. The first decision made by an apple grower is the choice of genetic material. At a minimum, this entails an interaction between the scion cultivar and the rootstock. After the trees are planted, early pruning and training decisions must be made. All these decisions have a major impact throughout the 20- to 30-year life of the orchard. Later factors to be considered include fertilization, subsequent pruning, and a whole spectrum of PGRs. Heavy pruning is much more likely to delay fruiting of a tree on a vigorous rootstock than the same cultivar on a dwarfing rootstock. Excessive nitrogen fertilization will be much more detrimental on a poor-coloring cultivar such as 'McIntosh' than on a highly colored 'Delicious' strain. Spreading of the tree is more necessary for 'Delicious' than for 'Golden Delicious' which naturally forms wide-angle crotches. Ethephon, which is helpful in bringing 'Delicious' into fruiting, is unnecessary with 'Golden Delicious' because they naturally begin to fruit at a young age.

CHAPTER SEVENTEEN

CROPPING SYSTEMS

When Europeans colonized the New World, one of their most immediate tasks was to establish a self-sustaining agricultural system. Resupply of foodstuffs across the Atlantic was not reliable, and many of the colonists came with the avowed purpose of breaking all ties with the Old World. Annals of several different settlements show that cropping was a top priority. Seeds and seedlings were brought with great care to the New World, and much effort and prayer went into their planting and tending. But the historical accounts show that many of the early efforts were dismal failures. Crops and practices that had worked successfully for centuries in England did not provide enough harvest even for seed to replant, much less to feed the settlement. Early years were marked by very lean times. It was often a hardscrabble existence that included foraging in the forests and fields for natural provender.

Many Native Americans took early compassion on the interlopers and provided them with food. (The Indians' compassion perhaps outweighed their judgment.) But even more valuable was the Indians' contribution of agricultural technology. New World crops like corn and squash were quickly adopted by the Europeans, as were New World practices such as fertilizing with a buried fish. Old World practices and crops (or the particular varieties brought over) were not well suited to the conditions of New England, Virginia, or Georgia.

Crops, the practices used to grow them, and the social, economic, and political factors that bear on crop production and utilization in a particular environment constitute a *cropping system*. Successful cropping systems are those which are adapted to all the environmental and human factors. Attempts to follow Old World cropping patterns and practices could not succeed in colonial America, because the physical environment, the European crops, and the old techniques were inappropriately matched. Nor could the settlers have adopted wholesale the cropping systems of the Native Americans. Only as the colonists adapted themselves to the New World did successful, self-sustaining cropping systems, based on European economic and social institutions, develop.

This chapter describes several different cropping systems, concentrating on their biological and technological aspects. The important economic and sociological aspects must be passed over for lack of space, but we encourage you to become more familiar with the effects of human institutions on agriculture through course work or readings in agricultural economics and rural sociology.

In discussing cropping systems we shall find some that are complex and others that are simpler. Some of the most biologically sophisticated cropping systems are found in less technologically advanced environments. As a generalization, the most complex and intensively managed cropping situations are associated with the tropics. By comparison, cropping patterns used in the United States and other more developed nations are relatively simple and straight-forward, with few crop components and rather direct relationships between the various noneconomic components.

SOME DEFINITIONS AND PRINCIPLES

In the final analysis, *agriculture* is a system for collecting the sun's energy and converting it into useful forms, which are then harvested as *crops*. Crops are usually plants or plant parts, but animals may also be considered crops. Cattle, sheep, and other animals can produce harvestable matter by transforming the plant matter they consume into meat, milk, wool, and other products (Fig. 17-1). In a plant science context, we concentrate on the plant crops, but we shall discuss some cropping systems that use animals as well.

Normally implied in agriculture and cropping is human manipulation of the system. People are the managers or manipulators, and the act (or art) is called *management*. As described in Chap. 13 on crop ecology, management involves imposing changes on the agricultural ecosystem or *agroecosystem*. Management also involves the economic aspects of agriculture, an important component of the total cropping system but beyond the scope of this text.

FIGURE 17-1
A Central American cropping system in which animals are the crop to be harvested. (*Courtesy of P. E. Hildebrand and Farming Systems Support Project.*)

FIGURE 17-2
A labor-intensive system of agriculture in which the tillage, planting, and harvest are done by hand. (*Courtesy of P. E. Hildebrand and Farming Systems Support Project.*)

The degree of management in cropping systems varies widely. *Intensity* is the word used to describe the level of human intervention; thus intense management implies that human inputs are considerable. Such inputs may consist of labor, machinery, fuel, chemicals, or know-how. *Labor-intensive* agriculture is built on a system that usually requires relatively large amounts of hand labor (as opposed to machines) for each unit of productivity (Fig. 17-2). As a general rule, a cropping system will be either labor-intensive or capital-intensive, requiring money for land, machinery, chemicals, etc.

One of the first exceptions to the general rule is *low-intensity management*. In some cropping systems, the manager may need to do relatively little to ensure periodic harvests. On the rangeland of the great plains and intermountain region, proper range management may involve placing the right number of cattle on a piece of land and then coming back at the appropriate time to "harvest" them. In some other cropping systems, the degree of management may be less. Wild or Indian rice, for example, is still harvested from naturally occurring stands of the plants.

A cropping system exists (and must be managed) in space and time. The land and other resources devoted to cropping over a period of time constitute the *management unit*. The land resource may be hundreds of hectares in some management schemes in modern large-scale, capital-intensive agriculture.

Even thousands of hectares may be included in a single management unit on western United States farms and ranches. In many world areas, however, the average farmer has control of 1 ha or less. Such small farms are the norm in much of the humid tropics. Obviously, a much greater premium is placed on efficiency of land utilization in such a space-limited situation, but the tropical environment generally offers more flexibility and management options, especially over time.

The concept of management over *time* as well as *space* is important in cropping systems. The good manager uses the available *growing season* to its fullest. In temperate areas, the growing season of many crops is bounded by the last freeze of the spring and the first freeze in the fall. Some crops can survive mild frosts, and their incorporation into a crop sequence can greatly extend the growing season beyond the frost-free period. Pea, radish, carrot, cole crops, and onion are a few of the cold-hardy species that permit the gardener to expand the period of productivity. Sugarbeet is important in some regions where it can be planted much earlier and harvested much later than most crops. Some crops, e.g., potato, can be planted before the last spring frost because they can begin to grow in cooler soils, but they do not fare well if frost occurs at the other end of their production cycle.

The growing season in some areas is regulated not by temperature but by rainfall. Many tropical and subtropical regions have seasons that are wet or dry, not hot or cold. These areas are often characterized by a *monsoon climate*, where three to eight consecutive months may be quite rainy but the remainder of the year sees very little or no rainfall. Here the wise manager works to take maximum advantage of the growing season afforded by moisture.

The key point in all cropping situations is to use the space and time available for crop growth to their economic maximum. The time may be no more than 3 frost-free months in parts of Alaska or Canada, but a good manager can produce as much per hectare in those 3 months as a poorer manager would in a 4- or 5-month season farther south. The space available may be only a few hectares, but good management can make them outproduce by severalfold a poorly managed larger holding. In dealing with some of the cropping systems in use throughout the world and especially in the United States, this chapter will focus on time and space management though man-agement of other resources will be brought in as well.

SINGLE-CROPPING SYSTEMS

On a global basis, the diversity in cropping systems is tremendous. Although cropping systems employed in the United States are numerous, many are simply variations on a few basic themes. One common basic theme is one annual crop per field. Whatever the crop, the general pattern of production and management over time and space is often repeated in all 50 states. To distinguish it from other, more complex management systems, this method is often called *single cropping* or *sole cropping*. In single cropping, only one crop species is harvested from a piece of land each *cropping year*.† The single crop may be corn, soybean, sorghum, tobacco, or any number of other species, but in every case, it is the only crop grown on the field over the year.

Management of single cropping is relatively simple compared with other patterns to be considered later. The field is prepared and planted to one crop at the appropriate time; all midseason management decisions will generally be based only on the one crop that is present; and harvests can usually be accomplished with a single pass through the field at the appropriate time. Thus management in sole-cropping systems is fairly straightforward. The manager attempts to provide the crop with an optimum "slice" of the growing season while manipulating the environment to keep the status of other controllable factors favorable. The results may still be highly dependent on factors beyond the grower's control, especially if the farmer cannot irrigate. A dry year in a normally rainy area can mean disaster.

In some areas where rainfall is typically rather limited and irrigation is not feasible, the grower can use the know-how input of management to advantage, e.g., choosing crops that require less water to produce a harvest. Water-conservation measures such as terracing or fallowing can often capture enough extra water to make the difference between profit and loss.

---

†A cropping year or *farming year* normally corresponds to the calendar year, but we shall consider one situation where a farming year lasts for two calendar years and another where a farming year overlaps two calendar years.

Alternate-Year-Fallow Systems

In the northwestern United States and other semiarid regions, annual rainfall is seldom abundant enough to carry a grain crop to maturity; typically only about half the necessary moisture is provided in precipitation each year. Some intelligent manager reasoned, however, that, since two halves equal a whole, 2 years' precipitation in these semiarid areas should provide enough moisture for one crop. The result is an *alternate-year-fallow* cropping system, in which crops are grown on a piece of land only every other year. (In this case, one farming year equals two calendar years.)

The alternate-year-fallow pattern of sole-crop production results in nearly optimum utilization of water, the most limiting resource in the agroecosystem. The manager grows crops on only half the potential cropland each year. The usual method involves planting crops in strips 100 m or so wide and perhaps 1 km or more long. Between the crops is the fallow ground. The strips in a well-managed farm follow the contours, producing sinuous bands of alternating green and brown.

The good manager takes pains to keep fallow fields free of weeds, either by cultivation or chemical means. The weeds must be controlled on the fallow strips to prevent their reappearance in cropping years and to conserve moisture. A field full of weeds can pull as much water out of the soil in transpiration as a field full of wheat. Only if the soil is left free of living plants will the fallow year's rainfall remain in the soil for the alternate year's grain crop. Other techniques to reduce evaporation from the soil surface include a *dust mulch,* produced by working the soil to shallow depth and leaving a protective layer of dissociated soil particles on the soil surface.

Some Shortcomings of Single Cropping

The alternate-year-fallow pattern of single cropping is an intelligent solution to a water-shortage problem that limits crop growth in an otherwise suitable environment. Not all single-cropping systems are so efficient, however. Some, in fact, may make poor use of time and space during potentially productive portions of the farming year.

In temperate humid areas, spring has many sunny, warm days suitable for growth of many species, but in a single-crop system, the land may be growing only weeds while waiting for the soil to get warm enough for the single crop, e.g., soybeans, to be

planted. Or the single crop may be put in early to take advantage of the front of the growing season and reach maturity well before the end of the growing season, as defined by frost or dry weather. In either case, the single crop fails to take full advantage of the time and space when temperature (or water) is not limiting.

Sugarbeet is a notable exception to the general pattern of poor coverage by a single crop of a temperate growing season. In northern areas, sugarbeet can generally be planted as early as the ground can be worked and is harvested well into the fall after several frosts that would have killed other crops. This exceptionally long period of growth means that in many temperate areas sugarbeet, a C_3 plant, can produce more harvestable carbohydrates as a single crop than corn or sweet sorghum, both C_4 plants.

Monoculture versus Rotation

Monoculture Certain crops are grown as single crops on the same piece of land year after year. This *monoculture* is commonly motivated by economic, not biological, considerations. For example, as soybean became an increasingly valuable commodity in the 1970s, some land was planted to soybean year after year. In the midwestern corn belt, land that had grown corn continuously was switched over to soybean monoculture. The prime motivation was economic. When corn and soybean were about equal in value to the farmer, the two would be planted in equal acreages and *rotated,* switched back and forth each year. Each is still a single crop, but monoculture is avoided by rotation.

Monoculture is a potentially counterproductive management practice; it can reduce the farmer's profits even when the price of the commodity is up. Disease, insect, weed, and fertility problems are magnified by monoculture. Populations of pests that thrive on a particular crop will increase if their crop host is continuously available. Weeds that cannot be successfully controlled while growing with a particular crop can go to seed and become more abundant each succeeding year they are not controlled. Nutrients needed in especially large amounts by the monocultured crop will be depleted more rapidly.

Monoculture of soybean in the southeastern United States in recent years has led to serious problems in all these categories. Soybean-specific diseases, whose spores persist in soil or plant residues, overwinter and reinfest beans planted in the same field. Nematodes, another serious pest, can do the same. Several perennial and annual weeds that defy chemical control may build up. Some of the worst weeds, e.g., sicklepod and cowpea, are also legumes. Their similarity to soybean makes chemicals that will kill the weeds but not the crop difficult to find. Potassium and magnesium fertilization is another concern in continuous soybean production, since both are extracted in relatively large amounts by the plants.

Rotation Alternating crops from one year to the next on the same section of land is an old-fashioned management practice that reduces the problems of monoculture. Crop rotation goes back at least as far as the Romans. Although often only a single crop is planted and harvested each year, the alternation of species from one year to the next is beneficial. Insects or diseases that prey on a particular crop will not be able to multiply in the years their host crop is not planted. Good management allows the grower to take advantage of the different nutrient demands of crops in a rotation. For example, peanut and soybean may need no additional phosphorus fertilizer if they follow an adequately fertilized crop, because these legumes are particularly efficient at extracting phosphorus from the soil. Crops that follow legumes may need less nitrogen fertilizer than when they follow nonlegumes.

Rotations may help in weed control also by permitting selective herbicides or other cultural practices to be used in alternate years without harming one of the crops in the rotation. For example, johnsongrass, a serious southeastern weed, can be controlled by herbicides, but many of them are lethal to corn and sorghum. On the other hand, soybean and some other crops are not sensitive to certain of the chemical control agents. Therefore, in fields where johnsongrass is established, nonsensitive crops such as soybean can be planted for a year or two until the weed has been controlled.

Another factor has been motivating growers to return to the practice of rotation in the management of their crops. As fuel and fertilizer costs have gone up, many growers have found their profit margin is so slim that they cannot afford or risk the high yearly investment for planting cash crops. The cost of nitrogen fertilizer in particular may be prohibitive when the farmer must borrow money to plant a crop. For that reason, some managers choose to plant a nitrogen-fixing legume in rotation with their cash crops. In many parts of the corn belt, alfalfa can provide as much as 100 kg of nitrogen per hectare to the crop that succeeds it. The result can be up to a 50 percent savings on nitrogen costs for the first corn crop after alfalfa. Of course, during its year in rotation the alfalfa also produces high-quality feed for livestock.

The particular cropping sequence used in a rotation varies with the climate, tradition, economics, and other factors. The first well-established rotation was developed in eighteenth-century England; the Norfolk 4-year rotation involved growing turnips, barley, clover, and wheat in turn. A common pattern in the United States corn belt today is to alternate the two major cash crops, corn and soybean. Longer rotations, taking more than 2 years to complete, might include insertion of a year each with a small-grain crop and a legume-grass mixture for a hay crop. Economics is often the major determinant of what crop to grow. When small-grain prices are down and the prospects are good for profitable soybeans, the rotational sequence is often suspended.

The sequence of crops within a rotation may be critical since some crops yield better or worse depending on the crop they follow. For example, Ohio wheat production typically is about 15 percent less following soybeans than following clover. A Rhode Island rotation study showed that onion yields vary by as much as 27-fold depending on the crop they follow. Yields of other crops may vary by 50 percent or more depending on the preceding crop.

The reason for such effects is not always known. Sorghum is a notoriously hard crop to follow. Yields of almost any crop after sorghum will be lower than after corn, soybeans, or wheat. It has been suggested that sorghum's effect on succeeding crops relates to the high carbohydrate content in sorghum's roots; decomposition of such roots stimulates soil microbial growth and "ties up" nitrogen and other nutrients in the soil microflora, the phenomenon known as *immobilization*. In other cases, the effect of one crop on the next may relate to chemicals left in the soil or generated by decomposition of the crop residues. Such *allelopathic* effects may be much more common than we realize. Wheat residues, for example, have been shown to inhibit the growth of several different crops that might follow. The allelochemicals are thought to be produced during decomposition of the residues by certain soil microbes.

Since crop rotation is practiced by many growers, including home gardeners on a miniature scale, our emphasis on monoculture and its dangers should not be misleading. Intelligent managers have never fully abandoned rotations, but by some estimates, perhaps 10 million additional hectares were switched to planting continuous cash crops during the 1970s. That trend has now been reversed—a healthy sign.

A Place for Monoculture

Single cropping, even monoculture, may in some situations be the only economically viable alternative for large-scale agriculture. The growing season may be so restricted by temperature or rainfall that it is theoretically or economically impossible to insert a second harvestable crop into the cropping year. For example, in the north central plains, spring and early summer are suitable for wheat or other small grains, but the summers tend to be too dry to support the growth of longer-season crops such as soybean or corn. Therefore, spring wheat is planted and then harvested by July to be followed by summer fallow. This has been the pattern for years. Sunflower and safflower are increasingly grown in some areas formerly planted exclusively to wheat but only if the land is not planted to wheat in the spring. There is not enough rain each summer to produce even a short-season crop like millet after a wheat crop matures.

Single cropping (with monoculture) is a logical management pattern for certain other crops or situations. Perennial crops may need to grow for a few to several years before they become productive; then they should remain at peak productivity for a number of years. Apple, avocado, blueberry, grape, peach, pecan, orange, and many other fruit and nut species are perennial. All are generally expensive to plant and maintain, and they may remain nonproductively juvenile for up to several years; so it is necessary to set aside a piece of land for their continuous monoculture (Fig. 17-3). Many orchards and groves are planted in the expectation that the land will be dedicated exclusively to that purpose for decades. Plantations of pine and other forest species fit into the same category.

MULTIPLE-CROPPING SYSTEMS

Unlike single cropping, with only one harvest on a piece of land each farming year, *multiple cropping*

FIGURE 17-3
An apple orchard is a single-crop, monoculture system of agriculture.

gives two or more crops or harvests from the same field annually. Multiple cropping is a way of intensifying crop production in time or space (or both), thus using the growing season and the land to better advantage. Multiple cropping takes many different forms since there are several ways of managing a piece of land to make it produce two or more crops in a year: planting two crops simultaneously, planting one after the other has been harvested, overlapping them in time, or alternatively, making two or more harvests from a single planting. These patterns and others have been used to great benefit in various cropping systems.

Multiple-cropping practices fit into one of two basic patterns. In *sequential cropping*, two or more crops are planted and harvested one after the other. In *intercropping*, two or more crops are grown side by side. Each form of multiple cropping intensifies crop production relative to single cropping. Sequential cropping tends to intensify production in time, often spreading production more nearly over the entire growing season. Intercropping tends to focus more on maximum utilization of space, planting more or different plants to take better advantage of all the available light, water, nutrients, and other resources. Nevertheless, intercrops can take greater advantage of time (growing season) than some single crops, and sequential crops may make better use of space than a sole crop.

Multiple-cropping systems are rather foreign to western agriculture and are the exception on large farms in industrialized countries. On the other hand, multiple cropping is the dominant form of agriculture in less developed countries and in areas where very small farms are common (Fig. 17-4).

In much of the world, the farm is less a place to make a living than a place to live—not a profit-making enterprise but a family's home and source of subsistence. On such holdings, labor and time for management are generally more abundant than equipment and money. Since family welfare and survival are paramount, management decisions on such holdings are generally not based on the profit motive. Cropping systems on such small farms are labor-intensive as well as spatially compact. In humid tropical areas, where most of these small-acreage farmers live, the growing season generally lasts year-round. With this combination of constraints (low capital) and advantages (cheaper labor and longer growing season), multiple cropping becomes a logical form of agriculture on small holdings in many tropical countries. Quite different patterns develop in other geographical or economic settings.

Many multiple-cropping systems may give the impression of casual or even totally unplanned management. To the person unfamiliar with the system, an intercropped patch where 15 to 20 different species grow apparently helter-skelter may look unmanaged, but nothing could be farther from the truth. The crops and their positions have been carefully chosen. Such biologically complex agroecosystems may approach a nearly optimal utilization of the land agriculturally. By carefully planting various crops together or sequentially the landholder can

FIGURE 17-4
A typical family farm in the tropics. An area of 1 ha or less will often yield all the food, feed, fiber, and fuel the family needs to survive. (*Courtesy of P. E. Hildebrand and Farming Systems Support Project.*)

take advantage of each plant's niche. The small landholder, say in Panama, who produces all the food, fuel, and fiber his family needs to subsist is managing a more complex biological system than the corn and soybean farmer in Iowa† who plants a thousand hectares. Obviously, each farmer has a different set of problems and management options, but the "peasant" farmer must be just as careful a manager and in some ways even more efficient than his tractor-driving counterpart.

On both small-scale, intensely managed multiple cropped farms and large-scale, intensely managed single-cropped farms, the intelligent manager takes into account the resources available and the possible limitations to productivity and then manipulates the resources to provide for an acceptable level of production. Intelligent management of cropping systems makes sense ecologically and economically and is the hallmark of good farming.

Sequential Cropping

Double, triple, and *quadruple cropping* are systems of sequential cropping in which two, three, or four crops are produced in turn on the same piece of land during the same cropping year. The number of crops involved in the sequence is determined by the length of both the local growing season and the life cycle of the individual crops. In a fourth type of sequential cropping called *ratooning* or *ratoon cropping*, crops are planted only once and harvested twice or more. Several different crops can be ratooned where the growing season is long enough to permit their regrowth.

In double cropping, the second crop, usually planted immediately after harvesting the original crop, may be the same or a different species. Monoculture is possible in a sequential cropping system. In some subtropical areas, the monocultural pattern of two crops of rice planted sequentially in the same paddies each year has probably been followed for centuries. In some humid tropical situations, the development of improved, earlier-maturing cultivars has made triple-crop rice common. Continued breeding work may eventually produce rice cultivars that can routinely produce four annual crops on the same land.

But there is no free lunch. Double, triple, or quadruple cropping results in greater productivity

---

†In Iowa, the greater complexity may be associated with the economic side of the total cropping system.

only at some greater cost and usually some reduction in labor efficiency. One can expect more rice from two crops than from one—but not twice as much. Three sequential crops should produce more in a year than double cropping would—but not 50 percent more.

Each planting in a sequential cropping pattern ideally makes optimum use of the environmental resources available to it during the growing season; but even if water and nutrients are fully available to a rice crop during the growing season, the number of days and the amount of sunlight are *always* going to be limited. This imposes a limit on photosynthesis and therefore productivity. The light limitation is caused by two developmental factors: canopy development and senescence.

Canopy Development Young seedlings do not have enough leaf area to capture all the sunlight falling in their vicinity (Fig. 17-5). Only as plants grow larger do they develop a leaf canopy that intercepts essentially all the sunlight. Until then, much light falls to the soil unproductively. *Canopy closure* results in essentially full interception of sunlight by crop leaves (Chap. 13). When two or more crops are planted in a single year, the period of wasted sunlight before canopy closure occurs twice or more.

Ideally one could plant a crop, allow it to develop an efficient canopy, and then make multiple harvests without disturbing the canopy. Keeping the leaves in place allows them to continue to "harvest" sunlight and convert it into something harvestable. This is the approach used with some crops; many home garden varieties of tomato, pepper, and other plants bear fruit continuously for the last half of the growing season in temperate areas. Careful harvesting preserves the photosynthetic leaf assemblage and prolongs production. For many crops, however, multiple harvests are not possible. The plants are genetically programmed to senesce and die after producing one crop.

Senescence Rice, wheat, corn, soybean, sunflower, and many other crops cannot be harvested more than once because they die after the harvestable crop has been produced. The senescence process inherent in most annual crops (Chap. 4) begins as grain or fruit reaches maturity and rapidly "kills" the whole plant. Although growers exploit this trait in mechanical

FIGURE 17-5
Young sorghum plants. Most of the sunlight falling on this field is not intercepted by sorghum leaves, and the energy is lost to the agroecosystem.

harvesting of dead, dried plants, it makes multiple harvests from the same planting impossible. It also halts sunlight harvesting in leaves, often in the midst of the growing season. Senescent canopies are yellow, without chlorophyll, and typically die when conditions are otherwise ideal for growth. If enough time remains in the growing season, a new planting can be made, but canopy making must begin again. Since the age and size of an annual crop plant when it produces fruit, senesces, and dies is at least partially under genetic control, breeders can manipulate the *maturity* of a plant so that it takes more or less time to reach a harvestable stage (Chap. 14).

There is a trade-off, however, in use of early-maturing, multiple-cropped plants. One cannot expect twice as much rice from double cropping an early-maturing rice cultivar as from a single planting of later-maturing types, because the faster-developing types generally have less leaf area and spend less time in the productive grain-development period before senescing. The combined double-crop yield will typically be 150 percent or more of the single-crop yield. But the "no free lunch" factor comes into play when considering the return on labor investment. The planting and tending of rice, which are very labor-intensive, must occur twice in the growing season; and less than a doubling of output is received for the doubling of input.

Fitting the Crops to the Season

Despite our emphasis on rice, sequential cropping is certainly not confined to monoculture of that crop, nor is it practiced only in the tropics and subtropics. Many other crops can be planted before or after rice in a sequentially cropped field, and many temperate-region cropping systems employ two and sometimes more crops in sequence.

Many temperate areas, where the growing season is fixed between the last freeze of spring and the first freeze of fall, may still offer an opportunity for sequential cropping. In Alaska and Canada, where the climate is more nearly arctic, the potential growing season is so short that only very rapidly maturing crops could be double-cropped; but productivity can still be intensified in some cases, especially for vegetable crops. For example, radish and some kinds of lettuce reach a harvestable stage quickly, allowing a second, more slowly maturing crop to be planted even where the frost-free period may last only 60 to 90 days. The very long days associated with subarctic areas actually favor some double crops.

In temperate areas, as the growing season increases to 210 days or longer, the opportunities for double cropping become correspondingly greater. It is possible to make two harvests of crops whose time to maturity is approximately half as long as the growing season. Of course, factors other than the time to reach maturity must be considered; the crop must be otherwise adapted to the local conditions. Rainfall patterns, availability of irrigation, soil factors, availability of planting and harvesting equipment, and economic considerations always should be weighed in deciding whether and what to double-crop. For example, there is no advantage in double cropping if the produce cannot be marketed within a reasonable distance and at a reasonable profit margin.

In the southern tier of the United States, where the growing season is favorable for many different sequential cropping enterprises, farmers often take nearly full advantage of their longer growing season. Planting early in the spring makes it possible to grow to full maturity two properly chosen sequential crops before the first fall freeze. In some areas of Florida and the southwestern United States, mild winters permit continuous sequential cropping, so that at any time of the year some crop is growing.

The particular crops used in a double or triple cropping sequence naturally vary from region to region and from farm to farm. In the irrigated, mild-wintered southwest, growers double-crop a number of different agronomic and horticultural crops. Although Florida's climatic conditions permit more flexibility, the southeastern United States generally shows less diversity in double-cropping systems, and some major crops are notable for their absence. Peanut, dent corn, tobacco, sorghum, and cotton are relatively long-season crops and do not fit well into double-cropping sequences. In contrast, many vegetable market crops, as well as sunflower, soybean, oats, and others mature rapidly enough to fit into the same growing season with another crop (Fig. 17-6).

Double Cropping with Winter Annuals

Winter annual crops offer distinct opportunities for double cropping. Crops such as the fall-planted small grains (primarily wheat, rye, and barley) are included in crop sequences extending over parts of two calendar years but considered as one cropping year. In

FIGURE 17-6
This field of sunflower was double-cropped following potato. (*Photograph by R. L. Harrison.*)

humid regions with growing seasons of 180 frost-free days or more and winters moderate enough not to kill winter annuals, winter small grains can be grown, harvested, and followed by a summer annual crop such as soybean or sunflower. In southeastern states, the double-crop sequence of small grains and soybean is very common.

The winter-annual–summer-annual double-crop sequence intensifies the time component of land utilization. The fall-planted crop captures sunlight and converts it into plant matter during a portion of the year not usually part of the growing season. The winter-hardy annuals such as wheat or barley can continue growing after the first frost of fall and begin growing again before the last frost of spring. (In the middle of the winter, they are more or less quiescent.) When winter small grains have been vernalized by extended exposure to the cold, they can respond to long days and begin to move rapidly into flowering and maturity in the spring. They are generally ready to harvest in mid-June, leaving 3 to 4 more months of good growing weather for a second crop.

In areas where the length of the growing season is marginal for double cropping and cool springs push small-grain maturity into late June or early July, cultivars that mature just a few days earlier are used to advantage. They can mean the difference between a successful double-crop season and a disastrous loss of the second crop to frost. Obviously the second crop must also be chosen with care with regard to maturity. A cultivar taking 120 days to reach a stage safe from freeze damage chosen when only 110 frost-free days are left after harvesting the small grain may doom the venture to failure.

Because the margin for error is sometimes narrow, and because the first frost of the fall is not consistent in its annual arrival, the good manager looks for ways to improve the odds of success in double cropping. The use of earlier-maturing cultivars for both crops in the sequence helps tremendously. Other approaches involve reducing the time it takes to get the second crop in the ground, which is one reason for the increased popularity of no-till planting.

Conventional planting typically includes some form of tillage for general seedbed preparation. Plowing, disking, and other tillage operations are used to make a smooth, noncloddy soil before passing through the field with the planter. All these operations take time and should be performed only when soil moisture conditions are appropriate; but when it is time

to plant the second crop in a double-crop sequence, the farmer is typically short on time and patience. Each day lost in tillage or waiting for suitable tillage conditions pushes the double crop a day closer to the profit-killing frost. One solution is simply not to till the field at all. *No-till planters* are implements that put seed in the ground without any previous seedbed preparation; they make a seedbed about 5 cm wide and plant the seeds in a single pass. These implements enable a farmer to plant soybeans on the same day and in the same field from which wheat or barley was harvested (Fig. 17-7).

Another way of hastening establishment of the second crop in a double-crop sequence involves planting it before the first is harvested, a technique that can also be considered to be a form of inter-cropping. In *relay intercropping*, a second crop is planted into an existing crop, usually near the time of maturity of the first. The aim is to let the second crop germinate and begin growth in the shade of the older crop and then to remove the older crop before it has any adverse effects on the younger. Although relay intercropping works well in some small labor-intensive cropping systems, its value in large imple-ment-intensive systems is questionable. It is difficult to plant one crop satisfactorily without damaging the other and equally difficult to harvest the first crop without damaging the second. Additional re-finement may make relay intercropping a viable

FIGURE 17-7
This double-crop no-till pattern allows the summer annual (soybean) to be planted immediately after harvest into the stubble of the winter annual (wheat). (*Photograph by G. R. Buss.*)

supplement or substitute for double cropping, especially where timing is critical.

Ratoon Cropping

Ratooning, or *ratoon cropping,* requires a single planting but allows multiple harvests. Typically, the aboveground portion of the plant is harvested at the appropriate stage and the plants allowed to regrow. Commonly after a second harvest, the crop is killed by frost or plowed under and replanted. The ratooned crop obviously must regenerate its canopy if it is to capture and transform more sunlight, but it can do so faster and more efficiently with a good root system already in place. The alternative of replanting takes time, and the young seedlings must start all over to generate both leaves and roots. In many cases, double cropping would not be feasible because it would take too long to reach maturity after the second planting. Of course ratooning saves not only growing time but also saves the time and cost of replanting.

Only nonsenescing perennials are suitable for ratoon cropping. When soybean, wheat, and field corn are ready for harvest, they are dead, roots and all. Among the field crops, important nonsenescing species are sugarcane and sorghums, which can be harvested for sugary stem juices (sugarcane and sweet sorghum) or for feed (forage and grain sorghums) and left to regrow.† The second harvest of most temperate-region crops tends to be smaller than the first, but still there is a yield of some value compared with the alternative of no yield or expensive replanting and days lost from the growing season.‡

Intercropping

In this form of multiple cropping, mixtures of plants are grown simultaneously. In sequential cropping, different crops in the same field are separated in time, but in intercropping, since the plants are separated only physically, the use of space at a given

time is intensified. The practice often leads to important—sometimes beneficial—interactions between crops.

Intercropping can be divided into four patterns:

Mixed Intercropping—Two or more crops are grown simultaneously without being arranged in distinct rows.

Row Intercropping—At least one of the two or more crops is arranged in a row. (One crop may be planted in distinct rows while the others grow randomly.)

Strip Intercropping—Crops are interplanted into strips, or rows, wide enough to permit cultivation of each but narrow enough for the crops to affect each other.†

Relay Intercropping—The second or third intercrops are planted after the initial crop has been growing for some time, usually when it approaches maturity.

As noted above, the relay intercropping pattern might equally well be considered a variation on sequential cropping, since there is decidedly some separation in time of the multiple crops. The idea in relay intercropping is to intensify both the use of space between the rows of a maturing crop and the use of the time available during the growing season.

History Intercropping is uncommon in modern, field-scale, mechanized agriculture; but it has a place and may find its way back into American farming systems. There was a time when intercropping was regularly practiced in this country. Native Americans used it extensively, and it was part of the Native American agricultural technology that Europeans adopted. Only as agriculture became increasingly mechanized did intercropping disappear from our fields. Not too many years ago, corn was commonly intercropped with squashes, pumpkins, and beans. With hand labor, these crops could easily be planted together and individually harvested. But as we began to produce corn in fields measured in tens and then hundreds of hectares, it became impractical to tend

---

†In tropical areas, where perennial sugarcane and pineapple are also ratooned, the year or more from one harvest to the next is an improvement over the approximately 18 months (sugarcane) to 24 months (pineapple) from planting to first harvest. In harvesting pineapple, since only the fruits are removed, the entire root system and canopy remain intact.

‡Despite similarities, forage crops are not generally considered to be ratooned and will be treated separately later in the chapter.

---

†The wide bands or strips of crops used in some soil-conservation practices do not constitute strip intercropping because the crops do not grow close enough to have any effect on each other. The wide strips are essentially long, narrow, single-crop fields.

multiple crops simultaneously. (There is also the very real consideration of what to do with 500 ha of intercropped pumpkins.)

Young trees are planted at spacings to accommodate their mature size, but, during early years of growth, they do not use the space and light effectively. Intercropping can be used to great advantage in such situations. Young orchards can be intercropped with strawberries or other crops. Walnut groves have been intercropped with wheat and hay during the first 10 years or so, giving an economic return on the land before the longer-term crop begins to produce.

Intercropping is still evident in two areas of American agriculture. On a large scale, pastures and rangelands, generally composed of mixtures of several species, satisfy the definition of a mixed intercropping system. The same systems can of course be considered a variation of ratoon sequential cropping, but in either case, it is an important multiple-cropping system, discussed later in the chapter. On a small scale, home vegetable and flower gardens certainly meet the criteria for intercropping systems. Most vegetable gardens fit the definition of strip intercrops (Fig. 17-8), and wise home gardeners use relay intercropping plus double and triple cropping to advantage. Many flower gardeners and some vegetable gardeners also employ mixed and row intercropping techniques artfully and productively.

FIGURE 17-8
A home vegetable garden constitutes a multiple-cropping system in which both intercropping and sequential cropping are practiced.

Advantage Intercropping is supposed to increase yield by intensifying the use of space. Why should adding bean plants to a cornfield increase total yield more than adding more corn plants? Why don't additional bean plants reduce corn productivity so that their combined yields only equal the single-cropped corn? If a crop has produced the maximum amount of grain possible for the conditions under which it is grown, it would be reasonable to expect a reduction in yield if additional plants are put into the field, since only so much light, water, and nutrients are available.

Consider a hypothetical situation where crop A is grown as a single crop. Under ideal conditions and at an optimum density of a plants per hectare, it produces a maximum yield of x kg ha^{-1}. Crop B grown under the same conditions and at a density of b plants per hectare produces y kg ha^{-1}. If A and B are grown together at $a + b$ plants per hectare, respectively, we should not expect a yield of $x + y$ kg ha^{-1}. In fact we might not even expect a yield as high as either x or y because we would anticipate that the two crops might compete to such a degree that yields of each would be drastically reduced.

On the other hand, if crop A is planted at slightly less than a plants per hectare and crop B is planted at a much reduced rate, what might the results be? Obviously there are many possibilities, but in many intercropping systems, the likely outcome is a yield of nearly x kg ha^{-1} for A *plus* a substantial yield from B. Sometimes one finds that A's yield at a plants per hectare is greater than x kg ha^{-1} and B's yield added to that gives an even greater combined yield.

In less abstract terms, this means that some intercropped plants can perform more productively than their single-crop components. More kilograms of beans and corn can be produced in 2 ha of a corn-bean intercrop than in 1 ha of corn plus 1 ha of beans. In fact, one can produce as much corn and beans in about 1.8 intercropped hectares as in 2 single-cropped hectares (Fig. 17-9). That translates to a 10 percent savings of land space. Other intercrop combinations can be much more productive.

In the tropics, sorghum and pigeonpea are commonly intercropped. Sorghum yields are typically reduced by only 5 percent as a result of the competition from pigeonpea, which produces 70 percent of what it would in a sole-crop situation. A common intercrop in India and West Africa is millet and peanut. The combination permits a 25 percent sav-

FIGURE 17-9
Intercropped maize, sorghum, and beans produced in Guatemala from the planting occurring in Figure 17-2. (*Courtesy of P. E. Hildebrand and Farming Systems Support Project.*)

ings in land; i.e., sole cropping of the two species would require 25 percent more space to produce the same yields.

Interactions Why certain crops can be more productive together than apart is not always clear. In many cases, the beneficial effect may be due to *annidation*, a situation where the intercrop components use environmental resources differently. Annidation implies the sort of peaceful coexistence discussed in Chap. 13, where each component of the intercrop uses the various growth requirements in complementary ways. One plant may be able to utilize lower light levels or nutrient levels. In some cases, one intercrop component may *require* the lower light level in the shade of a taller intercrop. Especially with legume-nonlegume mixtures, there may be annidation with respect to nitrogen utilization; the legume can fix its own nitrogen and allow the nonlegume to use the limited supply in the soil. Sometimes the legume may contribute nitrogen to the nonlegume. The sorghum-pigeonpea intercrop uses light more efficiently in that sorghum intercepts more light early in the season and pigeonpea develops a fuller canopy later.

Several physiological and physical effects may explain how productivity of one crop may be influenced positively by another. Sometimes chemicals produced by one plant help the growth of the other. Sometimes one crop provides physical support to

the other, as with bean vines on corn (Fig. 17-10). A taller, more robust crop may provide physical protection from wind or frost. (*Nurse crops* are sometimes coplanted with less vigorous species to provide earlier production and perhaps protect the slower-growing species. Oats are occasionally planted as a nurse crop for young alfalfa.)

Another way intercrops can excel single crops is in preventing or reducing yield losses from pest damage. Two different strategies may be used. In one, the attacked intercrop is lost, but the other intercrop compensates by using the additional resources to produce a greater yield than it would have. The intercrop is a form of insurance for the grower. (If the disease- or insect-killed plant had been growing alone, the loss of productivity would

FIGURE 17-10
A bean-corn intercrop. Used extensively in Native American agriculture, it is still employed around the world, as here in Indonesia. The vines of the bean find support on the corn, and the nitrogen produced by the bean can help corn growth. (*Photo by E. W. Carson.*)

have been absolute.) One intercrop may also protect the other by the so-called *flypaper effect*. The pests are repelled or otherwise attracted away from the crop sensitive to attack by another member of the intercrop. Cereals intercropped with cowpea or peanut in tropical settings often reduce insect attack and insect-carried diseases.

The yield-stability factor with an intercrop is critical to many small farmers. Where sorghum and pigeonpea are cash crops, one study estimated that sole-crop pigeonpea failed 1 year in 5 and sorghum 1 year in 8 but the intercrop only 1 in 36. For the farmer whose livelihood (and life) hinges on a narrow margin, the economic advantage of intercropping is clear.

In the Garden While a home garden represents an intercropping situation, it does not necessarily outproduce a single-cropped situation. Only if the gardener takes advantage of annidation, the flypaper effect, or other favorable interactions does the intercrop really intensify production. If all the rows in the garden are 0.8 m apart regardless of what is being grown in them, and if all crops are planted without regard to their neighboring crop, little benefit will come from the intercropping. To take advantage of the favorable interactions, one must plant the proper species at the appropriate times, densities, and distances. That may mean, for example, slightly increasing the distance between corn plants and interplanting climbing beans after the corn has gained some height. Radishes planted with carrots or lettuce can be a successful intercrop, since radishes mature rapidly and can be harvested before they interfere with growth of the other crop. In some situations, the vigorously germinating radish seedlings can improve emergence and establishment of less robust carrot or lettuce seedlings.

Companion planting in gardens is a popular idea with scientific merit. By choosing where plants are put in relation to other species, one can take advantage of potentially beneficial and avoid potentially harmful interactions. Although the exact basis for plant interactions is not always known, numerous observations confirm their existence. For example, plants that produce aromatic volatile compounds can often reduce insect damage to their neighbors. Marigold, nasturtium, and catnip have been documented to reduce cabbageworm numbers on intercropped cabbages (although the same study showed

that later competition from the aromatic species could reduce cabbage yields). Some plants that share disease or pest problems should not be planted together, e.g., potatoes and tomatoes.

Allelopathy In another important plant interaction that can occur in intercropping, *allelopathy*, chemicals produced by one plant affect another. The effect is usually harmful, but occasionally the released chemicals make another plant more productive. Harmful allelopathic interactions amount to a sort of biological chemical warfare. Several horticultural, agronomic, and tree crops are known to be allelopathic, as are many weeds.

Cucumbers and other members of the cucurbit family produce chemicals that inhibit the growth of some plants in their vicinity. The allelopathic response often results in natural weed control. The cucumber-produced "herbicide" suppresses many weeds that might otherwise compete with it for water, nutrients, and light. The stubble and residues from wheat and rye reduce the growth of several broadleaf weeds. Some sunflowers produce allelochemicals that have similar herbicidal effects. Black walnut trees release chemicals that are inhibitory to many other species. The allelochemicals in the leaves are washed into the soil by rain or released after the leaves fall and decompose. The consequences can be drastic for plants trying to grow in the vicinity of the tree. (Obviously one should never use black walnut leaves as a mulch or composting material.)

Some Characteristics Intercropping is very much the norm in the small holdings that constitute farms in less developed nations. There, where land availability and technological inputs are at a premium and labor is more abundant, the emphasis on all forms of multiple cropping to include intercropping is natural. In such economic and climatic environments, single cropping does not use resources efficiently or intelligently. The wise manager naturally intensifies utilization of space, time, and all other resources. For many such managers the alternative is starvation.

A common characteristic of *traditional farming systems*, handed down over centuries, is the use of many crops and many varieties of each species. In Columbia, a dozen different cultivars of bean may each have their special use in the kitchen and a particular niche in the family plot. In Zaire, the Medje people have cultivated 30 different crop species with as many as 27 different cultivars of one species. Each specific variety of each crop in an intercrop mixture has characteristics that suit it to a particular use in a particular place and time. The differences between cultivars may relate to maturity, growth habit, ecological situation (wet versus dry season or mixed versus row intercrop), taste, or cooking quality.

Another characteristic of traditional cropping systems is an arrangement of plants (Fig. 17-11) that seems haphazard to the uneducated eye. Strangers to the area might suppose that no thought is given to planting patterns. The arrangement looks purely random or even like an uncultivated and natural setting. The last supposition may be only half false. While traditional mixed intercrops are certainly cultivated and nonrandom, they have an element of naturalness about them. The systems' ecological balance often rivals those found in nature.

FIGURE 17-11
In this traditional intercropping system in Indonesia, although the plants may appear to be a random planting or even the result of natural succession, each was planted in a particular location in relation to the other plants. The result is highly productive. (*Photograph by E. W. Carson.*)

An African Case Study In the Abakaliki area of east central Nigeria, a unique cropping system in use for centuries provides an ingenious solution to food production in an unpromising environment. The region is characterized by dense tropical forests and water tables near the surface. Even when the natural vegetation is cleared to permit cultivation, the high water table prevents most crops from putting down a root system capable of supporting the top of the plant either mechanically or nutritionally. To compound the problem, the soils are inherently rather infertile, lacking a good supply of nutrients on which crops can feed.

Nevertheless, the farmers of this inhospitable region produce a variety of nutritious and energy-rich foods. By effort and ingenuity, they have brought from the land all their sustenance year after year for centuries. The age of the practice is highly significant, suggesting that the system is nonexhaustive and stable.† The Nigerian farmers have learned to live in equilibrium with their environment, and the environment has continued to provide their needs.

---

†Recall from Chap. 13 that some cropping systems in the United States are unstable or exhaustive. Soil and water resources have already been so depleted in some places that the land has passed completely out of agricultural production.

Such stability is crucial for any cropping system that is to be successful in the long run.

In the Abakaliki district, the land is made productive by clearing a portion of the jungle, piling the native vegetation into scattered mounds, and then burning it. Additional soil is placed on the mounds to raise them—sometimes taller than a person and more than 3 m in diameter. The mounds hold high levels of many nutrients provided by the burned and decaying vegetation; they also provide an unsaturated rooting environment above the water table. On the mounds a dozen or more different crops are interplanted. Between the mounds, on the flat waterlogged regions of the clearing, rice is planted, so that typical fields look very strange to the uninitiated (Fig. 17-12). Rice grows over the flat portions, but the mounds scattered all around the field are covered with a hodgepodge of peanut, corn, pumpkin, melons, pigeonpea, cassava, yams of several types, and various other root, fruit, and vegetable crops. Extended closer examination reveals a pattern of planting; the yams are clustered at the top, and peanut and pumpkin are nearer the base. Young cassava plants are interplanted with low-growing species, but in other mounds, larger cassava plants are associated with other crops. Harvests and plantings are being made continuously for some time once the field comes into production.

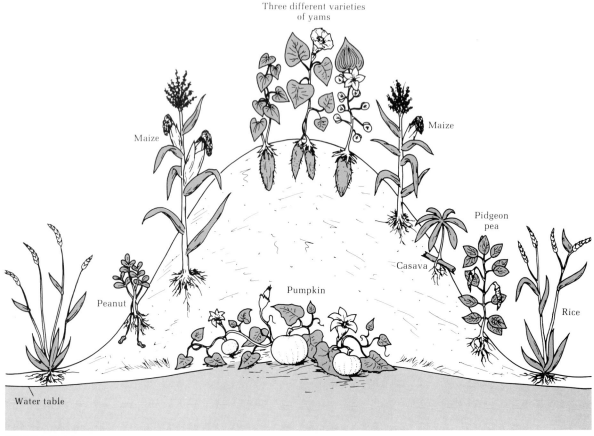

FIGURE 17-12
Diagram of the crop placement on and around an Abakaliki mound. (*Adapted from Multiple Cropping, American Society of Agronomy, Madison, WI, 1976.*)

Shifting Agriculture

While the mounding technique is productive for a time, it is not perpetually so. Since the supply of nutrients is eventually removed by leaching and by harvesting, the procedure must begin again. A new area of jungle is cleared, burned, mounded, and planted every few years. This method of *slash-and-burn, swidden,* or *shifting cultivation* is a common solution for low-technology farming in the generally infertile soils of the humid tropics. The nutrients potentially available to support plant growth are locked tightly in the nutrient cycles of the native vegetation (see Chap. 13). The cycle is broken by killing the natural vegetation and making at least some of the nutrients available immediately by burning. In some places, a section of jungle 3 or 4 times the size of the desired field will be cut and all easily

moved vegetation brought into the cropped area to be burned.

We stressed earlier that these cropping systems are notable for their stability and nonexhaustive nature, but now we seem to be saying that they are unbalanced, since soil nutrients are depleted and the farmer must move on. This is not the same as the depletion in American colonial and frontier agriculture, because the periodic depletion of nutrients in tropical rain forest slash-and-burn systems is only temporary. In these truly stable systems, not only is physical soil erosion less serious a problem but when cultivation shifts to another area, the jungle vegetation quickly reestablishes itself. With a succession of species, a few years later the former clearing can scarcely be distinguished from neighboring jungle. (We obviously are not dealing with a tropical rain forest with 70-m trees.) Species native to the area are adapted to the infertile soils and can quickly reestablish a nutrient cycle drawing on the soils' scant supply. Once the lush, native vegetation has been reestablished and a pool of nutrients is available again, the slash-and-burn process can start anew. To some degree, the same thing could happen in frontier America, but the cycle time was (is) much longer, partly because of slower regrowth in temperate regions.

In tropical areas, the interval between abandoning and then reclearing a piece of land is a fallow period, called *bush fallow*. (The jungle is commonly called the bush.) The fallow period may be from 2 to 15 years or more with intervening cropping periods of 1 to 5 years. The length of a cycle traditionally depended on climatic conditions that regulate rate of bush regrowth during the fallow period and the rate of nutrient depletion during the cropping period. Unfortunately, however, the cycle time is being reduced in some areas by population pressure. In parts of Indonesia, the cycle used to be 30 to 45 years, but the bush is being slashed and burned more frequently because more and more families are having to depend on the same land resource. Management of space and time has been intensified, not necessarily with any long-run benefit.

This discussion of multiple-cropping systems has not exhausted the subject or really done more than scratch the surface, but it has introduced the key principles of all intelligently managed cropping systems. By careful management of time and space and careful attention to full (but not exhaustive) utilization of resources, the land can be made more productive. Even some unlikely situations can be cropped to advantage. Whether we are considering a water-logged soil in Africa, a parched plain in Arizona, or a rooftop garden in New York City, with ingenuity, effort, and attention to their needs, plants can be made to produce. Higher levels of management generally mean higher levels of production. Management, however, controls how balanced a cropping system is in its use of resources; the more unbalanced the system the less stable and the higher the cost of maintaining it.

FORAGE CROPPING SYSTEMS

In some parts of the world, the predominant, natural vegetation consists of low-growing, primarily her-baceous plants that include especially the grasses (Fig. 17-13). The prairies on the plains of North America, the pampas of South America, and the savannahs of Africa occupy millions of essentially treeless hectares; but the land should not be considered unproductive. On the contrary, the African plains support tremendous populations of animals. The American plains once supported so many buffalo that trains could be delayed 2 days while a herd passed. The American cowboy once tended cattle that fed strictly on nature's bountiful *ranges*. Today, we are returning increasingly to grassfed beef, much of which is again being produced on our native *rangelands*.

The native or natural grasslands and ranges are the result of soil and climatic conditions that make grasses the predominant species. Moderately low rainfall is usually the major factor favoring grasses where species with large leaf area and low drought tolerance simply would not survive. Such natural grasslands can provide tremendous amounts of nutritious vegetation for native animals or domesticated livestock. We have also learned to produce and maintain grasslands in areas that would not otherwise be grassy, areas where the soils and rainfall would permit trees to grow and crowd out most low-growing species, especially the grasses. Many millions of hectares that would otherwise become forests are now managed as semipermanent *pastures*, very artificial ecosystems from a biological viewpoint. Such grasslands support much of our food- and fiber-producing livestock.

Ranges

Ranges used to be natural, casually managed areas of forage production; cattle were simply allowed to graze as long as there was forage. The vegetation of such ranges originally consisted largely of the native, adapted species whose physiology and morphology made them well suited to the environment. The native species were often the grasses (Fig. 17-14). In dry locations where sagebrush, mesquite, or chaparral were the original species—and even they could be sparse—very few cattle could be supported by the land. Overgrazing either lush grass or sparse vegetation resulted in sometimes drastic changes in the vegetation; productive, nutritious species might disappear and weedy species appear in their place, causing the value of the rangeland to fall drastically.

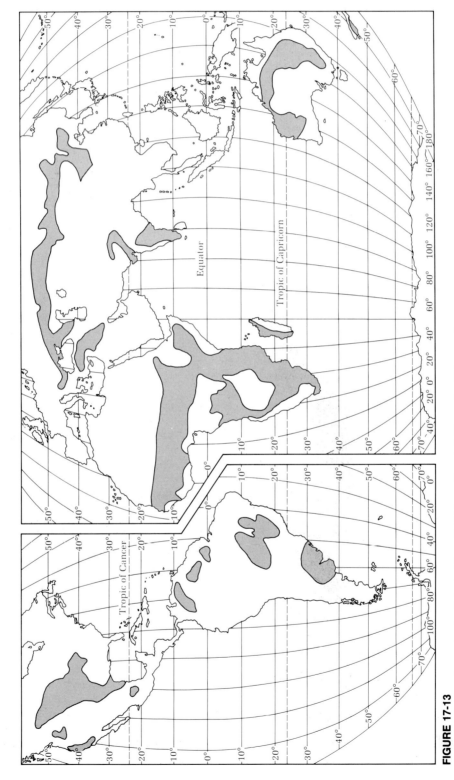

FIGURE 17-13
World distribution of grasslands. (*Adapted from G. T. Trewartha et al., Elements of Geography, McGraw-Hill, New York, 1967.*)

While demand for beef has been increasing, good rangeland in the west has been shrinking due to both overgrazing and human encroachment. Part of the solution has been an eastward shift of beef production, but part has been to manage the existing rangeland more wisely. By adjusting *grazing pressure*, i.e., intensity and duration of grazing, the more desirable native grass species have been preserved or restored on some ranges. In many places, it has been economically feasible to fertilize the rangeland or reseed it with native or introduced species.

Even with the best management, however, most western rangeland will support only a limited number of cattle. As a generalization, the *stocking rates*, i.e., number of animals stocked per land area, used in the pastures of the eastern United States are much higher than those of our western ranges. Water is usually the major limiting factor in growth of the forages, and the rainfall pattern of the Great Plains and intermountain area does not compare favorably with that in the east. Without irrigation, the stocking rate of much western rangeland is measured in hectares per animal. On the other hand, 10 or more head of cattle can sometimes be grazed per hectare on irrigated land or land receiving the moderately large amounts of rainfall common in the east.

Pastures

In contrast to the minimal cultivation associated with ranges, pastures can be thought of as cultivated forage crops (Fig. 17-15). Forages growing in pastures are there only through the activity and continued intervention of human beings. Pastureland must be cleared to permit forage growth and then kept free of undesirable species that would otherwise move back in successionally (Chap. 13). Annual field crops such as corn are usually managed more intensely,

FIGURE 17-14
A remnant of native prairie. The grass and forb species seen here once spread over hundreds of thousands of hectares.

FIGURE 17-15
Pastures in humid areas were generally created by clearing the native vegetation, planting, and managing introduced forage species.

but pastures are often just as unnatural an ecosystem as a cornfield.

Because pastures are not natural ecosystems, it is perhaps not surprising to find that most of the plants commonly grown in our pastures are not native to North America. The plants we use are generally well adapted for providing as much herbage as possible over as much of the year as possible. The manager must make some critical species choices and management decisions to ensure their survival and productivity.

The nutritional needs of the animals feeding on forages must be taken into consideration when choosing the plant species. Livestock need carbohydrate energy in their rations, but they also must obtain protein, vitamins, and minerals. In general, grasses can provide sufficient energy, available as cellulose to ruminants, for sustaining animals (other than milk cows and pregnant animals), but the protein intake may be less than necessary for good gain. On the other hand, legume forages produce feed that is relatively protein-rich. A mixture of grasses and legumes is often highly desirable for grazing, haying, or ensiling and is important enough to warrant a separate discussion.

Certain forages are preferred for particular animals. Timothy makes a highly desirable, clean hay for horses. Production of timothy fell dramatically as horses and mules passed from the agricultural scene. Kentucky bluegrass and common white clover make an excellent pasture for sheep in temperate humid areas. These two plants are particularly desirable

FIGURE 17-16
Sheep graze more closely than most other animals can.

FIGURE 17-17
A no-till planter especially designed for seeding into pastures that need renovation. (*Photograph by D. D. Wolf.*)

because they can tolerate the closer grazing of sheep (Fig. 17-16).

Grass-Legume Intercrops in Pastures

We have already mentioned one great advantage of a pasture or hayfield with a mixture of grasses and legumes: a balance of nutritional energy and protein. There are other benefits. One of the big economic values is the reduced need for nitrogen fertilizers. The nitrogen fixed in the nodules of legumes provides for the growth of both the legume and neighboring grass plants. Some of the fixed nitrogen appears to leak out of living legume roots, and more is released during the natural death and decay of legume leaves, stems, and roots. Table 17-1 indicates the nitrogen-fixing value of some pasture legumes. As much as 85 percent of the nitrogen in forages consumed by cattle passes through the animal and returns to the pasture. The nitrogen released in these ways stimulates growth of the grass intercrop component of

the pasture and saves the farmer the expense of buying nitrogen fertilizer, the cost of which is increasing.

Some grass-legume associations are particularly well matched for pastures. White clover and Kentucky bluegrass not only withstand close grazing but are also a good match in height, timing of growth, and general adaptation. Since their periods of greatest growth and height of growth generally coincide, one does not crowd out the other. In fact the two are so well suited for growing together that they frequently appear and even become dominant in eastern United States pastures where they have not been intentionally planted.

Haying of legumes is often easier or more effective when they are mixed with grasses. The *curing* or drying of the succulent legume tissues is faster when grass is interspersed with the mass of cut material to improve air circulation and facilitate drying. Grasses can increase the harvest of legume leaves, which are brittle and tend to shatter when baled alone, by retaining them during pickup by the baler.

For all these reasons, grass and legume associations or intercrops can work to the grower's benefit. The grower therefore strives to maintain an appropriate grass-legume balance by using good management practices (controlled grazing, proper fertilization, and weed control) and by *renovating* the pasture or range from time to time (Fig. 17-17). Renovation can reestablish a good ratio of legume and grass components and restore the highly beneficial, if artificial, ecosystem.

TABLE 17-1
Estimated Annual Nitrogen-Fixing Potential of Several Forage Legumes

Species	Estimated nitrogen fixation, kg ha^{-1}
Alfalfa	150–250
Crimson clover	80–160
Ladino clover	110–180
Blue lupine	60–135

PART FOUR

CROPS: CLASSIFICATION AND GROUPS

CHAPTER EIGHTEEN

TAXONOMY

The plant kingdom ranges from tiny one-celled algae to staunch trees over 100 m high. The exact number of different kinds of plants living on earth is not known and probably never will be. By various estimates, there are between 350,000 and 500,000 distinct plant species.

In 1886, De Candolle wrote in *Origin of Cultivated Plants,* "Science can make no real progress without a regular system of nomenclature." Standardizing plant names has always been a problem for plant scientists. That relationships exist between plants is often readily apparent, but the affinities are not always clear.

Even though plant classification is subject to constant refinement, and is sometimes a cause for confusion, it is critical in education and communication, both for the scientist and the layperson. Besides bringing order out of chaos, such classification can provide insight into fundamental relationships between individuals in the plant world.

The science that includes classification, nomenclature, and identification of plants is *plant taxonomy* or systematics, the oldest branch of botany. Although botanists have worked for centuries on the classification of plants, much remains to be done.

Taxonomy is vital to plant science. When the scientist identifies, names, and establishes relationships between plants through the use of taxonomy, plant scientists around the world can communicate in an exact and precise manner. When nomenclature is used properly, there is no confusion between individuals, even those using different languages, over what plant is being discussed. When a package of seed is mailed anywhere in the world, the recipient can know exactly what plant to expect when the seed is planted. If a scientific paper or popular article is written describing how to grow a specific plant, even a general understanding of taxonomy is enough to enable readers to understand that information.

The information base for plant taxonomy is continually increased by research. Plant taxonomists work to identify the different species of plants and to classify each plant by showing its true relationship to all other plant life. They group plants on the basis of anatomical, morphological, physiological, and genetic similarities. The analysis of these similarities and differences leads to various taxonomic systems.

Efforts continue to establish a system of plant classification that is both logical and natural, i.e., based on presumed evolutionary kinships. The more botanists learn about plants and the more they discover about natural relationships, the more natural the system can be made. All bases will be used in the future to reach, insofar as possible, the ultimate goal of a plant classification system that reflects genetic relationships between all species in the plant kingdom. Areas taxonomists explore to obtain data for classifying plants include the following:

Morphology—Traditional taxonomists use morphological characteristics to classify plants into defined groups. They reason that the more similar two plants are in form and structure, the more closely related they are. In most cases, this is the only type of information available.

Anatomy—The anatomical taxonomists study cell types of the vascular system to establish evolutionary patterns. For example, plants with similar stelar organization might be related. Much work is also now being done on pollen anatomy.

Embryology—Embryological taxonomists use the morphology and anatomy of embryonic development to search for phylogenetic, i.e., evolutionary relationships. Such characteristics as number and position of cells in the embryo sacs or the placement of the micropyle are used as bases of comparison.

Biochemistry—Biochemical taxonomists use such compounds as flavonoids, sugars, proteins (enzymes), oils, alkaloids, alcohols, terpenes, or phenols to establish relationships. The basis for classification is that the more complex compounds are believed to have evolved at only one time and in only one group of plants. Hybrids are classified by the presence of intermediate levels of the compounds present in both parental groups.

Ecology—Ecosystematics in botany isolate the natural biotic units resulting from natural barriers such as oceans or high mountain ranges. These are used to establish classification on the basis of divergent evolution, chiefly below the species level. Researchers today look at small populations of plants within the species and theorize by extrapolation the changes that occur in the plant population as a whole.

Statistics—Numerical taxonomists take statistical data from divergent pieces of evidence, instead of only one or a few criteria, analyze them, and then classify the plant. They use the findings from all methods and present a composite analysis. Classification schemes arrived at numerically often correlate very closely with those reached by more traditional means.

Cytotaxonomy—The cytotaxonomists study chromosomes to understand the history and future of plant life. The discipline has been significant in documenting evolutionary mechanisms.

CLASSIFICATION SYSTEMS

The different systems devised for classifying plants over the centuries are generally artificial, natural, or phylogenetic.

Artificial Systems Devised for the convenience of commerce, these are often based on gross similarities and economic or utilitarian characteristics. Artificial systems are usually based on the ultimate use of the plants being considered. For example, agricultural plants may be grouped into grain, cereal, pulse, forage, silage, sugar, timber, fruit, medicinal, or ornamental plants. Ornamentals, for example, may be further subdivided as herbs, greenhouse plants, foliage plants, garden plants, perennials, shrubs, or trees. Since the artificial classifications serve chiefly for convenience in communication, they are used primarily in nonscientific and practical circles and are not designed to indicate relationships between plants. There is so much overlap that other systems must be used for more exacting descriptions. For instance, rye can be a grain, a cereal, a forage, a cover crop, or a companion crop. Obviously, a tree can also be a perennial and corn can be considered a food, feed, or oil crop.

Natural System The natural system of classification attempts to show relationships between plants through the use of selected morphological structures. The thesis underlying this system is that morphologically similar plants are closely related. It is a method used to reveal the order that implicitly exists in nature by using all the knowledge available to the taxonomist.

Phylogenetic System By classifying plants according to their supposed evolutionary relationship, a phylogenetic system attempts to reflect genetic relationships between plants and to establish their progenitors. The chief drawback to this system is our limited knowledge of earlier plant forms. Cytogenetics, paleobotany, anatomy, biochemistry, and other sciences help taxonomists in their attempts to determine the intricate evolutionary relationships and interactions in the plant kingdom.

Systems of organized classifications have gradually shifted from purely artificial to natural or phylogenetic schemes. Early systems, which were strictly artificial and based on growth habits, were supplemented by systems based on the numerical aspects and sexual parts of the plant and later replaced by systems establishing natural morphological relationships as the focal point. The most recent systems use phylogenetic relationships to establish classification.

Artifical Classification Based on Habit

The early Greek philosophers and medical practitioners were first to leave written records of attempts to classify plants. The Greeks classified plants on the basis of such vegetative characteristics as growth habit and leaf structure. Theophrastus (c. 370–285 B.C.), a Greek naturalist and author of *Historia Plantarum*, the oldest existing botanical work, classified 480 species of plants into categories of herbs, undershrubs, shrubs, or trees. His book recognized and described families of flowering plants, such as the carrot family, and he perceived relationships between members of such groups as conifers, cereals, thistles, poplars, and birches. He recognized genera in the sense of a group of species, and he named species. Although he correlated the presence of one or two cotyledons with the occurrence of other characters, he did not make use of these characteristics in his system. He classified leaves and studied the arrangement of leaves on the stem. Many of Theophrastus' observations were so revolutionary that they were not used in a plant-classification system until 2000 years later; however, his work helped precipitate some of the later classification systems. The greatness of this early work has certainly earned Theophrastus the title of father of botany.

Artificial Systems Based on Numerical Classification

In 1737, *Genera Plantarum* by Carolus Linnaeus (1707–1778) introduced the era of systematics based on numerical classification. Linnaeus, a Swedish naturalist, physician, botanist, and teacher, founded this type of artificial system and laid the groundwork for the natural systems of plant classification that followed. Linnaeus' system attached special significance to the reproductive aspects of the plant and was known as the *sexual system*. He classified plants according to the numerical characteristics of flowers, e.g., the number of stamens and the number and organization of carpels. In 1753, he published the first edition of *Species Plantarum* which introduced many of the scientific names still used in the nomenclature of higher plants. His system divided all plants into 24 major groups, or *classes*, which were subdivided into *orders*. Approximately 7300 species were described, each group arranged according to an arrangement of reproductive parts.

The work of Linnaeus was the foundation of a consistent and extensive use of the *binomial*, or two-name, *system* of nomenclature. Linnaeus divided each order into genera and each genus into species. The binomial is comparable to the way people are named in the United States. The plant's generic name is comparable to a person's family name, and the plant's specific name to a person's given name. For example, the members of the Rogers family may be John Rogers, Mary Rogers, and Richard Rogers. The genus *Acer* (maple), includes *Acer rubrum* (red maple), *Acer saccharum* (sugar maple), and *Acer platanoides* (Norway maple).

Thus, new plants could easily be incorporated into such a system simply by adding new units at the necessary level, and units found not to be separate could be combined. Since the number of known plants was increasing tremendously during this period of increased exploration, the system was very popular with botanists. Nevertheless, Linnaeus realized the limitations of a system that was more convenient than it was natural and expected that increased knowledge of relationships between plants would eventually permit a more natural system.

The greatness of Linnaeus lay in his broad outlook and his organizational ability. His system was so revolutionary in its simplicity and workability that he is called the father of taxonomic botany. Much present-day work uses Linnaeus' fundamentals to establish classifications on the species level.

Natural Classification System

It was not until the latter part of the eighteenth century that other scientists realized that more was needed for plant classification and identification when naming plants than brief descriptions and illustrations. Between 1750 and 1800 an influx of tropical and exotic plants into European botanical gardens brought the realization that there were more differences in plants than the reproductive differences Linnaeus had emphasized. Increasing awareness of floral morphology led these later taxonomists to add it to the sexual characteristics to form the basis for a new system, referred to as the *natural system*.

Since the systems of this period predated Darwin, they failed to recognize the progressive changes that occur in all living things through the process of evolution. In this classification system, plants were treated as static objects and were arbitrarily grouped into categories established by the scientist rather than into groups relating to the plant's geneology.

Phylogenetic-Based Classification System

Darwin's *On the Origin of Species by Means of Natural Selection or the Preservation of Favored Races in the Struggle for Life*, published in 1859, presented scientific facts based on the theory of gradual development and evolution of both plants and animals. His theory was based on three principles:

1. In nature there is overproduction and variation.

2. Overproduction results in competition or struggle for existence.

3. Variation gives an opportunity for survival of the fittest.

Darwin's theory of the origin of species obliterated previous taxonomic theories, since its emphasis on variability and mutability of species was contrary to all the earlier beliefs. The acceptance of Darwin's theory of evolution led to changes in the plant-classification system and establishment of current taxonomic systems.

In 1875, August Eichler was the first to present a plant-classification system based on inherited relationships. This led to the publication of a 20-volume classic based on Eichler's theories, *Die Naturlichen Pflanzenfamilien*, by Adolph Engler, Karl Prantl, and others, beginning in 1892. Engler and Prantl divided all plants with seeds into Gymnospermae and Angiospermae. Angiospermae were further subdivided into the Monocotyledoneae and Dicotyledoneae classes.

Under this system, subclasses were divided into orders, which were further divided into families. This is the backbone of the taxonomic system used today (see Table 18-1). Since tables of this sort are not completely standardized, you may find different ones in other sources.

When designating maize scientifically, it is not necessary to give the whole family tree. Sometimes the binomial *Zea mays* is sufficient. This includes all types of maize from flint corn to popcorn. To be more exact the trinomial may be used, *Zea mays indentata*, to specify that dent corn is the plant being considered. Sometimes the name of the particular

TABLE 18-1
Units of Classification (Taxa) Illustrated by the Complete Classification of Cultivars *Malus domestica* 'Jonathan' and *Zea mays indentata* 'Silver Queen'

Kingdom: Plantae (plants)
  Division: Spermatophyta (plants that bear seeds)
    Subdivision: Angiospermae (seeds enclosed)
      Class: Dicotyledoneae (seeds with two cotyledons)
        Order: Rosales
          Family: Rosaceae
            Subfamily: Maloideae
              Genus: *Malus*
                Species: *domestica*
                  Cultivar: 'Jonathan'
Kingdom: Plantae (plants)
  Division: Spermatophyta (plants that bear seeds)
    Subdivision: Angiospermae (seeds enclosed)
      Class: Monocotyledoneae (seeds with one cotyledon)
        Order: Graminales
          Family: Poaceae
            Genus: *Zea*
              Species: *mays*
                Subspecies: *indentata*
                  Cultivar: 'VPI 648'

cultivar may be desired. The cultivar name, usually in single quotation marks, follows the binomial or trinomial. When the cultivar and varietal names are synonymous, the latter is omitted. For example, *Zea mays indentata* 'VPI 648' designates a specific inbred line of *Zea mays indentata* developed and described by corn breeders at Virginia Polytechnic Institute.

DEVELOPMENT OF SCIENTIFIC NOMENCLATURE

History and Background

The first organized effort toward global standardization and legislation of nomenclatural practices was at the First International Botanical Congress meeting in Paris in 1867. Up until then, nomenclature was controlled by a few botanists of prestige and influence. The rules officially accepted at the Congress in 1867 were published by De Candolle in *Lois de la Nomenclature Botanique*. The Congress has met irregularly since 1867, publishing the legislation agreed upon in several volumes of the *International Code of Botanical Nomenclature*. These rules apply only to scientific names.

The scientific names assigned to individual plants are derived from a variety of sources. For example, *Magnolia grandiflora* is named because of its large or "grand" flowers, *Geranium carolinianum* was described from the Carolinas and *Rosa wichuraiana* was named in honor of Wichuray, a Russian botanist. Not until 1930 did an International Botanical Congress adopt rules for the determination and coining of plant names, the one area where new rules are made at each International Congress.

The Organization of Scientific Nomenclature in Taxonomy

The plant kingdom is organized into divisions, classes, orders, families, tribes, genera, and species, as shown in Table 18-1. Each category is called a *taxon* (plural, *taxa*), and each taxon, except for the genus and species, is distinguished by having a characteristic name ending.

Most horticultural, agronomic, and forestry crops are found in the *division* of the plant kingdom called *Spermatophyta*, or seed plants. Division names always end in *-phyta*.

The Spermatophyta division has two *subdivisions*, *Gymnospermae* and *Angiospermae*. Subdivision names end in *-ae*. Characteristics of Gymnospermae are as follows:

1. Plants always woody but vessels not present in vascular system.

2. Seeds produced on naked surfaces of scales; scales arranged in cones.

3. Flowers absent.

Angiospermae are also vascular plants which have seeds, but they differ from the Gymnospermae in the following characters:

1. Vessels almost always present in the vascular system.

2. Seeds enclosed in fruits.

3. Flowers present, each usually containing four whorls of floral organs (calyx, corolla, stamen, and pistil).

The angiosperms are composed of two classes, *Dicotyledoneae* and *Monocotyledoneae*, contrasted in Table 18-2. Class taxa names end in *-eae*. Dicotyledoneae, commonly called the dicots, contains approximately 200,000 species grouped in more than 250 families, including the broad-leaved trees, roses, sunflowers, and legumes. Monocotyledoneae, or monocots, contains approximately 50,000 species grouped in more than 40 families including the cereals, lilies, orchids, irises, and palms.

Each class is divided into orders, with names ending in *-ales* (rhymes with Bailey's), e.g., Rosales. An order is a group of related families with some common traits but some marked differences.

Each order is divided into families. The family, the last of the major taxa, usually represents a more natural unit than any of the higher categories. It is composed of closely related genera or (rarely) a single genus. Its name is usually formed by adding *-aceae* to the stem of an included generic name, e.g., *Rosaceae* from *Rosa* for the Rose family. When the size of the family justifies it and the included genera can be naturally grouped, the family is divided into subfamilies. The name is formed by adding *-oideae* to the stem of an included generic name.

In the binomial system used today, the scientific name is represented by a latinized form of the genus and species. The basic unit is the species, which can be defined as an interbreeding, i.e., cross-fertile, group of individuals of common ancestry with similar structure.

In writing the scientific name, the genus and species are italicized or underlined. An abbreviation of the name of the person who first gave this name to the plant is often written after the binomial name, e.g., in *Viola canadensis* L. The L. signifies that Linnaeus gave this name to the plant.

The initial letter in the generic name is always capitalized while the initial letter in the specific name is lowercase. The names of the genera and species have diverse endings, *-a, -ea, -ia, -i, -is, -um,* and *-us* being common. The International Code has established that only one group of plants can carry a particular genus name and a genus name can be used only once per group of plants. For example, the genus name *Quercus* (oak) cannot also be used for any other group of plants, and if all the oaks are considered to be in a single genus, they must all carry the name *Quercus*. A species name, on the other hand, is often used more than once; for ex-

TABLE 18-2
Contrasting Characteristics of Monocots and Dicots

	Monocots	Dicots
Principal veins	Usually parallel	Branching from midrib or its base; not parallel; forming distinct network (Fig. 18-1)
Flower parts	Usually in 3s or multiples thereof (Fig. 18-4)	Usually in 4s or 5s or multiples thereof (Fig. 18-2)
Root system	Fibrous, no taproot	Taproot common
Vascular bundles	Scattered irregularly through pithy tissue (Fig. 18-3b)	In a single cylinder (Fig. 18-3a)
Cambium	Absent	Present
Growth pattern	No increase in girth by forming annual layer	Cambium in woody species adds new layer each year or growing season (secondary growth)
Embryo	One cotyledon, called a scutellum	Two cotyledons

ample, the name *vulgaris* (common) is widely used, as in *Phaseolus vulgaris* (bean) and *Beta vulgaris* (beet). Obviously, it can only be used once in a given genus.

The generic name is always a singular noun in the Latin nominative case. It may be descriptive, e.g.,

Liriodendron (tulip tree); an aboriginal name of the plant, e.g., *Quercus*, the old Latin name for oak, or a name honoring a person, e.g., *Linnaea* for Linnaeus. The species name may be any of a number of descriptive words indicating a growth habit, a color, a person, or place. Latin names for many plants are

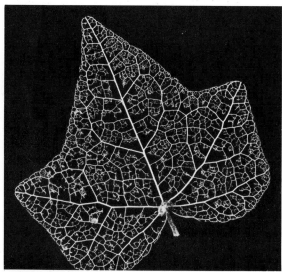

FIGURE 18-1
Surface view of cleared dicotyledon leaf of English ivy (*Hedera helix*) showing extensive pattern of venation. (*Carolina Biological Supply Company*.)

FIGURE 18-2
Dicotyledon flower of the peach (*Prunus persica*) showing the five-petal arrangement.

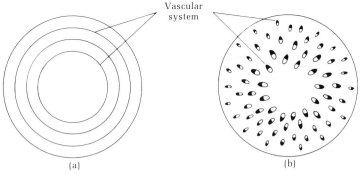

Vascular system

(a)

(b)

FIGURE 18-3
Cross-section of vascular system of dicotyledon and monocotyledon stems: (*a*) dicotyledon with continuous vascular system; (*b*) monocotyledon with discontinuous vascular system.

FIGURE 18-4
Typical monocotyledon flower of the lily (*Lilium longiflorum*) showing flower parts in threes. (*Courtesy of J. Martin.*)

TABLE 18-3
A Key to the Species of Oaks Occurring in the Mountains of the Southeastern United States

1. Leaves not widest at apex	2
2. No lobes in leaves. *Quercus phellos* (willow oak)	
2. Lobes in leaves	3
3. Bristle tips on leaves	4
4. 3–5 lobes/leaf, hairy. *Q. falcata* (southern red oak)	
4. 5–9 lobes/leaf, smooth. *Q. velutina* (black oak)	
3. No bristle tips on leaves	5
5. 7–11 lobes/leaf, smooth. *Q. alba* (white oak)	
5. 3–7 lobes/leaf, hairy. *Q. stellata* (post oak)	
1. Leaves widest at apex. *Q. marilandica* (blackjack oak)	

often descriptive of certain plant characteristics. In extended discussions where no confusion would result, the name of the genus is often abbreviated after it has once been spelled out, as in Table 18-3.

Using selection and breeding in many species, plant scientists have developed plants with specific characteristics for economic and cultural purposes. A cultivar is a cultivated variety that has originated and persisted under cultivation and is of agricultural importance, requiring a name. The term *cultivar* may be used interchangeably with *variety* to denote cultivated plants. New cultivars are constantly being developed in an effort to improve the quality and yield of crops and to permit wider climatic adaptation.

Cultivar names are now formed from not more than three words in a modern language and are usually set off by single quotation marks; e.g., the

'Jonathan' apple and 'Seneca Chief' sweet corn are written *Malus domestica* cv. 'Jonathan' and *Zea mays* cv. 'Seneca Chief,' thus making the nomenclatural system trinomial. Cultivars may differ from each other in color, quality, size, time of harvest, yield, disease resistance, or a combination of these or other agronomic or horticultural factors.

A cultivar may be any of several different types:

Clone—This is composed of identical material derived from a single individual and propagated entirely by vegetative means. Unless mutations have taken place, the plants of a clone are genotypically homogenous (or identical) rather than carrying the genetic variation inherent in plants propagated by seed. The abbreviation cl. is used.

Pure Line—This consists of plants of uniform appearance which, thanks to their horticultural or agronomic value, are reproduced as uniformly as possible by seed.

Assemblage—This consists of individuals reproducing from seeds showing some genetic differences but having one or more characteristics by which the plants can be differentiated from other cultivars.

Uniform Group—This is a first-generation hybrid (F_1) resulting from the cross of two pure lines. Such hybrids can be reconstituted whenever necessary by crossing two parental lines maintained either by inbreeding or as clones.

Hybrids formed between species are distinguished by a times sign between the generic or specific names, e.g., *Forsythia* × *intermedia*, or between the specific names of the two parents.

The term *form* is used to describe members of a population which differ genetically from other members but not enough to be separate cultivars. Such deviations are designated as forms only if they are beneficial and can be asexually propagated, thereby establishing a clone, to ensure that the desired characteristic can be retained. An example is the 'Starking Delicious' apple, a vegetatively propagated mutation from the original 'Delicious' cultivar.

HOW PLANTS ARE PLACED IN TAXA

There are always differences between individuals within a species due to environmental factors, to random genetic differences, or to hybridization, whether natural or induced. Since such variations mean that plants in an individual species can appear very different, careful study of any plant is necessary to determine whether the plant is correctly identified, whether as a member of an established species or as a new species.

The most common method of plant identification is to compare the plant in question with descriptions and illustrations from plant manuals, which list, in appropriate order, the plant families of the principal taxonomic divisions. Some of the manuals which are most useful in identifying plants are Bailey's *Manual of Cultivated Trees and Shrubs* and Gray's *Manual of Botany*. Books describing the floras of specific states or regions are also helpful and especially convenient, since they contain only the plants for that region and may give more information than books covering a larger area. Plants can also be identified by comparing them with herbarium specimens. A third method of identification of plants is by using a plant key (Table 18-3), which presents a sequence of choices, usually between two statements. By taking the correct choices in sequence, one arrives at the name of the unknown plant. In this type of key, each statement, called a *lead*, is identified by a letter or number, and each successive subordinate statement is indented under the preceding one.

Regardless of which method of identification is used, it is important to become familiar with the characteristics and variations in form and size exhibited by leaves, flowers, fruits, and seeds. These morphological structures are visible, whereas, reproductive structures often have to be dissected or examined with a hand-lens or dissecting microscope.

Other characters that may be valuable in distinguishing species are form, leaf apex, length of petiole, and pith characters of the twig or stem. Flowers and flower structures are frequently used to identify the family and genus of a specimen.

CLASSIFYING PLANTS ARTIFICIALLY

By Economic Value
A plant which has been found useful for food, clothing, shelter, fuel, feed, medicine, hallucinogens, weaponry, tools, transportation, or decoration fits the broadest definition of a *crop*. In the process of

discovering useful plants, people also found that some plants were thorny, foul-smelling, bad-tasting, poisonous, carriers of pests, or otherwise disagreeable. Plants which interfere with crops in any way were branded weeds. Most plants are neither crops nor weeds, and not all crops or weed plants fit neatly into one or the other of these categories. For example, in some places, bermudagrass is a valuable species for lawn and pasture, but elsewhere the home gardener spends hours digging it out of his potato patch.

By Harvest

A crop is any plant used by people. On the other hand, vegetation consumed directly by grazing livestock and not harvested per se is technically *pasturage*. This distinction is legally recognized in leases and contracts when it is stipulated that certain land may be cropped and other land grazed. In general agricultural terminology, however, crops include pasturage.

By Morphological Habit

All plants have structural features which enable us to divide them into separate groups. A tree, for example, has a main central axis (trunk) and an upright habit. A shrub has many stems arising from a single point or crown. Vines have long slender stems and trail on the ground prostrate unless supported by another plant or object. Plants may also be distinguished as being woody or herbaceous.

By Growth Pattern

Season It is the genetically fixed, physiological nature of some plants to survive only one season. Others remain active for many years.

Annuals—The life cycle is completed in 1 year or less. These plants germinate from seed; produce vegetative growth, flowers, fruits, and seed; and die within one growing season. Examples are marigold, pea, bean, corn, squash, pumpkin, ageratum, and garden cress.

Biennials—These ordinarily require 2 years or at least part of two growing seasons with a dormant period between growth stages to complete the life cycle. Only vegetative growth is produced the first year; the following spring and summer the plant produces flowers, fruits, and seeds, after which it

dies. Examples are celery, sweet william, and sweet clover. Biennials useful for consumption only in their vegetative state and therefore grown as annuals include cabbage, carrot, and beet.

Perennials—These plants do not die after flowering, but live from year to year.

Herbaceous Perennials—The soft, succulent tops are killed back by frost in many temperate and colder climates, but the roots and crowns remain alive and send out top growth when favorable growing conditions return. Examples are rhubarb, asparagus, alfalfa, timothy, chrysanthemum, and many cultivars of phlox. Sorghum is an example of a non-winter-hardy perennial that acts like an annual in temperate climates.

Woody Perennials—Woody stem tissues characterize these plants that are longer-lived and more durable.

Day Length Certain physiological processes are so important in the growth of some crops that they are used to describe the plants. Photoperiod or day length (Chap. 7) controls flowering in many crops, and they are described as *short-day* or *long-day plants.*

Hardiness Because of their resistance to frost and freezing, some plants are said to be *hardy*. Other plants are *tender*, sensitive to cold or cool weather. Winter wheat, winter oats, winter barley, and many other grasses are very hardy, being able to survive temperatures below $-18°C$. However, many of our crops such as soybean and corn are quite sensitive to frost or freezing weather and may be killed as seedlings or before reaching maturity. Landscape perennials must be purchased according to their hardiness in all areas having periods of cooler temperatures. Hardiness zone maps are distributed widely for this purpose.

By End Use

In the restricted agronomic sense, crops are often grouped according to their particular use, but this is a loose system of categories that are not fully inclusive or exclusive but overlapping. Some of the common agricultural crop categories are as follows:

Food Crops—for human consumption: wheat, apple, sweet corn, rice, soybean, potato, etc.; plant products fermented to produce ethanol (corn, barley, rice, grape, potato, etc.) may be included

Feed Crops—for livestock consumption: oats, barley, dent corn, millet, forages (alfalfa, bermudagrass, etc.)

Fiber Crops—used in making textiles and cordage: cotton, flax, hemp, jute, sisal, etc.; plants used in papermaking are sometimes included

Oil Crops—lipids extracted for human or animal consumption or industrial purposes: soybean, peanut, flax, sunflower, cottonseed, corn, tung, etc.

Ornamental Crops—juniper, holly, lily, azalea, dogwood, etc., used in landscaping.

Industrial and Secondary Crops—extracts for a great variety of personal and industrial uses: rubber, tobacco, etc.

By Discipline

The applied sciences in the field of agriculture are interrelated:

Agronomy—the production of grains, fibers, forage, and other field crops; soil science

Forestry—forest management and wood production and utilization

Horticulture—landscaping and the production, storage, and marketing of fruits, vegetables, floricultural products, and nursery crops

At one time agronomy, forestry, and horticulture were distinctly separated according to the purposes for which a particular crop was produced and the intensity of its cultivation, but as areas in plant science have expanded, the lines of demarcation have blurred. The relationship between horticulture and the other disciplines in the agricultural plant sciences can be shown best through examples. When grown for its nuts, the pecan tree is associated with horticulture, but when grown for its wood, it is a forest crop. Although grasses grown as turf are studied in both horticulture and agronomy, grasses used as pasture crops are strictly agronomic crops. Compared with agronomic and forest plants, horticultural crops are intensively cultivated and often have much higher economic returns per hectare of production; however, with expanded population and new production methods, the plantings of fruits and vegetables are increasing and production is becoming more extensive, making these factors vague in defining a particular agricultural area.

Within Agronomy

Some species are grouped together because their culture or management is similar, because they share a specific on-farm purpose, or because they are commonly recognized as having a natural (genetic) relationship.

A *cover crop* is planted to cover the soil and protect it from erosion. Wheat, rye, crimson clover, and other winter annuals are commonly planted in the fall and survive the winter in the field. When such plantings are plowed under in the spring to add nutrients and organic matter to the soil, they are also called a *green manure crop*.

A *companion crop* or *nurse crop* is one sown along with another to help the latter get established. For example, oats are frequently planted with alfalfa to help the alfalfa get a good start. The oats germinate quickly, provide shade for young alfalfa seedlings, and then are pastured off or harvested as hay or for grain before the alfalfa becomes dominant.

Row crops are *field crops*, i.e., plants grown on a large scale, planted with enough space between rows to allow for a cultivator to pass. Corn, soybean, peanut, tobacco, cotton, sugarbeet, etc., are commonly so grown. *Broadcast* or *drilled crops*, on the other hand, are sown evenly (broadcasted) across the field or in closely spaced furrows, or drills. Wheat, oats, barley, and alfalfa, usually seeded this way, produce dense stands which cannot be cultivated by mechanical means.

Although the terms cereals or cereal crops are often used interchangeably with grains or grain crops, there is a difference. Technically, *cereal crops* are grasses grown and harvested for small grains edible by human beings. Wheat, rice, corn, rye, oats, and barley are the six great cereals of the world, but in some places millet and sorghum may also be used for human consumption and are therefore cereal crops. *Grain* and *grain crops* are more inclusive terms. They are sometimes used to include the seeds of buckwheat, flax, amaranths, or even soybeans, but usually by grains we mean all the large-seeded grasses and their fruits whether eaten by people or not.

Small Grains—wheat, oats, barley, rye, and others

Feed Grains—corn, oats, rye, barley, grain sorghum, and others

Coarse Grains—in commerce, corn, oats, and barley

Bread Grains—in commerce, wheat and rye

Forage crops include many plants harvested (or grazed) essentially in their entirety for feed purposes. We have already mentioned pasturage, which is usually the vegetation of grasses and legumes grown for grazing by livestock. If the vegetation maintains itself (is perennial or self-seeding or both) and continues to be suitable for grazing year after year, it is called *permanent pasturage. Temporary* or *supplemental pastures* are seeded with plants which grow for only one season, e.g., sudangrass, or perhaps for 2 years, e.g., biennial sweet clover. When grasses and legumes are cut, dried, and stored for later feeding to animals, the product is called *hay*, a type of forage crop. *Silage* (or *ensilage*) is any forage crop cut before drying and stored moist in airtight silos, where it ferments, preserving much of its original nutritive value. It is often further distinguished as corn silage, sorghum silage, or grass silage, the latter usually being made from fine-stemmed species of grasses, alfalfa, or other legumes. *Green chop* is any forage crop cut green and fed immediately without drying or ensilage. Since *forage* as a noun is commonly used to describe any and all plant materials used as animal feed, "forage crop" has a much more restricted meaning.

Within Horticulture

Horticultural crops are commonly divided into four rather artificial groups: fruits, vegetables, floricultural crops, and nursery crops. Although traditional and convenient, the groupings are quite artificial and considerable overlap exists.

Horticulturists define a fruit as an edible, fleshy portion of a woody or perennial plant whose development is closely associated with the flower, and which is eaten as a dessert or snack (Chap. 21).

Fruit plants are subdivided on the basis of climatic requirements into tropical, subtropical, and temperate groups. The tropicals include avocado, banana, papaya, and pineapple. The subtropicals are represented largely by oranges and other *Citrus* species. The temperate fruits are further divided into tree fruits, tree nuts, and small fruits. The temperate tree fruits include apple, pear, peach, and cherry; tree nuts are represented by walnut, pecan, and almond. The small fruits are borne on low-growing plants and are also smaller than most tree fruits. Major small fruits include grape, strawberry, raspberry, and cranberry.

Vegetables can be defined as the edible portion of an herbaceous garden plant, usually eaten during the principal part of the meal, either in a raw or cooked state (Chap. 20). This definition is artificial and may lead to confusion in separating fruits and vegetables. If the product is of leaf, stem, or root origin such as spinach, asparagus, or carrot, respectively, there is no problem. However, the tomato and cucumber are grown and eaten as vegetables but are fruits from a botanical viewpoint (Fig. 18-5). These examples offer some insight into the confusion that can accompany artificial systems of classification.

Floricultural crops include cut flowers, flowering pot plants, foliage plants, and bedding plants (Chap. 24). Cut flowers such as rose, carnation, chrysanthemum, snapdragon, and orchid are grown mostly in greenhouses, whereas most gladiolus are grown outdoors in the summer or in mild climates. Flowering potted plants are sold as whole plants in bloom and include Easter lily, poinsettia, cyclamen, and chrysanthemum. Foliage plants are nonflowering plants grown in greenhouses or in protected outdoor areas in California and Florida. These plants such as philodendron, ferns, rubber tree, and weeping fig are used largely in indoor settings. Bedding plants are mostly annual flowers which are started in greenhouses and then sold for use in outdoor landscapes. Important bedding plants are marigold, petunia, and pansy.

FIGURE 18-5
Lycopersicon esculentum (tomato)—a perennial grown as a warm-season annual or in a greenhouse for its edible fruits (*esculentum*—edible). (*Harris Seed Company.*)

Nursery Crops In a very broad sense, nursery crops encompass some vegetable and many fruit plants, but the vast majority of the nursery business revolves around perennial ornamentals (Chap. 24). Certain nurseries specialize in fruit trees, others in strawberry or blueberry plants, and several grow and sell asparagus and rhubarb plants. The size of nurseries specializing in ornamentals is very varied as is the degree of specialization. The species grown differ markedly with the location but range from annual, biennial, and perennial herbaceous species to a myriad of shrubs and trees. They may be grown in the field or in containers but the intent is to provide ornamental plants for use in outdoor landscapes.

Within Forestry

A forest is a community of trees and undergrowth covering a considerable area. To facilitate forest management, foresters arbitrarily classify forest trees into several groups. One classification is based on the size of the individual tree (Chap. 25); others consider age and species. Trees in a forest that are essentially the same age, like those coming in after a devastating fire, are said to be *even-aged*. A forest with trees ranging from small seedlings to medium-sized trees and large veterans is said to be *all-aged*. A *pure forest* is composed mainly of one species; a *mixed forest* contains several. Trees which can grow in the shade of other trees are called *tolerant*, and species which cannot survive in shade are *intolerant*.

The stand density of a tract of timber indicates whether it is *well stocked*, *medium stocked*, or *understocked*. A medium-stocked stand has 40 to 70 percent of its canopy closed over. Another common classification of trees, based on the relative position of their crowns, will be discussed in Chap. 25.

CHAPTER NINETEEN

GRAINS AND PULSES

From an economic and nutritional standpoint, the grains and the pulses are undoubtedly the two most important groups of plants in the world. Between them they occupy more than three-quarters of the world's cropland (excluding land devoted to grazing), and they provide more than 90 percent of the human food supply. The grains are an invaluable dietary source of carbohydrates, the important metabolic energy component. The pulses are indispensible sources of protein, especially in countries where milk and meat are scarce.

Grain crops include those species whose harvested unit is called a "grain." Most are members of the grass family, but in commerce certain other crops may also be called grains. The nongrass grain crops include buckwheat and flax, and one even hears of soybeans being sold on the grain market. *Cereal grain* is often used to designate specifically the members of the grass family whose seeds are harvested for food or feed. Only members of the grass family are properly called cereals, cereal grains, or cereal crops.

Pulses are to the legume family what cereals are to the grass family; i.e., pulse crops are legumes whose harvested portion is the edible seed (usually a bean or pea). In the United States, soybean, peanut, and dry beans are the most common pulses, but others are important in world commerce or in certain geographic areas. As food or feed, they are critical protein sources, and some species are valuable for their edible or industrial oils as well. Our discussion of widely grown grains and pulses concentrates on those of the greatest economic and nutritional value.

Few crops grown in a particular geographical area are actually native to that part of the world. For example, from the region now occupied by the 48 contiguous states there have come only about a dozen truly native nonforest crop species: strawberry, pecan, cranberry, sunflower, Jerusalem artichoke, and a few forage grasses. (Many crops being cultivated by the Native North Americans when Europeans arrived, e.g., tobacco, maize, and squashes, were introduced from elsewhere in the New World.) All other crops now grown in this country came originally from some other part of the world. Their centers of origin and domestication cover almost the whole world, but a few areas have been more prolific than others. The most important centers of origin include central and western China (soybean, barley, and many others), India (rice, cotton, and more), and the Near East (wheat, rye, alfalfa, etc.). The following discussion of crops demonstrates how indebted each part of the world is to all the rest for most of its crops.

GRAIN CROPS (POACEAE)

Grain crops, especially the cereal grains, are efficient converters and storers of the sun's energy. Their photosynthetic products are stored largely as starchy materials in their seeds, which in corn, rice, wheat, and many other cereals are three-quarters or more starch. The starchy material in the *endosperm* of kernels of corn or wheat serves as an energy reserve and raw-materials storehouse for germination and growth of young seedlings. From an economic or nutritional point of view, the starchy grains are a valuable source of metabolic energy and an important industrial feedstock. Although we concentrate on the food and feed value of grains, their starches find many other uses, including gasohol and sizing for paper or cloth. (*Sizing* is the process by which a desirable finish or texture is given to paper and fabric.)

Wheat

The world's most widely cultivated crop, wheat, occupies more than one-fifth of the total cropland devoted to nonforage crops—almost twice as much as its next closest competitor, rice. Wheat is grown on a massive scale in the U.S.S.R., the United States, China, India, northwestern Europe (Fig. 19-1), Canada, Australia, and Argentina. Such widespread and abundant cultivation attests to its great adaptability as well as its role as the world's premier crop and most valuable cereal.

Classification and Origin One normally thinks of a crop as being a collection of cultivated varieties of a single species. The most notable exception to this generalization, wheat, is actually a collection of at least three distinct species in the genus *Triticum*. Some classification schemes list as many as 14 different species (Table 19-1).

FIGURE 19-1
A large field of common wheat, in this case soft red winter wheat, produced in West Germany. (*Photograph by M. M. Alley.*)

TABLE 19-1
Classification of Some Wheats (*Triticum* spp.)

Species	Common name	Number of chromosomes
T. aestivum	Common wheat	42
T. compactum	Club wheat	42
T. durum	Durum wheat	28
T. dicoccoides	Emmer	28
T. monococcum	Einkorn	14
T. spelta	Spelt	42

Common wheat (*T. aestivum*), as its vernacular name suggests, is the type most frequently grown, especially in the United States. Although in this country, emmer and durum wheats (sometimes listed as a subspecies of *T. turgidum*) are grown on a relatively small scale; in a few areas, such as North Dakota, where durum wheat is particularly well suited and productive, it is grown on large acreages. Durum wheat is also grown on a large scale in Russia and around the Mediterranean. Club wheat (sometimes listed as a subspecies of *T. aestivum*) is grown in this country primarily in the Pacific northwest, where climatic conditions favor its development. Most of the other types of wheats listed in Table 19-1 are grown on a very limited basis, especially in the United States. Spelt, for example, is grown occasionally as livestock feed. Einkorn is more or less a novelty crop, grown sparingly in Europe.

Common wheat has three distinct types:

Hard red winter wheat—is grown primarily in the central plains states, especially Kansas and Oklahoma, in southern Russia, and in Argentina. It is planted in the fall and overwinters as a short, immature plant. The grain head develops in the spring; grains are hard, relatively high in protein, and desirable in making bread.

Hard red spring wheat—is spring-planted in regions where the winters are too cold for fall plantings to survive, e.g., the northern tier of the United States, all of Canada, and parts of Russia and Poland. The grain, which is relatively high in protein, is the most desirable type for making bread flour.

Soft red winter wheat—is grown as a fall-planted crop in humid regions with relatively mild winters. In this country, the eastern half of the corn belt is

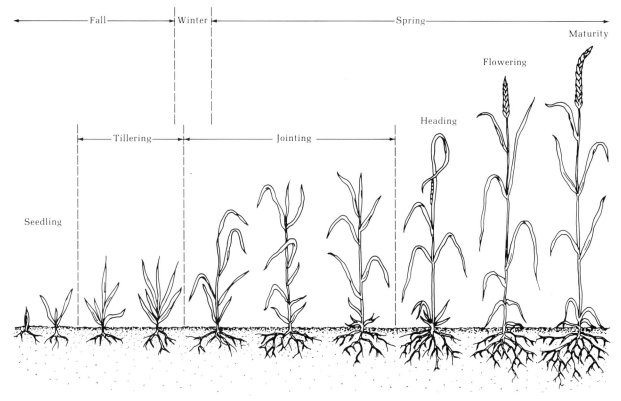

FIGURE 19-2
Development of winter wheat from seedling to harvest maturity.

the major production area. Western Europe also has a suitable climate, and France is a major producer. The kernels are noticeably softer and lower in protein than the types growing in the drier, often cooler, seasons of areas producing hard red wheat.

Common wheat is thought to have developed many centuries ago by hybridization and mutation from weedy ancestors resembling emmer, probably assisted by human intervention in selectively harvesting and then cultivating more desirable types. Wheats of various types were grown in prehistoric times and have been found by archaeologists in Egyptian, eastern Mediterranean, and European cultures that date back to the edges of civilization. In fact, cultivated wheat probably played a major role in the dawning of civilization—the development of human societies on a communal scale. For individuals to form the village grouping or civil unit, it was necessary to

support the larger population agriculturally. Simply gathering nuts, berries, roots, and grains from wild plants could not provide for higher-density communal living. An organized system of plant and animal husbandry was a prerequisite for producing food for many people congregated in one place. Wheat lent itself admirably to this need and has followed human migration extensively. It came to this country as early as 1602, thus preceding the Plymouth Colony and Jamestown.

Biology Wheat is a typical *small grain*† in its morphology and development (Fig. 19-2). The stems of young plants do not elongate during the production of the first several leaves and internodes; the visible plant is simply a tuft of leaves at this early stage.

---

†Wheat, oats, barley, and rye are the cool-season cereals commonly called small grains.

Branches, or tillers, develop below or at ground level and produce more leaves. For the fall-planted winter wheats, the seedlings enter the more or less dormant winter period in this form. When flower formation begins, each stem or tiller stops producing new leaves and starts to produce the grain head. Although the stem is quite short, within a few weeks, it shoots up to full height and raises the inflorescence nearly a meter or more aboveground. The florets open and pollination occurs, usually within a flower but sometimes by cross-pollination.

Use Most wheat is milled for flour used in baking. The hard red winter and spring wheats are preferred for breadmaking because of their higher gluten content. *Gluten* is a protein that increases the strength of dough by improving its water-absorption and gas-holding ability. This results in a moister, better-textured loaf and more 1-lb. loaves per barrel of flour. Soft red winter wheats, with lower gluten, are more desirable for cakes and pastries. Durum wheat flour is the most popular source of *semolina*, a milling product used for noodles, spaghetti, and macaroni. Club wheats are suitable for cakes, cookies, crackers, and cereals.

The starchy endosperm of a wheat kernel, or "berry," (Fig. 19-3) after milling and bleaching, produces the white flour commonly used in bread making or sold as household flour. The aleurone layer, which is the dark outer portion of the kernel (the *bran*), and the embryo (the *germ*) are normally removed to produce a white flour. Unfortunately this also strips away many vitamins, especially thiamin

(vitamin B_1) and tocopherol (vitamin E). Despite their lower inherent nutritional value, white flours, are preferred over whole-wheat flours by most consumers as a matter of cultural preference. (When only the rich could afford the more expensive white flour, white bread became a status symbol.) Also whole-wheat flours are more likely to become rancid due to the oils present in the germ. White flour will keep its quality longer than will whole-wheat flour when stored under identical conditions.

Rice

In some views, rice is the world's most important crop, because it is the chief food crop of perhaps half of the world's population. It ranks a distant second to wheat in total acreage, being grown on only about half as much land; but because of higher yields per hectare, tonnage of rice approaches that of wheat. China is the world's leading producer (and consumer) of rice (Fig. 19-4) with India, Indonesia, Japan, and other southeast Asian countries also producing large quantities. In the United States, rice is not a major crop, ranking behind five other grains in acreage and tonnage; it is grown only in a few southern or coastal states.

Classification and Origin Cultivated rice (*Oryza sativa*), apparently a native of southeast Asia, grows best in a warm, humid climate. Wild rice, a food source for Native Americans in the Mississippi River valley and an increasingly popular specialty item, is a different species (*Zizania aquatica*), resembling cultivated rice primarily in its submerged growth.

Most of the thousands of rice cultivars can be grouped into two major morphological categories: the short-grain (*japonica*) and the long-grain (*indica*) types, which appear to have developed from a common ancestor. Each became distinct after many centuries of cultivation in different geographical and climatic settings with different cultural practices. Short-grain rice has generally shorter, stouter stems (and grains) and is more resistant to lodging when fertilized heavily. Medium-grain rice, produced by crosses between the two major types, is now grown in many countries including the United States.

Rice cultivars can be classified according to their endosperm type. *Nonglutinous* (common or ordinary) grains, the preferred type, contain both amylose

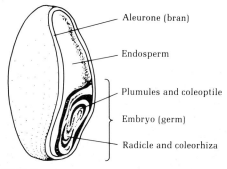

FIGURE 19-3
The wheat grain, also called a *berry* in commerce, and its commercially important parts.

Aleurone (bran)

Endosperm

Plumules and coleoptile

Embryo (germ)

Radicle and coleorhiza

FIGURE 19-4
Rice culture in China. (*Photo courtesy of J. Caldwell.*)

and amylopectin starches. *Glutinous* (waxy) rice has only *amylopectin*, which gives the endosperm a waxy, more translucent appearance.

Rices can also be classified by their requirement for water. Cultivars that grow on continuously flooded land until shortly before harvest are called *paddy rice* to distinguish them from *upland rice*, which can be grown without standing water. Upland rice cultivars are generally much lower-yielding and still require high rainfall and more or less continuously wet soils. Although such cultivars may be locally important, in many places incredible effort is invested in turning hillsides into flooded terraces so that paddy rice can be grown (see Fig. 13-13).

In tropical and subtropical humid regions, rice has been what wheat has been in temperate, humid areas—the staple food providing a reliable agricultural basis for civilization. Rice culture started in China perhaps 5000 years ago, when people discovered that husbandry and selection produced a bigger and more dependable crop than simply harvesting wild plants. Although rice cultivation spread to Europe before the time of Christ, it never gained the importance it has in the Orient, probably as a result of its poor adaptation to European conditions. Rice has been grown in the United States since colonial times; it was planted as far north as Virginia, but most of the acreage was in coastal regions of the deep south. After the Civil War, rice production shifted westward, till more than 90 percent of United States–grown rice now comes from Arkansas (Fig. 19-5), Louisiana, Texas, and California.

Biology Rice is an annual grass that resembles the small grains. It produces a number of tillers from a single seedling, and each tiller can develop a grain head. The number of tillers produced is affected by several factors, the most important being plant spacing, fertility, and cultivar. Plants that are widely spaced or well-fertilized or both may produce 20 or more tillers.

Rice is unlike most grains in its ability to germinate and thrive with seeds and roots submerged. Two almost unique characteristics are involved: seeds and young seedlings can grow using largely the energy obtained from anaerobic respiration (glycolysis), and when the stems and leaves emerge from the water, they can act as a snorkel and carry oxygen down to the submerged parts. Most plants do not have either capability; wheat and soybean, for example, are usually killed by 2 or 3 days of flooding. On the other hand, some of the most serious weed pests of rice have these or other adaptive characteristics, enabling them to compete with rice in the paddy.

Use When harvested, rice grains are still covered with the extrafloral parts, the *hulls*. Milling *rough rice* to remove the hulls leaves about 65 to 70 percent of the original weight as whole or broken kernels plus the bran. This *brown rice* is the equivalent of whole wheat in that the kernel with the pericarp and aleurone layer is still intact. Removal of the germ and the bran followed by polishing produces a shiny white grain.

Good-quality rice has only 7 to 8 percent protein. Most of the remainder is starch (75 to 80 percent). For people whose diet consists largely of rice, this

FIGURE 19-5
A terraced rice field in central Arkansas.

protein deficiency can be a serious problem. Brown rice is higher in vitamin content but has a shorter storage life because of the oil in its bran and germ.

Corn (Maize)

Although the acreage devoted worldwide to production of corn (*Zea mays*) is about half that of wheat, much higher yields per hectare mean that the annual tonnage of corn grain produced globally almost equals that of wheat. Corn can outyield rice on a per hectare basis, but since it grows on fewer hectares, global production of corn is slightly less than that of rice. In the United States, corn is indisputably the leading grain crop in acreage and total yields; and only in 1979, when soybean's total market value exceeded that of corn grain for the first time, has its place as the number one farm cash crop been challenged.

Almost half of the world's corn is grown in the United States, and that is done on less than one-quarter of the world's cropland devoted to corn. This significant statistic attests to the high technology (breeding, fertilization, irrigation, pest control, etc.) used in American corn production and the remarkable adaptation of *Zea mays* to the climate and soils of America's agronomic heartland, the corn belt (Fig. 19-6). Although maize is now grown almost worldwide, China and Brazil being major producers, the species, a native of the New World, seems to be particularly well suited to temperate North American conditions.

Classification and Origin Six distinct types of maize have been recognized, based primarily on certain characteristics of their kernels. Each type (Fig. 19-7) is so distinctive that it is often called a subspecies. The characteristics of the endosperm and certain extrafloral parts are usually diagnostic, but a number of other important cultural characteristics may be associated with particular types or subspecies.

Dent corn (*Z. mays indentata*) is by far the most commonly grown type, so that "field corn" is synonymous with dent corn. The endosperm of dent corn is made of two types of starchy material, one distinctly harder than the other. The *corneous* ("flinty" or "horny") endosperm forms a cup of hard, translucent material around the base and sides of a dent corn kernel; the rest of the endosperm is a more chalklike material that shrinks during drying, leaving a distinctive wrinkle or dent in the top. Dent corn is usually nontillering and selection for one ear per stalk is common.

Sweet corn (*Z. mays saccharata*) the common home garden type, is favored for the table because its endosperm has more free sugars and less starch than the other subspecies during early ear development. Some cultivars or hybrids are so high in free sugars that some people find them too sweet. Because they lack the corneous endosperm, mature sweet corn kernels shrink dramatically over their entire surface and produce a shrivelled or wrinkled seed. Sweet corn is typically harvested for table or market

FIGURE 19-6
Corn growing in the corn belt. (*U.S. Department of Agriculture.*)

FIGURE 19-7
The various types of corn. From left to right: dent, flint, flour, sweet, pop, and pod.

while the kernels are still in the succulent stage of development that precedes formation of starch. Sweet corn has a greater tendency to tiller than most other subspecies, and it may produce more than one ear per stalk.

The kernels of flint corn (*Z. mays indurata*) are rounded and uniformly surrounded by the flinty endosperm. The grains have less of the softer, starchy endosperm in their interior. Flint corn is seldom planted in the United States now except in northern, short-season areas, where its early maturity is advantageous. The northern flint varieties typically produce long, slender ears with only 8 rows of kernels (compared with the 14 to 18 rows common on dent corn). As expected from its rapid development and early maturity, northern flint corn is typically shorter (1 m or less) and has fewer leaves (7 to 8 leaves versus 10 to 15) than common dent varieties.

The endosperm in the kernels of popcorn (*Z. mays everta*) is distributed in much the same way as that in flint corn, but the kernels tend to be much smaller. The corn pops because the water trapped within the horny endosperm layer turns to steam when heated and causes the seed to explode. Popcorn plants frequently tiller, and the main stem may produce several ears.

Flour corn or Indian corn (*Z. mays amylacea*) is one of the oldest types and the one most closely associated with the Native Americans. The uniformly soft, starchy endosperm of the kernel makes this subspecies especially well suited for grinding into meal for flour. The kernels are often variously colored and the ears can be quite decorative.

In pod corn (*Z. mays tunicata*), individual kernels are surrounded by a pod or husk and the ears are also enclosed in husks. The pod is a genetic curiosity with no agronomic value. The subspecies designation is probably inappropriate, since the characteristic pod can be passed genetically to any of the other five subspecies. It is controlled by a single dominant gene giving rise to the extrafloral parts that form the pod. Nonpodded corn varieties are homozygous recessive for the trait.

Waxy corn is sometimes distinguished from the dent varieties, but its differences are not great enough to merit separate subspecies designation. The endosperm of waxy corn has a distinctive waxlike texture throughout with no white-starchy areas. The starch in waxy corn is almost entirely the branched, high-molecular-weight *amylopectin*. This type of corn starch is used in cooking and food processing and also in making adhesives.

Although maize is undoubtedly of New World origin, there is debate over just where in the Americas it first came under cultivation and began the almost automatic improvement that accompanies cultural selection. Favored are Peru, Bolivia, and Mexico. Regardless of its site of origin, several great civilizations were built upon its culture—the Incas, the Mayas, and the Aztecs. The large cities and cultural centers which were central to the flourishing of these remarkable, innovative pre-Columbian civilizations were quite likely able to develop only because of the adaptability and dependability of maize. It is no wonder that the religion of many of these societies was inextricably tied to corn culture.

When Columbus discovered the New World, corn culture had already spread from southern Chile to Canada, a remarkable feat of plant breeding and selection when one considers that the original plant was undoubtedly a tropical species. The Europeans' first efforts at colonizing Virginia and New England were near disasters when they tried to import English cultural methods and crops. It was due to the good will of certain Native Americans that the Europeans learned how to grow corn and avoid starvation.

The types most prevalent in pre-Columbian America were flour, flint, and popcorn. Dent corn is probably a relatively recent product.

Biology Corn is a remarkably variable member of the grass family. Variations in height (0.5 to 7.5 m), leaf number (4 to 44), ear number and shape, tillering, kernel characteristics, maturity (50 to 330 days), etc., produce plants which act and look almost as unalike as pears and palms. This great variation makes it crucial to pick a cultivar or hybrid with the desired cultural and physiological characteristics. Careful matching of the maturity group with local growing season is crucial. One generally wishes to use as much of the growing season as possible, but the corn must reach maturity before the first frost. A hybrid with a 130- to 150-day maturity should be planted in an area of at least 160 frost-free days, and if planting is delayed, switching to an earlier type is necessary. Because the timing of flowering (*tasseling* and *silking*) is partially regulated by day length, hybrid selection and development must be closely matched to the latitude.

Use As a forage crop, the whole plant can be grazed directly, chopped and fed green, chopped and ensiled, or harvested and shocked in standing bundles for later use. Only silage production is commonly practiced, however, and corn is the crop most frequently and abundantly ensiled. The corn is cut at almost full maturity and chopped into small pieces for packing into the silo. When air is properly excluded, fermentation preserves the material in a moist, palatable, and nutritious state. (Chapter 23 discusses corn as a forage crop.)

Corn grain is used principally (almost 90 percent) for livestock feed. It is a rich energy source (about 65 percent carbohydrate) but not particularly high in protein (about 10 percent). New *opaque*, or *high-lysine*, lines of corn are relatively enriched for that essential amino acid and therefore more nutritious. Animals fed on the opaque varieties show higher weight gains (Fig. 14-4), but these varieties are inherently lower-yielding.

Other uses for corn grain include such foods as breakfast cereals, alcohol and spirits, corn syrup and sugar, corn meal and flour, corn starch, popcorn, sweet corn, and hominy. Products and by-products of the milling process find use in the manufacture of adhesives, plastics, textile and paper sizing, laundry starch, and paints.

Corn residues have been used for a number of purposes from making corncob pipes to turning the harvested stems into cellulosic panels or burning them for fuel. The disadvantage to such total use of the crop lies in removing residues so efficiently that the soil is robbed of organic matter badly needed to maintain its tilth.

One not so new but newly popular use for corn is as a renewable source of liquid fuel either burned alone or in blends with gasoline. Much work has been done in improving the technology for converting corn and similar crops into alcohol. Serious questions to be resolved deal both with the energy efficiency of the process and the effect of diverting cropland now used to grow food into fuel production.

Sorghums and Millets

Some members of the grass family grown for their grain can be grouped with the sorghums and millets. This is a diverse group, many of whose members have in common robust growth, adaptation to warm seasons, a dense grain head, efficient use of water (Fig. 19-8), rapid growth, early maturity, and the C_4 photosynthetic pathway. Some also have value as forages (pasture, hay, or silage). Sorghums and millets grown for their grain take up about one-tenth of the world's nonforage cropland. Although certain members of the group can be found growing almost anywhere in the world, they generally do best in warmer locales.

Sorghum *Sorghum bicolor* is a highly variable species (Fig. 14-2) which includes grain sorghum (milo) (Fig. 19-9), sweet sorghum (sorgo), and broomcorn. The plant is thought to have originated in Africa and to have come into cultivation 4000 to 5000 years ago. The large number of different types suggests selective cultivation in many places for many centuries.

Sorghum is grown for grain on a large scale in the United States, India, Africa, Argentina, and Mexico. In the United States, most grain sorghum is used for feed; in Asia and Africa, food varieties are grown. The grain is generally slightly higher in protein than corn.

Millet The millets include four or five slightly to distantly related species. Pearl millet (*Pennisetum americanum*), the most commonly grown, is a valuable pasture species in the southern United States, but in India and some African countries, it is an important food crop. Pearl millet grows up to 4.5 m and bears a grain head resembling a cattail.

FIGURE 19-8
Grain sorghum, or milo, is a relatively short-stemmed plant with a dense grain head readily harvested by standard combines. Here it is growing on the plains of West Texas. (*U.S. Department of Agriculture.*)

FIGURE 19-9
A closeup of the head of grain sorghum, or milo.

Foxtail or Italian millet (*Setaria italica*), which probably originated in southern Asia, still finds its most extensive culture there. Its grain is used for food in China, Japan, and India. In the United States, it is sometimes used for hay, and some grain is harvested for birdseed.

Proso millet, or proso (*Panicum miliaceum*), finds use as a cereal for human consumption in some parts of the world and can be used in breadmaking, but its protein content is relatively low. In the United States, it is a highly desirable component of chicken feed and birdseed.

Other millets include finger millet (*Eleusine coracana*), browntop millet (*Panicum ramosum*), and teff (*Eragrostus tef*), each a specialty crop in some areas. (Teff is the staple food in some regions of Ethiopia.)

Barley, Oats, and Rye
The designation "small grains" usually means the four cool-season cereals wheat, barley, oats, and rye.† Of these four, wheat, of course, is the most extensively grown, but barley, oats, and rye are all in the

---

†Flax, a noncereal, is occasionally lumped in with the small grains because its adaptation and culture are similar.

top 10 or 15 world crops when ranked by the total area devoted to their culture or by total production.

Barley Barley (*Hordeum vulgare;* and *H. distichum* in some descriptions) does best in regions with cooler climates. Because much of the temperate region meets this criterion, barley is widely grown, being fourth among nonforage crops in total cropland devoted to its culture. Russia is the leading producer, the United States ranking third in acreage devoted to its production. The northern states (Minnesota, North Dakota, Montana, and Idaho) are particularly large producers in this country, and Canada ranks second in world production.

Most barley is of two general types (Fig. 19-10). The six-row barleys have six vertical rows of grain on each head, and the two-row barleys have a grain head which is flattened. Most barleys are bearded, i.e., have long, stiff bristles called *awns*. Cultivars without awns are often preferable as feed grain, since awns can damage the eyes and mouths of livestock. In some cultivars, the plant breeder has bred and selected for the complete absence of any extrafloral parts (palea and lemma). These "naked" or "hulled" varieties thresh out as free grains. Barleys are either fall- or spring-planted, depending on cultivar and climate. The biggest barley-producing areas in the north central states and Canada use six-row spring barley.

In addition to its value for feed, barley is used for malt in making beer and distilled liquors. In malting,

FIGURE 19-10
Two- and six-row barley spikes. Awns are present on both grain heads.

the grain is partially germinated, allowing the seed's own enzymes to begin breaking its starches down for later fermentation. Barley grain can also be pearled, a process that removes the hulls (the palea and lemma) and most of the germ (embryo). Pearled barley is processed into breakfast foods, baby foods, and other products.

Oats The decline of oats in the ranking of land devoted to its production parallels the passage of work animals from the agricultural scene, since the major use of oats has been as feed for such animals. The United States is still the world's major producer, however, suggesting that oats still have a place in a highly mechanized farm system. The relatively greater yields per hectare of some other feed grains, especially corn and sorghum, have also caused the diversion of many hectares from oats; but in cool climates, such as the north central states, Canada, and much of Russia, oats are still a valuable crop.

Oats are probably a relatively new grain crop, there being little indication of their culture before the time of Christ, whereas wheat, barley, and rye have been traced back 20 to 70 centuries earlier. The open panicle inflorescence of oats gives them a weedy look, perhaps indicative of a rather recent passage from the realm of weeds to that of crop.

Avena sativa is the common oat grown in the northern United States and Canada. There are a number of other species, but only a few have any positive agronomic value.† In addition to their value as feed for horses, sheep, cattle, swine, and poultry, oats are used in breakfast foods. Rolled oats are the usual form in commerce. The grain, or *groats*, are separated from the hulls following roasting and then steamed before being passed between rollers. The flaked product can be eaten as oatmeal or processed further.

Rye Rye (*Secale cereale*) is the least commonly grown small grain. Its popularity has declined, particularly as the preference for white breads has grown. The most adaptable and most tolerant of the small grains, rye is grown throughout the United States and from Switzerland to India. It does better on light, infertile, or acid soils than any of the other small grains. Improved varieties withstand colder

---

†Wild oat (*A. fatua*) is a serious weed pest in many parts of the United States and Canada.

winters than wheat, barley, or oats. Its culture requires less careful seedbed preparation and management than the others. Why, then, does it rank below the other small grains in its acreage? In addition to the declining preference for rye bread there are two reasons: with adequate fertilization, wheat and barley outyield rye; and rye grains shatter easily, so that volunteer plants can become pests.

Rye is a hardier and larger plant than the other small grains. When fall-planted, rye usually resumes growth earlier in the spring and heads out before the other species. It is often tall (sometimes exceeding 1.5 m), with a bluish-green tint and drooping, short-awned grain heads. The grains, or caryopses, thresh out when harvested.

Besides its use in bread making, rye is consumed as feed and in making whiskey and alcohol. Most of the rye produced in the United States goes to livestock. An increasing amount of the acreage planted to rye is fall-planted material that is turned under as green manure in spring or killed with a herbicide for no-till corn planting.

Two New Grains

Two of the newest crops in the world should be mentioned here. *Triticale*, an artificial hybrid of rye and wheat (Fig. 14-10), was developed by considerable genetic manipulation only within the last few years. Despite remaining genetic problems, there is growing hope that the good qualities of wheat, especially its high productivity, can be combined with the cultural, climatic, and high-protein characteristics of rye.

Naked-grain rice, an even newer crop, is a hybrid of rice and wheat recently developed in China. It is so new that we do not know whether it will be anything more than a scientific curiosity, but a crop combining the best characteristics of the world's two biggest crops would seem to have a sure place in agriculture.

PULSES (FABACEAE)

Many members of the legume family are grown for their edible seeds. Although their culture requires less than 10 percent of the world's cropland devoted to nonforage crops, their value as feed and food is far greater than that small figure would suggest. Their importance stems from the levels of protein present in many legume seeds, or their high oil content, or both. Protein, of course, is an essential dietary component for animals and people. Protein quality, the mixture of amino acids present, is especially important for nonruminant animals, including human beings, which have only a limited ability to transform one kind of amino acid into another metabolically. The *essential amino acids* cannot be produced in the body at all but must be obtained through the diet. Only plants can synthesize these amino acids. Meat is a source for them only because herbivores pass them along from the plants.

Legumes are especially valuable for their protein, something particularly deficient in the cereal grains. Since legume proteins are generaly deficient in sulfur-containing amino acids, however, it is possible to eat large amounts of legume protein and still be malnourished. People who rely on a strictly vegetarian diet must select protein sources carefully to ensure that they obtain sufficient quantities of all the essential amino acids.†

Soybean

The number of hectares devoted to the culture of soybean and the total tonnage of soybean production place it well down the list of the world's top crops, but its economic and nutritional value make it almost second to none (Fig. 19-11). It exceeds all other legumes in cropping area (over 50 million hectares) and production (around 80 million metric tons annually). Rapid and tremendous increases in soybean production make it impossible to predict this crop's potential.

Soybeans originated in the Orient and are now grown around the world, but they seem to have found a home in the western hemisphere, especially in the United States and Brazil, which now account for more than two-thirds of total world production. China, the site of origin of soybean, is another major producer and consumer.

Classification and Origin Soybeans of commerce all belong to the species *Glycine max*. The great number of cultivars (several hundred are registered in the United States alone) includes plants which

---

†The proteins in cereals are particularly deficient in the essential amino acid lysine. Combined with their generally low protein level, the lysine deficiency makes cereals an especially poor sole source of protein.

may look quite dissimilar and behave quite differently. There are black- and yellow-seeded types, plants with brown, black, or tan pods, purple- or white-flowered cultivars, and cultivars with broad or narrow leaflets. There are indeterminate, determinate, and semideterminate types. These characteristics and many others can be genetically incorporated into a cultivar that has the ability to respond to the appropriate day length for a particular latitude.

Because photoperiod adaptation is so critical, each soybean cultivar is classified into 1 of 13 *maturity groups* (000, 00, 0, I, II, . . . , X), based on the length of photoperiod which induces it to bloom. The lower numbers (except group 000, which is daylength neutral) are assigned to cultivars which are triggered to flower by the relatively brief summer nights of higher latitudes and which therefore bloom and mature more quickly (Fig. 19-12). Maturity groups with higher numbers contain cultivars that respond only

to the longer nights experienced later in the summer in the lower latitudes and therefore flower later in the year but still in time to mature before frost.

Soybeans were grown for food in China at least 5000 years ago according to Chinese writings of that time, but their culture almost certainly dates back many more centuries. Despite their prominence in Oriental agriculture, soybeans were still a minor crop in the United States as late as World War II. Until then, most soybeans grown in the United States were used as hay, but beginning about 1940, acreages and production skyrocketed. Between 1940 and 1950, annual production increased almost fivefold, and the next decade saw a near doubling. From 1960 to 1970, annual production in the United States more than doubled. Yields in 1980 alone exceeded total production for all the years before 1950. Brazil's record is even more phenomenal. From essentially no production in 1960, Brazil now grows (largely for export) over 15 million metric tons of soybeans annually. The increases in both countries are the result of greater acreages planted to soybeans and of greater yields per hectare. Brazil's average yield per hectare increased 28 percent in the 10-year period from 1965 to 1975 to the point where only the U.S. yields are higher. Argentina has also made great strides in soybean production.

Biology The distribution of soybean primarily in the temperate regions of the world suggests that the crop is best adapted to such climates, and its culture in the tropics and subtropics has met with little success until recently. Its prominence in temperate areas is not necessarily due to its native or original adaptation but to the work of plant breeders.

In the past 50 years, soybean breeders have produced many cultivars adapted not only to day length but also to the other climatic conditions and production practices of the central United States. It is largely due to their efforts that in 1975 three-quarters of the soybeans grown in the United States and one-half of those grown in the world were raised in eight states in the Mississippi River valley. There are many problems to solve besides adjusting photoperiod response, but soybean seems to be flexible enough genetically to fit into cropping practices almost anywhere between 50° latitude north and south. Certainly the areas of production have been increasing steadily for the last few decades.

FIGURE 19-11
Soybean is the most important legume crop grown in the United States.

FIGURE 19-12
Zones of adaptation within the United States of the soybean maturity groups. Seasonal changes in day length at different latitudes require careful matching of maturity group with latitude of planting. (*Adapted from Soybean Digest.*)

Use Soybean is included among the pulses because the seed is edible, but less than 10 percent is consumed directly. Instead, the beans are processed for their oil and protein-rich meal. Because the value of its oil is generally the major factor determining price, soybean is more often considered an oil crop than a pulse. The original use of the crop as an edible bean has been overshadowed by many other uses.

Soybean seeds are approximately 20 percent oil, 30 percent carbohydrate, and 40 percent protein, the remainder being largely water, cellulose, and various salts. The oil can be extracted with solvents from crushed soybeans leaving the protein-rich material. Most of the oil finds use in foods as shortening, margarine, and cooking or salad oils. The high degree of unsaturation in soybean oil makes it a welcome component of the human diet, but the instability characteristic of highly unsaturated oils requires edible soybean oil to be partially hydrogenated to prolong its shelf life. Industrial uses of soybean oil in paints, varnishes, plastics, and resins account for about 10 percent of the soybean oil produced in the United States.

The defatted meal which remains after the oil has been extracted from cracked soybean is used primarily as a protein source for animal feed and human food. When soybean flour is mixed with wheat flour to make protein-enriched bread, it also improves the handling quality of the dough. Texturized soybean proteins are used as ground-meat extenders; with the addition of flavors and binders, they make imitation chicken, ham, or beef. In pet foods, a major end use of soybean meal, the protein is texturized, flavored, and treated to produce a simulated meat. A soy milk product has been found useful for the approximately 7 percent of infants who are allergic to cow's milk. The nutritional value of soy milk is very nearly that of cow's milk but still falls short of that of mother's milk.

There are drawbacks to the use of soybeans as a protein source since soy protein contains relatively little of the essential sulfur-containing amino acid methionine. Animals feeding exclusively on soy protein sources do not grow as well as those consuming a complete protein, such as casein from milk. The problem is easily corrected, however, by the addition of methionine or the inclusion of other methionine-rich foods. A second potential problem is the presence of protein-digestion inhibitors in the seed. Unless properly treated, soy meal contains protein components which would inhibit their own digestion and absorption. Fortunately these inhibiting proteins can be destroyed by heating or steaming the meal before it is made into food or feed.

Neither of two other problems—the rather bitter or "beany" flavor soy meal develops after only brief storage and flatulence associated with its passage through the digestive tract of people and pets—is due to the proteins themselves. Undesirable flavors are caused by the chemical breakdown of unsaturated oils that have not been completely removed during solvent extraction. Despite some success in removing the undesirable flavors by additional extractions or additions of other chemicals, the problem remains with soy products meant for human palates.

The flatulence experienced by consumers of many legumes is caused by certain simple carbohydrates present in the meal or flour. When these carbohydrates are broken down by intestinal anaerobic bacteria, methane, hydrogen sulfide, and other gases are produced. Chemical and genetic efforts to reduce the content of these components have been not totally successful.

Peanuts

Peanut (*Arachis hypogaea*)—neither a pea nor a nut—is an atypical legume in several ways. Peanut (or groundnut or goober) is an important direct food source in some parts of the world as well as a valuable source of edible oil. About 20 million hectares annually are devoted to peanut culture worldwide. Although South America is its site of origin, most current production is centered in equatorial Africa, India, China, and the southeastern United States. Half of the world's total annual production comes from India and China alone. Peanuts

are best adapted to long, hot, sunny growing seasons, but cultivars have been developed that do well in shorter, cooler summers. The United States has three distinct regions of commercial peanut culture, each growing essentially only one type.

Classification and Origin The three major types of peanuts grown in the United States are Virginia, runner, and Spanish. Virginia varieties have either erect or prostrate stems, produce large seeds with russet seed coats, and are grown primarily in southeastern Virginia and northeastern North Carolina (Fig. 19-13). They are the preferred roasted or table type. The runner type is generally slower in maturing and has smaller seeds than the Virginia type. It has prostrate stems (runners), is grown extensively in Georgia and Alabama, and is used largely for oil and peanut butter. Spanish peanuts, grown almost exclusively in Texas and Oklahoma, are smaller-seeded with tan or red seed coats. They are erect-stemmed (bunch) types and are used primarily in peanut butter and confections.

Peanuts originated in the area of Brazil or Peru, were taken to Africa and Asia in the early days of exploitation by Old World explorers, and then came to the United States with Africans pressed into slavery. Although peanuts were grown in Virginia and elsewhere in Jefferson's time, they became important only when agricultural implements were developed that permitted machine planting, har-

FIGURE 19-13
The Virginia peanut produces large "nuts" valued for eating roasted and in candies. Pegs can be seen in several stages of development in this partially dug plant.

vesting, and processing and when George Washington Carver, at Tuskegee Institute, developed scores of processes for making both food and industrial products.

World production of peanuts increased steadily through World War II but has fluctuated in recent years with the demand for oil, food, and feed. The need for protein to feed the malnourished peoples of less developed and drought-stricken countries has periodically pushed peanut production upward.

Biology Peanuts have peculiar morphological and physiological characteristics. Germination of the seed is neither distinctly hypogeal, as in peas, nor epigeal, as in most other legumes, but intermediate. The cotyledons are pushed up slightly but seldom rise much above the soil surface. Peanuts are not unique in producing an underground fruit, but it is an agronomic and horticultural oddity. Peanut flowers have a typical appearance for a legume, but shortly after fertilization the ovary is pushed down into the soil by growth of a stalklike *peg*. Pegging is essential for production of the peanuts, since pods of fertilized flowers will not develop aboveground. The soils best suited for peanut growth are light-textured and stone-free, permitting ready penetration of the pegs and easy harvest.

A third peculiarity of peanut is its need for nitrogen fertilizer. While most legumes develop a symbiotic association with bacteria that permits them to satisfy all their nitrogen needs, peanut crops are meager unless provided with supplemental nitrogen. At least one other growth characteristic creates special problems for the grower: the vine does not senesce and die at maturity. The maturity of soybean, wheat, corn, rice, field beans, and many other agronomic species is readily determined by observing the dry down of the seeds when the plant senesces and dies. Peanut, like a few other annual crops, produces harvestable fruits while the shoot is still quite green and actively growing. The grower must therefore gauge the proper harvest time by experience. If the peanuts are dug too soon, many pods will not be completely filled; small, shrivelled nuts result. If the grower waits too long, however, crop quality deteriorates and older pods are lost in harvesting.

Use About 30 percent of the weight of the harvest is in the shell. Empty shells have shown promise as an organic mulch. The seeds contain 40 to 50 percent

oil and 25 to 30 percent protein. The oil is an important edible vegetable oil especially outside the United States. Peanut butter is the major end product in the United States, but many peanuts are eaten roasted and salted. Americans consume about 1 kg per person annually as peanut butter and more than ½ kg per person out of the shell or in candies. The protein in peanuts is complete, providing all the essential amino acids for nutrition.

Field Beans, Peas, and More

Several different species are grouped into the agronomic category of edible field beans or dry beans. Although many are members of the genus *Phaseolus*, other genera are also represented. One species, *Phaseolus vulgaris*, or common bean, is quite variable, with a large number of different-looking types. When all the dry bean species and types are considered as a single commodity (which they usually are), they become the world's second most important pulse crop both in acreage and production. Major producers are Brazil, India, Mexico, and the United States, where most field beans are grown across the northern tier of states, in the high plains, and in California.

Areas most favorable for growing dry beans have moderate temperatures during flowering and drier weather during the latter part of the season. In general, white dried beans are grown in northern, humid regions, while colored beans are grown in warmer, drier climates.

The common field bean (*P. vulgaris*) includes Navy or pea bean, pinto bean, kidney bean, and several other distinct types. Mung bean (*Vigna mungo* and *V. radiata*) is much less common in the United States but is an important pulse in parts of Asia and Africa. Mung bean is used for bean sprouts in oriental cooking. Lima bean (*Phaseolus lunatus*), a native of Peru and Central America, dries down at maturity and can be handled like other field beans but is commonly harvested in an immature, fleshy state for canning or freezing.†

Although chickpea, or garbanzo (*Cicer arietinum*), is a European native, modern cultivars grow best in warm, semiarid regions. Grown largely as a confection or snack food in the United States, it is a staple food in some areas. Broadbean (*Vicia faba*) was grown in ancient Egypt and still is a popular pulse there,

---

†Green, or snap, beans are likewise immature fruits of various types of *P. vulgaris*, the common field bean.

as well as in Europe and Latin America. It can be handled and prepared as a dry bean or eaten fresh. Lentil (*Lentilla lens*) is another pulse that dries down at harvest maturity and can be preserved until rehydrated for cooking. The small lens-shaped seeds of lentil have been found in the remains of bronze age cultures in eastern Europe. They remain important in several Asian and African nations but are grown only on a small scale in the United States.

The cowpea or blackeye pea (*Vigna unguiculata*) is also grown on a limited scale in the United States but is of some importance in central Africa, to which it is apparently native; it is best adapted to the long, hot growing season of such a climate. Cowpea is an important food in Niger, Nigeria, and other African nations. Grown extensively as a forage crop in the United States until the 1940s, the species has been largely replaced by more productive forage legumes.

Field pea or dry pea (*Pisum sativum*) is a variant of the common garden pea or English pea with characteristics making it suitable for field-scale cultivation. It dries to a form that is still edible (when rehydrated and cooked), permitting convenient harvest and storage. Pea is the only major legume preferring a cool growing season. In warm regions, it can be grown as a winter or early-spring crop. Peas are either prepared fresh for eating, canning, and freezing or allowed to go to full maturity, at which point they dry out. Dry, split peas are a nutritious, long-keeping source of protein, often made into soups. Pea is also used in some areas of the southeast as a winter cover crop, forage, and green manure.

A number of other field-grown legumes find use as pulses on a larger or smaller scale in various regions of the world. Some are locally very important, providing the major source of vegetable protein for a particular tribe or area. Tepary bean (*Phaseolus acutifolius latifolius*) is particularly drought-resistant and has been grown by the Native Americans of the southwest for centuries. Pigeonpea (*Cajanus cajan*) from India, still grown there and in Africa as a pulse food crop, is finding some acceptance as a forage species in the southeastern United States.

CHAPTER TWENTY

VEGETABLE CROPS

The vegetable industry is a large and complex part of plant agriculture. Despite the increase in home vegetable gardens, most vegetables consumed in the United States are produced commercially. The early colonists and many of our recent forebears grew their own vegetables and stored as many as possible for winter use. An underground area, called a root cellar, providing a moderate and more or less constant temperature was common in many areas for the storage of potato, carrot, beet, cabbage, and turnip. Home canning, drying, pickling, and (more recently) freezing are widely used for preservation of perishable vegetables.

As cities grew and the urban populace became unable to produce their own vegetables, commercial vegetable growing began. Early commercial growers produced a wide diversity of vegetables and often peddled them in a nearby city or town. This *market gardener* often had quite a diversified production. As transportation improved, *truck farming* developed. In the late 1800s, larger vegetable farms developed in areas particularly well suited for specific crops along the newly developing routes (railroads, rivers, and canals). Thus truck farming involved larger plantings of fewer crops for shipment to more distant markets.

The advent of refrigerated transportation for vegetables changed production areas again. Ice-cooled railroad refrigerator cars were succeeded by mechanical refrigeration of railroad cars and trucks.

Lettuce and other perishable produce can now be harvested in the Salinas Valley of California and be on a supermarket shelf in Chicago or Atlanta and in prime condition a few days later. The result has been the concentration of crop production in a few well-suited areas, so that leading vegetable states, California, Florida, and Texas, are determined by climate rather than by proximity to market.

Vegetable production for processing is an increasingly important part of the total industry. The volume of vegetables grown for the fresh market has been fairly stable since 1960, but processing vegetables has shown a gradual upward trend. Per capita consumption of vegetables is growing slightly, most of the rise occurring in processed vegetables.

In many areas of the country, increasingly important segments of the vegetable business are pick-your-own and roadside marketing systems. Although small compared with conventional vegetable farms, direct-marketing outlets have advantages for both the grower and consumer. The producer receives a larger portion of the consumer's dollar by saving marketing and shipping costs; the consumer not only gets farm-fresh produce but often pays less than in grocery stores. In more highly populated areas, farmers' markets, where farmers bring their produce for direct sale to consumers, are popular. The production of vegetables in millions of home gardens, although difficult to gauge, has an influence on the market. The tremendous sale of small-package vegetable seeds is one indication of the popularity of home gardening, and production of vegetable seed is a large, highly specialized business in some areas.

The wide diversity of vegetables makes a generally acceptable system of classification hard to devise. We shall use a combination of criteria—some genetic, some morphological, and some based on the plant part eaten. Terminology is further confused by the lack of clear distinction between fruits and vegetables. A *vegetable* can be defined as the edible portion of herbaceous garden plants, usually grown as annuals. With lettuce, celery, potato, and carrot, the edible portion is the leaf, petiole, tuber, and root, respectively. With tomato, melons, and cucumber, however, it is the fruit that is eaten. Most *fruit* plants are perennial, some woody, others herbaceous. A further distinction between fruits and vegetables is that fruits are generally eaten as a dessert or snack, a vegetable as a part of a main course.

Although the production systems used for vegetable crops vary with region, crop, etc., some general principles apply. For successful vegetable production, the soil should be of good tilth, well drained, and deep enough to accommodate the root system. Soil tests, standard procedure to determine the need for fertilizer or pH adjustment, should be made before soil preparation so that supplements can be incorporated into the field as it is plowed or otherwise tilled.

A basic requirement in establishing a vegetable crop is a relatively smooth, clodfree seedbed, prepared by plowing, disking, and sometimes further smoothing of the soil surface. A very fine seedbed is much more critical for small-seeded species, e.g., lettuce, than for large-seeded crops, e.g., snap bean and sweet corn. In areas where furrow irrigation is used, a system of ridges and furrows is formed before seeding the crop.

Vegetables are grown in rows to simplify the operations involved from seeding to harvest. With small species, e.g., radish and onion, the rows are quite close together (38 to 40 cm); with large ones, e.g., cucumber, rows are usually 1.2 to 1.8 m apart. Within-row spacing ranges from 2 cm for radish to 1 m or more for cucumber. The vast majority of vegetable crops are seeded directly in the field with various types of planters. The development of precision planters which space seed at even intervals has reduced the need for subsequent thinning. For early marketing of such high-value crops as tomato,

pepper, and cabbage, transplants are set in the field. The transplants are purchased from southern areas or are grown in greenhouses, cold frames, etc. The modified planting techniques for potato, sweet potato, and cassava will be described later in the chapter.

The vegetable industry has undergone major changes to reduce labor inputs. Herbicides have largely replaced hand hoeing and cultivation. Mechanical harvesters are standard for many of the major vegetables, particularly those destined for processing markets. However, lettuce, cantaloupe, and watermelon, are still largely hand-harvested because judgment is needed to evaluate the maturity of the product.

The need for irrigation is obvious in arid regions, but even in humid areas, irrigation is beneficial in ensuring rapid seed germination, seedling survival, and optimal product quality and yield.

The length of the growing season, climate, and types of vegetables grown determine the cropping system. Since radish, spinach, and snap bean mature quickly, successive plantings can be made. In other situations, a cool-season crop, e.g., spinach, may be followed by later plantings of warm-season species, e.g., snap bean, which in turn may be followed by a fall crop of cabbage. Many multiple-cropping combinations are possible and are selected largely on the basis of the anticipated market. Long-season crops, e.g., potato in the north, may occupy most of the growing season. A cover crop is often used to minimize erosion and build the soil during the off-season.

ROOT CROPS

Root crops generally include plants grown for various fleshy underground storage organs. In sweet potato, cassava, carrot, and beet, the usable part is an enlarged root; in others it is a tuber (potato) or a bulb (onion).

Potato (Solanaceae)

The potato (*Solanum tuberosum*) is a major food crop throughout the world and ranks with wheat and rice in world production. As a source of carbohydrates (mostly starch), the potato has become a staple part of the diet since early explorers introduced it into Europe from South America. The heavy dependence on the potato as a food staple set the stage for a tragic famine in Ireland during the mid-1840s (Chap. 14).

Since potato thrives under cool conditions, the species is ideally suited to more northern latitudes. Europe and the Soviet Union are heavy producers of potatoes both for human consumption and as livestock feed. The United States potato industry is of commercial importance in more than 40 states. As a cool-season crop, it is most productive at a temperature of 18 to 21°C. Varying cultivars and planting dates allows commercial production to range from Maine to Florida to California and Washington.

The potato is an annual which produces several upright stems 0.5 to 1 m high (Fig. 20-1). The fibrous root system is shallow, and multiple stolons develop and terminate in the tubers known as potatoes. Since the tubers are fairly shallow, when cultivation is required, soil is pulled up toward the plant both to hold up the stems and to keep the tubers well covered. Since it is a modified stem, the tuber readily forms chlorophyll when exposed to light, either before or after harvest.†

Commercially, all potatoes are vegetatively propagated by planting either small whole tubers or pieces of tuber.‡ The so-called seed potatoes are grown specifically for this purpose and are *certified* as being the indicated cultivar and free of major diseases and other pests. A potato tuber has several *eyes*, or buds, that correspond to the axillary buds of a more typical stem. Because of apical dominance, the terminal eye tends to inhibit the others, but when the tuber is cut into seed pieces, each bud will grow and develop.

Potatoes are harvested by machines which remove them from the soil and convey them to a truck or trailer (Fig. 20-2). Early potato tubers, harvested before the plants die, are immature and easily damaged; they must be handled carefully and shipped quickly. The main, or fall, crop is harvested after the tops of the plants have died or been killed by mechanical or chemical treatment to "set the skin" on the tubers, rendering them much less susceptible

---

†When exposed to light for as little as 24 h, the surface of the potato can turn green. Associated with the development of chlorophyll is that of solanidine, an alkaloid, which is toxic in sufficient concentration, rarely reached. Usually the worst result is an off-flavor.

‡Potatoes grown from seed, developed for the use of home gardeners, are not raised commercially.

FIGURE 20-1
A potato plant with some of the tubers exposed. Although the vines are still alive, the tubers are large enough for harvest.

to damage. The fall crop is usually stored and sold over a period of several months. After curing for 2 to 3 weeks at 13 to 15°C to heal injuries, the tubers are held at 3 to 4°C to minimize respiration and inhibit sprouting.

The marketing of potatoes has changed in recent years in that increasing proportions of the crop go to processing outlets. Trends show a marked increase in frozen potatoes, a moderate increase in dehydrated forms, and stable levels for chips, shoestrings, and canned potatoes. The continuing rise in eating away from home, especially in fast-food establishments, probably means growing use of potatoes in frozen french fries, which account for about 90 percent of the potatoes frozen.

Sweet Potato (Convolvulaceae)

The sweet potato (*Ipomoea batatas*) belongs to the morning glory family and originated in tropical America. The sweet potato is grown throughout the tropics and is a major food source. It was introduced to Europe by early explorers but was known in Polynesia long before.

World production of sweet potato, less than one-half that of potato, is heavily concentrated in Asia. China is estimated to account for more than 80 percent of the world crop, and other major areas include Africa and South America. The relatively high temperatures and the 4- to 5-month growing season limit commercial production to regions between 40° latitude north and south.

FIGURE 20-2
Modern potato harvester in operation. The tubers are lifted from the ground, gently shaken to remove soil, and conveyed to a truck or trailer which hauls them from the field. Normally the vines are allowed to die or are killed to make harvest easier. (*U.S. Department of Agriculture.*)

There is some confusion in the terminology used in describing sweet potato. *Dry-flesh* cultivars have a dry, firm, mealy texture after cooking; the flesh is varying shades of yellow, and in most cultivars the skin is light yellow to yellow-tan. *Moist-flesh* cultivars† are soft and moist after cooking; the flesh ranges from deep yellow to orange-red, and skin color is light tan to bright red.

The major sweet potato production area is in the southeast from New Jersey to Texas, with significant quantities also grown in California. The sweet potato is a perennial herb in the tropics but is grown as an annual in the United States and other areas where freezes occur. Its stem is a 1- to 5-m trailing vine which contains latex, as do the roots. The sweet potato of commerce is an enlarged root which usually develops in the upper 20 cm of soil (Fig. 20-3).

Propagation is by sprouts, cuttings, or a combination of both. To produce sprouts, or *slips*, true-to-name roots are bedded in sand or sandy loam—in

the open in very warm areas and in heated or unheated cold frames in cooler areas. A temperature of 20 to 26°C encourages good sprout growth without excessive succulence. When the sprouts are 20 to 30 cm long, they should be well rooted and can be

FIGURE 20-3
Sweet potato roots from one plant. The base of the trailing vine is still attached to the roots.

---

†The moist-flesh cultivars are sometimes marketed as yams; this is incorrect because true yams are monocots belonging to the genus *Dioscorea*.

pulled from the parent root for transplanting to the field. From one to four pullings can be made at 7- to 10-day intervals if the delay in time of setting is not detrimental. A bushel of good seed roots will produce 2500 to 3000 plants.

Sweet potatoes can be dug as soon as the roots reach marketable size. For years the standard method of harvest was to plow the roots out and have workers pick them up by hand. Careful handling is essential because the skin is easily injured and the flesh is tender. Increasing costs and the declining availability of field labor strengthen the trend toward mechanical harvest.

After harvest, sweet potatoes are cured at about 30°C and 85 percent relative humidity for 4 to 8 days. By encouraging wounds to heal, these conditions greatly reduce losses from rot and moisture loss during storage. After curing, the temperature is lowered to 14 to 16°C and relative humidity remains at about 85 percent. Temperatures should not go below 10 to 12°C because of chilling injury, which results in increased decay, internal breakdown, and lowered quality.

The major demand for fresh sweet potatoes is for medium-sized (5 to 8 cm diameter), well-shaped roots. Since prices for both jumbo and small sizes are low, many of these go to processing outlets. Most sweet potatoes go to market in wax-impregnated cardboard containers of various sizes. Polyethylene bags are not used because the high relative humidity and 22°C supermarket temperatures are ideal for developing *Rhizopus* soft rot, the leading cause of postharvest sweet potato losses.

Carrot (Umbelliferae)

The carrot (*Daucus carota*), an important vegetable throughout the world, is in the parsley family and is derived from wild forms native to Europe, Asia, and Africa. Its widespread use as food apparently dates back only to the sixteenth century. Before that, early forms of the carrot were used for medicinal purposes. It is thought that settlers introduced the carrot into Virginia in the early 1600s.

Of the total carrot crop, about 60 percent is grown for the fresh market, the remainder going to processing outlets. Carrots are also widely grown in home gardens and for local markets.

The plant is biennial; during the first year a rosette of feathery leaves and the storage root form. Temperature is particularly important for carrot produc-

tion; a range of 15 to 21°C is optimum for both shape and color of roots. If the plant is left unharvested, a short stem bearing flowers is formed during the second growing season. The edible part of the carrot, a taproot, varies in length and diameter but ideally is 13 cm or more long and 2 to 4 cm in diameter. Many small secondary roots serve as absorbing organs.

Carrots are direct-seeded in the field. The small size and slow germination of seeds make it difficult to get an adequate stand without excellent seedbed preparation. The high labor cost of thinning makes precision seeding mandatory.

Carrots are harvested in many ways, but the trend is toward more and more mechanization. Most fresh-market carrots are washed, sized, hydrocooled, and packed in polyethylene bags. Although carrots can be stored for up to several months at 0°C and 90 to 95 percent relative humidity, long-term storage is hardly justified since they are harvested in certain areas almost year-round.

Beet (Chenopodiaceae)

The garden or table beet (*Beta vulgaris*), which belongs to the same species as the sugarbeet and Swiss chard, is believed to be native to Europe and North Africa. The plant as we know it is of quite recent origin.

Most beet production is for processing; total tonnage is about one-fifth that of carrots. Climatic requirements are similar to those for carrots and other root crops. Although beets will grow in warm weather, cool temperatures improve sugar content, internal color, and texture of the roots.

Like most other root crops, the beet is a biennial. However, after extended cool periods, some cultivars tend to bolt even before the roots reach a marketable size. The edible root is made up of alternating layers of vascular and storage tissues. Ideally, the color is uniform, but conducting tissues are usually lighter red than storage tissues.

The seeds, instead of being individual, are small fruits containing two to six seeds, which make precision seeding difficult. In home gardens, the seedlings can be thinned and used for beet greens, but thinning is not economically feasible on a commercial scale. Fortunately, small-sized roots are preferred for processing.

Beets for processing are mechanically topped, dug, and conveyed to trucks or other large containers for hauling to the processing plant. For the limited fresh market trade, beets are handled much like carrots.

Miscellaneous Root Crops (Cruciferae)
Radish (*Raphanus sativus*) is of limited tonnage but is available year-round to the North American shopper. It is an annual or biennial plant which grows from seed to market size in about 3 to 5 weeks. Essentially all radishes are mechanically harvested, topped, and packed in small polyethylene bags, in which they are marketed.

Two other relatively minor root crops, turnip (*Brassica rapa*) and rutabaga (*B. napus*) are used mainly as a boiled vegetable or as an ingredient in soups, boiled dinners, etc.

Cassava or Manioc (Euphorbiaceae)
For millions of people in the tropics, an important food crop is cassava (*Manihot esculenta*), also known as manioc, yuca, and tapioca. It is native to the eastern equatorial region of South America and was under cultivation when the early European explorers arrived.

The plant is a herbaceous shrub growing 1 to 3 m tall. The edible part is the roots, which resemble those of the sweet potato except that they are 3 to 7 cm in diameter and up to 1.3 m long. Although the leaves are high in protein, the major dietary component of the roots is starch.

Cassava is grouped into bitter and sweet types, but the distinction between them is not clear. The major factor determining whether a particular plant is sweet or bitter is the concentration of a poisonous hydrocyanic glucoside. Sweet cassava requires no special treatment, but bitter types must be boiled or otherwise treated to destroy the hydrocyanic acid.

Although cassava is most productive when grown in a fertile soil, it also does quite well in marginal soils and with only minimal care. Probably because of this, it has been a staple food source in less developed countries. In the slash-and-burn agriculture of the tropics (see Chap. 17), stem cuttings are planted just before the rainy season. Some roots can be harvested in about 6 months but for maximum yields, 9 to 16 months are required.

Cassava may be boiled or baked like potatoes or shredded, dried, and ground into a flour. The major product of cassava in the United States is tapioca, made from the starch.

Yams (Dioscoreaceae)
True yams (*Dioscorea* spp.) are tropical climbing vines with large storage roots. Although not all yams produce edible roots, those which do are of major importance as a food source in many tropical regions. The edible roots are starchy and similar to potato, both in taste and use. Depending on the cultivar, flesh color may be white, yellow, or purple. Since yams require a growing season of 8 to 10 months, they are grown only in tropical climates.

Taro and Dasheen (Araceae)
Several genera in this family produce edible tubers or corms which have been an important food crop in southeastern Asia for over 2000 years. Taro and dasheen, the most important, are both members of *Colocasia esculenta*. Taro has very large leaves and petioles 1 to 2 m long. The edible corm (see Fig. 1-3) is high in starch and readily digested. Taro is baked or boiled to get rid of calcium oxalate crystals present in the raw corm. In Hawaii and other Pacific islands, taro has been used for centuries to make poi. Taro corms are crushed and cooked into a quite palatable starchy paste. The young leaves and petioles, also used as food, must be boiled to rid them of the calcium oxalate crystals. Dasheen tubers are generally used much like the potato—baked, boiled, or fried to make chips.

BULB CROPS (LILIACEAE)

The major bulb crop is onion but also included are garlic, leek, shallot, and chive. All the bulb crops are hardy and belong to the genus *Allium* in the lily family. Since most horticultural considerations, e.g., climatic requirements, are similar for the group as a whole, only onion is discussed in detail.

Onion
The onion (*Allium cepa*), a major vegetable crop in much of the world, apparently originated in Asia and has been used since the earliest recorded history. Early explorers and settlers introduced the onion into the New World, where its use is relatively recent.

The onion is a cool-season crop which is grown in the south in winter and in many other areas in summer. The winter and spring crops are sold soon after harvest, whereas much of the summer crop is stored for year-round sales. Since the onion is shallow-rooted, it needs a good supply of moisture, but a dry period during the harvest season facilitates curing of the mature bulbs.

Onions and the other bulb crops are used in many ways, depending on type and stage of maturity. The use of convenience products such as dehydrated onions, frozen onion rings, and frozen diced onions is growing.

The onion plant is very responsive to temperature and photoperiod. The critical day length for bulbing varies with the cultivar but is normally between 12 and 15 h. At shorter day lengths, no bulbs are formed. Even when the photoperiod is adequate, minimum temperatures must be met or bulbing is further delayed. An ideal climate would be cool weather early in the season with rising temperatures and decreasing moisture as maturity approaches.

Some onions do not form bulbs and are used strictly as green onions; others, called multiplier onions, form many small bulbs rather than one large one.

Onions are propagated from seed, sets, and transplants. Onion sets are small onions ranging from 1 to 2 cm in diameter which have been grown very close together and then dried, stored, and shipped just before planting.

Onions are harvested after the tops have dried and fallen over. Storage life varies widely with type and cultivar. The Bermuda type, for example, is a poor keeper, not stored over a month or so. For long-term storage, onions are cured, which is largely a drying process.

SOLANACEOUS FRUITS (SOLANACEAE)

The three major vegetables in this group, tomato, pepper, and eggplant, are tender perennials grown as annuals for their fruit. The family, known as the nightshade family, also includes potato, tobacco, and petunia.

Tomato

The tomato (*Lycopersicon esculentum*) ranks high among the important vegetables of the world and is grown in large quantities in most regions. It apparently originated in South America but may have been first cultivated in Mexico. Spanish explorers took the tomato back to Europe in the sixteenth century, but it was not widely used for many years. After its introduction into the United States in the eighteenth century, it was well into the nineteenth century before it was widely accepted as being edible.

Tomato production in the United States ranks second to the potato among vegetables. It is important to distinguish between the fresh market and processing industries, which are quite distinct. In spite of moderately declining processed-tomato acreages since the early 1950s, total production has almost doubled. The fresh-tomato industry accounts for about 10 percent of the crop and is quite different because growing at various locations makes fresh tomatoes available on a year-round basis. California and Florida each produce about one-third of the total fresh-market crop.

In some parts of the United States, particularly Ohio, sizable greenhouse ranges are used for vegetable production during cold weather. The tomato is the most important vegetable produced in greenhouses. Although this method of production was once profitable because the fresh produce commanded a high price, heating costs since the late 1970s have made it economically marginal.

In the tropics, the tomato is a herbaceous perennial; in temperate zones it functions as an annual. The growth habit of different cultivars varies widely but most common types grow as a highly branched stem 0.7 to 2 m long. The technique of ground production, in which plants are allowed to develop naturally, without support or pruning, is standard for processing tomatoes and is used in most fresh-market production. In some fresh-market production, however, all but one or two branches are removed as they start to develop and the remaining ones are trained to a short stake or to a string hung from above and tied loosely around the base of each stem (string-weave method). The additional expense of pruning and support is offset by better disease control, cleaner fruit, and continued production over a longer time. Another system involves the use of wire cages to support unpruned plants (Fig. 20-4).

Tomatoes are borne in clusters of four to eight. The fruit is a fleshy berry with many small seeds. Although the major commercial tomato is red and has a diameter of 5 to 8 cm, other types include the cherry tomato (1.5 to 3 cm in diameter), increasing in popularity for salads because it can be used whole;

FIGURE 20-4
Fresh market tomatoes grown in wire cages to keep the plant and fruit off the ground. Determinate cultivars are often used to suppress vine growth and concentrate the harvest.

yellow-fruited cultivars; and the plum tomato (Fig. 20-5) used for puree because of its high solids content.

Tomatoes are propagated from seeds, which germinate only after they have been removed from the flesh, cleaned, and dried. For early markets, transplants are grown in greenhouses or in southern areas and the transplants shipped in. In warm areas, they are now often direct-seeded in the field, particularly for processing.

Processing tomatoes are harvested almost exclusively by machine (see Fig. 4-14). The transition away from hand harvest has required the development of new cultivars specifically designed for once-over mechanical harvest with uniform fruit set, adequate but not excessive vine growth, uniform ripening, jointless fruit stems, and firm fruit which can withstand rolling, short drops, and the pressure in bulk containers (Chap. 14). Fresh market tomatoes are largely hand-harvested, but efforts have been underway to perfect harvest by machine. As with many agricultural crops, the substitution of machinery for labor has implications beyond the economic ones. This has become an emotional issue, likely to continue for many years to come.

Most fresh-market tomatoes are shipped at the mature green stage; they must be ripened and usually repacked before retail sale. They are put in ripening rooms at 19 to 21°C and 85 percent relative humidity, treated with ethylene, and held long enough to induce maximum color.

Pepper

Both the green, or sweet, and the hot pepper grown in the United States are *Capsicum annuum*, a species native to the New World tropics. Columbus apparently took pepper seeds back with him, and the pepper was much more quickly accepted than the tomato, both in Europe and eventually in the United States. This species includes a diverse group of peppers varying from 1 to 30 cm long, from green to yellow when immature, and from red to yellow when mature. The species includes all peppers of commercial value except the Tabasco pepper (*C. frutescens*).

The only pepper that is a major vegetable is the bell pepper, which tastes sweet rather than pungent, like most peppers. Sweet peppers are widely grown on small farms and in home gardens, but most commercial production is in Florida, California, North Carolina, and New Jersey. Fresh sweet peppers are a common part of tossed salads and relish trays; some peppers are pickled. The pimento pepper, which is canned in sizable quantities, is used to prepare pimento cheese and to provide the red

FIGURE 20-5
Processing tomato of the plum type. Because of its meatiness, this type of tomato is widely used in tomato paste and sauce. (*Asgrow Seed Company.*)

stuffing for green olives. The pungent cultivars belonging to *C. frutescens* are dried and ground into powder or used in the preparation of hot sauce.

The plant of *C. annuum* is an erect branched annual which grows 0.5 to 1.5 m high. The fruit, borne singly at nodes, are many-seeded berries.

Pepper, like tomato, is grown from seed. Plants started in greenhouses, hot beds, or open beds are transplanted to the field after the danger of freezing temperatures is past.

Green peppers are harvested when they reach full size but before the appearance of the red or yellow color. Harvest has traditionally been by hand. Once-over mechanical harvesters will undoubtedly be adopted as soon as cultivars are available which mature a heavy crop at one time and bear fruit which can withstand the rougher handling.

Harvested peppers are subject to chilling injury below 7°C, and temperatures above 10°C cause excessive rates of ripening and rots. Therefore, optimum storage conditions are 7 to 10°C with a relative humidity of about 90 percent. Under ideal conditions, peppers will hold for about 3 weeks.

Eggplant

The eggplant (*Solanum melongena*) is a native of India but gradually spread throughout the tropics. The type grown in the United States is *S. melongena* var. *esculentum*. The eggplant is not a major world crop except in Asia, where its production exceeds that of many other vegetables. Many types of eggplant are grown around the world, and the fruits vary not only in size and shape but in skin color, which ranges from white and yellow to the more familiar purple and black.

The commercial production of eggplant in the United States would rank it as a relatively minor crop, but it is widely grown for local markets and by home gardeners.

The eggplant most commonly grown in the United States is a bushy plant which functions as an annual although it is genetically a perennial. The usual size is between 60 and 90 cm in height; the base of the stem becomes woody. The fruit is a fleshy berry containing many seeds, which are eaten. The green calyx on the fruit is often spiny.

Eggplant is seed-propagated. Since the eggplant is a warm-season crop, even more readily injured than tomato by chilling temperatures, transplanting to the field is delayed later than for tomato. Ideal temperature ranges are 26 to 32°C days and 21 to 26°C nights.

The fruits are harvested when they reach one-half to two-thirds full size as both total yield and quality tend to decline if the fruits are left until they reach full size. Storage of eggplant should not exceed 10 days; optimum conditions are about 10°C and 90 percent relative humidity. At lower temperatures chilling injury is likely to occur.

Salad Crops

The salad crops include lettuce, celery, escarole, and a variety of miscellaneous green vegetables such as chicory and parsley. We shall consider only the first three, which rank among the more important vegetables.

Lettuce (Compositae)

Lettuce (*Lactuca sativa*) is in the sunflower family. It is widely accepted that *L. sativa* developed from *L. serriola*, the wild prickly form of lettuce native to Asia, Europe, and northern Africa. Lettuce was known and eaten by the early Greeks and Romans and even earlier by the Chinese.

The four major types of lettuce, all classified as *L. sativa*, are crisphead; butterhead; cos, or romaine; and leaf, or bunching. In China, stem, or asparagus, lettuce is widely grown; the fleshy stem is eaten fresh or cooked and the leaves are discarded.

Most of the commercial lettuce crop in this country consists of the crisphead type, characterized by firm heads and brittle texture of the leaves. For short-distance shipment, the butterhead and romaine types of lettuce are of some importance. The leaf lettuce cultivars are produced mostly in home gardens and in limited quantities in greenhouses during the winter.

Use of lettuce in this country is almost exclusively as a fresh vegetable, although in some parts of the world it is cooked like spinach. Lettuce is frequently the main constitutent of salads because of its attractive color, mild flavor, crisp texture, and low calorie content.

Lettuce is a shallow-rooted annual plant which forms a rosette of leaves at the base and (if left unharvested) a flower stem which may be 30 to 100 cm tall. Some cultivars form a definite head whereas others are merely a loosely formed rosette of leaves.

Lettuce is propagated by seeds which are very small, averaging over 100 per gram. In the major lettuce-growing areas, direct seeding in the field is standard practice.

Much hand labor is involved in the commercial lettuce harvest. Workers walk through the lettuce fields and cut heads which are mature, as indicated by both size and firmness (Fig. 20-6). After the lettuce is packed in fiberboard boxes, water is sprinkled over it, the top of the box is stapled, and the cartons are hauled to a *vacuum cooler*; this large vacuum chamber holds from 320 cartons up to a loaded freight car. The pressure is lowered from the normal 2.57 MPa to about 0.015 MPa, at which pressure water "boils" at 0°C. As the water vaporizes, it absorbs tremendous quantities of heat, so that the lettuce can be taken from a 24°C field temperature to 0°C in about 30 min.

If lettuce is harvested, handled, and cooled properly, it will store at 0°C for about 3 weeks. Lettuce is normally available every day of the year to consumers in the United States.

Celery (Umbelliferae)

Celery (*Apium graveolens* var. *dulce*) is the second leading United States salad crop, with production about one-third that of lettuce. Celery apparently developed from a wild marsh plant occurring throughout temperate Asia and Europe. Early use of celery was as a medicinal herb; later it was used as flavoring, but it has been accepted as a food since the seventeenth century. Early forms of celery were not particularly tender or tasty and were used mainly in soups. Not until mild, tender cultivars were developed did it become popular as a fresh vegetable. Like lettuce, celery is available year-round to the North American consumer.

Celery is normally a biennial plant, but certain stimuli can cause it to bolt. Ideal nonbolting temperatures are 15 to 18°C; temperatures above 21 to 24°C slow growth and lower quality; temperatures below 12 to 14°C for more than a few days may induce bolting.

The edible portions of the celery plant are the thickened petioles, or leaf stalks, which arise from a very compact stem. Celery is classified as an herb and grows to about 1 m. The dried fruits, often called seeds, although the actual seeds inside the fruits are minute, are used as flavoring, because they contain the same aromatic oils as leaves and petioles. Many other members of this family, e.g., anise, are used as herbs and spices.

Celery is seed-propagated but with difficulty because the seeds are very small and germinate slowly. In California, considerable direct seeding is done, but in other areas most celery is started in greenhouses, cold frames, or protected beds and transplanted (Fig. 20-7).

Celery has been hand-harvested until recently, when mechanization has increased. After packing, the celery must be cooled quickly by icing, vacuum cooling, or hydrocooling, i.e., flooding the produce with water at 0 to 2°C.

FIGURE 20-6
Lettuce is harvested by hand and packed in cardboard boxes in the field. It is then hauled from the field and placed in a vacuum cooler to remove field heat. (*U.S. Department of Agriculture.*)

FIGURE 20-7
Mechanical transplanters are widely used to set out young plants imported from southern areas or started in greenhouses. Some models apply water or a fertilizer solution to the young transplants. (Mechanical Transplanter Company)

Celery can be stored for up to a month if temperatures are at 0°C and the relative humidity is high. Polyethylene sleeves and shrink-film wraps are often used to provide mechanical protection and reduce water loss.

Escarole and Endive (Compositae)

Escarole (*Cichorium endivia*), in the sunflower family, is the broad- or straight-leaved form. The very curly or fringed-leaved form is called endive.† Escarole has been eaten since the days of the Egyptians and is thought to have originated in the region of East India. Escarole is used mainly as a salad ingredient although it can also be cooked as greens. Endive may be bitter unless it is blanched.

---

†Belgian endive is a second-year growth from roots lifted and replanted. Almost all sold commercially is imported.

The escarole plant is a moderately hardy annual or biennial which produces clusters of leaves, the margins of which vary from lobed to very cut and curled. Climatic requirements are similar to those for lettuce, so that escarole is grown as a winter or spring crop in the south and a summer or fall crop in northern states. Harvest, storage, and marketing are much like those for lettuce.

SWEET CORN (POACEAE)

Corn, or maize (*Zea mays*) was described in Chap. 19 as one of the major world grain crops. In this chapter we deal with *Z. mays* var. *saccharata*, sweet corn, which differs from the other types in that the conversion of sugars into starch is delayed or prevented as the endosperm develops. When harvested at the milky stage (a pricked kernel releases a milky juice) and cooked without delay, buttered sweet corn is a true delicacy. Plant breeders have recently developed supersweet cultivars which retain their sweetness considerably longer after harvest because the conversion of sugar to starch is much slower than in other cultivars.

The production of sweet corn for processing and for the fresh market are two distinct industries, accounting for three-fourths and one-fourth of the sweet corn crop, respectively.

Climatic requirements for sweet corn include adequate moisture and warm temperatures, although its wide adaptability is indicated by its production in every state as well as Mexico and Canada. Most processing corn is grown in northern areas because of cooler temperatures during the harvest season. Since corn matures more slowly at cooler temperatures, it can be harvested at optimum quality over a longer period of time and quality deteriorates more slowly after harvest in cooler regions.

Corn is grown from seed planted directly in the field. The harvest season is varied by selecting cultivars which mature in different lengths of time or by planting the same cultivar at intervals during the spring. Most of the sweet corn grown today consists of yellow hybrids. Hybrids are used because of vigor, high yields, and uniformity of maturity associated with heterosis. Increasing proportions of the sweet corn crop are being mechanically harvested. With improved cultivars, once-over harvest becomes practical and multiple hand pickings are eliminated.

The quality of corn begins to decline immediately after harvest, the rate depending on temperature; immediate chilling (usually by hydrocooling) and constant refrigeration are essential.

Much of the processed corn is grown on a contract basis, giving processors control of cultivars and planting dates, so that they receive an orderly supply at harvest.

COLE CROPS (CRUCIFERAE)

The cole crops of major importance in the United States are cabbage, broccoli, cauliflower, and Brussels sprouts. Of minor importance are kohlrabi and Chinese cabbage. All are considered to be members of the highly variable species *Brassica oleracea* in the mustard family.

Cabbage

Cabbage (*Brassica oleracea* var. *capitata*) is by far the most important member of this group. The early forms of cabbage apparently originated in Europe and parts of Asia and have been eaten since prehistoric times. The wild cabbage of antiquity was presumably a nonheading type, but the so-called hardheading cabbages are described in writings of the thirteenth century. Many types have evolved through mutation, selection, and breeding.

Cabbage is a cool-season crop and thrives on plenty of moisture. In the south, cabbage is grown in the fall, winter, and early spring; in the north, it is grown as a summer crop.

Cabbage is used as a fresh vegetable in cole slaw or cooked with meats, especially corned beef. Red cabbage adds both color and flavor to salads. Sauerkraut is cabbage which has been cut up, salted, and allowed to ferment.

The cabbage plant is a biennial but under some conditions will bolt during the first year. Cabbage is seed-propagated and can be direct-seeded in the field or seedlings can be started in greenhouses and transplanted. The head of cabbage is a large terminal bud, as can readily be seen by cutting a head longitudinally. As maturity approaches, stable moisture levels are desirable because a drought followed by rain can cause the heads to split.

Cabbage is usually hand-harvested; mechanical harvesters are being tested, but variations in the size and maturity of heads cause difficulties. Much less cabbage is stored than formerly, as it is available fresh year-round. Since fresh cabbage is green and stored cabbage is pale, consumer preference is strongly in favor of the freshly harvested product.

Broccoli

Broccoli (*Brassica oleracea* var. *italica*) is also called sprouting broccoli to differentiate it from heading broccoli, which is similar to cauliflower. Its early history is vague, and it seems to have been widely confused with cauliflower.

Climatic requirements are similar to those for cabbage. The broccoli plant forms a central head, which should be 15 cm in diameter or larger. The head consists of masses of flower buds borne on thick, fleshy stems. The entire head is cut, including 15 to 20 cm of stalk, before the flower buds begin to open. Once the main head is cut, many small heads, called *side shoots*, develop from buds in the axils of the leaves, so that multiple harvests are possible. Since currently available cultivars vary considerably in time of heading, fields are usually harvested several times to ensure optimum maturity.

Cauliflower

Cauliflower (*Brassica oleracea* var. *botrytis*) is another cole crop closely related to cabbage. Genetically and historically, there is great confusion about the development of broccoli and cauliflower and the distinction between them. Horticulturally, cauliflower differs from broccoli in that the white curd, or head, consists of malformed or hypertrophied flowers which form a dense cluster (Fig. 20-8).

Commercial cauliflower production is centered in California. Climatic requirements, even stricter than for other cole crops, include cool temperatures, plenty of water, and high fertilization.

Because it is not only white but tender and subject to discoloration, the heads are often covered by tying the leaves around them as soon as heads enlarge enough to spread the leaves and thus become exposed. From tying to harvest is normally from 5 days to 3 weeks. Self-blanching types do not require tying. Cauliflower is largely cut by hand because of uneven maturity in a given field.

Brussels Sprouts

Brussels sprouts (*Brassica oleracea* var. *gemmifera*) are grown in about half the quantity of cauliflower.

FIGURE 20-8
Creamy heads of cauliflower at harvest. Sometimes the large leaves are tied around the head as it matures to blanch it. (*Harris Seed Company.*)

Essentially the entire crop is frozen although small-scale local production may be marketed fresh.

Brussels sprouts form multiple small heads in the axils of leaves on a plant taller than other vegetables in the cabbage group (Fig. 20-9). The miniature, cabbagelike heads form continuously and have traditionally been hand-harvested repeatedly during the season. The cost of labor has led to the development of once-over mechanical harvesters, which cut the plant and then strip the sprouts from the stem. This procedure obviously reduces yields, but savings in labor costs make it economical.

VINE CROPS (CURCUBITACEAE)

The vine crops include cucumber, muskmelon, watermelon, pumpkin, and squash. Each is a tender annual crop grown for its fruit. Cultural requirements for all are similar.

Cucumber

The cucumber (*Cucumis sativus*) is a popular vegetable cultivated since very early times, presumably well over 3000 years. Its origin was probably in India; it was known by the Romans and was brought to the New World by Columbus.

Cucumber, like squash, thrives under somewhat cooler temperatures than muskmelons or watermelons, but germination will not occur unless temper-

atures are at least 13°C and germination is much faster at 30°C. Since the cucumber reaches harvest maturity in a relatively short growing season and will grow at relatively cool temperatures, it can be grown in most of the United States. However, for success in the far north it may need to be transplanted. In Europe, greenhouse cucumber production has been a large industry, especially in the northern areas. The English and Dutch developed cultivars which are long (often 0.6 m) and parthenocarpic for this type of production.

The plant is an annual trailing vine with a spiny or hairy stem. Cucumbers are monoecious, but recently introduced hybrids are gynoecious, producing female flowers only. For commercial plantings, a

FIGURE 20-9
Brussels sprouts are small heads which form in the leaf axils and resemble miniature heads of cabbage. (*Harris Seed Company.*)

small percentage of a monoecious type is included to provide pollen. Gynoecious cultivars are used because of earliness, concentrated ripening, increased yields, and suitability for mechanical harvesting.

Most cucumbers fit into one of two categories. Slicing cultivars grown for the fresh market should be 12 to 20 cm long with a diameter of 4 to 5 cm and dark green. To supply top-quality cucumbers, the fields are picked every 2 to 4 days. The fruit of the pickling cultivars are smaller, lighter in color, and more numerous (Fig. 20-10).

Cucumbers are usually direct-seeded in the field after the danger of freezing weather is past. For once-over harvest, high plant populations of gynoecious types are used. With the multiple harvests normal for the fresh market, plant populations are lower.

Muskmelon

The muskmelon (*Cucumis melo* var. *reticulatus*), or cantaloupe, is thought to have originated in Asia, whence it eventually spread throughout Europe and the New World. The muskmelon is distinct from some other melons for its netted skin and its much better keeping quality. Another member of this group, the Persian melon, also has netted skin but is larger. The muskmelon fruit is 11 to 14 cm in diameter and has thick orange flesh and numerous seeds in the internal cavity.

FIGURE 20-10
Modern cucumber cultivars are gynoecious and thus produce heavy, concentrated yields. This is a pickling-type cultivar. (*Harris Seed Company.*)

The plant is a tender, trailing vine which needs warm temperatures and plenty of sunshine. Low humidity and low rainfall make it easier to control fungal diseases, which are particularly severe on muskmelons in humid climates. Improved disease resistance is being incorporated into new cultivars. Most of our crop comes from irrigated regions in the southwest, where furrow irrigation is commonly used to avoid wetting the foliage.

For large plantings, muskmelons are direct-seeded but for small areas or where the growing season is too short, seed may be started in greenhouses and transplanted. In many areas muskmelons are grown on black plastic mulch, to which they are particularly responsive.

Maturity is critical if top-quality muskmelons are to be offered to the consumer. Because a muskmelon does not increase in sugar level after harvest, immature fruits are not only hard but insipid as well. The most widely used index of maturity is how easily the fruit can be separated from the vine. A full-slip melon breaks from the vine cleanly, leaving no stem attached (Fig. 20-11). This provides a melon of excellent quality if handled properly, shipped under refrigeration, and marketed promptly. Half-slip melons have some stem remaining on the fruit and are of inferior quality even after softening.

Although many machines to harvest muskmelons have been tested, most fruit is still picked by hand. Muskmelons should be hydrocooled or iced down as soon as possible and held at 4 to 7°C during shipment. They should not be stored for more than a few days.

Other melons closely related to muskmelon but classified as *C. melo* var. *inodorous* include the honeydew, a somewhat larger, smooth-skinned melon with a very pleasant, sweet, greenish-white flesh, and Crenshaw and Casaba; some of these are shipped but many are consumed locally in the west.

Watermelon

The watermelon (*Citrullus lunatus*) is usually considered to have originated in Africa but was also being grown by native Americans in the Mississippi Valley when early explorers arrived. Today watermelons are grown worldwide.

A long, warm growing season is required, but since leaf diseases are less troublesome than on muskmelon, high humidity is less detrimental. The plant (Fig. 20-12), a tender vine with deeply cut leaves, is

FIGURE 20-11
When a muskmelon is mature (the full-slip stage), the stem will separate completely from the fruit and fruit quality is excellent.

usually monoecious and is pollinated by bees. The fruits are round to oblong, ranging in weight from 4 to 25 kg and over. The small-fruited cultivars have been developed for the short growing season in the north; most of the commercial melons for the south are the larger-fruited cultivars. The commercial cultivars have red flesh, although yellow-fleshed types are available. Until recently only seeded types were on the market, but seedless cultivars have been introduced.

FIGURE 20-12
A watermelon plant growing on black plastic mulch. Note the deeply cut leaves and partially matured fruit.

Watermelons should be mature but not overripe at harvest. One of the best indexes of maturity is the change in color of the portion in contact with the soil from white to a creamy yellow. Watermelons are harvested by hand and loaded on trucks or trailers.

Pumpkin and Squash

These are major crops for home gardeners, pick-your-own, roadside marketing, and to a lesser degree commercial production.

The classification and nomenclature of the pumpkin and squash are very confused. All belong to the genus *Cucurbita* and are considered by many to be composed of five species, *C. pepo, C. moschata, C. maxima, C. mixta,* and *C. ficifolia.* The term *pumpkin* is generally used for the edible fruit of any of the *Cucurbita* species when used when ripe as forage, a table vegetable, or in pies. Pumpkins are coarse-textured and strongly flavored. *Summer squash* are edible fruit of *C. pepo* eaten in an immature stage as a table vegetable. *Winter squash* are mature fruit from any species of *Cucurbita* and are generally fine-grained and mildly-flavored compared with pumpkins.

BEANS AND PEAS (FABACEAE)

The two major types of bean used as vegetables are the snap bean (*Phaseolus vulgaris*) and the lima bean

(*P. lunatus*), both legumes. *P. vulgaris* is native to North and South America and was taken to Europe by early explorers.

Snap Bean

This group includes snap beans and beans used in a mature stage (dried beans; see Chap. 19). Some cultivars can be used in the immature stage as a snap bean or allowed to mature and used as a shell bean. Although most snap beans are green, there are also yellow cultivars called *wax beans*.

Snap bean ranks among the top 10 or 11 vegetables. More than 80 percent of the commercial crop is processed; the remainder is marketed fresh. Climatic conditions best for snap beans are mild temperatures and adequate but not excessive moisture. Since they require a relatively short growing season, several crops can be grown to provide continuous harvest in areas where temperatures do not become excessive.

The plant, a tender annual, is planted after danger of freezing has past and soil temperatures have reached at least 16°C to ensure germination. Beans are characterized as either bush or pole types. Bush beans grow on a low, erect plant which does not need support and seldom is taller than 50 cm. Pole beans grow as a vine, which must be supported as they grow to a length of about 2 m. Because of higher yields and distinctive flavor, pole beans are widely grown in home gardens.

Beans are direct-seeded in the field. From seeding to harvest takes 50 to 60 days for bush beans and 60 to 70 days for pole beans. Snap bean harvest is completely mechanized for bush types (Fig. 20-13). Pole beans are productive over a longer time but have declined markedly in commerce since they require hand harvest.

Maturity of the beans at harvest is critical for consumer satisfaction. Optimum quality is obtained when the pods are almost fully grown but the beans are only about 25 percent of mature size. Pods allowed to mature further tend to develop toughness and fibers or strings, a characteristic plant breeders have essentially eliminated if beans are harvested at the proper stage. For the fresh market, snap beans are hydrocooled and shipped in refrigerated trucks.

Lima Beans

Lima beans are of two distinct types, the small-seeded baby lima and the large-seeded standard lima.

FIGURE 20-13
Mechanical harvest of bush snap beans is standard commercial practice. The bin on the back of the harvester can be tipped to empty into a large trailer for transport. (*Chisholm-Ryder Company, Inc.*)

Although there is not complete agreement, most people include both types in *P. lunatus*. Each type has both vine and bush forms.

The plant resembles that of the snap bean in growth habit. Lima beans require a longer growing season than snap beans because they are grown for the seed rather than the pods. Nearly all commercially grown limas are of the bush type. All processing limas are harvested by machines, which mow off the vines and shell the beans mechanically so that only the seeds are hauled from the field.

Peas

The pea (*Pisum sativum*) is also called the garden pea or the English pea. Peas apparently originated in Europe or were at least well known to ancient Greeks and Romans; peas are widely grown in Europe today. Most cultivars in the United States are grown strictly for the seeds after shelling, but there are also many cultivars of sugar (edible-podded) peas.

Almost all commercial peas are processed. Production is centered in the north because a cool growing season is required and as maturity is reached, the cooler the temperature the more leeway available for harvesting with optimum quality.

Commercial pea cultivars are of the bush type; pole or climbing types are confined to home garden use (Fig. 4-9). The pea is one of the hardy vegetables and is direct-seeded 2 to 3 weeks before the average last freeze. Early planting is desirable to ensure maturity before the heat of midsummer, which reduces both yield and quality.

Peas for processing are all mechanically harvested by mowing and are shelled by machine, requiring minimal labor. For the fresh market, peas are picked by hand, a slow, expensive operation which no doubt contributes to the scarcity of fresh peas on the supermarket shelf. Pea quality depends on both sugar content and texture. For maximum tonnage per hectare, one would have to wait well beyond the stage of top quality. Since peas should be young, green, tender, and sweet, the grower must balance yield and quality for maximum returns.

GREENS

Among the vegetables used as greens or potherbs are spinach, Swiss chard, kale, mustard, collards, and turnip greens. Since spinach is the only green of major commercial consequence, it alone will be described in detail.

Spinach (*Chenopodiaceae*)

Spinach (*Spinacia oleracea*) is thought to have originated in Asia and spread through Europe only quite recently. Its widespread acceptance in the United States has occurred in only the past 60 years.

Of the United States spinach crop, about 85 percent is processed, and the processed spinach is split almost evenly between canning and freezing. Spinach is rich in vitamins and considered one of the most nutritious vegetables. Generally boiled as a potherb, it is now being used more and more as a salad green.

Spinach is a hardy, low growing, short-stemmed plant which produces a rosette of leaves; the leaves and petioles are the edible portion. One of the most serious problems with spinach is its tendency to bolt, which destroys the value of the plant. Bolting is encouraged by long days and heat and is also strongly influenced by cultivar. Spinach is direct-seeded, usually well before the last freeze, as it is one of the most hardy vegetables and also a cool-season crop.

Essentially all spinach is now mechanically harvested. For fresh market, it is cut just at the soil line so that the leaves remain in a rosette. For processing,

the plant is cut about 2.5 cm above the soil so that only loose leaves are removed and the growing point is left intact. The regrowth can be harvested later. Since fresh spinach is very perishable, it is put in cellophane bags and shipped immediately. Although spinach can be effectively vacuum-cooled, most is iced down when packed.

PERENNIAL CROPS

Asparagus (Liliaceae)

Asparagus (*Asparagus officinalis*), native to Europe and Asia, has been eaten for well over 2000 years; it was brought to the New World by the earliest settlers. Although asparagus is one of the most delicate, wholesome, and appetizing vegetables available to the American consumer, it is also one of the most expensive.

Commercial asparagus production in the United States is concentrated in a few states on the west coast and in the Great Lakes region. Asparagus thrives best in cool areas with adequate moisture.

Some asparagus growers produce their own plants from seed, but many are purchased from nurseries. The plant is a perennial and should remain productive for 15 to 20 years, depending on care. The harvested crop are spears (shoots) 15 to 25 cm long. Harvest is terminated after 6 to 10 weeks to allow the tops to develop and replenish reserves in the roots for the next year. When the plants develop after cutting stops, it becomes apparent that the plant is dioecious with the small berrylike fruits developing on about half of the plants (Fig. 20-14).

Harvesting asparagus by hand is expensive. Mechanized harvest has progressed considerably in recent years and may well save the asparagus business as an economically viable industry. Once harvested, asparagus is extremely perishable as quality declines very rapidly. Hydrocooling is necessary, and the vegetable must be held as close to 0°C as possible during marketing.

Artichoke (Compositae)

The artichoke (*Cynara scolymus*), also called the globe artichoke to distinguish it from the Jerusalem artichoke, is native to the Mediterranean region. Its use has been sporadic, but it was recorded as a delicacy over 2000 years ago.

FIGURE 20-14
Asparagus is a dioecious plant; note the fruits on the female plant on the left.

Commercial artichoke production is confined to low-lying coastal areas between San Francisco and Los Angeles. The artichoke must be grown in a frost-free region and thrives in coastal areas with foggy, cool summers.

Compared with the fresh market, the processed market is small, but artichokes are sold as frozen hearts or hearts canned in oil, water, or marinade. Sizable quantities of canned artichoke hearts are also imported from Italy, France, and Spain. Artichokes are eaten as appetizers, in salads, and hot or cold as a vegetable.

The artichoke plant resembles a thistle and is a herbaceous perennial with deeply cut leaves. It grows to a height of 1 m or more with a spread of up to 2 m. Although the artichoke can be propagated from seeds, the commercial grower uses offshoots or suckers arising from the base of the older plants. Often the old base or crown is divided, each piece having a sucker attached.

The artichoke is hand-harvested by cutting the stem 3 to 5 cm below the flower head. To maintain quality, the artichoke should be cooled to less than 4°C within 24 h.

CHAPTER TWENTY-ONE

FRUITS, NUTS, AND BEVERAGE CROPS

The early colonists had fresh fruit in season but for much of the year, they depended on dried fruit, fruit preserves, or fruit fermented to produce wine, apple jack, and other alcoholic beverages. Today rapid shipment, modern storage techniques, and large-scale food processing provide a year-round supply of a wide variety of fruits and fruit products.

Among the fresh fruits we can expect to see in a typical grocery store are strawberries from California, oranges from Florida, peaches from Chile, melons from Mexico, pineapples and papayas from Hawaii, grapes from Ecuador, bananas from Costa Rica, pears from Oregon, apples from Washington, and avocados from Puerto Rico. A newcomer is the kiwi fruit (Fig. 21-1). Some of these may have been in storage for several months, but others, e.g., the strawberries, will have been picked only recently. It is a tremendous credit to horticultural science and technology that such an array of fresh fruits is almost continually available to the consumer. When we also consider the tremendous choice of canned, dried, frozen, or otherwise preserved fruits and fruit products, the list becomes even more impressive.

From a botanical standpoint, a fruit is an expanded and ripened ovary with attached and subtending reproductive structures. As such, fruits include a range of products from apples and berries to beans, peas, peppers, tomatoes, and squash. More commonly a fruit is considered as a fleshy, usually sweet, plant product eaten out of hand, in salads, or as desserts.

Although fruit-producing plants can be classified in many ways, one of the most important distinctions can be made on the basis of climatic requirements:

Temperate fruits are deciduous and tolerate or even require a cool period each year.

Tropical fruits are evergreen, can withstand no temperatures below freezing, and may, like banana, suffer chilling injury at temperatures somewhat above freezing.

Subtropical fruits are intermediate between the temperate and tropical types, some being deciduous and others evergreen. They can ordinarily withstand temperatures slightly below freezing but are seriously injured by temperatures of −4 to −5°C.

With a few exceptions, the major fruit and nut crops are propagated vegetatively because they are open pollinating and do not breed true (Chap. 14). A seed from a 'Golden Delicious' apple will produce an apple tree, but its fruits will not resemble a 'Golden Delicious.' Although the seeds from a 'Bartlett' pear, a 'Redhaven' peach, a 'Parson Brown' orange, or a 'Hartley' walnut will produce a tree of the respective species, the fruits will in all likelihood be of very inferior quality. The same is true for the small fruits—strawberry, blueberry, grape, and cranberry.

Since the tree fruits, nuts, and grapes are budded or grafted on various types of rootstocks, the thousands or millions of plants of a particular scion

421

FIGURE 21-1
Kiwi fruit has become an increasingly popular fresh fruit in
recent years.

cultivar of apple, orange, or almond have the same
genetic makeup as the original plant (see Chap. 15).
The rootstocks may be of seedling origin or be
vegetatively propagated for a particular characteristic
such as dwarfing, pest resistance, or tolerance to a
specific climatic or soil factor. Most small fruits are
propagated by cuttings, runners, etc., so that the
entire clone is genetically identical. Some grapes are
grafted on rootstocks; others are propagated from
cuttings.

TEMPERATE FRUITS

The temperate fruits are subdivided into the *pome
fruits* (apple, pear, quince), *stone fruits* (peach, nectarine, cherry, plum, apricot), and the *small fruits*
(strawberry, blueberry, cranberry, raspberry, blackberry, and grape).

Pome Fruits (Rosaceae)

A *pome* is a type of fruit in which the pericarp forms
a papery core surrounded by flesh derived from
nonovarian tissues such as receptacle and calyx. The
only two pome fruits of great commercial concern
are the apple and pear. The quince is also a pome
fruit, but its production is minimal and it is used
almost exclusively for preserves. Apples are by far
the most widely grown temperate tree fruit, surpass-
ing the production of pears and the stone fruits

combined. Apples and pears are grown at similar
latitudes as their climatic requirements are similar;
in fact, they are often grown in the same orchard.

Apple Major countries that are producers of apple
(*Malus domestica*) include the United States, Italy,
France, and West Germany; Europe provides over
one-half of the world's crop. In the United States,
apple production is concentrated in Washington,
New York, Michigan, California, and in the Appalachian region from Pennsylvania to North Carolina.
Together these regions produce in excess of 80
percent of the United States apple crop, but apples
are grown commercially in at least 35 states.

The northern limit for commercial apple production is determined by minimum winter temperatures,
because it is usually impractical to grow apples
where winter temperatures dip below −35 to −40°C.
The southern extremity for United States apple production is determined by two temperature considerations: (1) apples need relatively cool nights during
the ripening season to develop good flavor, color,
and quality, and (2) most commercial apple cultivars
require 1200 to 1500 h below 7°C to satisfy the
chilling requirement, a condition not usually met in
the deep south.

Although there are thousands of apple cultivars,
the number grown commercially is quite small.
About 40 percent of the total United States apple
production is 'Delicious' and another 15 percent
'Golden Delicious.' Other cultivars of importance are
'McIntosh,' 'Rome Beauty,' 'Jonathan,' and 'York.'
These six cultivars combined account for about 80
percent of the total United States production. For
the country as a whole, about 50 percent of the apple
crop is processed into sauce, slices, juice, vinegar,
and such specialty items as jelly and apple butter.

There has been a recent trend toward intensifica-
tion of apple production systems. The traditional
orchards of large trees set 12 m apart and requiring
6-m ladders for harvest are disappearing. Hitherto
the desired scion cultivars have been propagated on
seedling rootstocks which produce tall, vigorous
trees; there is now, however, a trend toward trees
on vegetatively propagated, size-controlling rootstocks. The advantages of smaller trees include easier
pruning, spraying, and harvesting; greater production
per unit area of orchard land; and better fruit size
and color; but the greatest asset of the smaller tree
is that a new orchard can be in production 2 to 3

years earlier than one on seedling, i.e., nondwarfing, rootstocks.

Other techniques for controlling tree size are being tested. Many cultivars have *spur-type* strains, which are inherently compact in growth and grow to only 70 to 80 percent the size of standard strains. A wide range in tree size can be obtained by various combinations of standard and spur-type scion cultivars on the many available rootstocks. Interest in the possibility of chemical growth control is also widespread. Although many chemicals have been tested, none give effective control of tree size over extended periods of time. Several growth regulators induce early flower bud formation in young trees, which, thanks to the competition between fruit production and vegetative growth, can be helpful in holding vegetative growth in check. Growth regulators are widely used on apple trees for chemical thinning, increasing fruit firmness, increasing red color, accelerating fruit maturity, and delaying preharvest drop (see Chap. 4).

Several systems for mechanical harvest are being tested, but essentially all apples are still harvested by hand, as are pears. If mechanical harvest becomes a reality, it will probably be most useful for processing fruit, on which bruises are less critical.

Pear World production of the pear (*Pyrus communis*) is about one-third that of apples and is also concentrated in Europe, especially Italy and France. In the United States it is much more centralized than apple production. California, Oregon, and Washington produce about 95 percent of the total United States pear crop. On the west coast, 'Bartlett' accounts for about 70 percent of the pears produced; of those, about three-fourths are processed. Processing accounts for 60 percent of the total pear crop. The major products are canned pear halves, fruit cocktail, and mixed fruit. Limited quantities of pears are also dried. Important fresh fruit cultivars include 'D'Anjou' and 'Bosc' as well as 'Bartlett'. The limited pear production east of the Rocky Mountains is due to the devastation of pear trees by fireblight, discussed in Chap. 12.

Historically, pears have been propagated on rootstocks of several *Pyrus* species including *P. communis*, *P. ussuriensis*, *P. betulifolia*, *P. pyrifolia*, and *P. calleryana*. Pear decline, a disease caused by a mycoplasma, decimated pear orchards in the western United States in the 1950s and 1960s. Since then, most pear trees are grown on *P. communis* or *P. betulifolia*, which have resistance to pear decline. Until recently, the only dwarfing rootstock for pears was the quince (*Cydonia oblonga*). Because of incompatability with many pear cultivars and the need of support for the tree, quince rootstocks have not been used commercially.

Stone Fruits (Rosaceae)

The stone, or drupe, fruits include peach, nectarine, plum, cherry, and apricot. The fruit consists of the exocarp (skin), mesocarp (edible flesh), and endocarp (stony pit surrounding the seed).

Peach The peach (*Prunus persica*) is the most widely grown stone fruit in the United States and the world. The leading countries in peach production are the United States, Italy, and France. Peaches are characterized as *freestone* or *clingstone* based on whether or not the ripe flesh separates readily from the pit. The texture of the flesh of freestones is somewhat melting, whereas flesh of cling peaches tends to be very firm, making the latter preferable for canning. Peaches produced for the fresh market are generally freestone.

Freestone peach production in the eastern United States ranges from Florida to Michigan and New York; the major states are South Carolina, Georgia, New Jersey, Pennsylvania, and Michigan. Peaches are grown commercially in several western states but the only one of major importance is California, which produces most of the United States clingstone crop and about one-half of the freestone crop, or 65 to 70 percent of the total United States production. Of California's total production, about 75 percent is clingstone.

Peaches are not grown commercially as far north as apples and pears, since flower buds usually cannot survive temperatures below $-25°C$. The southern limit for peach production is determined by the number of hours of chilling during a typical winter. Until relatively recently, the southern extremity on the east coast was Georgia, because the chilling requirement of commercial cultivars ranged from 650 to over 1000 h. Newly introduced cultivars with chilling requirements of less than 200 h have encouraged the extension of peach plantings into Florida and other areas with minimal winter chilling.

Commercially important peach cultivars are more numerous than those of apples. Some, e.g., 'Redhaven,' are widely adapted, but others are regional. Although there are both white- and yellow-fleshed freestone cultivars, white-fleshed ones are grown only for local sales as they are not firm enough to ship well. Cultivar obsolescence is faster with peaches than with apples and pears, because many new peach cultivars become available, orchard life is shorter, it takes fewer years to bring a new peach orchard into full production, and peach consumers are much less cultivar-conscious.

Peaches have historically been grown on peach seedling rootstocks. Seeds from wild peach trees in Tennessee and the Carolinas were once used by eastern nurseries, but seedlings are now grown from pits collected at California canneries. The two major cultivars used as seed sources are 'Lovell' and 'Halford.'

Mechanical harvesting has recently become commercially feasible for processing peaches. A significant percentage of the cling peaches for processing in California is harvested by the shake-and-catch method, but most are still hand-harvested. Growers must balance the cost, availability, and dependability of hand labor against the capital investment in mechanical harvesters.

Fresh-market peaches are still harvested completely by hand, but experimental harvesters are under development for fresh fruit. With fresh-market peaches, not only must the producer be concerned with the balance between labor and machine expenses but fruit quality is of utmost importance.

Nectarine The nectarine (*Prunus persica*) which has increased in popularity in the past few years, differs from the peach in usually being smaller, having a smooth skin, and being more aromatic. Their close genetic relationship (they are in the same species) is demonstrated by the fact that peach seeds may produce nectarine trees and nectarine buds may mutate to produce a branch of peaches, or vice versa. For years, California has produced essentially the entire United States nectarine crop, but with new and improved cultivars and better disease control, nectarine production in the east is making moderate gains. The major disease problem is brown rot (*Monilinia fructicola*), to which nectarines are much more susceptible than peaches. Essentially all nectarines are sold on the fresh market.

Plum Among stone fruits, the plum (*Prunus* spp.) ranks second to peach in world production. Major

producing countries are Yugoslavia, Rumania, Germany, and the United States, where about 90 percent are grown in California, with small but significant quantities from Oregon, Washington, Michigan, and Idaho. In California, about 80 percent of the plums are dried to be sold as prunes; most of the others are sold fresh, although small quantities are canned. For the other states, about 50 percent are sold fresh, 30 percent are canned, 15 percent are dried, and a few are frozen.

Plums are much more diverse than other tree fruits. Several species native to North America are grown locally but have little impact on the market. Most important plums belong to either *Prunus domestica* (European plums) or *P. salicina* (Japanese plums). The *P. domestica* plums can be divided into four groups:

Prune—These are purplish-blue, oval in shape, and have a sufficiently high sugar content to allow drying without removal of the pit. Major cultivars are 'French,' 'Sugar,' 'Italian,' and 'Stanley.' Plums for the prune market are harvested by the shake-and-catch method.

Reine Claude (Green Gage)—These are roundish, sweet, and greenish-yellow or golden; the flesh is juicy and of high quality. Cultivars include 'Reine Claude,' 'Jefferson,' and 'Washington.' This group is used for the fresh market and canning.

Yellow Egg—This small group, which includes 'Yellow Egg' and 'Golden Drop' is used only for canning.

Lombard—This group is also relatively small but consists of oval plums which are red or pink. The quality is lower than either the Prune or Reine Claude group; most are sold on the fresh market.

Japanese plums (*P. salicina*) are characterized by conic or heart-shaped fruits with a more pointed apex than other species. Skin color ranges from a bright yellow to red and purplish red, but never blue. The flesh is yellow to red, juicy, and firm; overall quality varies widely with cultivar. Major cultivars include 'Santa Rosa,' 'El Dorado,' and 'Duarte.' The Japanese plums dominate the fresh market.

Cherry World production of the cherry (*Prunus* spp.) is about one-fourth that of peach; leading countries are Italy, the United States, Germany, and France. In the United States, cherry crops are about equally split between sweet and tart cultivars. Sweet cherries (*P. avium*) are predominately grown in the three Pacific coast states and Michigan, which together account for about 90 percent of the sweet cherries grown (Fig. 21-2). Sweet cherries are divided about equally between fresh-market and processing outlets. Most of the sweet cherries processed are brined for maraschinos, some are canned, and limited quantities are frozen. The tart (sour) cherry (*P. cerasus*) industry is concentrated around the Great Lakes, especially in Michigan, with much smaller plantings in New York, Pennsylvania, and Wisconsin. Almost all tart cherries go into processing channels; the most popular products are frozen, canned, and brined cherries, juice, and wine.

Using hand labor to harvest tart cherries is no longer economical, since from one-half to two-thirds of the total grower cost was in the harvest operation. Essentially all processing cherries are now harvested by the shake-and-catch method (Fig. 21-3), which has undoubtedly saved the industry from financial disaster.

FIGURE 21-2
Sweet cherries are borne laterally on previous season's wood. Each flower bud produces three to five flowers. This cultivar is 'Napoleon,' a white-fleshed type which is widely used to make maraschino cherries.

FIGURE 21-3
The shake-and-catch cherry harvester is positioned beneath the tree, has two hydraulic shakers to detach the fruit, which are collected on the catching frame. (*U.S. Department of Agriculture.*)

Apricot Production of the apricot (*Prunus armeniaca*) is slightly less than that for cherries and is centered in the Soviet Union, Spain, France, Turkey, Hungary, and the United States. Apricots have been grown in the United States since the early 1700s, but commercial production is limited to the west coast, where California accounts for 95 percent of the total crop. Apricots have a short chilling requirement, and since they bloom very early in the spring are highly susceptible to spring freeze damage. Apricot production is therefore confined to regions which seldom experience spring freezes. Attempts are being made to breed cultivars which bloom later and might be more widely adapted.

Apricot trees are grown on peach, apricot, or plum rootstocks. The major cultivars are 'Blenheim' and 'Tilton.' About 10 percent of the apricots are sold fresh, 65 percent canned, 20 percent dried, and 5 percent frozen. Shake-and-catch harvesters are used to some degree.

Although the almond is a stone fruit and genetically similar to the peach, we shall discuss almond under the category of tree nuts.

Small Fruits

The small fruits include a diverse group which are relatively small in size and are produced on bushes, vines, or other low-growing plants. Many are true berries but others are not, even though they are called berries.

Grape (Vitaceae) The grape (*Vitis* spp.) is classified horticulturally either as a small fruit or in a category by itself, but from a botanical standpoint it is a berry. Grapes rank first among the temperate fruits in tonnage produced in the United States although apples are a close second. In world production, however, grape production exceeds that for apples by about 2:1. Spain, Italy, and France are leading grape-growing countries.

Four distinct types are grown in the United States: the European grape (*V. vinifera*); the American bunch grape (*V. labrusca*); the French hybrids (*V. vinifera* × wild American species); and the Muscadine grape *V. rotundifolia*). Vinifera grapes have largely been confined to California because they are best adapted to climates with warm dry summers and cool wet winters. They have firm pulp, to which the skin adheres, and the flesh is sweet throughout. The 'Thompson Seedless' is the vinifera grape used in large quantities for both raisins and fresh fruit; about 25 percent of the total grape crop, or about 1 million metric tons, is dried annually (Fig. 21-4). Other fresh fruit cultivars include 'Emperor,' 'Flame Tokay,' and 'Ribier.' Many of the vinifera cultivars are grown exclusively for wine; about one-half of the total grape crop in the United States is crushed for wine. 'Zinfandel,' 'French Colombard,' and 'Cabernet Sauvignon' are well-known wine cultivars. The recent marked increase in wine consumption in the United States has generated interest in wine grapes in many parts of the country. The long-term success of the many relatively small recent plantings of vinifera cultivars in the eastern United States remains to be seen.

Most of the grapes grown in the eastern United States are *V. labrusca*, the American bunch grape (Fig. 21-5). The skin is rather loosely attached to the pulp, which tends to be soft and relatively acid around the seeds. In the Great Lakes area, more than 85 percent of the grapes produced are 'Concord' because of its wide adaptability, good productivity, and broad market acceptance for juice, preserves, frozen concentrate, and wine, as well as local fresh-market sales. Other cultivars used largely for fresh use are 'Fredonia,' 'Niagara,' and 'Delaware.'

The regions around the Great Lakes offer the best climate in the eastern United States for American

FIGURE 21-4
'Thompson Seedless' grapes are harvested and placed on paper trays between the rows where they dry to form raisins. (*California Raisin Advisory Board.*)

FIGURE 21-5
American-type grapes as they approach maturity. Grape flower clusters form at the first few nodes of the current season's shoots.

grapes. A strip of land about 8 km wide along the south shore of Lake Erie contains most of the grape acreage of Ohio and Pennsylvania and much of that for New York, which also has grape areas around the Finger Lakes. In spring, the cold lakes suppress temperatures and delay plant development, thus reducing the likelihood of freeze damage. In fall the lakes cool slowly, thus delaying the occurrence of freezing temperatures. Most cultivars of *V. labrusca* require a freeze-free growing season of 160 to 170 days for proper fruit and vine maturity.

The French hybrid grapes include many cultivars developed by breeders in France from crosses of *V. vinifera* with wild American species, which incorporated some of the inherent disease resistance and hardiness of the wild American species into the vinifera type. The increased demand for wine has revived interest in the French hybrids in the eastern United States; cultivars include 'Foch' and 'Seibel 10878.' How long this interest in French hybrid production lasts will doubtless vary with the region, cultivar, specific problems, and the market. Since these types are limited to wine production, some growers may prefer the flexibility of dual- or triple-purpose cultivars like 'Concord.'

Most of the muscadine grapes belong to *V. rotundifolia*, but there are two other closely related species. Muscadines grow wild in the southeastern United States and are adapted to areas where the temperatures remain above −12 to −15°C. Although limited

quantities of muscadine grapes are sold fresh, most are used for wine, preserves, and blending with other fruit juices. The older cultivars are generally harvested as individual fruits, but newer cultivars remain in a loose bunch. Of minor importance nationally, muscadine grapes are common in home gardens of the southeast. The 'Scuppernong' cultivar is the most widely grown.

Mechanical harvesting of grapes for juice, wine, and other processed products is becoming standard. In one system, an over-the-row harvester knocks the grapes off the vine, catches them below it, and conveys them to a truck or trailer (Fig. 21-6). For mechanical harvesting the individual grapes should separate readily from the bunch and resist bursting.

Strawberry (Rosaceae) Among the small fruits, strawberry ranks second to grape in total production in the United States and the world. The United States leads in strawberry tonnage, followed by Italy, Japan, and Poland. The commercial strawberry, which originated as a cross between *Fragaria chiloensis* and *F. virginiana,* is classified as *Fragaria × Ananassa.* In the eastern United States, commercial strawberry plantings range from Maine to Florida and as far west as Michigan and Louisiana. Heavy concentrations of strawberries are found in California, Oregon, and Washington.

FIGURE 21-6
An over-the-row grape harvester operating at Niagara Falls, Ontario. The grapes are knocked from the vines, collected below, and conveyed to a portable receptacle. (*Chisholm-Ryder Company.*)

The strawberry plant is a small herbaceous perennial with a very short stem or crown. From the axils of the leaves come branch crowns, runners, and flower stalks (Fig. 21-7). Strawberries are propagated by removing rooted daughter plants, which form on the runners.

The changes in strawberry acreage and production since the 1930s are amazing. In 1930 strawberry acreage was close to 81,000 ha with an average yield of about 2.8 metric tons per hectare; in 1975 the acreage was about 16,000 ha with an average yield of 15.7 metric tons per hectare. That is, in 45 years acreage was reduced 80 percent but yield per hectare increased by 560 percent, thus giving a slight increase in total production. Much of this increase in productivity results from the tremendous yields obtained in California, which averages about 45 metric tons per hectare. Florida averages 17 metric tons and the other states produce about 6 metric tons per hectare. The success with strawberries in California is the result of many improvements, including more productive cultivars, better pest control through insect and disease resistance (especially virus dis-

FIGURE 21-7
A flower stalk of strawberry. The terminal, or primary, fruit, which ripens first, is the largest berry. The secondary fruits ripen next, followed by the tertiary fruits.

eases), and improved cultural techniques, e.g., soil fumigation and use of plastic mulches.

Although strawberry production east of the Rocky Mountains has declined drastically, the interest in strawberries as a pick-your-own crop is likely to continue.

Blueberry (Ericaceae) Blueberry (*Vaccinium* spp.) has several species grouped into lowbush, highbush, and rabbiteye types, all native to the United States. Blueberries have small, soft seeds and are distinguished from huckleberry (*Gaylusacia*), which has 10 large, hard seeds. Blueberry production is limited to Europe and North America.

The most important *lowbush blueberry* is *V. angustifolium*. This low-growing bush or shrub, 15 to 45 cm tall, is harvested from the wild or semicultivation. Plants are not propagated and set out; instead their ecological niche between the field and forest is artificially maintained. By burning the fields in the spring when the ground is either frozen or wet, much of the competing vegetation is killed but the blueberry, which spreads by rhizomes, is pruned without killing. A typical program is to burn a third of the acreage each year so that the remaining two-thirds is harvested annually.

The United States lowbush industry is confined largely to the northeastern states, especially northeastern Maine, northern New Hampshire, and the upper peninsula of Michigan, with limited acreages as far south as West Virginia. Eastern Canada also produces sizable quantities of lowbush blueberries and ships many into the United States.

Most lowbush blueberries are sold to canning and freezing plants; limited quantities are sold fresh. The rough terrain on which they thrive makes mechanized harvest more difficult than with the highbush type, but machines are widely used.

Highbush blueberries consist of several species, the two major ones being *V. corymbosum* and *V. australe*. The plant can grow to 2 to 3 m but is usually pruned to a height of 1 to 1.5 m. Commercial production areas are limited because of the rather specific requirements for temperature, soil, and water. Temperatures needed are not drastically different from those for a peach cultivar with a chilling requirement of about 1000 h below 7°C. The top of the plants may be killed if winter temperatures go below −30°C, but a snow cover can provide protection. Since the southern limit is set by the need for adequate chilling in winter, they are not widely grown south of northern Georgia. Soil requirements include a soil pH of 4.3 to 4.8, which in many areas requires the addition of sulfur or other amendments to acidify the soil. The blueberry is shallow-rooted and needs ample water but cannot withstand submersion during the growing season.

Major production areas for highbush blueberries include North Carolina, New Jersey, and southwestern Michigan, with lesser acreages in Washington, Oregon, Massachusetts, New York, and Indiana. In Canada, commercial highbush blueberries are grown in both British Columbia and Nova Scotia.

As the result of breeding programs, many cultivars are available, most bearing fruits 3 to 4 times as large as wild berries. Important cultivars include 'Bluecrop,' 'Blueray,' 'Berkeley,' and 'Jersey.' Both highbush and rabbiteye blueberries (see below) are propagated by rooted cuttings.

Blueberries ripen over a period of several weeks and must be harvested repeatedly (Fig. 21-8). Although they are traditionally a hand-harvested crop, various mechanical innovations are now being used commercially. Hand-held vibrators shake the ripe berries from the bush onto a canvas catching frame which is moved from bush to bush. This system has increased output by 5 to 8 times over hand picking. Self-propelled harvesters straddle the row, shake the berries, and collect them beneath the bush. Large

FIGURE 21-9
Cranberry fruits are borne on upright shoots. Many are harvested by flooding the bog, knocking the berries from the shoots, and floating them to one corner of the bog, where they are collected.

quantities of highbush blueberries are sold fresh, many are frozen, and some are canned.

The *rabbiteye blueberry* (*V. ashei*) is important because it grows in warm areas where the highbush blueberry will not. Its short chilling requirement makes it suited as far south as northern Florida; its northern limit is central Alabama, Mississippi, and coastal North Carolina. Major advantages of the rabbiteye are excellent vigor, adaptation to wide variation in environment, and tolerance to heat and drought. 'Tifblue' is the leading cultivar.

Cranberry (Ericaceae) A plant native to North America, the cranberry, (*Vaccinium macrocarpon*) grows in acid, swampy areas, usually known as cranberry bogs. Commercial cranberry production is centered in Massachusetts and Wisconsin; other commercially important areas are New Jersey, Washington, Oregon, and some areas of Canada.

The cranberry is a low-growing, evergreen vine (Fig. 21-9). The fruits are borne on upright shoots 10 to 15 cm tall. Cranberries are propagated by stem cuttings. After an existing bog is mowed, the cuttings are spread over a new bog and disked lightly or pressed into the soil with rollers.

Cranberry bogs are expensive to develop because land preparation requires clearing, leveling, ditching, applying a layer of sand, and providing for both irrigation and drainage. Once established, bogs last

FIGURE 21-8
Clusters of blueberry fruits. Note that some fruits are mature and some are still green. Blueberries ripen over a period of 2 to 3 weeks and are harvested repeatedly.

for 60 years or more and are actually considered permanent if properly cared for.

Although once picked or "raked" by hand, most cranberries today are harvested by machine. Some machines use teeth to pull the berries from the plants; others operate in flooded bogs by knocking the berries free, after which they are floated to a corner of the bog and picked up by machine.

Large quantities of cranberries are sold fresh for use at Thanksgiving and Christmas. Major processing outlets are for cranberry sauce and cranberry juice.

Brambles (Rosaceae)

The bramble fruits (*Rubus* spp.) include red, purple, and black raspberries and the blackberry, as well as others of lesser importance. Brambles bear fruit on biennial canes which usually grow one year, produce fruit the second, and then die back to the ground. The fruit of the raspberry and blackberry is an aggregate consisting of many drupelets (Fig. 21-10). The core, or receptacle, remains in the blackberry fruit and is edible; the raspberry core remains on the bush.

Commercial bramble fruit production is limited in comparison to that of most other fruits, but it is important in certain areas of the United States, Canada, and several European countries. Much of the production in the United States is for processing into frozen fruit, jams, jellies, and preserves, but

FIGURE 21-10
A cluster of raspberry fruits, each an aggregate of many drupelets.

raspberries and blackberries are a high-priced item for local sales and are being increasingly planted for pick-your-own operations.

Major areas of production for raspberries include Michigan, Oregon, Washington, and New York. Blackberries are adapted further south and are found as far south as Texas. A major improvement in blackberries has been the release of thornless cultivars such as 'Smoothstem,' 'Thornfree,' and 'Black Satin.'

Because of viruses, poor yields, and especially high labor requirements, the brambles have gone through a decline in acreage, but better cultivars, virus-free plants, mechanization, and pick-your-own operations may restore some of their earlier prominence. Like strawberry, blueberry, and grape, the bramble fruits are well adapted to the backyard garden.

SUBTROPICAL AND TROPICAL FRUITS

Although subtropical and tropical fruits can be grown commercially only in limited parts of the United States, they account for a large part of the total fruit crop. Major production areas are confined to Florida, California, Hawaii, southern Texas, and Arizona, where temperatures seldom go below −5°C. Many of these fruits, especially mango, papaya, and pineapple, must be eaten where they are grown for the full aroma and flavor.

Citrus (Rutaceae)

Of all the tropical and subtropical fruits, the subtropical citrus fruits are by far the most important in the United States (Fig. 21-11). On a worldwide basis, citrus production slightly exceeds that of bananas.

Citrus fruits are borne on large evergreen trees reaching heights of 4.5 to 8 m. The fruit, a hesperidium, or modified berry, has a leathery rind with many oil glands. The juicy flesh consists of many juice sacs. Citrus fruit are valued for their high vitamin C content.

Although there are many citrus fruits, the sweet orange (*Citrus sinensis*) accounts for about 70 percent of the total United States citrus crop. Oranges are classified as common or navel, seedless or seeded, and by the season of ripening. In Florida the major cultivars are 'Valencia,' 'Pineapple,' 'Hamlin,' and

FIGURE 21-11
Citrus groves in Florida produce most of the oranges for frozen juice concentrate.

'Parson Brown.' Navel orange production in Florida accounts for less than 5 percent of the crop, primarily because of inconsistent productivity. In California, slightly more than 50 percent of the orange crop is navels, mostly of the cultivar 'Washington.' In Florida about 90 percent of the orange crop is processed, but in California most is sold as fresh fruit.

The second major citrus fruit in the United States is grapefruit (*C. paradisi*), about 70 percent of which comes from Florida. Grapefruit are classified as seeded or seedless and white-, pink-, or red-fleshed. Major cultivars are 'Marsh' (white, seedless), 'Ruby' (red, seedless), 'Duncan' (white, seeded), and 'Foster' (pink, seeded). Fresh-fruit markets take 35 percent of the grapefruit crop, about 30 percent is used for canned juice, 5 percent for canned sections, and the remaining 30 percent for other products such as frozen concentrate. For the fresh market, grapefruit should have symmetrical shape, smooth skin, and uniform yellow skin color and be seedless and sweet.

The lemon (*C. limon*) is grown mostly in California and Arizona, with limited quantities from Florida. Consumption is about half fresh and half processed, as juice concentrate.

Tangerine (*C. reticulata*), also called mandarin orange, is a specialty type of citrus which has a loose skin, peels easily, and has attractive orange-red flesh (Fig. 21-12). Because tangerines are less productive and do not hold quality as well as oranges after reaching maturity, they have never been widely grown in the United States. Tangerines are popular in Japan and other parts of Asia.

Other citrus species of lesser importance are limes (*C. aurantifolia*), tangelos (tangerine × grapefruit cross), and temples (tangerine × orange). There are many other citrus hybrids within the *Citrus* genus and even intergeneric hybrids.

Citrus trees are propagated by budding on rootstocks adapted to the specific area. Important rootstocks include sour orange, rough lemon, and sweet orange.

Citrus fruits do not exhibit a climacteric and are usually "stored" on the tree. Most cultivars can be harvested over a 2- to 3-month period, in sharp contrast to most deciduous fruit species, which must be harvested within a 1- to 2-week period after reaching maturity.

Avocado (Lauraceae)

The avocado (*Persea americana*) is an increasingly popular fruit in the United States, although it is not widely grown outside the western hemisphere. Mexico is the leading country in avocado production followed by the United States, Brazil, Colombia,

FIGURE 21-12
'Dancy' tangerine. (*Florida Department of Citrus.*)

Venezuela, and Ecuador. Commercial production in the United States is limited to southern Florida and southern California.

All commercial cultivars of avocado originated as chance seedlings. Trees are produced by budding or grafting of the desired cultivar on seedling rootstocks. Important cultivars include 'Fuerte,' 'Hass,' 'Booth 8,' and 'Lulu.' The avocado thrives only in tropical and subtropical climates.

The avocado fruit is a one-seeded berry, 10 to 15 cm long (Fig. 21-13). The flesh is green to yellow, buttery in texture, and rich in oils (10 to 30 percent). Compared with most other fruits, avocados are high in protein and vitamins. The skin may be smooth and thin or somewhat rough, thick, and almost woody. Skin color varies from green to purplish-black. The tree is an evergreen which grows 5 to 13 m tall.

Like citrus, avocados can be stored for several weeks on the tree; the fruits soften only after picking. Even though firm, mature fruits must be handled carefully to avoid bruising.

Olive (Oleaceae)

The olive (*Olea europaea*) has been a major crop for centuries around the Mediterranean Sea, where it has been used for oil as well as table use. Leading countries in olive production are Spain, Italy, Greece, Tunisia, Turkey, and Portugal. The entire United States crop, produced in California, accounts for less than 1 percent of the world crop.

The olive tree is an evergreen 7 to 12 m high. The leathery leaves are green-gray, and the fruit is a single-seeded drupe. The olive tree is subtropical and cannot withstand temperatures of $-10°C$; optimum fruiting is enhanced by winter temperatures of 6 to 8°C. Olives are readily propagated by cuttings; specific cultivars are grown for oil or table use. Important table cultivars include 'Manzanillo' and 'Sevillano.'

For table use, olives are harvested in the fall when the fruit is green to straw-colored. Postharvest treatment with sodium hydroxide (lye) solution is necessary to rid them of a bitter glucoside. If exposed to air, the olives turn black and are sold as ripe olives. Those kept from air during processing remain green and are sold as green olives.

Olives grown for oil are allowed to ripen on the tree and harvested in winter, when the maximum amount of oil has accumulated (see Chap. 23).

Date (Palmae)

The date palm (*Phoenix dactylifera*) is a tall, attractive evergreen tree which reaches heights of 15 m or more. It not only bears sweet, nutritious fruits but is also a stately ornamental.

Cultivated for many centuries around the Mediterranean, the date was a staple food crop in many early civilizations. For high yields of top-quality fruit, date palms are grown in areas with long, hot, dry summers and low relative humidity. Adequate irrigation is essential. Date palms are able to withstand some cold temperatures, but injury occurs at -6 to $-7°C$. Production is centered in Egypt, Iraq, Iran, and North Africa. In the United States, commercial date production is limited to hot interior

FIGURE 21-13
Avocado fruit ready for harvest.

regions of southern California and a few plantings in Arizona and Texas.

Since the tree is dioecious, commercial date production often requires growers to supplement natural wind pollination. Pollen collected from the male trees is dusted on the female flowers to ensure adequate fruit set.

As the large bunches of fruits approach maturity, they are frequently covered with paper wrappers open at the lower end (Fig. 21-14) to protect the fruit against rain and birds. In the United States, dates are eaten mostly as dried fruits, but in some Mediterranean countries dates are also eaten fresh.

Propagation of the date palm is unusual. They do not breed true from seed and as monocots, i.e., without a cambium, cannot be propagated by grafting or cuttings. Instead, rooted offshoots are removed from trees of the preferred cultivar and planted. One widely grown cultivar is 'Deglet Noor.'

Fig (Moraceae)

The fig (*Ficus carica*) is an ancient crop, which along with date, grape, and olive was an important food in early Mediterranean civilizations. Figs are grown throughout the world in subtropical regions. Leading producers are Spain, Italy, Turkey, and Greece; the United States grows only about 3 percent of the world crop. Of the United States production, about 80 percent is in California. Figs can be grown in the southeast from the Gulf states as far north as North Carolina, but most of the figs in this area are in small backyard plantings. Of the commercial fig crop in California, most are sold as dried figs, a few are canned, and even fewer are sold fresh.

The fig plant is a deciduous shrub or small tree with lobed leaves. Although various propagation techniques can be used, hardwood cuttings are most common. In California, fig plants are trained as trees, whereas in the southeast they are often grown as

FIGURE 21-14
Clusters of date fruits covered with paper for protection against rain and birds. The clusters were thinned to varying degrees to increase fruit size. (a. unthinned; b. moderately thinned; c. heavily thinned) (*U.S. Department of Agriculture.*)

shrubs. Winter temperatures of $-10°C$ will damage fig plants. The multistem plant in the southeast allows more rapid reestablishment of the fruiting surface after winterkill. Depending on the type of fig, one to three crops are produced each year. The first develops from buds produced the previous year; subsequent crops develop from axillary buds on the current season's shoots.

The fruit is a *syconium*, a hollow fleshy peduncle with female flowers borne on the interior surface (Fig. 21-15). At the distal end is a small opening called an *ostiole*, or eye.

Figs are divided into common figs, caprifigs, Smyrna figs, and San Pedro figs. *Common* figs have no staminate flowers, develop without pollination, and consequently are seedless. Cultivars include 'Mission,' 'Kadota,' and 'Brown Turkey.'

Caprifigs are wild figs growing in the Mediterranean region and of no value as fruit because of poor quality. They are unique in having male flowers

FIGURE 21-15
Mature 'Calimyrna' figs. If figs are not picked at this stage of ripeness for the fresh fig market, they are allowed to ripen fully and dry on the tree. (*Dried Fig Advisory Board.*)

which play a vital role in the production of Smyrna figs.

For Smyrna figs to develop and mature, the flowers must be pollinated from an external source. This is where caprifigs enter the picture, along with a fascinating symbiotic relationship between figs and an insect. The fig wasp (*Blastophaga psenes*) transfers pollen from inedible caprifigs to Smyrna figs, a process called *caprification*. The fig wasp completes three life cycles a year, each coinciding with a crop of caprifigs. Before emerging from a maturing caprifig, the females are fertilized by males, which hatch first. The female then leaves the fig fruit and finds a fruit of the next crop, where she lays her eggs.

The growers of Smyrna figs maintain a small planting of caprifigs. When the Smyrna fig flowers are receptive, caprifigs are collected, placed in wire baskets or perforated bags, and hung in Smyrna fig trees. The female fig wasps are coated with pollen as they emerge from the caprifigs. As they enter Smyrna figs in search of sites for egg laying, they pollinate the flowers. As the result of fertilization, seeds form and the Smyrna figs complete fruit development. The irony of the story is that the flower structure of the Smyrna fig is unsuitable for egg laying and the search is in vain. The leading Smyrna cultivar in California is 'Calimyrna.'

The San Pedro type is unique in that the first crop each year develops without fertilization, thus resembling the common figs described above, but the second crop develops only with caprification. Important cultivars include 'San Pedro' and 'King.'

Pineapple (Bromeliaceae)

The pineapple (*Ananas comosus*) is a low-growing perennial of the tropics. Although severely injured by temperatures of 0°C, the pineapple does not thrive at temperatures above 30 to 32°C. Features that make the plant quite resistant to desiccation include crassulacean acid metabolism, sunken stomata, narrow leaves, and funnel-shaped leaf bases which collect water.

The pineapple fruit is complex both in structure and method of development. The fruit is the result of about 100 individual flowers in an inflorescence at the apex of a vegetative shoot. Distal to the fruit a crown of small leaves grows. The fruit consists of the combined mass of the many individual fruitlets surrounding the inner stem tissue. When the fruit is prepared for use, the leathery outer rind is removed, as is the fibrous stem tissue in the center. Canned pineapple rings are the result of trimming the outer rind and removing the interior stem.

World pineapple production is in tropical regions such as South America, Asia, Africa, and Central America. Hawaii was the leading producer of pineapple, but much of the industry has shifted to areas where labor costs are lower.

Pineapples are grown in solid plantings of one cultivar to avoid cross pollination, which would result in seed formation. Fortunately, the fruits develop parthenocarpically.

Since the pineapple is seedless, it is propagated vegetatively by *slips* which arise beneath the fruit, *suckers*, which are shoots originating lower on the stem, or most commonly *crowns* at the top of the fruit (Fig. 21-16). To harvest pineapples for processing, workers wearing heavy chaps and gloves walk between the rows, harvest the fruit, and break off the crown. A week or two later the crowns are collected from the field and planted, often in fumigated soil using black plastic mulch. Newly set crowns normally take 18 to 24 months to flower, a process sometimes accelerated and commonly synchronized by the use of ethylene. The first crop consists of a single large fruit on each plant; the second crop, or first ratoon crop, consists of two smaller fruits per plant and is mature about 12 months after the first; the third crop, or second ratoon crop, consisting of four smaller fruits, is harvested about 12 months after the second crop. After this crop has been harvested, the field is prepared for replanting and the three-crop cycle repeated.

Most pineapples are harvested when they are fully ripe and processed immediately. The fruits ripen after harvest but since they lack starch reserves get no sweeter. Modern jet airplanes transport increasing quantities of fresh pineapples to the United States and Europe. Even these are harvested before becoming fully ripe on the plant and therefore are inferior to those available locally in pineapple-producing areas.

Banana (Musaceae)

The banana (*Musa* spp.) is a tropical fruit which is productive between temperature extremes of 10 and 40°C but is best suited to a narrower range of 15 and 35°C. The starchy cooking banana, or *plantain* (*M. paradisiaca*) is common in the tropics, but the table cultivars in world trade are *M. sapientum*. The only

FIGURE 21-16
Ripe pineapple ready for harvest. Note the slip arising below the fruit. For the fresh market, the crown would be left on the fruit; for the cannery, the crown would be knocked off and left in the field.

state that produces bananas commercially is Hawaii. Most bananas sold in American markets are imported from Central and South America. Leading countries in banana production are Brazil, Ecuador, Costa Rica, Mexico, and several African nations.

The banana is a fast-growing, herbaceous, monocotyledonous plant which develops suckers (pseudo stems) from underground rhizomes and reaches a height of 1.5 to 6 m. The leaves are large, up to 3.5 m long; the pseudo stem is composed of compressed leaf sheaths. After growing for several months, the flower stalk emerges from the top and usually bends downward (Fig. 21-17). Female flowers produce bunches, or *hands*, of bananas parthenocarpically.

FIGURE 21-17
Fruit stalk of banana. Bananas are produced parthenocarpically by the flowers at the base of the stalk.

Since a sucker fruits only once, it is cut back to the ground after harvest. The underground rhizome is perennial, and new suckers develop while the one or two mature suckers are fruiting. New plantings are established by cutting and planting young suckers, 0.3 to 1 m tall, with part of the rhizome attached. Suckers develop rapidly, and the first harvest begins about 9 months after planting.

Bananas are harvested when mature but hard and green and develop excellent quality when ripened. Ripening is done at the distribution center by exposing the fruit to 1000 ppm ethylene for 24 h at 14 to 16°C. The ethylene treatment triggers natural ripening and also makes a shipment of bananas ripen more uniformly. Bananas are subject to chilling injury at temperatures below 10°C and so are not held in cold storage like most fresh fruits and vegetables.

Papaya (Caricaceae)

The papaya (*Carica papaya*) is a large tropical herbaceous plant which grows to a height of 8 m or more. The normally single-stemmed plant grows rapidly; fruits are borne in the axils of the leaves, which are progressively shed from the base upward. Although the papaya plant continues to produce fruit, the planting is usually cut down and replanted when the fruit can no longer be harvested from the ground. Best production is limited to frost-free regions between 32° latitude north and south. Leading areas in papaya production are Central and South America and Asia. Hawaii is the only state with large acreages.

The fruit is a fleshy berry which may be round or pear-shaped and contains many small, round, blackish seeds (Fig. 21-18). The skin changes from green to yellow or orange as papaya ripens. The texture, flavor, and color of the flesh resemble those of a muskmelon. Papaya is widely used as a breakfast fruit, in salads, and as a dessert as well as in juices, ice cream, and jams.

Papaya plants are staminate, pistillate, or monoecious. To establish a new planting, three to five seeds or young transplants are set per location, as the sex of the plants cannot be determined until they flower. At that time, pistillate and staminate plants are removed, leaving the monoecious plants to fruit. The fruits from monoecious plants are preferred because they are pear-shaped whereas the fruit on female plants are round.

FIGURE 21-18
Papaya fruit; note the many black seeds.

The enzyme papain, used as a meat tenderizer and in medicine, is gathered from the sap exuded from scratches on immature papaya fruit.

Mango (Anacardiaceae)

The mango (*Mangifera indica*) is an important tropical fruit. The broad-leaved evergreen tree often reaches heights of 25 m. Although it is grown throughout tropical regions, India leads all other countries by far. United States production is limited to Hawaii and southern Florida.

The inflorescence is a long, branched panicle with several hundred individual flowers. With hundreds of inflorescences per tree, the potential number of fruits per tree is tremendous, but only a small percentage of the flowers develop into fruit.

The fruit range in weight from less than 100 g to over 2 kg, depending largely upon cultivar, but crop load and growing conditions also influence fruit size. The mango is classified as a drupe; the edible flesh is mesocarp. Some cultivars have many fibers extending from the pit into the flesh which do not affect the flavor but make it difficult to eat the fruit.

Mangos are harvested when the skin has partially changed from green to yellow but before the flesh has softened. After receipt at a distribution center, mangos may be treated with ethylene, much like bananas.

When fully ripe the mango is enjoyed as fresh fruit; it is also frozen, canned, dried, and used in chutney.

TREE NUTS

The major edible tree nuts are almond, walnut, pecan, macadamia nut, and filbert.

Almond (Rosaceae)

In the United States the almond (*Prunus amygdalus*) is grown commercially only in California because of its climatic requirements. European production is concentrated in Spain and Italy. The almond tree is deciduous and can survive moderately cold winters. Because of a short chilling requirement, the almond blooms early in the spring and is therefore subject to spring freeze damage.

The almond is a drupe, much like the peach and apricot. In the almond, however, the mesocarp is leathery; the edible portion is the true seed, or nut

(Fig. 21-19). As the fruit matures, the mesocarp dries and splits. Long, hot, dry summers are required for proper maturation and drying. Since the major almond cultivars are self-sterile, three cultivars are usually planted together to ensure adequate pollination by bees. Most almonds are mechanically shaken from the trees, allowed to dry a few days, and picked up by machine. Before harvest the soil is smoothed and packed by a specialized piece of equipment.

English Walnut (Juglandaceae)

The Persian or English walnut (*Juglans regia*) is a large deciduous tree. The male flowers (catkins) and female flowers are borne separately on the same tree, which is wind-pollinated. Although commercial cultivars are self-fruitful, an additional cultivar is often included in the orchard. Pollen of a particular cultivar may be shed before or after the female flowers of that cultivar are receptive, a phenomenon called *dichogamy*. A thick husk encloses the nut, inside of which is the edible seed.

Persian walnuts are widely grown in Europe, the Soviet Union, and China. United States production

FIGURE 21-19
'Peerless' almonds, as shelled meats, and still in the shell. (*California Almond Growers Exchange.*)

is primarily in California, with limited acreage in Oregon. California cultivars such as 'Hartley' withstand winter minimum temperatures of about − 10°C. A more cold-hardy type, the Carpathian walnut, hardy to − 40°C, is grown in limited numbers in the midwestern and eastern United States.

The numerous other walnut species are not widely grown commercially, e.g., *J. nigra* (eastern American black walnut) and *J. cinerea* (butternut). The black walnut produces edible nuts but is most prized for its beautiful wood, used for making fine furniture and gunstocks.

In California, walnuts are harvested much like almonds, and, like the other tree nuts, are sold in the shell or as shelled kernels.

Pecan (Juglandaceae)

The pecan (*Carya illinoensis*) is one of about 20 species of *Carya* native to North America, all except the pecan being classified as hickories and having much harder, rougher shells. Commercial pecan production occurs (in order of decreasing importance) in Georgia, Texas, Louisiana, Alabama, and Oklahoma and is about evenly divided between native seedlings and improved cultivars grafted on seedling rootstocks. The improved cultivars have thinner shells and are often referred to as papershells. Cultivars grown vary markedly with locality and climate. Since the pecan requires some cold weather to satisfy the chilling requirement, it is unsuited to subtropical and tropical climates. Severe winter cold is detrimental, and northern production is further limited by too short a growing season.

Filbert (Betulaceae)

The filbert (*Corylus avellana*) is but one of several edible species in this genus; two other species native to eastern North America are often called hazelnuts. Commercial filbert production is centered in Washington and Oregon. About 70 percent are sold in the shell, often as a part of bagged mixed nuts; much of the remainder is shelled and used in canned mixed nuts. Turkey and Italy are large producers of filberts.

Chestnut (Fagaceae)

The American chestnut (*Castanea dentata*) has been an important edible nut in Europe and Asia and, until the chestnut blight early in this century, was of considerable value in the United States from the Appalachians north to Canada. The European or Spanish chestnut (*C. sativa*) is widely grown in southern Europe and is sometimes grown in America. Although widely consumed, the European chestnut is considered to be of only moderate quality.

Macadamia (Proteaceae)

Macadamia nuts (*Macadamia ternifolia*) have become a major crop in Hawaii, with production unable to keep pace with demand. The nut grows inside a fleshy husk and has a hard shell (Fig. 21-20). Since nuts mature and fall from the trees over an extended time, they must be harvested several times, often by collecting them from nets hung under the trees. The nuts are dried, sized, and cracked between rotating drums. They are usually vacuum-sealed in small containers and command a high price because of their high quality and excellent flavor.

Other Nuts

Other important nuts in world trade are the cashew (*Anacardium occidentale*), pistachio (*Pistacia vera*), and Brazil nut (*Bertholletia excelsa*), which are largely imported into the United States from India, Turkey, and the Amazon basin, respectively.

BEVERAGE CROPS

The main nonalcoholic beverages of world trade are coffee, tea, and cocoa, all grown in the tropics. In the United States, we drink much more coffee than

FIGURE 21-20
Macadamia nuts approaching maturity.

tea, whereas the reverse is true in the world as a whole. Both coffee and tea contain caffeine, an alkaloid which serves as a stimulant. Cocoa contains theobromine, which is similar to caffeine.

Coffee (Rubiaceae)

Coffee is made from the ground seeds of the genus *Coffea,* which contains about 25 species. The commercial coffee industry is based primarily upon *C. arabica* (Arabian coffee), *C. canephora* (Robusta coffee), and *C. liberica* (Liberian coffee) with Arabian coffee accounting for about 90 percent of the world crop. *C. arabica* is a small tree from 4.5 to 9.0 m in height when unpruned but often held at 2 m by pruning to facilitate harvest. The fruits are two-seeded drupes called cherries. The fleshy berries change from green through yellow to red. Harvesting of the mature berries has been by hand, but attempts to mechanize the harvest are under way. The two greenish-gray seeds are surrounded by several layers of tissue which must be removed by a combination of machine separation, fermentation, curing, milling, and polishing. The coffee beans must be roasted at a high temperature to develop the characteristic aroma, flavor, and color. Many commercial coffees are blends of different types and sources to provide a particular character. The final step is grinding the beans.

Coffee production is limited to tropical and subtropical regions and is heavily concentrated in South and Central America and Africa; Brazil is the leading producer. Coffee is best adapted to an area with temperatures of 15 to 25°C, 200 to 300 cm of annual rainfall, and also a 2- to 3-month dry season, during which flowers are initiated. World coffee production is from 3 to 4 million metric tons. The only state in the United States growing coffee commercially is Hawaii, with annual crops of about 1000 metric tons.

Tea (Theaceae)

The most widely used caffeine-containing beverage is tea, made from the dried leaves of *Thea sinensis.* Allowed to grow to its full size, the tea plant is a small tree, but commercially it is kept pruned to a shrub about 1 m tall to make harvesting easy. Harvest involves plucking young tender shoots consisting of two to four leaves and the terminal bud. This has been a hand operation for centuries, but mechanical harvesters are being tested. The number of pickings per season varies from three or four in areas where

growth stops in the winter to 25 to 30 in warmer areas.

The harvested leaves are wilted, rolled under pressure, and then further dried. At this stage of preparation, the leaves are completely dry but still green and can be used as *green tea.* For *black tea,* the major tea of world trade, the leaves are rolled once, allowed to ferment, and rolled repeatedly during fermentation, so that the green color is lost and the flavor changes, the result varying with the type of leaf used and the length of time fermentation is allowed to proceed. The grade of the final product depends on which part of the shoot is used, 'orange pekoe' being the smallest leaf, 'pekoe' the second leaf, etc.

Tea production is centered in India, China, Japan, and parts of Africa. Best growth is obtained in a tropical to subtropical climate at elevations of 1 km in the tropics to as low as sea level in cooler regions.

FIGURE 21-21
Cacao fruit split to expose the seeds. (*U.S. Department of Agriculture.*)

Tea has been propagated from seed, but now cuttings and budding are being used to increase uniformity.

Cacao (Sterculiaceae)

The cacao plant (*Theobroma cacao*) is native to tropical America, where it is widely grown today. Parts of Africa also provide an excellent climate for cacao. Commercial production is limited to a zone of 20° north and south latitudes at low elevations with 150 to 200 cm of evenly distributed rainfall per year. Leading countries of production are Ghana, Nigeria, and Brazil. The cacao tree is quite small, usually 5 to 8 m high, and thrives best when partially shaded by larger trees. Flowering and fruiting occur continuously. The fruits are large football-shaped pods or capsules from 15 to 23 cm long and 7 to 10 cm in diameter. The fruits are harvested by hand when fully ripe, usually at weekly intervals, and split open (Fig. 21-21). The contents are removed and fermented to eliminate the pulp surrounding the seeds. After drying, the seeds or beans are dried and bagged for shipment.

The beans are cleaned, sorted, roasted, and ground into an oily paste called bitter chocolate. It may be hardened and sold, or converted into sweet chocolate by adding sugar and spices or into milk chocolate by adding milk as well. Cocoa is made by extracting most of the fat (cocoa butter), leaving a residue which is ground into powder.

CHAPTER TWENTY-TWO

OIL, SUGAR, AND FIBER CROPS

The oil-, sugar-, and fiber-crop industries are large, both in the United States and the world as a whole. This chapter discusses the major crops in each category and provides an introduction to their botany, geography, and use.

OIL CROPS

Vegetable oils have been used by people since very early times. Long before petroleum was discovered, plant oils were used in lamps and candles, as a source of heat, and for anointing the skin. The realization that the world's store of petroleum is finite and dwindling has led to increased interest in oils produced by plants.

Oils are universally present in plants; in most species, the greatest concentration of oil is in the seeds, the major exception being the olive, where the oil is contained in the fleshy fruit.

The production of vegetable oils is a large and increasing part of American and world agriculture. Not only is there an increasing demand for oil for conventional uses, but there is considerable interest in vegetable oils as fuel. Crops covered elsewhere, e.g., corn, soybean, and cotton, will be discussed here primarily in relation to oil.

Some oil crops, e.g., the olive, have a long history; for others, e.g., sunflower and soybean, large-scale commercial production is of relatively recent origin.

The cultivation of new oil crops is being commercially explored.

About 40 plant species are used as sources of oils but many others are potentially useful. About 90 percent of the plant oils in commerce come from soybean, sunflower, peanut, cottonseed, coconut, rapeseed, palm, linseed, and castorbean.

Oils are made up of fatty acid esters of glycerol (Fig. 5-24). When the fatty acid components are saturated, each carbon atom is linked by single bonds to other carbon atoms. In unsaturated oils, at least one carbon atom is linked to another by a double bond. Fats and oils differ in their melting point; at room temperature, fats are solids and oils are liquids. Fats consist largely of saturated fatty acids whereas oils have a higher proportion of unsaturated fatty acids. This difference in the degree of saturation has received special attention in recent years because of the apparent link between saturated fats and cholesterol levels in the blood. With modern technology, chemists are able to transform vegetable oils into solid fats by hydrogenation, the basis of the margarine industry.

Another feature by which vegetable oils are classified is their reaction to air. Fatty acids that are largely saturated, e.g., peanut and olive oil, do not react with the oxygen in air and are called *nondrying oils*. Highly unsaturated oils, e.g., tung and linseed oils, form a tough elastic film upon exposure to air. These *drying oils* are widely used in paints and varnishes. Intermediate types, e.g., cottonseed, corn, and sunflower oils, are called *semidrying oils.*

Most vegetable oils are used as cooking oils, salad oils, or for the production of margarine. Modern oil chemistry makes it possible for many oils to be used interchangeably, a fact of economic importance since it allows oil and margarine manufacturers to buy oils on the basis of supply and price.

In spite of the large size of the vegetable oil market in the world, much of the oil available is actually a by-product. For example, the demand for high-protein soybean meal for use as a meat substitute and as animal feed has made soybean oil one of the most widely available vegetable oils. Other examples include cottonseed oil, an important by-product of cotton, and corn oil, a by-product of the cornstarch industry.

The extraction of oil from plant sources normally involves two major processes, one mechanical and the other chemical, although in some cases only one is used. Before extraction, the product is usually cleaned to remove foreign matter such as weed seeds as well as broken or contaminated seeds. Contaminants can lower both oil quality and the value of the oil cake, the residue left after extraction. Oil-bearing seeds are usually crushed and then cooked at 85 to 95°C to disrupt the oil cells and inactivate enzymes. Seeds with a high oil content are usually crushed in large presses. Although these presses can leave as little as 10 percent oil in the oil cake, under such great pressure the quality of both the oil and oil cake decrease. Commercial practice is therefore to leave about 20 percent oil in the press cake and to extract it with organic solvents, often hexane. The hexane-oil mixture and the oil cake are distilled to remove the solvent, which has a low enough boiling point to be removed completely at temperatures that preserve the quality of the product.

Soybean (Fabaceae)

Soybean (*Glycine max*) has already been discussed in Chap. 19. The tremendous demand for soybean meal as a protein source has made soybean oil the leading vegetable oil. About 90 percent of it is used to make edible products such as cooking and salad oils, margarine, and mayonnaise. The rest is used for industrial purposes such as paints, soaps, and ink.

Although the major use of soybean meal has been livestock feed, increasing amounts are being used in

food as meat substitutes, flour, and in the preparation of specialty foods for those who have food allergies. As meat prices continue to increase, it seems likely that the market for soybean meal will grow and soybean oil will continue to dominate the vegetable oil market.

Corn or Maize (Poaceae)

Although corn (*Zea mays*) is grown primarily for livestock feed (see Chap. 23), significant portions of the crop are processed for cornstarch, syrup, protein, and oil. Corn oil comes from the germ (or embryo), and although oil constitutes only about 5 percent of the corn kernel, it represents a major source of vegetable oil because of the tremendous quantities of corn processed. The seed components are separated in milling. The endosperm makes up about 75 percent of the kernel and yields starch, which is used for starch, corn syrup, or corn sugar. The germ contains 25 to 30 percent oil and contributes significantly to the vegetable oil supply. Corn oil is used largely as salad and cooking oil and in the manufacture of margarine.

Sunflower (Asteraceae)

The sunflower (*Helianthus annuus*) is one of the relatively few commercially important crops originating in the New World. It grows wild in the Great Plains from northern Mexico to Nebraska and was used as a food source by the Native Americans.

The native sunflower is a tall (1 to 3 m), usually single-stemmed, annual plant with single or multiple seed heads 8 to 15 cm in diameter. Oil content of wild sunflower seeds is about 25 percent. Plant breeders, particularly in the Soviet Union, have developed superior cultivars for commercial production with single seed heads 15 to 30 cm in diameter (Fig. 22-1) and oil content of seeds approaching 50 percent. These cultivars are also much more suitable for mechanical harvest.

Although not a major agricultural crop, the sunflower has been used for several purposes. In the Soviet Union, sunflowers are a significant food crop. In the United States, the seeds are often baked like nuts and eaten as a confection but are not important as food. In some areas, sunflower plants are used for silage, especially where the growing season is too short for corn. The seeds are also grown for bird feed.

FIGURE 22-1
An immature sunflower head. Each head produces many seeds, which yield up to 50 percent oil.

Sunflower production has increased because of the demand for the oil, which is similar to corn oil and can be used to manufacture salad and cooking oils and margarine. New cultivars with smaller seeds but higher oil content have been developed for the

vegetable oil market. Perhaps the greatest concentration of sunflower production in the future will be in areas where the growing season is too short or too cool for major competing crops such as corn. Sunflowers are more tolerant of freezing temperature than corn and mature in 90 to 120 days.

Safflower (Asteraceae)

Although a fairly new crop in the United States, safflower (*Carthamus tinctorius*) has been grown for thousands of years. The origin of safflower is unknown but the Near East, Middle East, and Asia have all been suggested. Until synthetic dyes were developed, the flowers were used as a source of red, orange, and yellow dyes for cloth and food.

The plant is an annual with multiple stems growing to a height of 0.6 to 1.5 m. Leaves are sessile, toothed, and spiny. Like other thistles, each branch produces a head 1.3 to 3.5 cm in diameter (Fig. 22-2) with numerous flowers, each of which produces a seed. Flower and seed number per head range from 20 to 100. The seed is really an achene, which is small (1 to 1.3 cm long), smooth, shiny, and wedge-shaped.

Although still grown in some parts of the world as a source of dyes, safflower production is largely for oil. Modern cultivars have been developed for

FIGURE 22-2
Head of safflower before flowering, in flower, and after flowering. (*U.S. Department of Agriculture.*)

higher oil content, which ranges from 25 to 35 percent. Safflower oil is a high-quality, light-colored product used for salad and cooking oils and margarine and as a drying oil in paints and varnishes. Production in the United States is mostly in the Great Plains and California.

The minimum frost-free growing season for safflower is about 120 days. In many areas, it is in competition with small grains for cropland. Any increase in production in the United States may hinge on the development of cultivars which produce more oil per hectare. Safflower is harvested with small-grain combines.

Cottonseed (Malvaceae)

As discussed later in the chapter, cotton (*Gossypium* sp.) continues to be a major fiber crop despite the inroads made by synthetics. A major by-product of cotton-fiber production is the oil from the seed, which contains 16 to 20 percent oil. The oil is used for salad oil, shortening, and margarine. The press cake left after oil extraction contains up to 35 percent protein and is used as a feed supplement for livestock. A substance which is toxic to people can now be removed, so that cottonseed meal may become important as a low-cost protein supplement for human consumption.

Peanut (Fabaceae)

Peanut (*Arachis hypogaea*) was described in Chap. 19. Most of the United States harvest is used for peanut butter, salted peanuts, candy, and livestock feed. In other parts of the world, however, peanut is grown mainly for the oil.

Peanuts contain from 40 to 50 percent oil and are processed much like the other oil seeds. Both press and solvent-extraction methods are used. Peanut oil is a nondrying type widely used in cooking oils, shortening, and margarine. The oil cake remaining after removal of the oil contains about 50 percent protein and is used as a protein supplement in livestock feed.

Olive (Oleaceae)

Among the oil crops, the olive (*Olea europea*) is unique in that the oil is contained in the flesh of the fruit rather than the seed. About 90 percent of the world's olive production is for oil.

The tree is a large, subtropical evergreen most widely grown in Spain, Italy, Greece, Portugal, and Turkey (Fig. 22-3). A moderately cold winter and a long, hot, dry summer constitute an ideal climate for olive production.

For oil extraction, olives are left to ripen fully on the tree, allowing the oil content to reach its maximum of 20 to 30 percent. After washing, the olives are ground up and the oil removed in large hydraulic presses.

Olive oil is most widely used as a salad oil although some is used as a cooking oil. In Mediterranean countries, the olive has been a major crop for centuries. Its production in the United States is limited to parts of California.

Coconut (Palmaceae)

The coconut (*Cocos nucifera*) is the fruit of a large palm tree grown throughout the tropics (Fig. 22-4). Its origin is unknown because it has been grown for many centuries and because the mature fruit, which floats, is readily transported by ocean currents. Production in the United States is limited to southern Florida and Hawaii. Although the coconut palm can withstand light freezes, growth and production are best at mean annual temperature of 25 to 26°C. Since the coconut needs ample moisture and is highly tolerant of salt, it thrives along tropical seacoasts.

The large fruit is 25 cm or more long and somewhat triangular in cross-section. Inside the outer husk, or exocarp, which is hard and fibrous, is a fibrous

FIGURE 22-3
Olive branch bearing small white flowers.

oils, shortening, soaps, and cosmetics and for various industrial purposes.

Oil Palm (Palmaceae)

The oil palm (*Elaeis guineensis*), closely related to the coconut, is also a major source of vegetable oil. This plant is sometimes referred to as the African oil palm, because of its African origin and because most production is still in Africa.

Both the shell (pericarp) and the kernel contain significant amounts of oil. The oil from the husk is a deep yellow to reddish-brown and is used to manufacture soaps, candles, lubricants, and margarine. The kernel oil is a lighter yellow and is used for a variety of products including margarine and soaps.

Flax (Linaceae)

Besides being a source of fibers, as discussed later in the chapter, the flax plant (*Linum usitatissimum*) is the source of linseed oil. Cultivars used for seed production have shorter stems and produce more seeds than those used primarily for fibers.

Flax seed contains 35 to 45 percent oil; it is a drying oil widely used in paints, varnishes, and lacquers. Other uses are in the manufacture of linoleum, printer's ink, and imitation leather. The development of latex paints has decreased the use of oil paints, of which linseed oil is a major component.

The oil cake contains up to 35 percent protein and makes an excellent livestock feed supplement. The straw from seed flax is used in the manufacture of high-quality paper such as that used for cigarettes.

Most seed-flax production is in Canada, Argentina, India, the United States, and the Soviet Union. Although not one of the leading oils in total production, linseed oil has played a major role in certain industries.

RapeSeed (Brassicaceae)

Rapeseed (*Brassica napus*) and turnip rape (*B. campestris*) are closely related to the turnip, which is in the same family. Although rapeseed is sometimes grown as a forage crop, our interest is in rapeseed as a source of oil.

Rapeseed is a herbaceous annual or biennial growing 30 to 90 cm tall with heavily branched stems and long, lobed leaves. Rapeseed produces a rosette

FIGURE 22-4
Coconut palms. (*Courtesy of E. Pehu.*)

mesocarp, which is sometimes used for the fibers, called *coir*. When one buys a coconut in a store, the exocarp and mesocarp have usually been removed to leave the endocarp, the hard shell which is comparable to a peach pit. It contains the true seed, which is surrounded by a thin brown seed coat. The white "meat" inside the seed coat is part of the endosperm and contains from 35 to 50 percent oil. Coconut "milk" is liquid endosperm.

For the production of coconut oil, the mature nuts are harvested, the husk is removed, and the nuts are split in half. The milk is discarded and the halves spread out to dry. After sun-drying for a few days, the meats are pried from the shells and further dried. The dried coconut meat, known as *copra*, is chopped and shipped to processing plants for oil extraction.

Coconut oil is more highly saturated than most vegetable oils. It is widely used in margarine, cooking

of leaves, after which an indeterminate, racemose inflorescence develops. In biennial types, an extended cold period is required for bolting of the flower stalks. Each fruit (silique) holds 15 to 40 very small round seeds, ranging from yellow to black and containing up to 40 percent oil. Rapeseed production is largely in relatively cool areas with short growing seasons, such as Canada and northern Europe. Many rapeseed cultivars mature in growing seasons as short as 90 days, which is marginal at best even for safflower production.

Rapeseed is a source of edible oil, sometimes called *colza oil*. Other uses for rapeseed oil include the manufacture of lubricants and soaps. The meal left after oil extraction contains as much as 35 percent protein and can be used both for human consumption and as a supplement for animal feed.

Tung (Euphorbiaceae)

The tung (*Aleurites fordii*) plant is a small tree which produces fruits with two to five seeds. The kernels contain up to 60 percent oil. Tung oil is considered to be one of the finest drying oils and was highly prized for use in paints and linoleum. A sizable tung-oil industry which developed in the Gulf Coast region of the United States in the 1940s on the basis of these markets has now declined because of decreased demand.

Since tung oil is not edible and the press cake is toxic to livestock, the poor competitive position of tung oil in industrial markets makes a continued decline in importance likely.

Castor Bean (Euphorbiaceae)

Although hardly a major oil crop in the United States or the world, castor bean (*Ricinus communis*) has been grown for centuries, largely in the tropics but sporadically in the United States.

In the tropics, the plant is a short-lived perennial often reaching a height of 10 to 13 m (Fig. 22-5); since it is killed by freezing temperatures, it is grown as an annual in temperate regions. Most castor bean production in the United States is in the southern Great Plains region and California.

At maturity, the thorny seed pods (capsules) split into three segments, each with one seed. Oil content of the seeds ranges from 45 to 55 percent. Since the seeds contain a highly toxic compound, ricin, the oil cake cannot be fed to livestock and is used largely as fertilizer.

FIGURE 22-5
Castor bean plant. Plant grows 1 to 13 m high. (*U.S. Department of Agriculture.*)

Castor oil is a nondrying oil of high viscosity which is in demand for the manufacture of lubricants, greases, and hydraulic fluid. Other uses are in plastics and printing ink. Its use in medicine, formerly widespread, is now minimal.

Sesame (Pedaliaceae)

Sesame (*Sesame indicum*) is an ancient crop not widely grown. Its seeds are an excellent source of oil suitable for salads and cooking. Oil content of the seeds is as high as 50 to 55 percent.

New Crops

New crops are being explored as sources of oils, waxes, and hydrocarbons. Two promising ones will be discussed briefly.

Jojoba (Buxaceae) The jojoba† plant (*Simmondsia chinensis*) is native to Mexico and the southwestern

---

†Pronounced ho-ho'-ba.

United States (Fig. 22-6). Its seeds contain up to 50 percent of a liquid wax, which is sometimes called an oil.† Jojoba wax is an ideal substitute for sperm oil and makes an excellent lubricant. It can be used in the manufacture of high-quality hard waxes for floors, furniture, etc.

Commercial production of jojoba is in its infancy but as improved cultural and harvesting practices are developed, it may well become an important crop. Jojoba can be grown in arid areas because of its drought tolerance. As an excellent substitute for sperm oil, jojoba may help preserve dwindling populations of sperm whales.

Guayule (Asteraceae) Like jojoba, guayule‡ (*Parthenium argentatum*) is native to Mexico and the southwestern United States. A low-growing shrub, its latex was used to make rubber during World War II. When rubber from Asia again became available after the war, interest in guayule dropped. With concern about dependency on foreign sources of vital resources, interest in guayule has revived. Like jojoba, it can be grown in areas unsuited to production of agricultural crops.

SUGAR CROPS

The two major sources of sugar are sugarcane and sugarbeet, both of which yield sucrose. As the sugar of commerce, sucrose is in great demand, especially in areas of the world with a high standard of living. The annual per capita consumption of sugar in the United States averages about 40 kg.

Less important sources of sugar include maple syrup, honey, and sweet sorghum. When sugar prices have risen markedly, increasing amounts of corn syrup have been used as a sweetener, especially in processed foods. Cornstarch is hydrolyzed by acids to produce glucose (dextrose). Since glucose is not as sweet as sucrose, about 15 percent sucrose is often added to corn syrup.

---

†As noted above, oils consist of three fatty acid molecules attached to a glycerin molecule. The plant waxes are esters of a fatty acid and a long-chain alcohol.

‡Pronounced wa-yoú-lee.

Sugarbeet (Chenopodiaceae)

The sugarbeet is *Beta vulgaris*; although table beet and sugarbeet belong to the same species, they differ markedly. Table beets are usually red and contain only 2 to 4 percent sugar; sugarbeet is white with a white or creamy skin and contains up to 20 percent sugar (Fig. 22-7).

Not until the early nineteenth century was sugarbeet grown commercially. When a naval blockade

(a)

(b)

FIGURE 22-6
The wax from the jojoba plant can be substituted for sperm whale oil and has generated much interest as a cash crop in the southwestern United States: (*a*) 5-year-old plant; (*b*) fruit approaching maturity. (*Courtesy of D. A. Palzkill.*)

FIGURE 22-7
Sugarbeet roots.

cut off the supply of sugarcane from the tropics during the reign of Napoleon, a domestic sugar industry in central Europe evolved around the sugarbeet. Early types contained only about 6 to 7 percent sugar; selection and breeding have increased sugar content to more than 20 percent. In 1840, sugarbeet supplied only 5 percent of the world's sugar, but by 1890, the percentage had risen to almost 50 percent.

World production of sugarbeet is about 300 million metric tons; leading countries are the Soviet Union, the United States, France, West Germany, Poland, and Italy. Since sugarcane production is confined to tropical or subtropical areas, large-scale domestic sugar production in high latitudes is limited to sugarbeet.

The sugarbeet is a biennial grown as an annual except for seed production. During the first year, the large succulent 1- to 2-kg root forms. The long taproot requires a deep, well-drained soil. Yields vary widely but in the United States average about 45 metric tons per hectare (more than 80 percent of that yield is water). The four major sugarbeet-producing regions in the United States are the humid area in the north central states, the Great Plains, the mountain states, and the Pacific coast. Climatic requirements include an ample supply of moisture, and a temperature of 19 to 22°C is ideal because the combination of warm days and cool nights encourages good growth. The cooler nights of late summer, combined with the depletion of available nitrogen, slow vegetative growth and lead to increased sugar storage in the roots.

For many years, a major problem in growing sugarbeet was getting the desired stand of plants. Like the garden beet, the sugarbeet produces a seedball containing two or more seeds called a *multigerm* seed. During the 1950s, cultivars were developed which produce a single-seeded fruit, called *monogerm* types. Precision planting of these seeds ensures plant spacing which allows the minimal thinning necessary to be done mechanically.

Sugarbeets are typically grown by agreement with processors, the contracts specifying acreage to be grown, seed to be used, planting date, cultural practices, and harvest. The processor provides the seed and agrees to pay a specified price for the crop if it meets minimal quality standards. Sometimes delivery dates to the processor are specified to reduce the need for storage before processing.

Sugarbeets are harvested by machines which cut off the leaves, lift the roots, and load them onto trucks for delivery to the factory. The leaves are often collected and used as feed for livestock. Harvest is delayed as long as possible to allow maximum sucrose accumulation in the roots; it is usually started in late September or early October and finished as soon as possible to get the last beets out of the ground before the soil freezes. Modern harvest methods are so rapid that processors receive sugarbeets faster than they can handle them. To avoid heating and spoilage in the storage piles, ventilation and cooling equipment are often used.

In the processing plant, the roots are washed, sliced, cooked and the juices are extracted, purified, and crystallized. The solid material remaining after extraction of the juices is dried and used as livestock feed. A by-product is molasses, which is used for livestock feed, alcohol production, and other purposes.

Sugarcane (Poaceae)

Sugarcane is a member of the grass family and is a tropical perennial. Some cultivars are a cross between *Saccharum officinarum* and related species, especially *S. barberi* and *S. spanteneum*. The last two have been used in breeding programs to incorporate disease resistance.

Sugarcane is an ancient crop. Although its origin is uncertain, sugarcane is probably derived from wild species in Melanesia or northwestern India.

There is evidence that sugarcane was under cultivation in India as early as 1000 B.C. and in China by 760 B.C.. Sugarcane gradually spread to Java, the Philippine Islands, and Hawaii and westward to Iran and Egypt. Columbus brought sugarcane to Dominica in 1493, where it became established in the early 1500s. Imported sugarcane was processed in New York as early as 1690, and it was introduced into the United States as a crop in 1751, when the Jesuits brought it to Louisiana.

Annual world production of sugarcane is about 700 million metric tons, and leading countries include India, Brazil, Cuba, China, Mexico, and the United States. In the United States, sugarcane is grown commercially only in Hawaii, Louisiana, and Florida.

Sugarcane is considered to be one of the most efficient plants in fixing solar energy. As a C_4 plant, sugarcane thrives in a hot climate and can make use of the high light levels common to the tropics.

The plant grows up to 5 m tall and consists of multiple stems or stalks 2.5 to 7.5 cm in diameter (Fig. 22-8). The stalk is solid, and the cortex and pith of the 5- to 15-cm internodes consist of parenchyma storage cells. Leaves originate at each node and often reach a length of 2 m; leaf structure is similar to that of corn.

Sugarcane is propagated vegetatively by ''seed pieces,'' sections of stalk 0.7 to 1 m long that are buried in a shallow furrow. Plants emerge when axillary buds begin to grow at each node. The first shoot to emerge is the primary stalk, which is usually surrounded by secondary stems, the whole being known as a *stool of cane*. From planting to harvest requires 8 to 30 months depending on climate. In the United States, most sugarcane is harvested before flowers appear. In tropical areas, the silky panicle or arrow appears at the shoot apex in response to short days. Few seeds are fertile, and even those quickly lose viability.

Various systems are used in harvesting. In the past, large numbers of workers cut the stalks with special knives, stripped off the leaves, and topped the canes. In some areas today, the field is burned to remove the leaves and the canes are cut, windrowed, and loaded mechanically. Removal of leaves and tops reduces loading and hauling time and makes sugar mill operation easier.

After the first crop (*plant crop*) has been harvested, new shoots from the stubble give rise to one to three *stubble* or *ratoon* crops. A newly planted sugarcane field can therefore be in use for up to 8 years.

Sugarcane grows best in warm to hot areas with plenty of sunshine, e.g., in Hawaii, Cuba, and Puerto Rico. Since temperatures above 38°C are detrimental, the ideal climate is where mean monthly temperatures are between 21 and 38°C. Air temperatures of 10°C are harmful to the leaves, and freezing temperatures kill the plant. The effects of temperature on sugarcane growth are very apparent in Hawaii: at low elevations, sugarcane is fully mature in 2 years, whereas at an elevation of 1000 m an extra 12 months is required. Maximum production of sugarcane requires ample water and fertilization. A minimum of 115 to 130 cm of rainfall or irrigation must be uniformly available throughout the year.

After harvest, sugarcane is hauled to mills, where the sugar is extracted by a series of crushings with the addition of hot water between them. The water dissolves additional sugar, which is removed by subsequent pressings between heavy metal rollers. The fiber (*bagasse*) is separated from the juices; after drying it is burned as fuel for the sugar mill or used to make paper or fiberboard. The juice is clarified by precipitating impurities, and much of the water is evaporated. After the sugar has crystallized, large centrifuges are used to remove molasses, leaving ''raw sugar,'' which is about 96 percent sucrose. Molasses is used for livestock feed, alcohol production, and other purposes. The raw sugar is shipped to refineries, where it is redissolved and

FIGURE 22-8
Sugarcane field in Hawaii.

impurities removed by further processing. The final product is the white sugar of commerce.

Juice from sugarcane contains from 12 to 17 percent sucrose and 2 to 3 percent impurities. A metric ton of cane (1000 kg) yields 75 to 100 kg of raw sugar. Yields of cane average about 90 to 100 metric tons per hectare.

Sweet Sorghum (Poaceae)

Several types of sorghum are grown worldwide for grain, forage, and, to a minor degree, for syrup; all belong to *Sorghum bicolor*. In Asia and Africa, large amounts of sorghum grain are used for human food, but sorghum is most widely grown in the United States as feed for livestock and poultry.

We are dealing here only with sweet sorghum, often referred to as *sorgo* or *cane*. Although a relatively minor crop, sorgo is important in the south central and southeastern United States. Even there, however, production has declined markedly since 1920, probably because other inexpensive syrups and sugar products have become available.

The juicy pith of the stem contains considerable amounts of sugar—in a good sorgo cultivar 13 to 17 percent, of which 10 to 14 percent is sucrose. Since crystallization of the sugar is poor because of the presence of excessive amounts of starch and glucose, the major use of sweet sorghums is for syrup.

A yield of 35 metric tons per hectare is considered a good yield of fresh sweet sorghum; when topped and stripped of leaves this provides about 23 metric tons of stalks. A metric ton of stalk yields about 550 kg of juice. As the juice is heated in an evaporator, impurities coagulate and are skimmed off the surface. The yield of syrup from 1 metric ton of stalk is about 60 L.

Most sweet sorghum is grown in small plots, usually less than 0.5 ha. Sorghum syrup or molasses used to be made in small on-farm operations, but now one larger plant may manufacture the syrup for an entire locality. In areas where sweet sorghum is grown, the light-colored, mild syrup is popular for its characteristic flavor.

Maple Syrup (Aceraceae)

The manufacture of maple syrup is an important though localized industry. The sap from the sugar maple (*Acer saccharum*) is a clear, almost tasteless liquid. The product left after heating in an evaporator to drive off most of the water is a delicacy.

Historically the industry has been centered in northern New England, but it has some importance wherever sugar maples grow. Native trees (instead of a planting) are used as the source of sap. In early spring, holes are drilled into the outer sapwood, from which the sap exudes. Formerly, hollow wooden taps were driven into the holes and a pail hung on the end to collect the sap (Fig. 22-9), but since emptying the pails requires much time and labor, other systems are being used. In one, plastic tubing conveys the sap to a main collection point by gravity flow. The maple syrup is made in *sugar houses*, where the sap is boiled, usually on a relatively small scale. The yield of sap depends on the weather in early spring.

Conditions which maximize sap flow are moderately warm days and cold nights. The cold nights lead to hydrolysis of starch into sugars in the xylem parenchyma. These sugars are transported to the xylem elements. As temperatures rise, carbon dioxide released from the solution generates a positive pressure in the xylem, forcing the sap upward in the tree just before growth starts. "Sugaring" off lasts for several weeks in a good year, but temperatures that warm up quickly shorten the season.

Most people who make maple syrup do so only as an income supplement. Although it is profitable in a good year, the vagaries of weather make it an undependable business.

Most Americans have probably never tasted pure maple syrup since table syrups usually contain only

FIGURE 22-9
Traditional system of collecting sap from maple trees. Sap is boiled down to yield maple syrup.

small amounts of it combined with large amounts of artificial flavoring.

FIBER CROPS

The most widely grown fiber crop is cotton, which has been traced back over 6000 years. Fibers from flax were used by the ancient Egyptians to wrap mummies. Other fibers of some importance include jute, hemp, ramie, and kenaf.

Synthetics have offered increasing competition for natural fibers since the beginning of the twentieth century. Rayon, the first, appeared commercially in the 1890s. It is made from natural cellulose dissolved in a solvent. The thick solution is forced through tiny openings to form a strand, which is subsequently hardened, spun into thread, and woven into cloth. Today a wide range of synthetic fibers is available, e.g., nylon and the polyesters Dacron and Orlon. Fortunately for the cotton growers, many of the best fabrics are blends of cotton and a synthetic fiber.

Cotton (Malvaceae)

Cotton and kenaf, another fiber crop, are the only important agricultural crops in the mallow family. Most commercially grown cotton is either *Gossypium hirsutum* or *G. barbadense*, but there are several other species. Almost all cotton grown in the United States is upland cotton (*G. hirsutum*), with a fiber length of 2 to 3 cm, depending on cultivar. The long-staple cottons, with fiber lengths of 2 to 5 cm, belong to *G. barbadense*.

Cotton has been an important crop since the dawn of civilization. It is claimed to have evolved in various parts of the world, e.g., Central and South America, Africa, and Indochina, but the evidence is inconclusive. Archeologists have uncovered proof of cotton production in Arizona over 500 years ago, so that cotton appears to have been in North America long before it was brought to Virginia in 1607. Columbus found Native Americans using cotton cloth.

As the cotton industry developed in the southeastern part of the United States, with its dependence on slave labor, it played an enormous role in the economy, politics, and sociology of that region. Much of this development occurred in the latter part of the eighteenth century when machines became available to spin and loom the fibers and Whitney invented the cotton gin.

Although grown as an annual, some cotton species are perennial in the tropics. The plant is herbaceous, reaches a height of 60 to 150 cm, and has a long taproot. Many branches arise from axillary buds on the main stem. The large hairy leaves are simple with three, five, or seven lobes. At the base of each leaf petiole are two buds; one is vegetative and gives rise to an axillary shoot, and the other is a flower bud. The flowers are complete and have three leaflike bracts. The corolla is made up of five sepals, and the five petals range from white to purple. The fruit is the enlarged ovary, consisting of three to five carpels, each usually contains nine seeds. The fruit is a capsule, or boll (Fig. 22-10).

The cotton fibers, or lint, are elongated single cells growing from the seed coat. They start to elongate as the flower opens and continue to grow for about 3 weeks. For the next 3 weeks, the fibers thicken by additional deposits in the cell walls. The time from flowering to opening of the bolls is about 6 to 7 weeks, depending on growing conditions.

On reaching full ripeness, the boll splits along the sutures where the carpels meet and bursts open, displaying the cotton, now expanded into a white fluffy mass.

Production of cotton is in areas with a minimum frost-free season of 200 days and warm temperatures. The cardinal temperatures for germination and growth are minimum 16°C, optimum 34°C, and maximum 39°C. Since cotton requires a great deal of sunshine, production is usually confined to areas where less than 50 percent of the days are cloudy. Cotton is sensitive to the amount of light but not responsive to photoperiod. Water requirements are high, especially during flowering; minimum annual rainfall or irrigation needs are 50 cm, but yields are much higher with 150 cm.

In the eighteenth and nineteenth centuries, cotton was concentrated in the southeastern United States, where slave labor was plentiful. Cotton was grown as a monoculture and the land was abandoned after becoming so poor that cotton was unprofitable. Production spread into the Mississippi Valley and then further west to escape the boll weevil. Cotton is a major crop in Arizona and south central California, where the hot climate is ideal when irrigation is practiced.

In early days, cotton was harvested by hand, initially by slave labor. The worker carried a cotton sack 3 to 4 m long and in a good day could harvest

FIGURE 22-10
Open boll of cotton. (*U.S. Department of Agriculture.*)

from 70 to 100 kg. Still hand-harvested in many parts of the world, cotton in the United States is over 99 percent mechanically harvested. The evolution of the harvester was accelerated by the shortage of labor during World War II.

Modern machinery harvests one to three rows at a time. Rotating spindles pull the fibers and attached seeds from open bolls, leaving unopened ones for a second harvest. A two-row spindle harvester can harvest at the rate of 1000 kg h^{-1} whereas a worker can average about 10 kg h^{-1}.

Mechanization of the harvest required modification and standardization of various cultural practices. Both yield and ease of harvest are affected by plant population. Freedom from weeds is a necessity, not only to maximize yields but because grasses picked up by the harvester are difficult to remove from the lint. The plants are often chemically defoliated before harvest to improve efficiency and reduce contamination (see Fig. 4-18).

Flax (Linaceae)

Flax (*Linum usitatissimum*) is grown as an annual. The stem fibers are used to make linen cloth, cigarette

and book paper, and paper currency. Linseed oil, discussed earlier in this chapter, is the product of flaxseed.

Flax was used by the Egyptians to make cloth. Although linen use has declined as a result of competition from cotton, wool, and synthetics, for certain uses, it is still one of the finest kinds of cloth.

The major source of flax for fibers is the Soviet Union; lesser amounts are grown in France, Poland, Belgium, Rumania, and Czechoslovakia. North America no longer grows flax for fiber, but some is grown in Asia.

Linen is made from the long phloem fibers in the stem. The plant grows 30 to 120 cm tall, depending on cultivar, climate, and planting density; for fiber production, tall unbranched plants are preferred. Leaves are narrow and short and are alternate on the stem.

Flax, a cool-season crop, thrives at temperatures of 21 to 27°C. Cultivars grown for fiber require moist, cool weather during growth, but a dry fall is necessary for the plants to cure. It is a long-day plant and continues to flower until the occurrence of freezes in the fall.

Harvesting may be mechanized, but in many parts of the world it is still pulled by hand to obtain maximum fiber length. The stems may be left in the field or tied in bundles and submerged in water. To allow fiber removal, most of the other stem components must be eliminated by *retting*. In the field, dew collects at night and encourages decay of the softer tissues by action of microorganisms. A similar process occurs much faster with submersion. After retting, the stems are passed between rollers which break up the remaining tissues while leaving the fibers intact. Finally, the fibers are shaken to remove any woody stem pieces and are baled for shipment to spinning mills. After further cleaning and combing, the individual fibers are spun into thread. Whereas most linen is woven on large looms, in Ireland and certain other countries traditional hand weaving is still practiced.

Minor Fiber Crops

Jute (Tiliaceae) On a world scale, fiber production from jute (*Corchorus* sp.) is second only to cotton. Although fiber quality is low, jute is widely used in making rope, burlap, backing for carpets and linoleum, and packing in electrical cables.

Jute is a herbaceous annual grown in warm areas with plentiful rainfall. From seeding to maturity takes about 5 months. At maturity, jute ranges in height from 2 to 4.5 m. The fibers occur in bundles just beneath the bark; individual fibers are only 2 to 3 mm long.

At harvest, the stems are cut at ground level and allowed to dry until the leaves abscise. The stems are tied into small bundles and submerged in ditches or pools for 10 to 30 days of retting. After microorganisms have broken the soft cementing substances down, the stems are beaten with paddles. The exposed fibers are pulled from the stems, dried, graded, and baled for shipment to a weaving center.

Most of the world's jute is grown in India and Pakistan, where labor is plentiful and cheap and production and harvest methods go back for centuries.

Hemp (Cannabaceae) Hemp (*Cannabis sativa*), which produces fibers used for the manufacture of twine, rope, and in the Far East for cloth, is now better known as the source of marijuana. World production of hemp fibers is only about half that of flax. Major producers are the Soviet Union, India, and China. Growth, harvest, and retting resemble those for flax.

Kenaf (Malvaceae) Kenaf (*Hibiscus cannabinus*) is used to a minor degree as a fiber crop. The fibers are coarser and less flexible than those of jute but stronger. Major uses are for twine, rope, and fishing nets.

Kenaf produces an unbranched stem which ranges from 1.5 to 4.5 m. From planting to harvest, kenaf requires 3 to 4 months with plenty of rain or irrigation. Production, harvest, and retting are similar to those for other stem-fiber crops.

Ramie (Urticaceae) Ramie (*Boehmeria nivea*), or chinagrass, produces fibers that are long strands (30 cm) in the slender shoots. Since the plant is perennial, double cropping of the land is impossible. Ramie is often less profitable than jute or kenaf even though three harvests can be made per year.

Two major problems have prevented large-scale use of ramie. Although the cells constituting the fibers are long and the fibers very strong, the fibers are difficult to separate from the surrounding gummy tissues and are too smooth for use with conventional spinning machines.

Major producing countries are in the warm, humid regions and include China, Indonesia, and Japan. Attempts to introduce ramie into the United States have met with little success.

CHAPTER TWENTY-THREE

FORAGES AND TURF

For several reasons, *forages*, i.e., plants grown to be fed in their entirety to livestock, could be considered the most important crops in the United States. More land is devoted to their culture than to all other crops combined; about 400 million hectares in the United States are used as *rangeland* or permanent *pasture*. Although the forages grown on such land are usually consumed on the farm and do not reach the marketplace, their real-dollar value exceeds that of any other crop. The true value of the forage crops lies in their support of the human diet through meat and milk from forage-fed animals (Fig. 23-1). This nation's whole agricultural economy is critically dependent upon a strong forage-livestock production system.

The value of forage crops to people is directly linked to the ability certain domesticated animals have to digest and utilize forages. The cellulose that makes up a large portion of forages cannot be digested by human beings, dogs, chickens, or pigs, whose stomachs lack the necessary enzymes. For us, cellulose is *fiber*. The multiple stomachs of *ruminants*, on the other hand, contain enzymes, produced by bacteria living there, that can break cellulose down into simple sugars. Ruminants like cows, sheep, goats, buffalo, and camels thrive on grasses and other cellulose-rich forages and can provide sustenance for nonruminants, especially human beings.

Their value in feeding food-providing animals causes forages to be widely planted and maintained

as crops, but establishing and maintaining grasses and other low-growing plants is desirable in many other situations. The low, matted growth makes many plants desirable as *ground covers*, vegetative shields against erosion by wind and water. Roadsides, drainage ditches, and deforested hilly areas are often planted and managed to maintain a protective grassy cover.

Aesthetic and recreational considerations are also responsible for the growth of a few million hectares of unnatural but desirable grassland. Home lawns, golf courses, athletic fields, cemeteries, and the like are maintained as *turf*. The most intensively managed land on the earth's surface is that portion devoted to lawns; no other fraction of land receives as much labor, water (both irrigation and sweat), fertilizer, and pesticides as the American home owner's yard.

This chapter discusses range, pasture, ground covers, and turf and how to manage them for optimum use, whether the use be feed, erosion control, or bragging rights to the prettiest lawn in the neighborhood.

FORAGES

Fresh Forage

Forages are, by definition, plants used essentially in their entirety as feed for animals, but the condition or form of the forage when fed varies tremendously.

When forage is *grazed*, animals feed directly on fresh plants in the pasture or on the range. This is an efficient method, since the animals harvest their own feed, saving the time and expense of mechanical harvests. Most rangeland and many millions of hectares of pasture or meadowland are grazed. In favorable situations, livestock can graze year-round with no other feed needed.

Forages can also be fed fresh after being harvested mechanically. When freshly harvested forage is brought to animals in confinement, it is called *greenchop* or a *soiling crop*. Such crops provide forage to the animals in its freshest, most nutritious form—still alive and full of all its valuable vitamins and other dietary components. Confined animals generally gain weight faster than those on pastures or range because they spend less energy acquiring food. Green chop forage has the disadvantages of requiring mechanical or human harvest, and when it is being transported, 80 percent or more of the weight is simply water.

A major problem in fresh-fed forage production is matching supply with demand. In temperate regions, growth of forage species is seasonal, but cattle, sheep, and other ruminants must be fed year-round. In some cases, supply and demand can be kept in balance by timely adjustments of herd size; animals are purchased when forage production is on the upswing and then sold at the end of the grazing season. Only as many head are kept as can be supported by grazing the current forage production.

Forage Preservation

Another solution to the problem of the uneven production of forages is to harvest and preserve them for use during the off-season. Three different methods of preservation are generally used, one of drying and the other two essentially processes of pickling. It should be stressed that forages, like human food, are most nutritious when fresh; their preservation is only partial at best, and improper management during preservation can result in total loss of the plants' nutritional value.

Hay The most common method of preserving a forage is to dry it and store it as *hay*. In haymaking, plant tops are cut, dried in the field, compacted, and stored. Sunny, warm weather is essential for drying or *curing* hay. (The expression "make hay while the sun shines" dates back at least to 1546.) Haymaking is the oldest method of forage preservation and has

FIGURE 23-1
Beef cattle feeding on good forage can produce useful products on land otherwise too hilly or rocky for cropping.

changed very little since it was described by the Romans in the first century.

During curing, the moisture content of the plant material drops from 80 percent or more to 25 percent or less. Rapid drying is needed because the longer the material lies exposed to the elements the poorer the quality of the hay. Thorough drying is necessary to prevent microbial activity, but forages such as the legumes must not be overdried; when too dry the plants become brittle and the leaves, which contain most of the vitamins and minerals, *shatter* or fall off as the material is picked up.

In the United States most hay is stored in bales, the shape and size of which have recently been changing. When labor was relatively cheap, bales light enough (10 to 40 kg) to toss around by hand were more common. They were moved to a barn for dry storage and used as needed. High-quality, high-value hay may still be handled in this way, but balers or stackers that produce round bales (Fig. 23-2) or loaf-shaped stacks of several hundred kilograms are becoming more common. The hay is often left in the field, because it cannot be moved readily. The outer portion of field-stored hay loses quality as it takes up moisture, but the bulk of the forage remains nutritious and can be fed as needed in the winter.

Silage and Haylage Instead of drying forages to preserve them, many growers store and feed moist material. Many of the same forages that are dried to make hay can also be preserved in a high-moisture form but obviously under different conditions. The

FIGURE 23-2
Large, round bales hold 10 to 20 times as much hay as the traditional rectangular bales.

normally wet-stored forages are protected against spoilage by a procedure similar to pickling.

Silage and *haylage* are distinguished from each other by the amount of moisture they contain. Silage frequently contains 60 to 75 percent water while haylage generally has a moisture level between 35 and 50 percent. The two are produced and handled in much the same way except that silage is made from wetter materials. The plants used include the whole range of forages to be described shortly.

Silage and haylage are made by harvesting the aboveground portion of plants after they have matured to a suitable moisture level or have been allowed to wilt briefly after cutting. The material is usually chopped into small pieces and packed into some type of container that can be made airtight, commonly the familiar tower *silo*. Silos are designed to receive large quantities of the chopped forages and maintain them in an *oxygen-free* environment. Horizontal trenches or bunker silos are also used, and some growers are using plastic sheeting made into gigantic (30 to 40 m) sausagelike casings into which the chopped material is packed (Fig. 23-3).

Important biochemical and microbiological processes occur after the material goes into the silo. When the oxygen has been used up by respiring tissues inside a properly designed airtight silo, naturally occurring organisms begin to carry on other critical processes. Sugar in the plant material is converted into lactic acid and the pH of the mass drops. If the silage or haylage has been properly handled and the silo is airtight, lactic acid concentrations may reach 3 to 5 percent. The low pH thus produced inhibits further bacterial activity and preserves the silage or haylage. The ensiling process is complete in about 2 or 3 weeks, and the product is

FIGURE 23-3
A low-cost, disposable silage system using a plastic film casing into which properly wilted forages are packed for ensiling. (*From D. A. Miller, Forage Crops, McGraw-Hill, New York, 1984.*)

then ready for feeding. As long as oxygen is excluded, the ensilage can be held for years.

The feed value of ensilage is less than that of fresh forage. There is nothing magical about the silo; definite losses of nutrients occur during the ensiling process, but silage or haylage is generally a better-quality feed than the same material made into hay. The major advantages of ensilage lie in its storability and nutritional value. (Of course, dry grains such as corn or barley store more easily than silage.) The so-called *high-energy silages* are made from grain-bearing plants, especially corn, and contain more of the energy-rich carbohydrates. The disadvantages or limitations of ensilage as a feed include its bulkiness and rapid deterioration once removed from the silo.

Dangers

Some plants that provide excellent forage for one animal may be unhealthy or even fatal for another. For example, subterranean clover causes reproductive failures in sheep. Flatpea, an erosion-control species and occasional forage for cattle, has caused fatalities in sheep feeding on its foliage, and its seeds are toxic to other animals as well. A number of forage-related disorders take a high toll of livestock each year as a result of poor forage management, poor animal management, or just poor luck.

Bloat One of the most serious problems as measured by economic losses and the discomfort caused in animals is *bloat*. It occurs when the gasses normally produced in the rumen of cattle become trapped in a foam and cannot be eructated, or belched out. The rumen swells up (bloats) and puts pressure on the other internal organs. Without treatment, a painful death is imminent. Bloat losses in the United States alone exceed $100 million yearly and 0.5 percent of the cattle population. New Zealand has had bloat death rates of 1 to 2 percent annually in its dairy cows.

Bloat occurs primarily in cattle grazing on legume-rich pastures. Fresh material (grazed or green-chopped) is more likely to cause bloat than hay, and feeding exclusively on legumes also increases the likelihood of bloat. This suggests as preventive measures grazing animals on pastures with a suitable grass-legume mixture or feeding grass hay before or during pasturing on legumes. Another approach has been to administer antifoaming agents to reduce or eliminate the buildup of bloat-causing foam. Breeding legumes with less of the foam-causing component or with some built-in antifoaming agent may also be successful. Until we have such a solution, bloat will continue to be a deadly problem.

Grass Tetany *Grass tetany* is another serious malady experienced especially by animals grazing on fresh material, in this case, grasses, although occasional difficulty arises with haylage or grass silage. The problem tends to be sporadic, but it can be devastating, losses in affected herds running as high as 20 percent or more. The total number of animals affected by grass tetany may equal that affected by bloat, but fatalities are fewer. The visible early symptoms in cattle and sheep include nervousness, muscle twitching, and a staggering walk. In later stages, the animals fall over and become spastic. The telltale internal symptom is low levels of magnesium in the blood, hence the technical name *hypomagnesemia*, which means "low magnesium."

The condition seems to be related to an insufficiency of magnesium in the diet. Grasses are relatively low in magnesium compared with legumes and other forbs and tend to be especially so during cool, wet periods. Since cool, wet soils seem to inhibit the uptake of magnesium by grasses, it is not surprising that grass tetany is much more common in the spring and fall or in areas with cool, rainy summers. The Netherlands has a high incidence of grass tetany, presumably because the climate and soils are so likely to produce low-magnesium grasses.

The prevention or treatment of grass tetany involves getting magnesium onto the feed or into the animal. Fertilization with magnesium or spraying magnesium solutions onto pastures or into silos can often eliminate the problem before it starts.

Toxic Forages Under certain conditions, a number of compounds that occur in plants may accumulate to such a level that the plants become toxic to animals feeding on them. Under conditions of high nitrogen fertilization and often in conjunction with drought or a series of cloudy days, *nitrate* may accumulate in plants to the point of being harmful or even fatal to animals feeding on them. Certain grasses seem especially prone to nitrate accumulation, e.g., orchardgrass and tall fescue. Care must be taken, and laboratory testing for nitrate content may be advisable in some situations.

Prussic acid, or cyanide, highly poisonous to all oxygen-requiring organisms, may be produced in lethal concentrations in a number of different species. Cyanide and chemicals that form cyanide in the rumen can become dangerously high in the sorghums and their close relative sudangrass. Some types of white clover have also caused cyanide poisoning in New Zealand. Stress conditions, especially drought and frost, combined with high nitrogen fertilization favor formation of the cyanide. Grazing of these species should be avoided at such times.

Biology

Forages are such a large and diverse group of plants that few generalizations about their biology can be made and exceptions to those generalizations will be many. Nevertheless, the discussion of important forage species will be simplified by first categorizing them taxonomically, morphologically, and ecologically.

Taxonomy From a taxonomic standpoint, most of our important forages belong either to the grass or the legume family, the same two families that provide most human food. The forage grasses are more numerous in total species and generally more productive than the forage legumes; but some of the best forages are legumes. Alfalfa, a legume, has been called the "queen of the forages," and the clovers, also legumes, are of tremendous forage value.

Some important forages are neither grasses nor legumes. Sunflower, turnip, mustard, and other nongrass, nonlegume plants are valuable forage crops in some places. On the rangeland and prairies of the Great Plains and intermountain areas, the grasses are by far the predominant native species; all other forage plants (including the legumes) are called *forbs*. The designation is not taxonomic but a convenient, loose term like "vegetable."

Morphology A major morphological characteristic shared by almost every forage species is the scarcity of woody tissues; i.e., the plants are *herbaceous*, with the following advantages:

Palatability—The tendency for livestock to prefer a forage is naturally greater when the plant is easy to chew.
Digestibility—The tendency for forage matter to be digested and absorbed as it passes through an animal is greater in herbaceous plants than more woody materials. Ruminants can digest the cellulose in woody tissues, but the lignin that also occurs in wood interferes with cellulose digestion.
Nutritiousness—This naturally goes hand-in-hand with greater digestibility, but it also refers to availability of key dietary components such as proteins, vitamins, and minerals.

Although forage grasses do not actually become woody, they can produce stems relatively low in palatability and digestibility. The stems of grasses are of distinctly lower feed value than the leaves. The *straw* that remains after small grains are harvested is essentially all stem. Livestock do not feed on straw as readily or gain weight as rapidly as with good hay. When a stand of grass is left too long before grazing or cutting for hay, the crop becomes stemmy. Earlier harvest produces a crop with less lignified material and greater palatability and digestibility.

While overmature grasses may pose a quality problem, overmature forage legumes are even more likely to be of poor forage value. Most legumes and other forbs have an active vascular cambium and can make a fairly large amount of woody tissue. Although alfalfa is rightfully considered herbaceous, the older portions of the stems can easily get as thick as a pencil and almost as tough. This obviously is a consideration in managing any such species for forage (Fig. 23-4).

Ecology The period of greatest growth for a forage plant varies sharply from species to species; no forage grows at a constant rate throughout the season. For example, *cool-season species* grow well in cool and moderately warm weather but essentially cease growing in the middle of a hot summer. The cool-season forages are C_3 plants with the photorespiratory pathway. Hot conditions accelerate photorespiration and depress net photosynthesis to the point where C_3 plants may not produce enough photosynthates to make new stems and leaves. During spring and fall, cooler conditions provide for rapid photosynthetic rates and maximum growth rates.

Some forages, including the C_4 species, are much better suited metabolically to hotter, drier weather. Thanks to the more efficient photosynthetic pathway and absence of the heat-accelerated, photosynthate-robbing photorespiratory pathway, C_4 plants grow

FIGURE 23-4
As forages become too mature (foreground), they are more stemmy, with less and less nutritious leaf material.

more rapidly in hot weather. Although they and other *warm-season species* grow rapidly in warmer weather, their activity is often drastically reduced in the cooler spring and fall. In areas subject to long, hot summers, warm-season species are obviously the preferred forages. In more temperate regions, cool-season species will be the most desirable.

The unevenness of forage production during the growing season poses the greatest problem for producers living in a climatic *transition zone*. This uneven band of territory stretches across the midline of the United States and is found in disjunct areas elsewhere. It is the region where summer conditions often inhibit growth of cool-season species but winter temperatures frequently kill the not too hardy warm-season species. There are some solutions to this problem, or at least some good compromises, but each must be applied to the particular situation. Some of those cultural approaches to forage management in transition zones will be mentioned below. Forage breeders are also making progress in improving the cold hardiness of some warm-season species.

Another major climatic factor besides temperature affecting forage growth and survival is precipitation or moisture availability. Some forage species are able to thrive in regions where rainfall is too limited to support the growth of other species, while some can tolerate soil moisture levels that would "drown" others. These differences in moisture requirements are the result of a unique combination of physiological and morphological features within each species.

The characteristics that increase drought resistance are quite diverse. Forages with deeper-penetrating root systems are usually less susceptible to drought. The special photosynthetic machinery of C_4 forages permits them to photosynthesize even when moisture is so limiting that the stomata close. Under such wilting conditions, many C_3 plants essentially stop photosynthesizing. Certain forage plants can become partially dormant when water is less available and resume growth when rainfall returns. Some grass plants have specialized cells in their leaves that cause them to curl up when they wilt, reducing further moisture loss. Species with fewer stomata per unit of leaf area tend to lose water via transpiration less rapidly. These and other genetically determined physiological and morphological features are adaptations to locations with limited rainfall or soils with little ability to hold the water that falls. Obviously, a different set of adaptive characteristics permits other forages to tolerate or even to require high moisture levels.

Whatever the causes of their differences, the forage species differ tremendously in the climatic conditions to which they are adapted. The grower must take into consideration the inherent suitability of a particular species for the particular situation. Kentucky bluegrass makes an excellent forage in temperate regions with moderate rainfall but would be a poor choice in an area of long, hot summers and low rainfall. Conversely, little bluestem, which has great merit as a forage on the semiarid, hot south central plains, cannot compete with Kentucky bluegrass in the humid, cooler northeast.

Of course, other environmental factors also enter into the choice of forage species. Such soil conditions as fertility, pH, depth, texture, pans, and shallow water tables affect the growth of different species differently. For example, alfalfa will do poorly in a soil with low pH, while alsike clover grows fairly well under such conditions. Nevertheless the major environmental factors are still temperature and water. Because they are so critical, in the following discussion we group the important United States forages into just four basic ecological situations or categories: warm and dry, cool and dry, warm and wet, and cool and wet.

Adaptation and Introduction

In the United States, the vegetation of western rangeland is usually a mixture of native species, but some ranges have been improved by the introduction of forage species from elsewhere. The *introduced spe-*

cies are typically plants that have been used as forages for centuries in their land of origin. Any forage crop that has been highly productive in a warm, semiarid region is a likely candidate for introduction onto our southwestern rangelands. Similarly, a species that is adapted to the Russian steppe and productive under the cool, dry conditions existing there would probably be well suited to the northern Great Plains or the intermountain region and perhaps even more productive there.

While natural grasslands are common to most of the western United States, they are not common in the east. Pastures are essentially unnatural ecosystems in the humid eastern United States, so that it is not surprisingly that few native species from eastern North America are good forages. Most valuable pasture forages found in that area have been introduced.

FIGURE 23-5
Big bluestem. (*United States Department of Agriculture.*)

FORAGE SPECIES

For Warm, Dry Areas

The forages adapted to warm, arid locations would be the species of choice in the central plains and southward. These plants can withstand drought and remain productive in the scorching summers common to that portion of the United States. Most of the species now cultivated are native to the Great Plains, but some have been introduced.

Grasses The bluestems include two important native grass species, big bluestem (*Andropogon gerardi*) and little bluestem (*A. scoparius*) (Fig. 23-5). These two species once covered huge areas of the Great Plains before the land was disturbed by overgrazing and plowing. The bluestems are naturally well adapted to the region, little bluestem being able to tolerate somewhat warmer and drier conditions. The bluestems are being replanted in many places. They can provide nutritious pasturage and hay and are relatively productive in a region where rainfall seriously limits the growth of most species.

The grama grasses are native forage species that fed buffalo, deer, and antelope for centuries before cattle and sheep appeared on the plains. Two species, sideoats grama (*Bouteloua curtipendula*) and bluegrama (*B. gracilis*), are important range species that usually grow in association with other grasses. Bluegrama is lower-growing and more drought-tolerant

than sideoats grama. Its shortness makes it unsuitable for hay, but it is excellent for grazing and is extremely valuable in holding soil. In many places where farmers have plowed stands of bluegrama or ranchers have permitted it to be overgrazed, virtual deserts have appeared.

Buffalograss (*Buchloe dactyloides*) can remain green and productive in hot, semiarid conditions that shrivel up most other grasses. Its dried leaves also provide nutritious forage in the winter. It is too low to be a hay species, but it forms a dense soil-protecting sod that withstands close grazing.

Switchgrass (*Panicum virgatum*) is another naturally occurring, drought-tolerant, warm-season forage that has been brought under cultivation because of its productivity and easy establishment. It yields good quantities of hay and summer pasture and provides good erosion control. All these native range species are available as improved cultivars with greater disease resistance, better forage quality, and wider adaptability.

Legumes A number of native forbs are of minor importance on ranges, but few have ever been cultivated. Plant breeders have worked primarily on grasses, the predominant species on the range. Alfalfa is grown widely in the southern plains and the arid southwest but only with irrigation, since alfalfa is productive only with more rainfall than is normally available there.

Biennial sweetclovers (*Melilotus alba* and *M. officinalis*) are common in the central plains and upper Mississippi valley, where they are valuable in improving soils. They provide nitrogen, create air and water channels into and through the subsoil, and increase the soil organic matter as they decompose. They produce good hay if managed properly, provide excellent pasture in some regions, and are one of the most valuable sources of nectar for honey.

A native of temperate Eurasia, sweetclover does not tolerate the acid soils common in the eastern United States. Largely for this reason, it is essentially restricted in the United States to the regions of higher soil pH, and when grown in the east must be adequately limed. Of the two species, *M. alba*, biennial white sweetclover (Fig. 23-6), is usually preferred, because it is less likely to lodge and produces more total growth.

For Cool, Dry Areas

Many different species of grasses and forbs are adapted to the cool, semihumid to arid conditions of the northern Great Plains and intermountain region. Only a few are sufficiently productive, palatable, and nutritious to be of great value as forages. The most important include a few domesticated native species and a few introductions.

Grasses The wheatgrasses, members of the genus *Agropyron*, include many cool-season species, some of which are highly valued forages and erosion-

FIGURE 23-6
White sweetclover, here being visited by a bee, is an excellent source of nectar for honey.

preventing ground covers in regions of low rainfall. Two or three native species and one or two introductions from Eurasia now are the principal vegetation on millions of hectares of rangeland in the central and northern Great Plains and the northern intermountain regions.

Western wheatgrass (*A. smithii*) (Fig. 23-7) occurs naturally across vast portions of the northern plains and northwest. Its distribution extends southward into the areas commonly occupied by gramas and buffalograss. The association with the two warm-season species makes an especially desirable range, because the wheatgrass provides early cool-season growth and the gramas and buffalograss become productive as wheatgrass goes into its midseason slump. In more northern locations and especially on higher-pH soils, western wheatgrass may form pure stands and provide good pasturage, satisfactory hay yields, and excellent erosion control.

Slender wheatgrass (*A. trachycaulum*) was the first native American grass to be domesticated. Seeds of an especially productive native type were distributed in Canada and throughout the western United States at the end of the nineteenth century. It is a valuable rangeland grass but somewhat less drought-resistant than western wheatgrass or crested wheatgrass.

Crested wheatgrass (*A. desertorum*) originated on the cold, dry plains of Siberia and eastern Russia. It is particularly well suited to the coldest regions of the Great Plains and the western mountains. A deep root system and the ability to become semidormant in hot, dry weather permit crested wheatgrass to persist in rather inhospitable environments. The early spring growth of this species makes it especially valuable in mixtures with other grasses that may produce more heavily later.

Intermediate wheatgrass (*A. intermedium*), another introduction from Eurasia, is less drought-tolerant and winter-hardy than western or crested wheatgrass, but it is highly productive as a pasture and hay species in the eastern plains, where rainfall patterns and winters are more favorable.

Smooth bromegrass (*Bromus inermis*), also introduced from Eurasia, is a valuable hay and pasture species all across the northern tier of states from Maine to Washington. Although it is a cool-season species, it does not fit neatly into either the dry or wet category. It is most productive in regions of higher rainfall but still able to survive drought and temperature extremes. Smooth brome produces large

FIGURE 23-7
Western wheatgrass, here being grown for seed production. (*United States Department of Agriculture.*)

amounts of nutritious, leafy material for grazing and hay early in the spring and again in the fall. It makes an excellent grass for growing in association with the all-purpose alfalfa.

Two species of wildrye are of some value as cool-season forages in the drier areas of the plains and western mountains. Russian wildrye (*Elymus junceus*) and Canada wildrye (*E. canadensis*) provide forage later into the summer than most other cool-season grasses. The introduced Russian wildrye is considered one of the best cool-season pasture species to supplement warm-season grasses.

For Warmer, Wet Areas

The natural climax vegetation for most of the southeastern states was primarily forest before man began clearing larger and larger areas. Few native southeastern grasses meet the forage criteria of adaptation, productivity, palatability, and nutritiousness. The

millions of hectares of pastures and meadows in the eastern United States today support their millions of cattle and sheep largely with introduced legumes and grasses domesticated in Europe or Asia hundreds of years ago.

Grasses The humid southeast, with its hot, sometimes droughty summers, is hospitable to a limited number of forage grass species. The most prominent, bermudagrass (*Cynodon dactylon*), has been a favorite of southern cattlemen since it was introduced in colonial times. The origin of bermudagrass is uncertain, but it is undoubtedly from the tropics, perhaps Africa or India. It is a warm-season species, adapted to, and productive in, hot weather. It grows well across the southern tier of states from Florida to southern California wherever rainfall is adequate or irrigation is available. The northern limit of its productive range extends across Kentucky and Oklahoma. The cold winters of northern states kill even the more winter-hardy cultivars. In its usual range, the tops of bermudagrass die back to soil level after the first frost and make no more growth until the ground warms up the following spring.

'Common' bermudagrass, the type first widely distributed, is a low-growing, aggressively spreading plant that provides good pasturage but may become a serious weed pest in row crops. The popular name *wiregrass* describes the persistent rhizomes and stolons by which bermudagrass spreads.

The most important cultivar of bermudagrass, 'Coastal,' a much taller plant than the common variety, can be used for pasture, green chop, hay, or haylage. Whether grazed or harvested and preserved, the plants must be cut closely and frequently (every 3 to 6 weeks) during the most active growing season. Otherwise the growth becomes stemmy and much less palatable and nutritious. Properly managed bermudagrass can provide enough highly nutritious, young forage to maintain good milk production in dairy cows and rapid weight gain in beef animals.

Bermudagrass is often grown in association with various legumes, especially white clover (*Trifolium repens*) and crimson clover (*T. incarnatum*). White clover is a low-growing perennial adapted to both warmer and cooler regions of moderate rainfall. Crimsom clover, an annual, is quite sensitive to hot weather. It is grown with bermudagrass as a winter annual, planted in the fall and disappearing by midsummer.

Other forage grasses for the southeast include dallisgrass (*Paspalum dilatatum*), bahiagrass (*P. notatum*), and johnsongrass (*Sorghum halpense*). Johnsongrass is native to North Africa or southern Asia, but the other two come from Central or South America. The two New World species are similar in their region of adaptation (the cotton belt) and use (pasture and occasionally hay). Since johnsongrass has become a noxious weed pest in row crops it is not commonly planted now. It is so vigorous and so difficult to eradicate, however, that many cattle raisers have learned to use it to advantage, for it provides large amounts of high-quality pasture and hay if properly managed.

Sudangrass (*Sorghum sudanense*), a close relative of johnsongrass, behaves like an annual and is less of a threat as a weed. It is a tall, fast-growing species native to northern Africa. It provides abundant amounts of nutritious forage for hay and haylage if properly managed. As a C_4 species, its maximum productivity occurs during the middle of the summer season.

Italian or annual ryegrass (*Lolium multiflorum*) fits into a rather special situation in the southeast. When bermudagrass and the other warm-season species die back to soil level in the fall, the pasture becomes nonproductive. But winters in much of the southeast are generally warm enough to provide for the growth of cool-season species. Annual ryegrass, a European introduction, can be planted into a dormant bermudagrass sod and provide nutritious, fresh forage all winter long. The ryegrass tends to die out as the warm weather returns, but bermudagrass will have resumed growing by that time anyway. Perennial ryegrass (*L. perenne*), from the Mediterranean area, is sometimes used in the same way to provide quickly developing, temporary pasturage. Since neither of the ryegrasses is very drought-tolerant or winter-hardy, they find little use outside of the southeast and the west coast; but, in those locations, they can be a valuable short-term forage.

Legumes Of the forage legumes adapted to southeastern conditions, the lespedezas, a few clovers, and (to some degree) alfalfa are the most valuable. We have already mentioned crimson clover and the special role it plays as a winter annual and the value of the low-growing white clover in a pasture. Neither species is very productive in hot weather, however. The most important warm-season forage legume species are three members of the genus *Lespedeza*.

All three of the cultivated lespedezas, two annuals and one perennial, are introductions from the Orient. Striate, or common, lespedeza (*L. striata*) and Korean lespedeza (*L. stipulacea*) are small-stemmed, low-growing, leafy annuals that provide nutritious hay and midsummer pasturage. They appear naturally in most southeastern pastures now, because they were once so widely sown and because of their tendency to reseed themselves each year. Sericea (*L. cuneata*), a tall-growing, potentially stemmy perennial (Fig. 23-8), has some value as a forage but is primarily used in erosion control and soil building. Sericea can be established on badly depleted or infertile soils such as cuts and fills along roads and disturbed soils around mines and other excavations. It is a major component on roadsides and medians throughout the southeast.

FIGURE 23-8
Sericea lespedeza tends to grow in dense erect stands.
(*Courtesy of D. D. Wolf.*)

Kudzu (*Pueraria lobata*), a vigorous, viny legume native to Japan and China, has a history similar to that of johnsongrass. Introduced shortly after the Civil War as a "dream plant" that could provide effortless tons of protein-rich forage for a region badly in need of it, kudzu proved so vigorous and fast-growing in the southeast that it often produces nightmarish tangles of growth climbing over everything it touches, including 20-m trees, telephone poles, and farm buildings (Fig. 23-9). Its growth exploits resemble science fiction stories of city-eating plants. While kudzu is out of control in many areas of the southeast, it can be controlled and produce good-quality pasturage and hay as well as stop erosion.

For Cool, Wet Areas
The humid, moderately cool summers and very cold winters of the northeastern states and eastern corn belt represent another geographical area with special forage requirements. Again, there are few native species of any forage value, but several forages used for centuries in central and northern Europe are equally well adapted to the conditions in this important dairy and beef- and sheep-producing area.

Grasses Kentucky bluegrass (*Poa pratensis*) is to the northeastern states what bermudagrass is to the southern. It is one of the few agronomically important species that appear to be North American natives. Its arrival on this continent probably predated human beings by a few million years, but some botanists argue that it actually originated in Eurasia. Regardless of its exact origin, it has long been indigenous in North America and is well adapted here. It is a cool-season grass that tends to go dormant in hot weather but flourish in spring and fall. It produces highly digestible and nutritious pasturage, but it is too short to be a good hay species.

Perhaps up to 90 percent of the Kentucky bluegrass pastures in America are volunteer stands, where the highly adapted grass has simply invaded and spread after land was taken out of cultivation. Management to maintain or improve such stands requires an understanding of the morphological and physiological characteristics of bluegrass and its desirable associates. The goal of a weed-free sod of high productivity can be achieved by liming, fertilizing, grazing and/or cutting to maintain both the grass and some desirable legumes.

When overgrazed or cut too frequently and closely (or both), bluegrass roots and rhizomes develop poorly and the sod may become weedy and unproductive. When undergrazed or not mowed, lower-growing legume species may be shaded out, so that weedy herbs and shrubs can invade the sod. The bluegrass itself can tolerate short periods of over- or undergrazing. Like its common associate, white clover, it is especially tolerant of very close grazing by sheep.

Tall fescue (*Festuca arundinacea*), a vigorous, sod-forming grass, is probably the most widely adapted, agronomically important grass species. This European introduction is both an excellent forage and a

FIGURE 23-9
Kudzu produces vines that can cover the ground or climb trees and other objects.

highly desirable turf in areas of heavy use, such as pathways or recreational areas, because of its ability to withstand considerable traffic.

As a forage crop, tall fescue has much to recommend it. In the area for which it is best adapted (the Ohio and Tennessee valleys and the middle Atlantic states), tall fescue makes excellent growth, is drought-resistant, and remains green through most of the winter. Growth, palatability, and nutritional value of this cool-season species increase in the fall as days grow cooler. These characteristics make tall fescue a valuable species for *stockpiling*, in which late-season growth is allowed to accumulate for grazing in the winter. Because it can even provide some growth on warmer winter days, it is especially attractive for stockpiling.

Despite its potentially high nutritional value, cattle feeding exclusively or largely on tall fescue sometimes do poorly or show adverse effects. *Fescue foot* is a disease of cattle that causes lameness and, in severe cases, gangrene and loss of the extremities of legs or the tail (Fig. 23-10). Fescue foot is not common, but it can be economically devastating when an entire herd is affected. The problem appears to stem from toxins produced by a fungus growing in the fescue. Fungus-free tall fescue does not cause the symptoms.

Timothy (*Phleum pratense*) (perhaps named for Timothy Hanson) was also known as herd grass (for John Herd) in colonial America. An introduction from Europe, it is especially well adapted to the cool, wet seasons of the northeast. It is shallow-rooted and quite drought-sensitive. Timothy hay is bright and clean and especially desirable for horses. The plant is highly palatable and nutritious as silage, hay, or pasturage. It suffers from droughty weather or close grazing and is susceptible to uprooting during grazing and frost heaving.

A distinctive and important morphological feature of timothy is that one of the lower internodes on its culms remains short and becomes swollen. This *haplocorm* serves as a site for storage of carbohydrate and produces new growth from its buds. The new shoots can develop roots and produce new plants vegetatively. An individual shoot is biennial but the haplocorm's ability to regenerate makes a stand of timothy perennial.

Orchardgrass (*Dactylis glomerata*) is another European cool-season, perennial forage grass. It is quite well adapted to northeastern conditions (so much so that it can become a weedy pest) and can provide nutritious forage for much of the growing season if properly managed. It is more tolerant of shade than many other forage grasses, even growing well under

FIGURE 23-10
Early symptoms of fescue foot include lameness and a coarse coat. (*From D. A. Miller, Forage Crops, McGraw-Hill, New York, 1984.*)

orchard trees, the characteristic from which its common name derives. It can be used for pasture, hay, greenchop, or haylage.

Orchardgrass is tall and leafy but does not produce a dense sod. It begins to grow early in the spring and will provide fair amounts of growth through the warm summer months if moisture is sufficient (being more drought-tolerant than Kentucky bluegrass or timothy) and if it is well fertilized. Its management, like that of other forage grasses should take into account its tendency to become stemmy with advancing maturity.

Legumes Alfalfa (*Medicago sativa*) is an introduction from the region of Iran, an area somewhat warmer and definitely drier than the northeast. The varieties suited to growth in the colder climates become dormant with the shorter days of fall. This ability was introduced genetically into the more productive Near East species from a similar plant native to Siberia.

Soil acidity can be a problem, as one would expect for a plant that developed in, and is best adapted to, the alkaline soils of the Near East. Soils that are neutral to slightly alkaline are generally most productive for alfalfa, which responds especially well to high fertility and pH adjustments with lime.

Alfalfa can produce large quantities of highly nutritious forage in the form of pasturage, hay, or haylage. Its high protein content makes it an especially valuable feed, and it is grown essentially throughout the United States. We include it here, among the humid, cool-season species, because it tends to be most productive under such conditions. It is also highly productive in the dry, hotter areas of the Great Plains but only with irrigation.

White clover (*Trifolium repens*) can be found from the Arctic Circle southward into temperate climates. It is sensitive to drought, heat, and competition from other plants, but given moisture, moderate summers, nonalkaline soil, and some sunshine it will grow and even thrive. White clover is a prostrate or creeping plant. Its stems, or stolons, grow on the soil and produce highly nutritious erect leaves and flowering stalks.

White clover is a complex of three different types distinguished primarily by size, minor morphological differences, and/or length of life. Two types, *common* and *ladino*, are domesticated and agricul-

turally important; the *wild* type is small, long-lived, indigenous, and not usually cultured, although it has some real value. (The Native American called it "white man's foot grass" because it seemed to appear spontaneously wherever the colonists went.) Ladino clover (Fig. 23-11) is a giant type that developed in the Po valley of northern Italy. It is the most valuable of the three, determined by hectarage devoted to its culture.

Red clover (*Trifolium pratense*), which probably originated on the northern side of the Mediterranean, is best adapted to areas where summers are moderate and moisture is plentiful. It is a winter-hardy, shade-tolerating perennial that is very sensitive to hot weather. It can be used only as a winter annual in the southeast. Its leafiness and edible stems make it a nutritious pasture and hay species and the most important legume hay crop in the northeastern United States.

Medium red clover produces enough growth for two or three cuttings of hay each year. *Mammoth red clover* is less productive, yielding only one cutting and moderate aftermath. Neither type is particularly long-lived; diseases and insects tend to reduce stands drastically after 2 or 3 years, making rotation to some other crop desirable.

Alsike clover (*Trifolium hybridum*) is believed to have originated in the Alsike area of Sweden. It tolerates wetter and more acidic soils and more severe winters than any other clover. On the other hand, while it is above average in tolerance to alkaline

FIGURE 23-11
Ladino clover can produce an abundance of low-growing, leafy material.

soils, alsike clover is sensitive to drought and heat. The stems of alsike are erect but rather thin and weak, so that the plants tend to become prostrate unless grown with another species, such as timothy, that gives them some support. This is especially important when the alsike is to be cut for hay.

Birdsfoot trefoil (*Lotus corniculatus*), a fairly recent introduction from the Old World, was not a cultivated crop even in its area of origin before this century. Until then, it was simply a naturally occurring plant that provided good forage when found in pastures.

Trefoil is adapted only to regions of cool summers and moderately good rainfall. It is like alsike clover in its tolerance of wet, shallow soils, but it tends to be more productive. Although the mature plants of trefoil are very leafy and well-branched, most are weak-stemmed. The species provides ample growth and is a valuable pasture, hay, and silage crop. It is being used widely in the northeastern United States to replace alsike and red clovers.

The natural or original distribution of crownvetch (*Coronilla varia*) appears to be centered around the Mediterranean. It is best adapted to fertile, well-drained slightly acidic to alkaline soils. It has been used primarily for ground cover and slope stabilization on highway banks, mine spoils, and other disturbed areas. Its value and advantages as forage have yet to be shown conclusively, but its worth in erosion control on road banks in the northeast is unequaled.

TURF

Turf, or sod, refers to the matted mass of roots, stems, leaves, and soil formed in the growth of certain grasses and other low-growing herbs. Strictly speaking, *turf* refers to the plant and soil mat in place, and *sod* refers to plugs, blocks, or strips of the turf that have been removed for planting elsewhere (Fig. 23-12). *Turfgrasses* are members of the grass family particularly well suited for forming a sod and producing a dense, usually close-cut ground cover. Turfgrass management is the professional domain of those who maintain golf courses, athletic fields, cemeteries, and other large expanses of closely cut grass for sporting or aesthetic purposes. Many homeowners spend a lot of time (and money) in culturing their own piece of turf as well.

FIGURE 23-12
Sod can be cut, rolled, and stacked for shipment.

Morphology

In most areas, the common turf species are grasses, but not just any grass will do. Many grasses are no more suitable for lawns than corn—a member of the grass family too—would be. The morphological, physiological, and ecological characteristics that make some species much more desirable for turf include the ability to spread and to tolerate frequent close cutting. The second is obvious, but the first may need some explanation.

Sod-forming grasses usually form a continuous, dense ground cover (the turf) by horizontal development of stems. Whether the horizontal branches develop aboveground as *stolons* or belowground as *rhizomes*, these leaf-bearing structures can spread and fill in gaps between plants. The ability to spread permits the plants to fill in bare spots left by disease, a dead weed, or mechanical damage. A plant with the spreading habit will be especially valuable, for example, on a football or soccer field, where heavy use may tear out patches of sod. The spreading habit can generate a dense stand without gaps that might give weeds a toehold.

Adaptation

Relatively few grass species are considered turfgrasses, but not all of them will suit a particular situation. As with the forages, the species chosen must be matched both to the environment and the functional, recreational, or ornamental use to which it will be put—home lawns, putting greens, athletic fields, grass walkways, roadway margins, or erosion control. Although in theory the environments are potentially just as diverse as for the forages (from hot and dry to cool and wet), in practice, moisture is seldom a limiting factor; in semiarid areas homeowners and greenskeepers usually provide water as needed to keep a green healthy turf (Fig. 23-13). In drier areas where supplemental water is unavailable or its use unfeasible, a few species can still fill the turfgrass role.

For Cool, Wet Areas

The cool-season grasses adapted to areas of moderate to high rainfall are the species of choice for the northeast and north central regions of the United States. As with the forages, the most valuable turf species for this area are introductions from Eurasia. In fact, several plants originally introduced as forages in the sixteenth and seventeenth centuries later became turf species. The culture of lawns and ornamental grass stands developed in the eighteenth and nineteenth centuries. Before the lawn mower was invented in 1830, sheep were the primary means of keeping grass short; it is not at all surprising then that forage species became turfgrasses. The same characteristics that made them tolerant of close grazing also made them suitable for frequent, low cutting. In fact, it has been suggested that millions

FIGURE 23-13
Irrigation of a golf course may provide needed moisture or cooling or both. (*Courtesy of J. R. Hall.*)

of years of grazing by the ancestors of our forage animals caused a natural selection for grasses with prostrate or underground stems and meristems at the ground or belowground—characteristics that make grazing and mowing feasible and the spreading habit possible.

Kentucky bluegrass (*Poa pratensis*), used worldwide as a cool-season turfgrass, is the most important turf species in the northern United States. Although this is the same Kentucky bluegrass that is such a valuable forage species, different cultivars are normally used in lawns. Kentucky bluegrass is also a favorite for parks, athletic fields, fairways, roadsides, and airfields. Its rhizomes make for good recovery when it is used on athletic fields and other heavy-traffic areas. It can produce an excellent quality turf when cut to 3 to 5 cm and otherwise properly maintained. Several other species of *Poa* have some use as turfgrasses, but none is as important as *P. pratensis*.

Three different European species of *Agrostis* find use as turfgrasses in certain situations. Creeping bentgrass (*A. palustris*) is a fine-leaved, stoloniferous plant that tolerates extremely close cutting to produce a dense, carpetlike turf. It is used most extensively on putting greens in the humid northern states, but new, heat-tolerant cultivars are being developed for golf courses in the upper south. The creeping or spreading habit and the ability to withstand cutting as low as 1 cm or less make creeping bentgrass almost ideal for forming smooth, living putting surfaces (Fig. 23-14)—"almost" because culture of creeping bentgrass is time-consuming and demanding. Careful attention to mowing height, fertility, irrigation, and disease control is essential. Professional greenskeepers live in fear of an outbreak of some disease right before a big tournament. Fungicides can control most of the diseases, but some can discolor or disfigure greens almost overnight.

Colonial bentgrass (*A. tenuis*) also lends itself to close mowing and produces fine, dense turf under such management. Since it lacks the vigorous stolons or rhizomes of creeping bentgrass, it does not recover quickly from damage. Its primary use is therefore not on putting greens but in high-quality, fine-textured lawn turfs. It is a rich-man's turfgrass because its culture is expensive and time-consuming. It must be cut, fertilized, and watered frequently to maintain a good quality turf. It is also susceptible to a number of unsightly but controllable diseases.

FIGURE 23-14
A fine putting surface created by close mowing of creeping bentgrass.

Nevertheless, where one can afford the time or money to provide the proper management, colonial bent makes a beautiful, luxurious turf.

Velvet bentgrass (*A. canina*) exceeds its two cousins in producing fine-textured lawns. It has the low-spreading habit, thanks to stolons, and narrow, almost needlelike leaves, a combination that can produce a close-cut turf unexcelled in beauty. It is rather restricted in its adaptation to certain soil types and its culture is tedious; but where it is adapted and its management is adequate, it rewards the homeowner or greenskeeper with an exceptional lawn or putting surface.

Fine fescue (*Festuca rubra*), or red fescue, is a fine-leaved, shade and wear-tolerant, slowly spreading plant that can form a dense, high-quality turf. It is one of the most widely used cool-season turf species, being more drought-tolerant than Kentucky bluegrass or the bentgrasses and requiring much less attention. It is especially useful in shady, dry areas, where it will make more growth than any of the other fine-leaved turfgrasses.

Tall fescue (*Festuca arundinacea*) is a coarser-leaved plant than fine fescue, with little tendency to spread. It is most valuable as a forage, but its ability to withstand intensive traffic makes it useful on athletic fields, airfields, and playgrounds. New cultivars have narrower leaves and produce a finer-textured turf gaining popularity in lawns (Fig. 23-15).

Perennial ryegrass (*Lolium perenne*) and Italian or annual ryegrass (*L. multiflorum*) find some use for turf situations. Each is used often in mixtures with other turfgrass seeds to provide rapid establishment. The rapid early growth of the ryegrasses may reduce erosion and make a more hospitable environment for slower-germinating species, and keep the homeowner happy that his or her efforts are succeeding. The slower-germinating species, such as Kentucky bluegrass and fine fescue, will ultimately be more valuable in the turf than the short-lived, nonspreading, drought-sensitive ryegrasses. But the ryegrasses do have their place. Inexpensive lawn seed mixtures found in most stores usually contain high percentages of the ryegrasses. In the south they are often over-seeded onto bermudagrass to provide green lawn during the winter months, when bermudagrass is dormant and its tops are an unattractive brown.

For Cool, Dry Areas

Two species of *Agropyron*, the wheatgrasses, have found some use as turf species in the cool, dry areas of the north central and northwestern states. Fairway wheatgrass (*A. cristatum*), an introduction from Siberia, is shorter, denser, and slower-growing than the wheat grasses used as forages in the semiarid northern plains. In its region of adaptation, it is a valuable species for planting on roadsides and, as its name suggests, golf course fairways.

Western wheatgrass (*A. smithii*), a North American native, is a little more tolerant of warmer and wetter conditions than fairway wheatgrass. It finds occasional use in lawns, roadsides, fairways, and general-purpose, low-maintenance turfs.

FIGURE 23-15
A lawn of tall fescue.

For Warm, Wet Areas

The grasses that produce dense, close-cut stands in the humid southeastern states are led by bermudagrass (*Cynodon dactylon*), the species so valuable as a forage in the same area. As usual with species that do double duty, some cultivars are more suited for turf situations, while others have been selected for their forage qualities. The improved turf-type bermudagrass cultivars form vigorous, dense, aggressively spreading stands that tolerate heavy wear and recover rapidly when damaged (Fig. 23-16). The narrower-leaved cultivars produce a fine-textured, high-quality, persistent turf in areas with hot, humid summers and mild winters. The discoloration of the leaves that follows the first frost of fall can be overcome either by using turfgrass colorants or by overseeding cool-season species such as the ryegrasses.

Two zoysiagrasses, members of the genus *Zoysia* and natives of southeastern Asia, are used extensively in the upper south as lawn species. Their rather slow regrowth makes them generally less desirable than bermudagrass for athletic fields, playgrounds, and other areas of heavy wear. Japanese lawngrass (*Z. japonica*) is faster-growing and more winter-hardy than manilagrass (*Z. matrella*); both turn brown soon

FIGURE 23-16
This football stadium has a bermudagrass turf; such a playing surface is particularly tolerant of heavy wear. (*Courtesy of J. R. Hall.*)

after the first freeze and remain dormant until warm weather returns. The prostrate stoloniferous and rhizomatous growth habit of the zoysias allows them to tolerate and even thrive on low mowing; 2 cm or less is quite acceptable.

Saintaugustinegrass (*Stenotaphrum secundatum*), a native of the West Indies, is a popular lawn species in the deep, humid south, especially in shady areas. It is not a fine-textured plant but requires relatively little maintenance except for frequent mowing, and it fills the important role of providing truly warm-season turf in shaded environments. It is a stoloniferous, aggressively spreading plant that looks best when mowed to a height of 4 to 7 cm.

Three or four other grasses see limited use as turf species in the southeast. Centipedegrass (*Eremochloa ophuroides*) is somewhat like Saintaugustinegrass in its area of adaptation and in its general culture, but it is less shade-tolerant. Carpetgrass (*Axonopus affinis*) is another lawn species for the deep south that can be used to produce a dense, rather coarse-textured turf. Bahiagrass (*Paspalum notatum*) is of significance primarily in low-maintenance turf situations in the deep south, where it is used extensively to provide a ground cover on roadsides, airfields, and other areas where low maintenance costs are more important than high-quality turf.

For Warm, Dry Areas

Three grass species are used to produce nonirrigated turfs in the hot, semiarid regions of the southwestern and south central states. In areas where enough water has been available, many homeowners have planted and irrigated drought-sensitive turfgrasses, but seriously declining water supplies are forcing a reversion to the three naturally adapted species, each of which is a native of the southern Great Plains or the southwest. Each is also a forage species whose tolerance to close grazing, drought, and high temperatures makes it equally at home on the range or in a turf situation. Buffalograss (*Buchloe dactyloides*), bluegrama (*Bouteloua gracilis*), and sideoats grama (*Bouteloua curtipendula*) are all forages that have simply been brought in from the range to produce lawns (primarily buffalograss) and turfs on roadsides and other low-maintenance areas. Their major value lies not in the high quality of turf that they produce (irrigated bermudagrass or zoysiagrass makes a much prettier lawn) but in their ability to produce a turf at all under dry conditions.

CHAPTER TWENTY-FOUR ▬▬▬▬

FLORICULTURAL AND NURSERY CROPS

Modern commercial floriculture includes the production and distribution of cut flowers, foliage and bedding plants, and container-grown flowering plants. Companies specializing in indoor landscaping, now a significant part of the floricultural industry, install and maintain plants in interior areas of offices, stores, and other public places. The nursery business involves the production and distribution of plant materials for outdoor use. Most commercial nurseries grow and sell ornamental trees, shrubs, and vines for use in the landscape; specialized nurseries produce and market fruit plants.

FLORICULTURE

Commercial floriculturists must find solutions to several problems. Higher costs of energy have stimulated research to find reliable and economical substitutes for petroleum fuel. The use of solar energy for heating greenhouses is being extensively studied, as are methods to reduce heat loss, particularly at night. Rapid transportation has led to increased competition from foreign countries, where labor costs and environment are more favorable for production. In recent years, sales of cut flowers have declined as customer preference changed; a combination of new crops and improved marketing strategies may help to reverse this trend.

The floricultural industry consists of wholesale florists who are production specialists, commission merchants or jobbers, and retail florists who provide a wide range of floricultural services and products to the public. Retail growers often cannot produce all the flowers and plants they need, so they purchase from wholesale growers or commission houses. Wholesale growers mass-produce merchandise for the market in a highly specialized business, producing one or at most a few crops in large greenhouse ranges. Because of the great cost of heating in the cold season, the trend is toward greenhouse production in the milder climates. The temperature advantage, however, must be balanced against the transportation costs to determine the most economical growing area. Greenhouse crops are produced in all areas of the country, but California, Pennsylvania, New York, Ohio, and Illinois account for one-half of the total United States production. Flowers can be produced in outdoor areas under cloth, plastic screen or lath houses, in frames, or in totally open areas, as they are in Florida, Georgia, and California. Large areas of foliage-plant production are centered in Florida, California, and Puerto Rico. The leading states for the production of bulbs are Washington, Michigan, Illinois, and Oregon; California leads the United States in production of seeds for bedding plants. Bedding-plant and pot-plant production is not geographically concentrated because of high shipping costs.

The retail florist industry mainly encompasses flower sales and floral design along with such customer services as delivery, gift wrapping, and a reminder service for special dates when flowers would be appropriate. With the expansion of the floricultural industry, came a need for the services of the middleman, or wholesale commission florist. Retail florists purchase almost all flowers, plants, and greenery used for arrangements from growers or wholesalers, making everything from wedding bouquets and corsages to wreaths and funeral arrangements. Although the retail florist industry consists mostly of small businesses, many belong to national organizations which coordinate advertising and marketing.

FLORICULTURAL CROPS

Floricultural crops are classified by use as cut flowers, potted plants, foliage plants, and bedding plants. The development of plants, especially the initiation of flowering, is determined by a number of factors, the most important usually being environmental cues like day length and temperature. Since flowers of many species are produced naturally only once a year at a particular time, the manipulation of environmental factors to produce a marketable pot plant or cut flower out of its normal season has become a highly sophisticated skill called *forcing*. The principal requirement for forcing is exacting environmental control, especially of light and temperature. Forcing may be accomplished by using naturally occurring climatic conditions in combination with artificially controlled ones. Sometimes the plants are controlled by artificial means, both environmental and chemical. The exact technique depends on the species being forced, how and when it is to be marketed, and climatic conditions in the forcing locality during the critical period.

The public interest in foliage plants and a greater demand for bedding plants has aided the recent growth of the floricultural industry. The demand for certain floricultural crops changes rapidly. All these factors make the industry highly specialized and exacting.

Cut Flowers
Some of the most beautiful floral material is specially adapted and grown to be sold after being cut from the main plant. California produces 30 percent of all cut flowers sold in the United States, and Florida produces another 10 percent. Colorado, which spe-

cializes in carnation production, produces 7 percent of all cut flowers. Increasing numbers of cut flowers are being imported.

Flower production in general and that of any species in particular is influenced by market demand, the general state of the economy, customs, fashions, availability, and to some extent the biology of the plant. Flower sales are irregular and seasonal, and since flowers are perishable, the florist bears a degree of risk. The popularity of chrysanthemum has greatly increased because they can be produced year-round. The decreased use of corsages has lessened the demand for camellia, gardenia, and orchid, while carnation, chrysanthemum, and rose are the largest sellers of all the cut flowers.

Carnation, chrysanthemum (Fig. 24-1), orchid, rose, and snapdragon are usually grown in a greenhouse because temperature control is essential. The photoperiod, which is critical in the flowering of some species, can be shortened by using black cloth or lengthened with electric lights, and irradiance can be decreased by using shade materials. Nutrients can be supplied in solution through irrigation water or as granular fertilizers.

Some cut flowers can be field-grown, e.g., gladiolus, Easter lily, and snapdragon. Outdoor production takes place throughout the United States. Chrysanthemum and aster are often grown in protective cloth houses to keep out leafhoppers and other insects which can transmit disease. These houses also lower light levels and leaf temperature.

FIGURE 24-1
Greenhouse production of chrysanthemums for corsages. (*Courtesy of J. W. Love.*)

The life or keeping quality of cut flowers is determined by the physiological factors affecting water uptake, transpiration, and respiration and by the genetic makeup of the species or cultivar. The water-holding capacity of the tissues, the temperature, relative humidity, air movement, and the absorptive area of the cut surface help determine the amount of water absorbed and transpired. Lower air temperature, high relative humidity, and little air movement will decrease transpirational losses. Since cut flowers can absorb water only through the stem, the absorption area is small compared with the transpiration area.

The rate of respiration is influenced particularly by availability of stored sugars and tissue temperature. Flowers with high sugar content usually last longer; respiration can be decreased by lowering the temperature to extend the life of cut flowers. Sugar in the solution supplies energy for the flowers, and the addition of special chemicals controls the bacterial and fungal growth which would otherwise result from the presence of sugar in the water. Other chemicals are being used to suppress blockage of xylem in the stems.

Chrysanthemum (Asteraceae) Among cut flowers, the chrysanthemum *(Chrysanthemum morifolium)* ranks first in importance. Native to China and Japan, where it is the national flower, it was first brought to America about 1795. It was grown outdoors until the middle of the nineteenth century, when it began to be produced as a greenhouse crop.

The amazing versatility of the chrysanthemum accounts for its importance. The many cultivars offer variation in size, flower type, color, and use. The standards (single large flowers) or pompoms (groups of small flowers) offer variety and contrast. Chrysanthemums are also noted for their keeping quality. From the grower's standpoint, the ability to produce the desired grades and types at any time during the year is an advantage.

A large part of the industry is devoted to developing new cultivars, producing disease-free stock, and selling rooted cuttings. Terminal stem cuttings are used to propagate chrysanthemum. Vegetative propagation maintains the genotype and the particular desirable phenotypic characteristics, which would be lost in sexual reproduction by seeds. The young plants are grown under long-day conditions until the stem approaches the desired length. Then, short

days are provided for flower initiation, which occurs when the night is 9.5 h or longer. Depending on the cultivar, 9 to 15 weeks is needed for full flower development. Temperatures at night should be 16 to 18°C during vegetative growth and 13 to 16°C after flower initiation. Plants can be grown as a single stem or pinched to allow two, three, or four stems per plant.

California and Florida lead the nation in chrysanthemum production under shade-cloth houses. Since high summer temperatures limit midsummer production, most of the production is between October and June. Plastic cloth is the standard shade material; although expensive, it can be used for several seasons. Production parallels that of greenhouse-grown flowers, but disease can be more of a problem.

Rose (Rosaceae) Roses are produced and purchased year-round, but the demand is greatest at certain holidays. The Chinese were the first to cultivate the rose; today, all commercial roses are hybrids *(Rosa hybrida)*. The prevalent form is the hybrid tea rose (Fig. 24-2), usually with one large terminal flower per stem, lateral buds being removed. Floribunda roses, also grown commercially, are multi-flowered and generally keep longer than hybrid teas.

Roses are planted as 2- to 3-year-old field-grown budded plants during the spring, and flowers appear 2 to 3 months later. The plants are pinched in early stages of development to encourage branching before the shoots are allowed to flower. This results in more branches and therefore more terminal flowers.

It is crucial that precise light, temperature, and moisture requirements be met. Most rose growers are specialists and do not grow other crops. Night temperatures should be maintained at 16°C for most cultivars. On sunny days, temperatures should range between 20 and 21°C. Humidity should be raised during the day by wetting the walks and watering mulched beds. Roses are generally cut back in summer, when flower production is very low. From late May through August it is desirable to shade the plants to lower temperature.

Dutch Iris (Iridaceae) Iris bulbs used for forcing are usually the Dutch iris *(Iris tingitana)*. Produced largely in Holland, Japan, and the Pacific Northwest, bulbs can be forced from December to June.

Each year the iris produces a new dormant bulb and the old one disintegrates. After the bulbs are

FIGURE 24-2
Hybrid tea roses. (*Courtesy of G. E. Rose.*)

in demand for corsages, wedding bouquets, and floral designs. Most of the orchid family's 20,000 recorded species are tropical or jungle exotics. Only a few genera are grown commercially, the most common being *Cattleya, Cymbidium,* and *Phalaenopsis* (Fig. 24-3). In a temperate climate, all must be grown in greenhouses. Orchids require 4 to 8 years from seed to produce flowers.

Some species respond to photoperiod, and photoperiodic control is generally used to regulate flowering to coincide with an optimum marketing period. For example, cattleya flowering can be delayed by long days and promoted by short days, while cymbidiums do not respond to photoperiod. Each cultivar has exacting light and temperature requirements, the range of which is quite wide. Cattleyas require about 16°C and between 15 and 17 klx, while cymbidiums require 10°C and 60 klx. All growing media must be well aerated, and uniform moisture must be maintained.

Snapdragon (Scrophulariaceae) Native to the Mediterranean area, snapdragon (*Antirrhinum majus*) is especially valuable as one of the few commercially produced flowers with a spike inflorescence (Fig. 24-4); nevertheless it presents difficulties since the inflorescences shatter easily. They also have a strong *geotropism* or response to the earth's gravitational pull. Spikes placed horizontally slowly

dug, they are cured at relatively high temperatures and then held at carefully regulated warm temperatures to control flower development. The bulbs must be vernalized at low temperatures to cause flower initiation. Temperature control during forcing is also critical to avoid excessive stem length and ensure proper flower development.

The iris can be cut and shipped when buds are only beginning to show color; putting the stems in water allows flower development to continue. When a flower is not initiated during vernalization, the disorder is called *blindness* and may result from premature digging or improper handling. Common causes of *bud blasting*, when flowers are initiated but do not develop, are night temperatures over 16°C, overcrowding, or insufficient light or water.

Orchid (Orchidaceae) Orchids, produced worldwide in many forms and colors, are long-lasting and

FIGURE 24-3
Phalaenopsis orchids are long-lasting cut flowers, excellent for corsages and contemporary floral designs.

FIGURE 24-4
Flower beds of snapdragons at different stages of development. (*Ball Seed Company.*)

start to bend upward.† Snapdragons must be carefully packaged, standing upright. Most snapdragons are grown in greenhouses, but outdoor production is possible in southern California.

Summer-grown plants flower within 7 weeks after planting from seed. Nutrient levels are usually kept lower than for other floricultural crops. Photoperiod, irradiance, and temperature influence snapdragon growth and flowering. Snapdragons are long-day plants. Different cultivars require various light conditions; some require high light and will flower only in summer, while others tolerate low light and can be grown in winter. Flowers are cut when expansion of the lower florets is complete and the top florets are still tight. They are graded on the basis of spike and stem length and placed in water immediately. After cutting, a 4°C air temperature is desirable.

Carnation (Caryophyllaceae) A native of southern Europe, the carnation (*Dianthus caryophyllus*) has been a major floral crop in the United States

since the late nineteenth century. It is produced chiefly in Colorado and California. Carnation prices have actually decreased slightly over the last 20 years while fixed production costs have doubled; research has led to production of more flowers per unit of bench area with an improvement in flower quality and grade. Large numbers of carnations are imported from Colombia.

Commercial production often requires a 1-year cycle: plants may be benched in the spring, flowers cut in the fall, and plants removed the following spring after the second cutting. High light and cool temperatures favor more rapid development of the plants.

Demand for shorter stems, a variety of colors, and multiple blooms should increase in the future. Since the carnation lends itself to mass market sales, production in the United States and importation from Latin America are likely to increase. Cost reductions should create a greater sales potential and more profit.

Narcissus (Amaryllidaceae) Narcissus (daffodil or jonquil) (*Narcissus* spp.) is found in many sizes,

---

†This is not related to phototropism, the tendency of all plants to grow toward the light.

shapes, and colors, but only the cultivars grown in the Pacific northwest or the Netherlands are used extensively for forcing. Bulbs shipped to growers for forcing are distributed as early as September. Temperatures around 9°C are required for forcing, but bulbs not ready for planting can be held at 13 to 16°C.

Cold and forcing temperatures influence stem length; temperatures of 10 to 13°C cause the longest stems. Forcing procedures must be carried out carefully for quality flower production. Losses from basal rot, a serious fungal problem during forcing, can be reduced by dusting the bulbs with fungicide before planting.

China Aster (Asteraceae)
Asters (*Callistephus chinensis*) are grown chiefly in cloth houses during the summer. Supplemental lighting is necessary during short days for proper stem elongation. Night temperatures should be kept at 10°C to produce high-quality flowers with strong stems. Greenhouses are generally used for aster production in the spring, when high light and low temperatures are prevalent.

As soon as they can be handled, the seedlings are benched. No pinching is required since they branch naturally; 8 to 10 flowers can be produced per plant.

Gladiolus (Iridaceae)
The gladiolus (*Gladiolus grandiflorus*), a native of South Africa, is usually field-grown. Florida produces the winter crop and has the greatest overall production. Other states grow the summer crop; North Carolina is second in production.

The gladiolus produces a *corm*, a thickened underground stem often incorrectly called a bulb. Gladiolas are day-neutral and have a rest period that can be broken by 4°C storage for 8 weeks before replanting. High light is necessary during forcing in order to produce tall, thick flower spikes.

Statice (Plumbaginaceae)
Although the genus *Statice* has been reclassified as *Limonium*, the common name statice remains popular. Dried statice is used commercially in winter bouquets and wreaths. It is also used fresh in arrangements as filler material. Statice is grown mainly in Florida and California. The hardy perennial types are grown throughout the United States and Europe.

Statice is sown in late July or early August in Florida. Flowers that develop in January or February are sold to tourists or sent to northern markets. Later sowing produces successive crops of flowers until spring. Some statice is sown in the midwest in late May to produce late-summer flowers.

L. latifolium, the most popular perennial type, is quite hardy. Blooms which are produced the second or succeeding years can be dried successfully, after which they are often dyed.

Gerbera (Asteraceae)
Gerberas (*Gerbera jamesonii*), or Transvaal daisies, are excellent cut flowers. They are usually field-grown in California and Florida in ground beds, although a superior crop is also grown in greenhouses. Gerberas are most productive with a night temperature of 16°C and they require more ventilation than most greenhouse crops.

The flowers are long-lasting and ship well if cut at the proper time, i.e., when the outer row of staminate flowers releases pollen. Premature cutting will cause wilting; they do not require refrigeration.

Other Cut Flowers
The decline of stocks (*Matthiola incana*) and sweet pea (*Lathyrus odoratus*) is due to the development of cut flowers that last longer. The labor required for their production has also contributed to their decline. Stocks are largely grown outdoors in California and Arizona. Some retail growers still produce sweet peas on a small scale during the Christmas and Valentine seasons. Freesia (*Freesia* spp.) is produced in southern California fields. These three crops along with peonies (*Paeonia* spp.), dahlias (*Dahlia* spp.), and baby's breath (*Gypsophila* spp.) are grown on a small scale for a limited market.

Potted Flowering Plants

Potted flowering plants are grown in containers to be sold while in bloom, often for a particular season or holiday, e.g., Easter lilies for Easter and poinsettias for Christmas. Nevertheless, potted plants like chrysanthemum, gloxinia, African violet, and azalea sell well year-round. Poinsettia, Christmas cactus, and Christmas begonia are usually produced for December sales. Hydrangea, calceolaria, and lily are usually forced for the spring holiday sales. Since these are all short-term crops, greenhouse space can be rotated, cutting costs by continuous use of the facilities.

Since pot plants are in containers that limit root growth, the volume of soil for root contact is small compared with that of plants growing in a bed or outdoors. The consequent restricted root growth makes water and nutrient supply critical. The frequent waterings necessary to provide consistent moisture tend to leach nutrients from the medium, making frequent applications of soluble fertilizer necessary.

The soil mix must be selected for optimum moisture and air relationships and contains a high quantity of organic matter, usually peat. Adjustments in the pH of the soil or in its phosphorus or calcium content, if necessary, are best made when the potting medium is mixed.

The fertilizer program for potted flowering plants should begin as soon as new root growth is visible, often within a week after potting. The most labor-efficient method is constant fertilization in the irrigation water, or *fertigation*. The soil must be uniformly moist so that the fertilizer salts will not harm the roots. Frequency of irrigation with or without nutrients is modified by the water-holding capacity of the soil, temperature, light, air movement, and the kind and size of plant. Small growers may find slow-release fertilizers useful.

Although most pot-plant production is for local markets, improvements in shipping have widened the range of the market. The cost of marketing the plants amounts to approximately 25 percent whether the plants are sold through a commission house or (more common) directly to the retailer.

Azalea (Ericaceae) All forcing azaleas are derived from either *Azalea indica* or *A. obtusum*. Azaleas bloom naturally in the spring but can be forced to flower year-round by manipulation of day length and temperature. Breeding programs have produced hybrids with distinctively different physiological characteristics. The length of cooling period required for flower-bud induction and development leads to the classification of cultivars as early, midseason, or late.

Commercial producers usually buy plants in the fall which have already had flower buds formed and precool them at 4 to 10°C before forcing. Plants grown in Oregon, Washington, and many southern areas do not require precooling since they have been precooled naturally. Cultivars of *A. obtusum* (Kurume) require about 4 weeks and those of *A. indica* (Indica) about 6 weeks of uniform cold storage to break the rest period before forcing.

Potted azaleas must be pinched to give the compact vegetative growth that will later yield many flowers.

After initial manual pruning to shape the plant, pruning to increase branching can be done with chemical pinching agents.

Easter Lily (Liliaceae) The pot lily *(Lilium longiflorum)* is generally the most profitable flowering pot plant on a unit area basis produced in the greenhouse, but since it is in demand only at Easter, marketing can be risky (Fig. 24-5). Bulb production takes place on the west coast, in the south, and in Japan. In a state of rest when dug in the fall, the bulbs are precooled for 5 weeks or more at temperatures 0.5 to 3°C, depending on the cultivar, to break the rest period and promote uniformity of size and synchronous flowering.

Poinsettia (Euphorbiaceae) When first introduced into the United States from Mexico by J. R. Poinsett, the poinsettia *(Euphorbia pulcherrima)* was grown as an exotic plant in conservatories and botanical gardens. Later it was grown as a cut flower. Today, it is chiefly raised in greenhouses for Christmas flowering. Short days are required to produce the colorful bracts which are often mistakenly called the flower. (The true flower of the poinsettia is inconspicuous.)

Propagation is by stem-tip cuttings taken from stock plants and rooted during the summer or early fall for Christmas sales. Plants are grown for the Christmas market as a single stem, producing one large flower; pinched to produce three or four "blooms"

FIGURE 24-5
Pot-plant production of Easter lilies. *(Courtesy of J. W. Love.)*

per plant; grouped to form a huge hanging basket with a multitude of blooms; or even as "trees."

Flowers of this short-day plant are initiated along with the colorful bracts when nights are 12 h or longer and the night temperature is between 13 and 18°C. These 12-h intervals must not be interrupted by any light source; even street lights or several passing automobiles can supply enough light to delay flowering. In warmer night temperatures, a night of more than 12 h is necessary for flower initiation. The higher the temperature the longer the night requirement. For bract development, 16°C is ideal. The plants become reproductive in early October and flower for Christmas. Extremely early Christmas sales can be made by black-cloth long-night treatments in mid-September.

New short-stemmed cultivars have reduced the need for growth regulators, used until recently to keep stems from excessive elongation.

Begonia (Begoniaceae) Begonias are ideal flowering pot plants because they can survive in the poor light of most homes. They may be tuberous *(Begonia tuberhybrida)* or fibrous-rooted *(B. semperflorens)*. The Christmas begonia *(B. socotrana)* is a semituberous type. The Rieger begonia, developed in Germany by Otto Rieger, is fibrous-rooted and a valuable addition to the flowering group.

The fibrous-rooted or ever-flowering wax begonias *(B. semperflorens)* are sold both as flowering pot plants and as spring bedding plants. These day-neutral plants are reproduced by seed or propagated readily from stem and leaf cuttings. Tuberous begonias, often used for summer hanging baskets, can be produced from seed in 6 months. Early December planting in high humidity with light shade beginning in February will result in June flowering.

The semituberous Christmas begonias are propagated by leaf cuttings. These cultivars flower in late fall and for Christmas. Available daylight is normally all that is necessary, since the short-day plants will naturally flower around Christmas, the time of greatest demand. Norwegian and Scandinavian cultivars exhibit sturdiness, flower retention, and longevity in the home, making them especially desirable.

Rieger begonias can be made to flower year-round (Fig. 24-6). Since they are slightly responsive to photoperiod, lights can be used in winter to stimulate growth; 3 weeks of black-cloth covering in the summer to induce flowering will shorten plant height. Crops can be produced in 10 to 16 weeks.

FIGURE 24-6
Rieger begonias can flower year-round. Note the tubular irrigation system for efficient and economical watering. (*Courtesy of J. W. Love.*)

Chrysanthemum (Asteraceae) Chrysanthemum, (*Chrysanthemum morifolium*) one of the major flowering pot plants, is desirable because of its lasting ability in the home, a wide variety of colors, and year-round availability (Fig. 24-7). Demand for these plants reaches a peak at Easter and Mother's Day, although their color range has made them popular Christmas plants in recent years.

Chrysanthemums require careful control of temperature and day length to produce the desired amount of vegetative growth and then flower at the appropriate time. Flowers are initiated with 9.5-h nights, but full flower development requires at least 10.5-h nights. From the start of short days, it takes 8 to 15 weeks for the plant to flower, depending on the cultivar. Sometimes classification is by the length of time from flower initiation until flowering, e.g., 11-week cultivars.

Standard cultivars and spray types are used for commercial production because of their large flowers; pompoms are undesirable for pot plants. When se-

FIGURE 24-7
Chrysanthemum pot plants grown and flowered in the greenhouse using an automatic black-shade-cloth treatment to provide short-light and long-dark periods. (*Courtesy of J. W. Love.*)

lecting a cultivar for pot production, one should look for short stems, large flower size, and good branching. When large-flowered cultivars are used, stems are disbudded to produce only one large flower per stem. Growth regulators are often used to help obtain the desired height, which can also be controlled somewhat by day length.

Narcissus (Amaryllidaceae) Narcissus, or daffodils (*Narcissus* spp.), can be used for pot-plant production from December through April. Upon receipt of the bulbs, the grower ventilates, inspects, and stores them, just like those used for cut-flower production. Forcing temperatures of 15 to 17°C should be used because lower temperatures tend to produce taller plants. The actual forcing procedure for container-grown narcissus is very precise.

Kalanchoe (Crassulaceae) The kalanchoe (*Kalanchoe blossfeldiana*), a long-lasting plant with showy heads of bright flowers (Fig. 24-8), was introduced by Robert Blossfeld, a German hybridizer. Grown worldwide, kalanchoe is popular at Valentine's Day but will flower any time with the prescribed day length.

Temperatures of 16°C and full sun are necessary to prevent the stems from becoming overly elongated.

Under natural light in a greenhouse, kalanchoe will flower in the spring, but using black cloth from 5 P.M. to 7 A.M. to provide short days allows the grower to produce flowering plants in proportion to demand. From the start of short days, it takes from 9 to 14 weeks for kalanchoe to flower.

African Violet (Gesneriaceae) The African violet (*Saintpaulia ionantha*), native to tropical Africa, is easy to grow as a houseplant because it is tolerant of the low light, warm temperature, and low humidity found in the average home. Optimum growth occurs at a relatively low light level (11 klx), making commercial production economical. Day-neutral, the violet flowers year-round, although plants grown in long days generally produce more flowers. Flowering African violets can be produced from seed in about

FIGURE 24-8
Kalanchoe plants are produced worldwide and flower year round if day length is controlled. (*Pan American Plant Company.*)

10 months; usually, however, they are propagated from leaf-petiole cuttings.

Leaf spotting is caused by watering with cold water; water at 21 to 24°C prevents this unsightly damage. Nights of 21 to 22°C should be maintained. Commercial growing of African violets has increased considerably due to demand by hobbyists and collectors and the wide variety of cultivars now available.

Calceolaria (Scrophulariaceae) Calceolaria *(Calceolaria crenatiflura)*, or pocketbook plant, has showy, pouch-shaped flowers in brilliant yellows, reds, and bronzes (Fig. 24-9). New cultivars with better keeping qualities will probably increase the demand for this old favorite. Most cultivars tend to wilt under normal home conditions unless careful attention is paid to watering.

Seeds are planted in late summer to produce plants for spring sales. Vegetative growth occurs at night temperatures of 16°C and 3 months of cooler temperatures is necessary for flowering. At present, calceolarias are not suitable for the south, but improved cultivars may eventually make production possible in warm areas.

Cyclamen (Primulaceae) The cyclamen *(Cyclamen persicum)* is a showy plant which blooms in midwinter in normal greenhouse culture. It is popular in Europe, where 40 percent of the sales of cyclamen are for cut flowers.

FIGURE 24-10
Gloxinias are grown in the greenhouse in spring and summer. *(Ball Seed Company.)*

Plants may be produced from seed or corms, though most are planted from seed in September or October and sold 13 to 15 months later. Newer cultivars can be produced in 8 to 9 months. Flowering date can be figured roughly as 5 months from potting. Forcing temperatures from 10 to 15°C are necessary for flowers to develop.

Gloxinia (Gesneriaceae) The gloxinia *(Sinningia speciosa)* is produced as a greenhouse potted plant in the spring and summer, although year-round demand is increasing. An elegant short-stemmed plant with large, velvety leaves and bell-shaped flowers, the gloxinia has an extremely long flowering period (Fig. 24-10). Plants are day-neutral and unresponsive to temperature for bud formation. Gloxinias can be produced quickly from the tuberous stem in 2 to 3 months; if large quantities are needed, production from seed, which takes 5 to 7 months, is more economical.

FIGURE 24-9
Calceolarias coming into bloom. Favorable growth at low temperatures make this a good crop to conserve energy in greenhouse production. *(Ball Seed Company.)*

Cineraria (Asteraceae) The cineraria *(Senecio cruentus)* is a colorful and inexpensive pot plant to produce (Fig. 24-11). Plants are grown from seed for sale January through April. Cinerarias are cool-temperature plants and will not set buds at night temperatures above 16°C. Temperatures of 13°C or lower for 3 to 4 weeks are required to set buds. Although they develop and mature regardless of temperatures, a night temperature of 7 to 10°C is ideal.

Cinerarias produce a large leaf area and should be watered frequently. In the intense spring sunlight plants may wilt even if the soil is moist and will need some protection from the sun; during short days, however, the shading should be removed. Since the plants grow rapidly, the cost of production is comparatively low. Several compact strains producing numerous flowers are available.

Other Flowering Pot Plants Rose *(Rosa spp.)*, primula *(Primula malacoides)*, geranium *(Pelargonium × hortorum)*, hydrangea *(Hydrangea macrophylla)*, Christmas cactus *(Zygocactus truncatus)*, and others are produced commercially as flowering pot plants but are of less economic importance. Pot roses are usually sold in the spring so that they can be replanted in the garden. Polyanthas, or baby ramblers, hybrid teas, hybrid perpetuals, and climbers are often sold as pot roses. The development of cultivars of primula that are easier and quicker to produce has increased interest in them for the mass market. Crossandra plants have a glossy, gardenialike foliage and flower spikes with overlapping light-orange florets. Geraniums are usually sold as bedding plants, but the demand for potted geraniums is increasing, especially in the south. Geraniums can be grown from seed or cuttings and flower year-

FIGURE 24-12
Geranium grown from seed. *(Pan-American Plant Company.)*

round (Fig. 24-12). Christmas cactus often survives unfavorable conditions. These hardy plants grow best when they are watered only when dry, and fertilized lightly in the early fall.

Foliage Plants

Foliage plants have been used indoors for many years as points of interest, screens, or elements of design. It has been documented that workers feel happier and healthier when green foliage plants occupy a part of their work space. There has been a steady increase in production of tropical foliage plants since 1945.

Our most useful foliage plants generally come from the tropics or other mild climates similar to those occurring in some areas of the south and California. Florida produces about 45 percent and California about 15 percent of the foliage plants. Many foliage plants are propagated and begun outdoors or in a protective structure in the south and shipped to northern greenhouses for further growth. In the

FIGURE 24-11
Cineraria plants ready for marketing. *(Ball Seed Company.)*

southern parts of Florida and California, *Sansevieria*, and palms are grown outdoors, but *Philodendron*, *Ficus*, *Dracaena*, *Dieffenbachia*, and *Pothos* must be grown in plastic screen houses (Fig. 24-13). In central Florida, foliage plants are grown in slat sheds, plastic screen houses, and greenhouses. The stock plants of *Dieffenbachia*, *Maranta*, *Schefflera*, *Peperomia*, *Philodendron*, and *Pothos* are grown in heated slat sheds or plastic houses and transferred to greenhouses for propagation and finishing (Fig. 24-14). California growers are the chief suppliers of ivies, cacti, ferns, and succulents.

Although about 10 klx is necessary for most foliage-plant production, exceptions include *Sansevieria* and *Peperomia* (20 klx) and *Aglaonema simplex* (only 7 klx). Most foliage plants need a daily minimum of 300 lx for 12 h in order to produce enough photosynthate for their successful maintenance. Artificial light may be necessary to meet this requirement.

Temperatures of 21 to 24°C and humidity levels of 75 to 80 percent are necessary for optimum growth of most indoor plants, but healthy plants can be maintained, i.e., surviving but with little to no new growth, for long periods in our too dry, overheated, or overcooled houses. Newer cultivars are better adapted to difficult conditions than their earlier counterparts; many do best with a certain amount of neglect. (Plants kept indoors usually require less water due to the lower light levels.) Many people tend to over water foliage plants.

FIGURE 24-14
Production of large foliage plants in 15- and 20-cm hanging baskets. (*Courtesy of C. E. Bell, Hines Nurseries.*)

As breeders develop more useful and widely adapted foliage plants, it is the grower's responsibility to help ensure the buyer's long-term success. A special holding area for plant acclimation is often installed in production greenhouses, where plants are slowly withdrawn from the highly growth-promoting light and moisture levels and eased into the hardship of the average indoor environment. Acclimated plants are much less likely to undergo severe shock when moved from greenhouse conditions.

Customer education is an important part of foliage-plant marketing. The happy buyer will be a repeat buyer. The foliage-plant producer or supplier can help purchasers achieve success by providing such educational devices as care labels, flyers, booklets, and direct information.

Bedding Plants

Bedding plants are used for flower gardens, window boxes, hanging baskets, and miniature gardens. Although some biennials and perennials are grown as bedding plants, annuals generally predominate and are replanted each spring when the danger of frost has passed. Important bedding plants include petunia, marigold, and impatiens. Commercial growers must plan carefully so that the plants are at the right stage of development and size for transplanting. Commercial growers generally propagate bedding plants by seed because it is the most economical method.

FIGURE 24-13
Production of *Pothos* in a fiberglass house where heating, cooling, fertilization, and watering are automated. (*Berisford Photography, Apopka, FL.*)

Seeds are usually sown in vermiculite, ground sphagnum moss, or peat-lite mix. Liquid fertilizer can be applied after germination. Seedlings are transplanted to a growing medium consisting of peat moss and perlite, vermiculite, or fine sand in a variety of containers (Fig. 24-15).

Most bedding plants require full sun. Crowded plants and those grown in low light have weak stems and flower more slowly. Artificial light may be used for growing young seedlings. Fluorescent light of 1 klx obtained by spacing lights at 1-m intervals 15 cm above the plants is effective. At least 16 h of light per day should be used until the plants are moved into full sunlight in the greenhouse (Fig. 24-16). Although warm temperatures, 21°C or higher, are required for germination, most bedding plants grow better under cool temperatures. Night temperatures of 18 to 21°C are needed for compact growth and stems of the desired diameter.

FIGURE 24-16
Greenhouse production of bedding plants. (*Courtesy of W. H. Carlson.*)

Careful attention must be paid to watering since bedding plants are grown in small containers under light and temperature conditions which quickly dry the soil. Overwatering must be avoided since it often leads to root rot. Growth retardants can help produce uniform, compact plants especially if environmental conditions cause etiolation.

NURSERY CULTURE

The American Association of Nurserymen defines the nursery industry as

> the production and/or distribution of plant materials, including trees, shrubs, vines, and other plants having a woody stem or stems, and all herbaceous annuals, biennials, or perennials generally used for outdoor planting by companies whose major activities are agricultural or horticultural.

A nursery is a place where trees, shrubs, vines, and other plants are grown and maintained (Fig. 24-17) until they are placed in a permanent planting. The nursery may propagate the plants from seed or cuttings, purchase rooted cuttings from a wholesale nursery and grow them to a salable size, or purchase plants of the desired size and retail them.

The first commercial nursery in America was established in 1730 in Flushing, New York. Early

FIGURE 24-15
Containers for production and marketing of bedding plants. (*Pan-American Plant Company.*)

FIGURE 24-17
Wholesale nursery producing a wide range of nursery crops. (*Monrovia Nursery Company.*)

nurseries specialized in growing fruit trees and other food plants. Beginning in 1750, the industry expanded westward from the Atlantic seaboard. Evergreens were first grown commercially in Illinois in 1844.

The industry has grown into a multi-billion-dollar business. Technology and increased interest in the environment and aesthetics of the landscape have changed production from chiefly fruit trees for home and farm orchards to shade trees, ground covers, and shrubs. The technology needed for nursery crop production has been provided largely through university, government, nursery associations, and industrial research programs. New and better cultivars of plants have been introduced. The methods and materials for controlling plant pests have improved. Labor-saving devices have been designed and incorporated into all phases of nursery operations from planting to shipping. Improved methods of transportation have made longer shipments less expensive, allowing firms to locate in the area with the best resources for production.

Garden centers became popular in the mid-1940s because they allow consumers to buy seeds, fertili-zers, hand tools, and plants at one location. Nurseries were only seasonal businesses at that time, but with improved technology and intense interest in gardening and landscaping, nurseries have become year-round operations. High-density living seems to make people realize how important plants are for maintaining a pleasant environment.

Nursery Crops

The several hundred species of plants produced in nurseries are classified into three broad groups based on growth characteristics and similarities in cultural requirements. The classifications are coniferous or needled evergreens, broadleaf evergreens, and deciduous plants. The species and cultivars within a group can be classified according to habit of growth, i.e., trees, shrubs, ground covers, and vines.

Needle Evergreens The conifers generally have thin, narrow foliage, the leaves being extremely slender in relation to their length. Coniferous evergreens include the pines (*Pinus*), the firs (*Abies*), and the spruces (*Picea*). Some coniferous evergreens have small, dense scalelike leaves which overlap on the stem, e.g., junipers *(Juniperus)* and false cypress *(Chamaecyparis)*.

Most coniferous evergreens are produced from seed collected from parent plants with the desirable characteristics. Many of these seeds will not germinate without stratification or scarification (or both) to break dormancy. Once seedlings have germinated and are established, they are either lined out (planted in a row) in the field or transplanted into containers.

The exception to seed propagation is the perpetuation of desirable sports, i.e., mutations in form (pendulous or prostrate as opposed to upright), in color (blue-green or yellow instead of green), or size (dwarf shrub instead of large tree). When a sport is worth perpetuating, cuttings are taken from the mutant branch or plant and the plant is propagated vegetatively. Most ornamental junipers are propagated by cuttings—over 1000 different species or cultivars to date.

Coniferous evergreens can be produced throughout the United States wherever temperature requirements for the particular crop are satisfied. There are coniferous evergreens for almost any location.

Broadleaf Evergreens These plants, which generally have a higher sales value than coniferous evergreens, have leaves varying in size and shape but always with considerable width and length. Variations in leaf size occur both in the general group and individual genera. Thus the broadleaf evergreens are a large and diverse collection of some of our most important landscape material. One can find plants to fit every color, textural, and size requirement.

Most broadleaf evergreens are produced from cuttings. Once rooted, the plants are lined out in the field or placed in containers. Most broadleaf evergreens, e.g., hollies and camellias, prefer slightly acid soil with a pH from 5.5 to 6.5. Daphne and cotoneaster prefer a more alkaline soil, and the heath family, which includes azalea, rhododendron, and heather, prefers a pH of 4.5 to 5.5.

Deciduous Plants These plants, produced for sale throughout the United States, are propagated by vegetative methods and from seed. Important examples are apple *(Malus)*, propagated by budding; grapes *(Vitis)*, propagated by dormant cuttings; and *Viburnum*, produced from seed. Each crop has specific cultural requirements for production.

Types of Nurseries

Wholesale The wholesale nursery, which produces plants in quantities sufficient for sale to retail outlets, is usually either a field operation or a container operation. Many nurseries propagate their plants and grow them to salable size. Some wholesalers specialize in producing liners† for sale to other wholesalers, who then grow them for the market.

Wholesale nurseries are usually located in rural areas where land prices and taxes are lower and where it is possible to hold some land for later expansion without incurring too much additional tax burden. Important criteria for locating a nursery include a plentiful source of clean water, a good labor supply, and proximity to a dependable means of transportation such as a highway system or airport.

Retail Since the retail nursery or garden center depends largely on homeowners for trade, it must be located close to a community or city, where land prices and taxes are usually higher. Like the wholesaler, the retail nursery should have sufficient space to expand as necessary.

The retail garden center sells plants grown by itself or bought from a wholesaler. Also available are fertilizer, seeds, tools, and similar supplies. Such a center often contains an office and sales building with showroom, a shade house or other areas where plants are displayed (Fig. 24-18), and possibly a small greenhouse. Ample parking is important, as customers will usually not stop if parking appears jumbled or confused. Garden centers may also offer landscape services, such as planning, installation, and maintenance.

Landscape The *landscape nursery* is normally located near a population center with a high percentage of homeowners, since these will be its major clientele. The nursery can be located on the outskirts as its clients are more likely to travel a short distance if necessary to obtain landscaping services. Most nurseries conduct business within an 8-km radius.

The landscape nursery derives the major portion of its income from landscape jobs, which include drawing plans, implementing or installing the de-

---

†Liners are rooted cuttings or seedlings which are essentially self-supporting when planted in the field or in containers.

FIGURE 24-18
A garden center for marketing nursery plants and landscape services.

signs, and maintaining the plantings. The landscape nursery may grow its own plants, but more often it buys from wholesale nurseries or even retail garden centers.

Mail-Order The *mail-order* nursery is a specialized retail nursery depending primarily on catalog sales. It is usually situated in the country, where land is relatively inexpensive, near a good source of water, labor, and transportation. Like the local retailer, it sells to individual customers, such as homeowners, and may also carry supplies.

Agency The agency nursery sells stock through agents or salespeople. Such nurseries, though comparatively few, are a highly specialized part of the industry.

PLANT PRODUCTION IN A NURSERY

Nursery stock is grown according to two basically different methods—in the field or in containers. Commercial nursery production of plants in containers on a large scale is only about 25 years old and is due to the need for mechanization resulting from higher cost of labor and resources.

Field Production
Field production involves planting liners in nursery rows, growing them to a salable size, and digging

and selling the mature plants. Once the only way of producing landscape plants, this method has been partly replaced by container production; nevertheless, some growers are returning to field production of large specimen landscape plants. Technology has also improved and streamlined the operation, making it much more efficient. Successful operations employ a combination of production methods.

Lining Out The plants, either as rooted cuttings or as seedlings, are first lined out in nursery rows by hand or transplanting machines. Liners may be left in one location until they reach the desired size if spacing permits, or sequentially shifted to beds with larger and larger spacing until they are of the desired size. In the field, the plants must be fertilized, pruned, and watered according to the species being grown and the environmental conditions.

Fertilizing Fertilizers are applied broadcast or as side dressings down the rows of the plants. A soil test should be conducted on all parts of the field to determine how much of what kinds of fertilizer to use. Environmental conditions, soil differences, and drainage can make the requirements in a field vary greatly.

Pruning The type of pruning and whether it is used at all depend on the use or purpose of the plant. The lower limbs are usually removed from shade trees, whereas shrubs are pruned to promote lateral growth. For high-quality shrubs and trees, pruning (Chap. 16) should begin at an early stage. Root pruning should also be practiced in a field nursery to force growth of a more compact, fibrous root system, which helps reduce transplanting shock. Root-pruned plants have a higher recovery and survival rate when transplanted than plants which have not been root-pruned. For plants with naturally coarse roots, such as pine, juniper, and Burford holly, a fibrous root system is encouraged by frequent transplanting, beginning when the plant is young, root pruning each time.

Large trees and shrubs are often root-pruned the year before they are to be transplanted giving the plant several months to develop new fibrous roots in the root ball before being moved. The root pruning can be done manually or mechanically by a root-pruning machine, which is a large, hydraulically controlled semicircular knife pulled through the soil with a tractor. The knife enters the ground on one side of the plant and moves under it, severing roots at a predetermined depth.

Weed Control Weed control is just as necessary in the field nursery as elsewhere and for the same reasons. Weeds not only compete with plants for minerals, water, and light but make removing the plants difficult. Herbicides for weed control are cheaper and more efficient than hand labor, but chemical control is complicated by the fact that many ornamentals react differently to the same herbicide. What works well with one plant species may be toxic to even closely related species. Also, a herbicide can be legally used only on the plants listed on the label of the herbicide.

Irrigation Irrigation may be used to supplement natural rainfall or it may be the main source of water, with rainfall supplementing irrigation. Application may be through various overhead methods, trickle, or in furrows (Chap. 9).

Plant Removal Removing the plants from the field before selling them may be done by hand or machine. Most deciduous evergreens are balled and burlapped (B and B); i.e., the root ball and its surrounding soil is wrapped and secured with burlap. Machine removal is becoming increasingly important as the lack and cost of skilled labor prohibit hand removal. Digging by machine is also much faster. One type of machine digger consists of four large triangular blades, which open, encircle the plant, and then close (Fig. 24-19). The blades are driven into the ground hydraulically one at a time until they meet under the plant.

The American Association of Nurserymen (AAN) has established standards for B and B plants specifying the size of the root ball for certain types of plants. For landscape jobs, these specifications should be followed carefully.

Many deciduous plants, fruit trees in particular, are dug while dormant by large machines moving down the row. They cut the roots at a certain depth and gently lift the plant, which must be in a relatively loose, sandy soil so that the roots can easily be cleaned. Bare-root plants are cheaper to dig as they require less time and of course are considerably cheaper to ship.

FIGURE 24-19
A hydraulically powered machine for digging and balling plants in the field.

Plant Storage Plants are stored before selling or planting by several methods. Plants may be *heeled in* by covering the root ball or roots with sawdust or planting in a sandy soil, so that plants can be removed any time the ground is not frozen. Storage houses provide the proper environmental conditions of -0.6 to $1.7°C$, relative humidity of 85 to 90 percent, and adequate ventilation. These temperatures not only minimize respiration rates but help woody plants satisfy their chilling requirement. High relative humidities minimize transpiration. The ventilation supplies sufficient oxygen and carries away the carbon dioxide and heat of respiration.

Container-Plant Production

Growing plants in containers in commercial nurseries began in the 1950s in California and spread gradually to other areas. Virtually every part of the country enjoys some local nursery products. Container growing, which has gained in importance in the last 20 years, consists of planting liners in containers then carefully watering, fertilizing, pruning, and weeding in order to produce healthy plants. Transplanting as needed into larger containers until the plant has reached a salable size is a critical step in container production.

The important advantages of using container-grown plants in the landscape are as follows:

1. The planting season is extended with less transplanting loss. This is of special importance in warmer areas of the country since both B and B and bare-root materials are highly susceptible to shock when transplanted in summer heat.

2. Container plants are easier and cheaper to harvest.

3. Transportation costs are less than for B and B.

4. There is greater production per hectare.

5. Even though container-grown plants require precise cultural attention, they are easier to adapt to new production programs than field-grown plants.

6. Quality of nursery stock has increased with the use of containers.

There are, however, several disadvantages:

1. The plants may become *pot-bound,* and such a plant usually will not grow vigorously when it is transplanted into a larger container or into the landscape.

2. Container-grown plants may prove difficult to adapt to the landscape site and soils after being grown in artificial media, although appropriate site preparation will usually take care of this problem.

3. Many landscape companies want to buy large trees and shrubs, but these large specimens must be grown in oversize containers, which are expensive and hard to handle. The 100-L containers do not fit into a standard operation and require a nursery specially equipped to handle them.

4. Container-grown plants are much more dependent upon proper and frequent irrigation than field-grown plants (Fig. 24-20). Careful inspection must be made to determine that all containers are being watered properly.

5. Repotting is still a disadvantage in container production. Much work is being done to make the process more economical, and potting machines have helped in larger businesses. Planting the liners in larger containers to avoid further repotting results in a lower-quality root system than when the plant is progressively potted up.

Containers The perfect allround container for plant growth does not exist. An ideal container should have a neat appearance and be rustproof, lightweight, and reusable with a relatively long life. It should be structurally strong enough to protect the root ball from physical damage and to withstand rough handling and should be insulated to protect the roots from heat and cold. It should not be adversely affected by herbicides, fertilizers, other chemicals, or weather conditions, nor should it be brittle or have a tendency to crack. A perfect container should be stackable for storage and available to growers at prices they can afford.

The first containers were used food cans. Although they rusted easily and required a large storage area, they were inexpensive, strong, and available. Next, a steel can with tapered sides was developed. It was strong enough to protect roots against crushing; tapered sides made it stackable and permitted easy removal of plants. The steel, however, readily conducted heat and thus offered the roots no protection against adverse weather conditions. Plastic containers, developed next, make stacking and plant removal easy. Since they conduct heat more slowly, they offer some protection against adverse weather conditions. Although they are thin and may allow root damage, especially in transit, they are relatively inexpensive and economical. Improved plastic containers, despite a greater initial investment, are more durable and easier to use.

Polyethylene bags are being tried in California. Choosing a soil mix that will prevent sagging, determining the proper amount of tamping, deciding on the number of holes to put in the bottom, and developing proper irrigation techniques are the major problems associated with the use of plastic bags. Consumer acceptance is also a problem.

For moderately large, fast-growing plants, treated bushel baskets have been a success. The continuous-stave type with no seam at the bottom is the only satisfactory kind. The chief advantage is that they

FIGURE 24-20
Production of nursery plants in metal containers.

1. It should be well-drained but moisture-retentive and must have adequate aeration.

2. The components should be easily obtainable, lightweight, easy to work with, and clean.

3. It must be adaptable to mechanical mixing.

4. It should be easily reproducible from one batch to the next.

5. It should withstand heat pasteurization without breaking down.

6. It should be retentive of nutrients.

Ideally, only one basic mix should be used throughout the nursery, slight modifications being made for species or cultivars with other requirements.

Since natural soils vary in texture, structure, and nutrients, any mix using soil would be nonuniform from batch to batch. The clay in soils also has a tendency to cake in a container. The most common components for soil mixes are sphagnum, or peat moss, sand, vermiculite, perlite, bark, and sawdust (Fig. 24-21). Quarry dust, rice hulls, or shavings may also be used, as well as many other locally available types of organic matter.

The English were the first to develop a standard mix for growing a wide variety of plants, but it used composted organic matter, which is not uniform and therefore not reliable for exact duplication of the mix. It also requires a great amount of storage space

are relatively inexpensive, but they deteriorate quickly and give little protection to the roots. Wooden baskets do not conduct heat readily and give the roots some protection from heat or cold, but they have little structural strength. Drainage is good, and the plants can be planted still in the basket, which soon rots when buried in the soil.

Soil Mixes Since plants in containers are not grown in field soil, a growing medium or soil mix must be supplied which has the following properties:

FIGURE 24-21
Soil-mixing operation at a large container production nursery. (*Monrovia Nursery Company.*)

in the nursery, and the composting action is relatively slow.

The University of California developed mixes that contain sphagnum moss as the organic matter and fine sand as the inorganic matter in varying proportions, depending on the use, the most common ratio being 3 parts of fine sand to 1 part sphagnum peat. These mixes are designed to have the advantages of clay-base soil with none of the disadvantages. Cornell University has also developed a soilless mix, which contains sphagnum moss and either perlite or vermiculite; it is called the *peat-lite* mix.

The mix can be prepared in several different ways, scaled to the size and scope of the nursery operation using it. The ingredients for large operations can be mixed mechanically by tractors with front-end loaders, or a converted cement mixer can be used. For large-scale operations, a specially designed soil mixer is available (Fig. 24-22).

Pine and hardwood bark, a by-product of the timber industry, is widely used by container nurseries, sometimes mixed with soil or sand but often used alone. Both its water-holding capacity and nutrient retention are excellent.

Fertilizing Essential elements can be applied either in granular form or in solution. When dry fertilizer is used, it can be applied on the surface of the soil or incorporated into the growing mix. One method of applying dry fertilizer is to use measuring spoons to apply the correct amount to each plant. A fertilizer dispenser enhances uniformity and speeds up the operation. Tablets of fertilizer can also be applied to the soil surface of the containers.

The trend in the nursery industry is toward long-lasting, or slow-release, fertilizer. These encapsulated pellets, or coated granules, are formulated to break down gradually. Each time the plant is watered, some fertilizer is released. This method saves labor, since the fertilizer lasts so long.

Potting and Bedding Out Potting consists of filling the containers with soil mix and planting the liners in them. Again, the potting operation must be tailored to the size and scope of the nursery using it. There are machines designed specifically for potting (Fig. 24-23), or a machine can be built to fit the needs of the nursery. Plants are also potted manually. After the containers are filled with soil mix, a hole is

FIGURE 24-22
A soil-mixing machine. (*Monrovia Nursery Company.*)

FIGURE 24-23
A potting machine. (*Monrovia Nursery Company.*)

prepared and the plant placed in it. The mix is watered and the container loaded to be taken to the growing beds (Fig. 24-24).

Pruning Pruning is used to produce compact plants of the proper shape. Root pruning is not necessary as the container acts to restrict the root's outward development; however, care must be taken to see that the plants do not become pot-bound. This is prevented by transplanting to larger containers as necessary. At that time, the root ball can be cut or cracked to encourage new root formation and to discourage pigtailing, or encircling growth of roots.

Irrigation Watering is absolutely essential with a container nursery, since the containers hold so little water. The source of water must be ample for the demands of the nursery. Water can be delivered to the plants by overhead risers or the tubular system, and either system can be operated manually or automatically through a system of clocks and solenoids. The tubular system consists of a series of thin flexible pipes which feed off a main pipe to each individual pot. Water goes only where it is needed and work can be going on in the beds during irri-

gation. Also, if fertilizer is injected into this irrigation system, there is less waste. Fungal disease is held to a minimum because the foliage of the plant is not wet. However, the cost may be prohibitive for the smaller containers. Setting up a tubular irrigation system is expensive and requires a high level of maintenance.

FIGURE 24-24
Bedding-out containers in the field. Note the gravel base for drainage of the beds.

FIGURE 24-25
View of lath shade house showing section of container production of *Camellia japonica.*
(*Monrovia Nursery Company.*)

Weed Control The smaller growing space in a container makes competition from weeds potentially even more severe than in the field. For this reason, weed control is essential in container nurseries. As a precaution, the beds where the plants will sit should be fumigated to kill weed seeds, soil insects, nematodes, and disease organisms and then covered with black plastic, which will act as a physical barrier to any germinating seeds that are not killed or that are carried in before the beds are covered. Herbicides should be used to control weeds in the containers themselves, using the same precautions as in the field. Herbicides are an inexpensive method of weed control, often amounting to only 5 to 10 percent of the cost of control by hand or machine. Generally, a combination of chemical and mechanical methods is used because labor costs have become prohibitive.

Winter Protection Winter, or cold, protection, which varies with the location of the nursery and the anticipated severity of the winters, is especially important for container-grown plants. Where the winters get cold and stay cold, as in the northern areas of the country, unheated plastic houses may be the best and safest form of protection. Such houses are expensive but may well be worth the price. In areas such as the southeast and California, where it is not consistently cold throughout the winter months, such houses may not be worth the expense. Lath or shade cloth may be sufficient to protect plants (Fig. 24-25). Containers may be moved close together, especially under a stand of pines, which act as a canopy and windbreak. After placing the containers close together the outer edges can be ridged with sawdust or other insulating material to provide additional protection.

CHAPTER TWENTY-FIVE

FORESTS AND FORESTRY

Forestry is the art and science of managing and using woodland. Although forests are the source of many services and products, timber production is the major source of income to finance forest management (Fig. 25-1). This complex subject encompasses both the biological aspects of tree production and socioeconomic considerations. It usually includes related areas of range management, recreation, watershed management, and wildlife biology as well as such business skills as marketing and accounting. The complexity of forest management is illustrated by the fact that forestry includes 20 or more fields of study.

There are hundreds of uses for wood, from whole-tree stems for poles to small pieces of hardwoods and fruitwoods such as cherry for fine furniture. The list of forest products includes building lumber, mine timbers, railroad ties, veneers, shingles and shakes for roofing and siding, wood containers, wood composition board, wood flour (a filler used in explosives), by-products such as shavings and sawdust, wood fuel, and such secondary products as pencils, packaging, and even toothpicks and clothespins.

Wood is also essential for the manufacture of other products. Much of the lower-grade softwood in the United States is grown as pulpwood for paper production. Some synthetic materials, e.g., rayon and acetate, are derived from wood cellulose, as are some explosives. Wood is carbonized or otherwise processed to yield charcoal. Finally, certain types of wood can be hydrolyzed to sugars and then fermented to alcohol, which is growing in importance to meet our energy needs. The demand for reconstituted and derived products (mostly paper products) has almost reached that of solid wood. The per capita consumption of paper products exceeds 272 kg annually, encompassing a vast array of products.

FORESTS AND FOREST BIOLOGY

Extent of Forests

Forests cover over 30 percent of the land surface of the United States (305 million out of 918.6 million hectares). The eastern part of the United States has extensive forests, mainly as a result of adequate rainfall. The Great Plains region is virtually treeless, and in the west there are extensive softwood forests consisting of pine, spruce, fir, hemlock, and the magnificent redwood. The eastern and western forests merge in northern Canada.

The United States has six major forest regions, three east and three west of the Great Plains. The three eastern forests include the northern, the central hardwood, and the southern region. The west includes the Rocky Mountain, the Pacific coast, and the Alaskan regions. In addition, there are subtropical to tropical forests in Hawaii and parts of Florida and Texas.

FIGURE 25-1
Timber production is the major source of income to finance forest management. A group of fine specimens of ponderosa pine (*Pinus ponderosa*) in Montana. (*U.S. Forest Service.*)

Forest Ownership

About one-third of the American forests are not used for commercial forest purposes. These lands are in parks, wilderness areas, or protected watersheds or are unsuitable for commercial production. Of the remaining 200 million hectares of forests, the federal government is the biggest single owner, holding about 55 million hectares, or about 28 percent of the total area. Almost 60 percent of the commercial forest area, some 120 million hectares, is owned by about 4 million private individuals. Only about 27 million hectares, or about 14 percent of the commercial forest, is owned by companies of the forest products industry.

Principles of Forestry

Since the primary task of the forester is centered on producing trees, understanding trees and their growth is essential. Although certain special terms may be

used for convenience, the principles underlying the growth, development, and reproduction of trees of the forest are the same as those for other crop species. Thus, forestry must be based on a sound understanding of botanical principles. One critical point to keep in mind is that trees are generally managed for wood production, i.e., stem tissue. Compared with most horticultural and agronomic crops, forests represent long-lived, perennial plants. (Orchards are somewhere between forests and most other crops.)

Classification

In United States forests, all trees are spermatophytes, or seed plants. They are classified formally as either gymnosperms (cone-bearing plants) or angiosperms (flowering plants). This formal classification is accepted scientifically, but others are routinely used by foresters. The most common general classification identifies a tree according to its leaf form as a hardwood or softwood tree. *Hardwoods* are broad-leaved, generally deciduous, flower-bearing trees (Fig. 25-2). *Softwoods* are needle-leaved, mostly evergreen, cone-bearing trees (Fig. 25-3). This broad classification is far from perfect: the southern long-needled pine is classified as a softwood species although its wood is harder than that of many of the hardwoods, and several deciduous broad-leaved spe-

FIGURE 25-2
Hardwood forest trees are broad-leaved, generally deciduous, and flower-bearing. (*U.S. Forest Service.*)

FIGURE 25-3
Softwood forest trees are needle-leaved, usually evergreen, and cone-bearing. Thinned six times over the past 30 years, these red pines have maintained a high level of production. (*U.S. Forest Service.*)

cies yield softwood, e.g., basswood, willow, and aspen. This leads to the confusing terms hard hardwoods and soft hardwoods. Some sense can be made from this when it is remembered that the terms were established early, when the demand was for the soft white pine and the very hard white oaks only. Later

use of numerous species of widely varying hardness or softness could not erase the old established usage.

Morphology and Anatomy
Although a forest may contain annual, biennial, and short-lived perennial plant species, trees are gener-

ally considered to be the major vegetation; they are long-lived perennials that usually produce a single central stem and attain a height of 6 m or more. Rarely does a forest tree reach maturity in less than 15 to 20 years; some trees may grow continuously for centuries though generally very slowly after the first hundred years. In considering the growth of trees, increases in both height and diameter, or girth, are important in determining the yield of lumber or other products. In terms of gross morphology, as a rule, trees growing in a typical forest environment are taller and have smaller root systems than those growing separately or under widely spaced, ornamental settings. The apical or top portion of the tree, the *crown*, is also smaller under crowded conditions. Note that the word crown has a different meaning in forestry than in horticulture and agronomy. In forestry, the crown is that portion of the tree which has branches (Fig. 25-4). Since shade decreases the ability of branches to survive, the trees in crowded forests tend to shed their lower branches and thus have less crown in proportion to clear stem.

Roots Roots serve the same general function as for other plants: anchorage, absorption of water and essential minerals, and storage of photosynthate. The storage function is less important in forest trees than for most perennial field crops since stem tissue takes over much of the storage function in trees.

Tree species differ markedly in the extent of their root systems, differences that may affect how the species are managed in a forest. For example, most oak, hickory, and walnut trees have deep and extensive taproot systems which provide extremely solid anchorage. Thus, these and other trees with similar root systems are not as prone to uprooting by severe winds. In harvesting a forest, wind damage is a minor concern with these trees. Spruces and balsam, on the other hand, have shallow root systems and can be uprooted even by moderate winds. In planning the harvest, such trees must be removed or allowed to remain in a pattern that will ensure adequate wind protection. Most forest trees fall somewhere between these extremes. Generally, the root system provides sound anchorage except in relatively extreme conditions. Of course, even deep-rooted species can be uprooted under severe conditions, such as prolonged heavy rains before or during high winds. Hurricane conditions can level extensive areas of forest. In addition, trees that normally produce taproots may develop shallow root systems because of poor soil conditions, e.g., only a thin layer of soil above the bedrock. In such cases trees may be not only stunted but also easily blown over.

Stems The trunk, or *bole*, of a tree is its central stem (see Fig. 25-4). In most species, mature trees have few if any branches on the lower portion of the bole near the soil surface because they are shaded out and die. Various types of damage, however, may cause basal branching or the development of suckers. This tendency is most obvious in the growth of numerous shoots from the stump of a cut tree. In terms of wood yield, the trunk is the most important part of the tree. Forestry and the wood sciences require a detailed understanding of the nature, growth, and development of the trunk. All trunks, regardless of species, length, and straightness, have similar basic anatomy. The nature of the tissues determines many of the uses for the wood.

The general anatomy of a trunk is easily seen in the cross section of a stump or freshly cut log. Generally, four or five distinct regions can readily be identified from the outside to the center of the trunk: the outer bark, the inner bark, the sapwood, the heartwood, and, on close examination, a tiny core of pith in the center. The new wood formed from the cambium is the light-colored *sapwood*. This xylem tissue transports water and minerals upward throughout the tree; the phloem is produced toward the outside of the cambium, as the inner bark, and functions in food transport. As the sapwood ages, it becomes darker and harder, due to deposits of various chemicals called *extractives*. It ultimately ceases to function in solute transport and becomes *heartwood*. The time required to darken, the extent of darkening and hardening associated with heartwood development, and the chemical composition and color of the extractives deposited all vary with the species.

Even casual observers are familiar with the annual rings clearly visible in the cross section of tree trunks. Actually, these rings have two parts: an inner layer of early wood formed in the spring, which is lighter in color and generally softer with larger thin-walled cells, and an outer layer of late wood, which is darker, with smaller thick-walled cells formed later in the summer. These growth rings provide a record of past growing conditions. Good conditions favor rapid, extensive growth, which results in wide bands; narrow bands suggest poor growth conditions such

FIGURE 25-4
The crown is the top portion of the tree that has not yet shed its branches and the trunk is the central stem. In terms of wood yield, the trunk is the most important part of the tree. Sugar pines (*Pinus lambertiana*). (*U.S. Forest Service.*)

as crowding or drought. Where trees are isolated with little or no competition, these rings represent a record of climatic conditions.

The cambium layer lies between the inner bark layer (the active phloem tissue) and the newest or outer sapwood (xylem). Recall that this meristematic tissue produces the vascular tissues, phloem on the outer side and xylem on the inner side. The cambium is not visible to the unaided eye, being only one cell layer thick; cells on either side begin to differentiate into either phloem or xylem.

Hardwood trees also have vascular rays extending laterally from the bark through the active conducting tissue of the xylem to the pith. These lateral transport tissues, which are quite obvious in oaks, birch, and sycamore trees, contribute to the splitting of logs as they dry. Some conifers, e.g., pine and spruce, have a system of ducts or channels containing resin, which appear as minute pores on freshly cut cross sections or as fine streaks on longitudinal surfaces.

Effective production management requires detailed understanding of how a crop is produced, whether it is a forest or any other crop. Because lumber is the principal product, the trunk growth will be considered in more detail although since every plant is an integrated whole, stem growth cannot truly be separated from root, shoot, and leaf growth. Trees grow in two directions—height and diameter. Increases in height are from apical meristematic tissue (primary growth). This produces primary xylem and phloem, which then elongate. Production and elongation of the primary cells is the only way length or height is increased. In contrast, recall that in members of the grass family, the pattern of elongation is intercalary, from the base of internodes. With trees, removal or death of any shoot apex will stop the elongation of the apex of that shoot and encourage the development of buds that occur lower on the stem. This leads to branching. Understanding these phenomena is of great importance in forest management, particularly timber management, where excessive branching is undesirable.

For lumber production, growth in diameter is more important than growth in height. Increases in diameter ultimately come from the production of secondary vascular tissue, secondary phloem and secondary xylem. Secondary xylem is more or less synonymous with wood (90 percent or more of a log is secondary xylem). The nature of this vascular development in a stem explains the appearance of annual rings.

The overall growth habit of a tree is important in determining the type of lumber it will yield and in its management in a forest setting. The growth habit depends largely on how the tree branches and subsequently on the growth, development, and spread of its crown. Trees with a central, major stem and much smaller lateral branches are described as having an *excurrent* branching habit; many common conifers, e.g., spruce, pine, fir, and cedar, display a typical excurrent branching pattern. Hardwoods (generally deciduous dicots) may have a *deliquescent* branching pattern, in which the trunk branches and rebranches repeatedly. In this case, it is difficult to identify the bole in the crown. Obviously, the branching habit determines the length of usable stem that can be cut from a given tree or type of tree. It also clearly affects the grade of lumber in terms of whether it is free of knots and straight-grained. A *knot* in a piece of wood is the cross section of a branch which has been engulfed by the trunk as it added secondary xylem.

The crown region is the foliar canopy of the tree, which includes limbs and branches from the bole, leaves, buds, flowers, and fruits. The structure of the crown determines how leaves are exposed to sunlight. This is the portion of the tree in which all photosynthesis occurs and from which most water is transpired. The size, shape, and density of the crown region also determine the potential competitive position a given plant may have for light and the density of shade it creates for nearby plants. Understanding these phenomena is important in managing a forest in terms of thinning and reforestation. Finally, characteristics of the crown are commonly used in identifying certain major forest species.

Trees grown in the open normally have a larger crown that starts close to the ground; the crown will tend to be symmetrical as well as wide and deep. Four common crown shapes under these conditions are conical, vase-shaped, rounded, and oval. In a forest setting, trees are taller and crowns smaller as a result of crowding or competition, and crown shapes are more irregular.

PHYSIOLOGY AND NUTRITION

Trees are among the largest living organisms; indeed, the giant redwoods of California *(Sequoiadendron giganteum)* are the largest living organisms known. In spite of their size, trees have essentially the same physiological and nutritional requirements as other crop plants. Although as far as is known, all forest trees follow the C_3 photosynthetic pathway, it is not unreasonable to suppose that some tropical tree species could follow the C_4 pathway, much as sugar cane and corn differ from most other grasses.

Since, to some extent, the amount of water transpired by a plant depends on the size of the plant, particularly the amount of leaf surface and depth of the root system, it is not surprising that forests transpire immense amounts of water. The very nature of a forest in its natural setting precludes detailed study of the amount of transpiration, but forests may exert a substantial influence on the depth of the water table in wet sites.

From a nutritional point of view, trees in the forest require the same minerals as cultivated crop species. As part of a relatively undisturbed ecosystem, however, forest trees are amazingly efficient in recycling nutrients. As leaves and branches fall to the forest floor, the nutrients are gradually released through decomposition and most are reabsorbed by the trees. This is quite different from crops, where much of the plant tissue is harvested and removed annually. When timber is cut, only the trunk is removed, and since xylem tissues contain relatively low concentrations of nutrients, the loss of nutrients to the forest is minimal. The undisturbed layer of plant residue on the soil surface, plus the canopy that reduces raindrop impact, effectively prevent the erosion of soil and nutrients common on cropland (Fig. 25-5). Mycorrhizae are also important to forest trees in increasing their nutrient uptake. Because of the efficiency of the forest's recycling system and the vast areas involved, little fertilization has been carried out. Now, however, as harvesting of smaller trees means more frequent removal, and since the whole tree is often removed (branches as chips for fuel, for instance), more thought and research are being devoted to the loss of nutrients. A few sites with known deficiencies are being routinely fertilized.

Forests trees are susceptible to various fungi, bacteria, and viruses. Insects also cause periodic, significant losses. Insect control has depended more on agricultural chemicals than either disease control or the improvement of soil fertility. As in all of agriculture, serious efforts are being made toward finding natural means of controlling both insects and diseases, to minimize the use of chemicals. So far, chemical control has been the most effective method

of halting devastating insect attacks, but there is hope for introduced predators and pathogens.

Fire

More than other agricultural resources, forests are subject to nonbiological hazards, both natural and those caused by people. Tens of thousands of hectares of timber are lost to fire every year. Some result from lightning; others are caused by human carelessness and even arson.

A forest fire causes significant losses of several kinds; it is difficult to place a value on most of them. All too frequently there is loss of human life and heavy loss of private property. Obviously, there is significant loss of timber. In addition, there is habitat destruction and the loss of wildlife. Sometimes these losses are irreversible. Damage to watersheds due to destruction of protective ground cover leads to severe surface runoff of rains, flooding, erosion, nutrient losses, and landslides like those which have devastated parts of southern California in recent years.†

For many years the philosophy of forest managers has been that any fire in a forest is bad. As a result, fire control has been a major management consideration and expense in many forest areas and has been more or less equated with forest management in the mind of the public. As unburned fuels build up through leaf fall and brush accumulation, the likelihood of a fire increases; thus fire is a natural event in a forest. There is good evidence that periodic, small fires have always removed some of the understory plant growth and much of the accumulation of branches and leaves in our forests, thereby removing fuel for a major fire. This includes evidence that Native Americans in many areas recognized the importance of ridding the forest of this rough brush and either allowed fires to burn unchecked or even started them before the accumulation of understory material could provide adequate fuel for a disaster. The character of the forests found here by the early settlers was shaped to a large degree by fire. In recent years, the benefits of fire have been recognized more by foresters, wildlife biologists, and range scientists, and prescribed fire has become a major tool of forest management (Fig. 25-6). It has been extended, for example, to wilderness areas in Yellowstone National Park, where naturally caused fires are sometimes

---

†The California mud slides have been largely from brush-rather than timberlands but illustrate the results of the loss of vegetation due to fire.

FIGURE 25-5
The undisturbed layer of plant materials shows salmonberry, blueberry, ferns, and mosses to prevent erosion of soil and nutrients. The trees are Sitka spruce and hemlock. (*U.S. Forest Service.*)

allowed to burn unchecked in an attempt to restore a more natural, or at least a less hazardous, balance.

Loss of nitrogen due to burning is a serious problem with wild fires, which tend to burn with great intensity. Prescribed fires, on the other hand, appear to recycle nitrogen faster with little loss to the ecosystem. Available nitrogen is generally increased, as these fires create conditions favorable to nitrification. To a large degree, other nutrients in the burned material are returned to the soil in ashes.

CLASSIFICATION SCHEMES

A forest is an ecologically complex unit consisting of a community of trees and undergrowth covering a considerable area. Management may consider many elements other than trees, e.g., watershed, wildlife, and recreation, but in this discussion, only the trees will be covered. There are a number of terms used to describe forests or parts of a forest:

1. A *stand* is an aggregation of more or less uniform trees in terms of species, age distribution, etc. (comparable to a field or farm).

2. A *type* is similar to a stand but more extensive. It is a group of similar stands, not necessarily adjoining but defined by the species composition wherever it is found.

3. A *site class* considers the physical factors affecting tree productivity, e.g., soil, slope, exposure, etc.

4. Pure versus mixed stand; at least 80 percent of the trees in a *pure stand* are of the same species. *Composition* considers the species present.

5. Trees are classified by age or by size, which reflects age. In determining tree size, both height and diameter are considered. Diameter is expressed as diameter at breast high (dbh)—1.35 m from the soil surface; i.e., diameter in centimeters at 1.35 m.

    a. *Seedling*, a tree grown from a seed, not yet 1 m high.

    b. *Shoot*, a sprout not yet 1 m high.

    c. *Small sapling*, a tree 1 to 3 m high.

    d. *Large sapling*, a tree over 3 m high and up to 10 cm dbh.

FIGURE 25-6
An almost pure stand of slender bluestem (*Andropogon tener*), in a longleaf pine forest. This area is burned practically every year and is moderately grazed. (*U.S. Forest Service.*)

e. *Small pole*, a tree 10 to 20 cm dbh.

f. *Large pole*, a tree 20 to 30 cm dbh (Fig. 25-7).

g. *Standard*, a tree 30 to 60 cm dbh.

h. *Veteran*, a tree over 60 cm dbh.

The most important of several broad ecological classifications of trees for forest management is shade tolerance. A *shade-tolerant tree* can generally withstand closer planting or more competition for light than a nontolerant type. This becomes a major consideration in planting, thinning, or harvesting operations and schedules. Forest trees are classified under three broad headings with regard to light requirements:

Shade-tolerant species—e.g., balsam fir, hemlock, redwood, basswood, spruce, birch, and maple

Intermediate—e.g., Douglas fir, ash, elm, and many oaks

Shade-intolerant—e.g., cypress, eastern red cedar, larch (tamarack), pine, aspen, black cherry, black walnut, cottonwood, hickory, locust, red gum, sycamore, yellow poplar, and willow

Species in these three groups presumably differ in basic physiological processes related to photosynthesis. The most common explanation is that the differences in shade tolerance are related directly to differences in the amount of light required to reach light compensation, or the amount of light a plant

FIGURE 25-7
Longleaf pole-size pine stand. (*U.S. Forest Service.*)

needs for net photosynthesis to occur, as well as light saturation (see Chap. 7). It is important to note that these classifications are broad. Within any species are genotypes which may be more or less tolerant to shading than the average. Since a common cause of shading is crowding or dense planting, trees that tolerate shading are, within limits, also expected to tolerate more crowded conditions than nontolerant species. Tolerance is obviously a major consideration in deciding whether to manage trees as even-aged stands or as uneven-aged, stands where regeneration and growth must take place in the shade of several older age classes.

The pattern of crown development is important in identifying the species that will succeed in uneven-aged conditions; related to, or an expression of, shade tolerance, it can be a significant factor in thinning, harvesting, and forest-management decisions. In a given stand, some trees, as they go through their life cycles, grow above all others and appear to dominate a stand. Others, at the opposite extreme, may never emerge at the upper surface of the forest crown or foliar canopy. Many fall somewhere in between, not truly dominant but not obviously occurring only at the lower levels of the canopy. Obviously trees that do not reach a dominant crown position must be shade-tolerant or else they would not survive and might as well be cut and used.

Recognizing crown types, shading, and spacing requirements are crucial in making harvest and thinning decisions. Four types of crown development and position are recognized:

Dominant Trees—The crowns extend above the general height of the forest canopy and receive full sunlight from above and partial sun from the sides. They experience minimal shading because they reach beyond competitors.

Codominant Trees—Crown development is at the general level of the forest canopy. Trees receive full sunlight from above but lateral shading from adjacent trees. Crowns are medium-sized and not as extensive as those associated with dominant trees.

Intermediate Trees—Crowns are formed below the general level of the forest canopy but extend up into it. Trees that develop in this manner must tolerate shading.

Overtopped Trees—Crowns are formed below the canopy level, receiving only filtered light.

Species can also be classified according to other ecological factors. Three species of pines, white, red, and jack, are adapted to the dry, sandy soils of the north. On heavier, well-drained soils in the same region, mixed hardwood forests are found, along with other pines, spruce, fir, and hemlock. Of course, in both situations, trees are adapted to the typical northern climatic conditions.

Other examples of widespread habitat are the extensive southern pine forests (see Fig. 25-7), typical of the southern coastal plains, the gums (Fig. 25-8) and cottonwoods found in association with southern bogs and swamps, and the cedars and balsam firs found in association with the colder, northern swamp areas.

KEY TREES OF AMERICAN FORESTS

Even the casual observer must be impressed by the number and diversity of trees which constitute the forests of the United States and Canada. North America has about 850 species of at least some significance to the total forest industry. The Society of American Foresters recognizes 147 forest types, about two-thirds of which are in the eastern United States and one-third in the west. Obviously, even the forest types are too numerous to describe individually. To simplify matters, the forest types have been condensed into major groups, each recognized by the prevalence of certain key species.

Pines

Eastern white pine *(Pinus strobus)*, one of the common tree species of eastern forests (Fig. 25-9), is found in forests stretching from Newfoundland to Lake Winnipeg to the Great Lakes states and New England, extending in the southern Appalachians as far south as northern Georgia. It is best adapted to humid, cool northern latitudes, where trees can reach a height of over 30 m and a diameter of 65 cm in 80 years, and some trees tower over 60 m with a girth of up to 2 m. The eastern white pine yields softwood that is widely used in the lumber industry today and was an important part of colonial life and commerce.

Western white pine *(P. monticola)*, also a source of softwood, is a valuable and important species in the lumber industry on the west coast. The wood is used in millwork (window sashes, frames, doors,

FIGURE 25-8
Typical stand of swamp gum (*Nyssa biflora*). (*U.S. Forest Service.*)

etc.) and is prized in the construction industry for framing and similar basic construction.

Western white pine is rarely found in pure stands; it occurs commonly with hemlock, Douglas fir, and several western firs. Significant quantities of this species occur in forests from British Columbia, along the west coast south to central California and inland to Idaho and Montana. The largest trees are found in Idaho and Montana, where they thrive on the deep porous soils of that region.

The largest of all the white pines is the sugar pine (*P. lambertiana*) (see Fig. 25-4), which lives up to 500 years. Trees with diameters of 0.6 to 1 m, reaching 48 to 55 m in height, are not uncommon. Trees 76 m tall and 3 m in diameter or even larger have been reported.

Sugar pines are confined mostly to the western mountains (Fig. 25-10) at elevations of 300 to 2100 m in the Cascade Range of southern Oregon and in the Sierra Nevadas of California extending into scattered stands in lower California. The largest trees are found in the north central to northeastern corner of California. Sugar pine is seldom found in pure stands, commonly growing in association with ponderosa pine, California incense cedar, white fir, and Douglas fir. Sugar pine requires the relatively high moisture typical of the mountain environments, particularly during the seedling and sapling stages.

The wood from sugar pine is similar to that of eastern white pine, and uses are about the same. It is fairly soft, moderately weak, and not particularly stiff. It is easy to paint and does not tend to split

FIGURE 25-9
Eastern white pine (*Pinus strobus*) is one of the very extensive tree species of Eastern forests.
(*U.S. Forest Service.*)

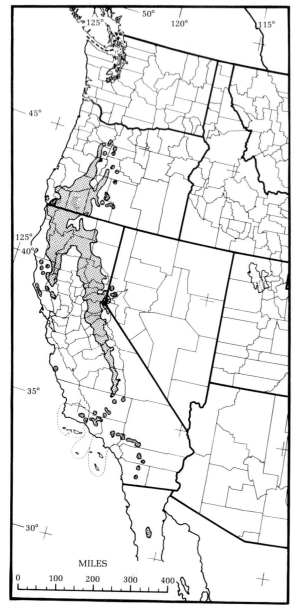

FIGURE 25-10
Natural range for sugar pine, *Pinus lambertiana*. *(U.S. Forest Service.)*

when nailed. A major characteristic of the white pines is stability, with little tendency to warp, shrink, or swell.

The most widely distributed of the western pines, ponderosa pine *(P. ponderosa)*, also called yellow pine, is found generally west of the Great Plains from British Columbia into Mexico (Fig. 25-11). It is the most valuable forest species of the southwest in terms of volume marketed. Given their wide distribution, it is not surprising that these trees vary markedly in size depending on where they are grown. Average measurements of mature full-grown trees are heights from 46 to 70 m with corresponding diameters of 1.5 to 2.4 m. Larger trees are found in California and Oregon than in Arizona and New Mexico.

The lumber is soft but of high grade and resembles lumber from white pine. Large quantities are used for planing-mill products, such as doors and sashes. Ponderosa pine is also used for manufactured products, including furniture and toys.

Shortleaf pine *(P. echinata)* is widely distributed and one of the four most important pine species in southern forests (Fig. 25-12). Its other common names, yellow pine and southern yellow pine, are also applied to longleaf, slash, and loblolly pines. Shortleaf pine trees are relatively small and mature fairly rapidly. Under average conditions, trees are about 10 m tall and 15 cm in diameter in 20 years. In 70 years, trees may reach heights of 25 m with diameters of 30 cm or more on the better sites.

Shortleaf pine may be found in association with loblolly pine, red oak, hickory, and sweet gum, but best growth is usually in dense, pure stands. This species is widely distributed and grows quite abundantly in 24 states. The northern limits seem to be Staten Island and Long Island, New York, into parts of Pennsylvania. The distribution is southward to northern Florida and westward to eastern Texas and Oklahoma. Where it occurs in abundance, shortleaf pine is frequently considered to be the most valuable conifer for timber.

The wood from shortleaf pine is strong, straight-grained, and moderately hard and rigid. It is dark yellow to light brown and relatively free from resinous matter. The physical properties of the wood make it well suited for house construction, particularly frames, interior finishing, floors, and sashes, although it is an excellent all-purpose lumber. Some

FIGURE 25-11
Forest road passing through a stand of virgin ponderosa pine (*Pinus ponderosa*). (*U.S. Forest Service.*)

is used for low-grade furniture, wood pulp, and specialty items.

Fir

Douglas fir *(Pseudotsuga menziesii)*, an important native species of western forests of North America,

is widely distributed in mountainous areas from central British Columbia south to Mexico and from the Rocky Mountains to the Pacific (Fig. 25-13). Trees thrive under a wide range of climatic conditions, but most rapid growth is favored by deep soils and abundant moisture (100 cm annually). Rapid growth

FIGURE 25-12
Shortleaf pine (*Pinus echinata*) is one of the four most important pine species in southern forests. (*U.S. Forest Service.*)

is also favored by a long growing season, as in the regions of Washington and Oregon between the coast and the Cascade Mountains, where trees are commonly 60 m or more tall with diameters from 1 to 2 m. Ancient giants more than 1000 years old have been reported to be over 90 m tall and up to 5 m in diameter.

Lumber from Douglas fir is valuable in construction because it combines strength and rigidity with moderate weight. The relatively large size of the trees makes Douglas fir a source of very large cuts of lumber, up to 0.6 m wide. Douglas fir is the major source of wood for construction plywood and is also used in boat construction, as mine timbers, and for staves for casks and kegs.

Hemlock

Although the wood of western hemlock (*Tsuga heterophylla*) has been considered inferior, it is used in some furniture and for pulp. It is also used as rough-sawed siding for houses and roofing. Western hemlocks may reach heights of up to 70 m and diameters of over 1.5 m and live for more than 500 years. They are best adapted to the cool moist Pacific coast climate from Alaska to northern California (Fig. 25-14).

Eastern hemlock (*T. canadensis*), found in eastern Canada and the eastern United States, is a native of the northeastern United States and the Great Lakes states. It is an important species in the forests of New England, New York, and Pennsylvania. The range of this species extends south in the Appalachian Mountains to northern Georgia and Alabama.

Eastern hemlock trees are medium-sized, 24 m tall and 0.6 to 1 m in diameter. The largest trees are found in the southern Appalachians of North Carolina and Tennessee, where heights may exceed 30 m with diameters of 1.2 m. In the east and Great Lakes states, trees are smaller. In general, trees grow slowly, reaching maturity in 250 to 300 years and surviving for 600 years or more. Occasionally pure stands of eastern hemlock are found, but generally

trees are found in mixtures with white pine, red spruce, sugar maple, beech, and yellow birch. The wood from eastern hemlock is used for general building construction and is a major source of pulp for newsprint, wrapping paper, and some high-grade printing paper.

Spruce

The Sitka spruce (*Picea sitchensis*), the largest of the North American spruces (Fig. 25-14), is comparable in size to the huge Douglas fir. Trees yield wood of

FIGURE 25-13
Mature stand of Douglas fir (*Pseudotsuga mengiesii*). (*U.S. Forest Service.*)

FIGURE 25-14
Sitka spruce (*Picea sitchensis*) and Western hemlock (*Tsuga heterophylla*) adapted to the cool moist Pacific Coast climate. (*U.S. Forest Service.*)

high commercial value. Because of their size, clear lumber of nearly any length and width can be obtained. Trees may be from 1.2 to 1.8 m in diameter and over 60 m tall, with no branches in the first 12 to 24 m of the trunk. The wood is prized because it is relatively light but strong for its weight.

Production of Sitka spruce is limited to a relatively narrow band along the Pacific Coast from northern California to beyond Cook Inlet on the Alaskan coast. It is rarely found more than 64 km inland. Although Sitka spruce may occur in pure stands, it usually is found in mixtures. In Washington and Oregon, it is found with Douglas fir, grand fir, western hemlock, and western red cedar.

Redwood

Redwood (*Sequoia sempervirens*), also known as coastal redwood (Fig. 25-15), is a giant of the forest, taller than the giant sequoia (*Sequoiadendron giganteum*) but not as massive. Redwoods are limited to a narrow foggy coastal band from the southwestern corner of Oregon to Monterey, south of San Francisco. The moderate, cool, moist climate is apparently essential for their growth. Redwoods are rapid growers; in 20 years a tree may reach a height of 15 m and diameter of about 20 cm. Mature trees average 60 to 75 m in height and 3.0 to 4.6 m in diameter. Some trees are over 100 m tall and nearly 6 m in diameter with estimated ages of 2000 years.

Redwood forests have been heavily cut. Conservation efforts have been successful in protecting certain remaining forest areas. The wood is desirable because it resists decay and is valuable for pilings and posts. It is also used in heavy construction and rough-cut for natural siding and roofing materials, as well as finished into specialty items ranging from musical instruments to caskets. The wood is very soft but exceptionally resistant to insects because of chemicals that accumulate in the cells of the heartwood and give it its distinctive color and durability.

Maple

Sugar maple (*Acer saccharum*), also called rock maple for its hardness, is one of the most common, highly valued hardwoods. The wood from sugar

FIGURE 25-15
Redwood (*Sequoia sempervirens*) is a giant of the forest and is taller than the giant sequoia (*Sequoiadendron giganteum*) but not as massive. (*U.S. Forest Service.*)

maple is used in many types of furniture. If properly dried, the wood is excellent for floors because of its hardness, beauty, and resistance to cracking. Sugar maple is used for one unique purpose in the United States: virtually all bowling pins are made from this wood, as is the first half of every bowling alley.

Sugar maple trees are vigorous and persistent and range in height to over 30 m and in diameter up to 1.4 m. The distribution of this species is from eastern Minnesota into Canada, the Great Lakes states east throughout New England, and south to northern Georgia, Alabama, and Mississippi.† Significant stands are found in northeastern Texas. Beyond doubt, the sugar maple is the most important of all the maple species; even though the sugar and syrup industry for which it was named is now of minor importance, the sap is still used to produce these luxury products.

Ash

White ash (Fraxinus americana) is another valuable hardwood species. Not common in pure stands, the trees are most frequently found in mixtures with maple, oaks, and hickories. The distribution of this species is similar to that of sugar maple, except that white ash trees are found farther south, into northern Florida. White ash wood is used for a number of specialty products, including furniture fittings such as drawer pulls and some parts of small wooden boats. Its most common use seems to be in athletic equipment, e.g., bats, rackets, and hockey stocks, and almost exclusively for handles of nonstriking tools such as shovels and hoes.

Oak

White oak (Quercus alba) is one of 60 or more species of oaks (Fig. 25-16). White oak is a valuable source of hardwood used for a variety of purposes—from early furniture manufacturing and flooring to railroad boxcars, railroad ties, whiskey barrels, and mining timber. The wood is strong, fairly heavy, and durable. Typical forest-grown trees have relatively long, straight trunks with little branching except at the top.

Kentucky is the center of the distribution of white oak, but this species is widely spread through the eastern United States and extends into the southwest. Stands are found from southeastern Minnesota to Maine and from southeastern New Brunswick to

†This species has been recognized as the official state tree in New York, Vermont, West Virginia, and Wisconsin.

northern Florida and westward into southeastern Texas (Fig. 25-17). In Kentucky, stands have been harvested that had trees over 400 years old reaching heights of 40 m with diameters of over 1 m. White oaks are also valuable ornamentals in spite of their relatively slow growth. Given adequate space and good soil, they form massive trees.

Basswood

American basswood trees (Tilia americana) are valued as ornamentals as well as for their wood. Trees are symmetrical and may be quite compact, although they range in height to 40 m and in diameter to 1.3 m. American basswood trees are found in forests widely spread in the northern half of the states east of the Mississippi. The wood is used for a variety of specialty purposes from furniture and moldings to shade rollers and beehives, where a soft, nonwarping, easily worked wood is required.

Beech

Like American basswood trees, American beech (Fagus grandiflora) is valued both for its wood and as an ornamental. Trees are well-shaped with delicate, graceful branches and twigs and smooth, blue-gray bark that make them attractive even in fall and winter. American beech trees tolerate shading very well; in dense plantings, they may displace trees of less tolerant species. In general, beech is found in the same regions as sugar maple. Beech usually occurs in mixtures with other hardwoods and is a typical part of the northern hardwood forest. The growth habit is branchy with a very short bole, which makes it undesirable for lumber production because of the cost of handling compared with the low yield of clear wood.

The wood from the American beech is hard and because it leaves essentially no odor or taste is ideally suited for manufacturing food containers, including cooperage for aging distilled liquor. The wood is also used for toys, brush handles, chair rounds, some veneers, as shavings serving as a support for bacteria in vinegar generators, as a source of pulp, and even for railroad ties.

Birch

In the northeast and Great Lakes states, yellow birch (Betula alleghaniensis) is the most common and most valuable native birch. The most productive

FIGURE 25-16
White oaks (*Quercus alba*) are a valuable source of hardwood. Typical forest-grown trees have
relatively long, straight trunks with little branching except at the top. (*U.S. Forest Service.*)

stands are near the Canadian border, where trees thrive on moist, cool, north-facing slopes. Trees grow slowly and may take up to 150 years to reach adequate size for sawtimber. At maturity, trees may range to over 18 m in height and 1 m in diameter.

Yellow birch is a common species of the northeastern hardwood forests. It is frequently found in association with sugar maple, beech, eastern hemlock, red spruce, balsam fir, and eastern white pine. Lumber production is declining because of poor reproduction on cutover areas, although numerous seedlings can be found on moss-covered logs and stumps and where the forest floor has been disturbed. It is the most common veneer for hardwood doors, cabinets, and paneling; other uses of the wood are railroad ties and for distillation products.

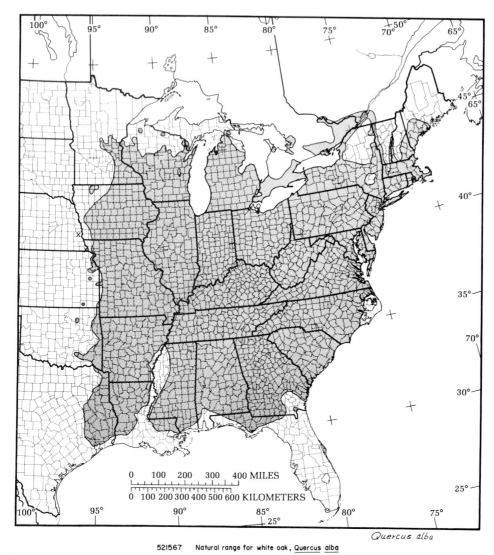

521567   Natural range for white oak, <u>Quercus</u> <u>alba</u>

FIGURE 25-17
Natural range for white oak. (*U.S. Forest Service.*)

Elm

American elm *(Ulmus americana)* is the most widely distributed of the six elm species native to the United States. Like the other species, it occurs naturally east of the Rocky Mountains, but it can be grown successfully throughout the west. The American elm is the largest elm grown in the United States; under favorable conditions, heights may exceed 30 m and diameters 2 m, but averages are closer to 20 m and 1.3 m, respectively. In addition to a significant role in the lumber industry, American elm trees are among the most beautiful shade trees with extensive ornamental uses in the northern United States. Unfortunately, many mature plantings have been destroyed by the Dutch elm disease, caused by a fungus, *Ceratocystis ulmii*, which is transmitted by the native elm bark beetle and the European elm bark beetle.

Wood from American elm trees is used for the manufacture of furniture, plywood, boxes, crates, and pallets and for a range of special products from sporting equipment and musical instruments to caskets. Elm wood is also used for casks and barrels not intended to hold liquids, known as slack cooperage.

Poplar

Yellow poplar, or tuliptree *(Liriodendron tulipifera)*, is one of the most valuable hardwood trees grown in the United States. It is common in farm woodlots and forests and has been planted extensively as a shade and ornamental tree. The species is native to the states east of the Mississippi River except Maine, New Hampshire, and Wisconsin. The distribution extends westward to Missouri, Arkansas, and Louisiana. Indiana, Kentucky, and Tennessee recognize the yellow poplar as their state tree.

Yellow poplar is one of the fastest growing hardwood species. Trees may reach heights of over 60 m, but the average is 24 to 36 m with diameters of up to 2 m; some giant trees may be up to 3.6 m in diameter. Yellow poplar does not tolerate shade well and seldom constitutes a portion of the forest understory. Most frequently this species comes up as even-aged stands in abandoned fields, clear-cuttings, or burns.

Wood from yellow poplar trees is used for furniture, fixtures, millwork, and cabinets. Large planks can be cut from larger trees, and some wood is used for pulp, particle board, and similar products.

Sweetgum

Sweetgum *(Liquidambar styraciflua)* is one of the most beautiful hardwood trees. Foliage in the fall is brightly colored. Although found as far northeast as Connecticut, sweetgum is recognized as a species of

FIGURE 25-18
Black walnut (*Juglans nigra*) grow rapidly but they are intolerant of shade and do well only on deep, rich, well-drained soils. (*U.S. Forest Service.*)

the south and southeast. Trees tolerate flooding and in the south commonly attain heights of up to 36 m and diameters of over 1 m. Forest-grown trees yield straight, clear trunks. The prized lumber, with beautiful grain and texture, is easily finished. As a result, this wood is used in interior finishing and furniture as well as in musical instruments.

Walnut

Black walnut (*Juglans nigra*) can attain immense size—1.2 to 1.8 m in diameter and heights of 45 m (Fig. 25-18). Average sizes are about 0.8 m and 30 m in diameter and height, respectively. The trees grow rapidly but are intolerant of shade and do well only on deep, rich, well-drained soils, the kind suited to farming. Although a valuable species, it cannot compete economically with most agricultural crops.

Black walnuts are native to the eastern United States, from western Massachusetts to eastern Nebraska. Their range extends to western Florida and eastern Texas, although plantings may be found on the west coast.

The wood is used for many purposes. Its fine, clear grain makes it excellent for specialties, such as rifle and shotgun stocks. It is also used for furniture and cabinets, and has been used as railroad ties and fence posts. Defective or low-quality wood is even used for fuel. The nut is a favorite food and condiment.

A problem with walnut trees as ornamentals is that their roots and leaves contain a chemical that has an inhibitory effect on many other plants including tomato and pine.

CONVERSIONS

TABLE 1
Common SI Prefixes

Multiple	Prefix	Symbol	Multiple	Prefix	Symbol
10^9	giga	G	10^{-9}	nano	n
10^6	mega	M	10^{-6}	micro	μ
10^3	kilo	k	10^{-3}	milli	m

TABLE 2
Conversion Factors

To convert	To	Multiply by†
Mass:		
metric ton	U.S. ton	1.102
kilogram (kg)	pound	2.205
gram (g)	ounce	0.035
Volume:		
liter (L)	U.S. gallon	0.264
Length:		
kilometer (km)	mile	0.621
meter (m)	foot	3.281
centimeter (cm)	inch	0.394
Area:		
hectare (ha)	acre	2.471
square meter (m²)	square foot	10.764
square centimeter (cm²)	square inch	0.155
Pressure:		
megapascal (MPa)	bar	10
	lb/in²	145
	mmHg	295.3
	atm	9.87
Yield:		
metric ton per hectare	ton per acre	2.242
Energy:		
joule (J)	calorie	4.184
	Btu	9.478×10^{-4}
Illuminance		
lux (lx)	lumen	
	(= footcandle)	0.093

† To convert in the other direction, from customary unit to SI equivalent, simply use the reciprocal; for example 2 lb × 1/2.205 = 2 × 0.435 = 0.907 kg. This method is particularly simple on hand calculators with the 1/x function.

TABLE 3
Conversions between Celsius (°C) and Fahrenheit (°F) Temperatures†

°C	Either	°F	°C	Either	°F	°C	Either	°F
− 73.3	− 100	− 148.0	− 6.1	21	69.8	16.1	61	141.8
− 70.6	− 95	− 139.0	− 5.6	22	71.6	16.7	62	143.6
− 67.8	− 90	− 130.0	− 5.0	23	73.4	17.2	63	145.4
− 65.0	− 85	− 121.0	− 4.4	24	75.2	17.8	64	147.2
− 62.2	− 80	− 112.0	− 3.9	25	77.0	18.3	65	149.0
− 59.5	− 75	− 103.0	− 3.3	26	78.8	18.9	66	150.8
− 56.7	− 70	− 94.0	− 2.8	27	80.6	19.4	67	152.6
− 53.9	− 65	− 85.0	− 2.2	28	82.4	20.0	68	154.4
− 51.1	− 60	− 76.0	− 1.7	29	84.2	20.6	69	156.2
− 48.3	− 55	− 67.0	− 1.1	30	86.0	21.1	70	158.0
− 45.6	− 50	− 58.0	− 0.6	31	87.8	21.7	71	159.8
− 42.8	− 45	− 49.0	0	32	89.6	22.2	72	161.6
− 40.0	− 40	− 40.0	0.6	33	91.4	22.8	73	163.4
− 37.2	− 35	− 31.0	1.1	34	93.2	23.3	74	165.2
− 34.4	− 30	− 22.0	1.7	35	95.0	23.9	75	167.0
− 31.7	− 25	− 13.0	2.2	36	96.8	24.4	76	168.8
− 28.9	− 20	− 4.0	2.8	37	98.6	25.0	77	170.6
− 26.1	− 15	5.0	3.3	38	100.4	25.6	78	172.4
− 23.3	− 10	14.0	3.9	39	102.2	26.1	79	174.2
− 20.6	− 5	23.0	4.4	40	104.0	26.7	80	176.0
− 17.8	0	32.0	5.0	41	105.8	27.2	81	177.8
− 17.2	1	33.8	5.6	42	107.6	27.8	82	179.6
− 16.7	2	35.6	6.1	43	109.4	28.3	83	181.4
− 16.1	3	37.4	6.7	44	111.2	28.9	84	183.2
− 15.6	4	39.2	7.2	45	113.0	29.4	85	185.0
− 15.0	5	41.0	7.8	46	114.8	30.0	86	186.8
− 14.4	6	42.8	8.3	47	116.6	30.6	87	188.6
− 13.9	7	44.6	8.9	48	118.4	31.1	88	190.4
− 13.3	8	46.4	9.4	49	120.2	31.7	89	192.2
− 12.8	9	48.2	10.0	50	122.0	32.2	90	194.0
− 12.2	10	50.0	10.6	51	123.8	32.8	91	195.8
− 11.7	11	51.8	11.1	52	125.6	33.3	92	197.6
− 11.1	12	53.6	11.7	53	127.4	33.9	93	199.4
− 10.6	13	55.4	12.2	54	129.2	34.4	94	201.2
− 10.0	14	57.2	12.8	55	131.0	35.0	95	203.0
− 9.4	15	59.0	13.3	56	132.8	35.6	96	204.8
− 8.9	16	60.8	13.9	57	134.6	36.1	97	206.6
− 8.3	17	62.6	14.4	58	136.4	36.7	98	208.4
− 7.8	18	64.4	15.0	59	138.2	37.2	99	210.2
− 7.2	19	66.2	15.6	60	140.0	37.8	100	212.0
− 6.7	20	68.0						

† To convert a given temperature locate it in the center column and read to left or right as appropriate. To find the reading on the Kelvin scale, add 273 to the Celsius reading.

GLOSSARY

ABSCISIC ACID A phytohormone involved in many physiological processes including dormancy, drought resistance, and abscission.

ABSCISSION The natural separation of leaves, flowers, and fruits or buds from the stems or other plant parts by the formation of a special layer of thin-walled cells.

ABSORPTION SPECTRUM The various wavelengths of the spectrum absorbed by a plant part or a pigment.

ACCESSORY BUD Lateral buds occurring at the base of a terminal bud or in an axil at the right or left of the axillary bud.

ACHENE An indehiscent dry fruit with only one seed, which is separable from the walls of the ovary except where it is attached to the inside of the pericarp. Typical of fruits in the sunflower family.

ACTINOMORPHIC See radially symmetrical.

ACTINOMYCETE Microorganisms similar to fungi and bacteria and active in the decomposition of organic matter.

ADAPTATION A characteristic of survival value for plants or animals; survival in a specific environment.

ADVECTIVE FREEZE A freeze associated with the invasion of a large, cold air mass; accompanied by windy conditions and hence no temperature inversion.

ADVENTITIOUS STRUCTURE Structures arising from places other than the usual, e.g., roots growing from leaves or buds developing at locations other than in leaf axils or shoot apices.

AEROBIC Requiring oxygen or occurring only in its presence.

AFTERMATH The residue and regrowth of forage plants after harvesting by animal or machine.

AGGREGATE Soil particles which are clumped together and thus improve the structure of a soil; also called *crumb, granule, ped.*

AGROECOSYSTEM The collection of physical, environmental, economic, and social factors that affect a cropping enterprise.

AGRONOMY The study of soil and crops grown on large scale.

AIR-BLAST SPRAYER A sprayer which has a large fan to produce a high-speed, high-volume air flow to break spray particles into small droplets and carry the spray to the plant.

AIR CONDITIONING The use of overhead irrigation to cool a crop through the evaporation of water and the associated uptake of the heat of vaporization; in turfgrass management, also called *syringing.*

AIR DRAINAGE The flow of cold air downhill. Freeze-sensitive crops are planted on hillsides so that on calm spring nights the cold air will drain down and away from the crop.

AIR LAYERING Propagation procedure in which a ring of tissue about 2 cm wide is removed from the stem below the tip of the plant, and the stem is surrounded with a moist medium such as peat.

ALEURONE The outer protein-rich layer of cells of the endosperm of grass seeds.

ALKALI SOIL A soil with a pH of 8.5 or higher, a high sodium content, or both.

ALKALOID Any of a large number of nitrogen-containing, naturally produced, alkaline biochemicals, e.g., caffeine, nicotine, and cocaine.

ALLELE (1) A given form of a gene which occupies a specific position (locus) on a specific chromosome. Alternative forms of a gene occur and can thus occupy a specific locus. These variant forms are said to be alleles or to be allelic to one another. (2) One of a pair of genes.

ALLELOPATHY An interaction between different plants or between plants and microorganisms in which substances (allelochemicals) produced by one organism affect the growth of another (usually adversely).

ALTERNATE LEAF ARRANGEMENT An arrangement characterized by one leaf per node on alternating sides along the stem.

AMINO ACIDS The biochemicals that serve as the building blocks of proteins; 20 different naturally occurring amino acids are present in plants and animals. Essential amino acids are those which animals cannot produce but must rely upon plants for.

ANAEROBIC Not requiring oxygen or occurring only in its absence.

ANALOG A compound chemically similar to another.

ANAPHASE The stage of nuclear division during which the chromosomes move to opposite poles.

ANGIOSPERM A plant that bears enclosed seeds in fruits formed by development of the pistil of the flower.

ANNIDATION An ecological relationship in which two or more organisms grow side by side and avoid or reduce competition by using environmental resources differently.

ANNUAL A plant that completes its life cycle in 1 year or less.

ANTHER The upper part of the stamen, in which pollen is produced.

ANTHESIS The stage in floral development when pollen is shed. In grasses, the period during which anthers are extended from the glumes.

ANTITRANSPIRANT A material applied to plants to reduce the rate of transpiration. It is usually a plastic or wax formulation which is sprayed on and dries to form a relatively impervious film.

APEX The tip of a stem or root.

APICAL MERISTEM The growing point of the plant. The tissue at the tip of the stem or root in which cells undergo cell division.

APOMIXIS Form of reproduction in which new individuals are produced without nuclear or cellular fusion. The embryo develops from an unfertilized egg or from tissues, such as the integument, which surround the embryo sac.

AQUIFER A naturally occurring underground reservoir of water.

ARABLE Land suitable for the production of crops requiring tillage.

ARID REGION A geographical region receiving less than 25 cm of annual precipitation. Most crop production in arid regions requires irrigation.

ARTIFICIAL SYSTEM Any classification system of plants devised for convenience. Artificial systems are often based on arbitrary, variable, and superficial characteristics.

ASEXUAL REPRODUCTION The production of a whole plant from any somatic cell, tissue, or organ of that plant.

ASSEMBLAGE Group of individuals reproducing from seeds that show some genetic differences but have one or more characteristics by which the plants can be differentiated from other cultivars.

ATP Adenosine triphosphate, an important energy-storage form in cellular metabolism.

AURICLE Appendage projecting from the collar of a grass leaf.

AUTOTROPH Organism which obtains energy by oxidizing inorganic substances or using the sun's energy in biochemical processes.

AUXIN A phytohormone involved in many physiological processes including stem and root elongation, fruit set, cambium activity, and sex determination. Indoleacetic acid (IAA) is the major naturally occurring auxin.

AVAILABLE WATER Soil moisture between field capacity and the wilting coefficient (-0.03 to -1.5 Mpa soil-water potential); generally considered to be available for plant uptake.

AXIL The angle formed between the junction of a leaf and the stem.

AXILLARY BUDS Buds borne laterally on the stem in the axils of leaves.

BACKCROSS The cross of a hybrid to one of its parental types; a method of incorporating traits into otherwise adapted cultivars. May include multiple crosses to a recurring parental type.

BARK All the tissues of a root or stem from the cambium outward.

BERRY Simple fruit in which the entire pericarp is fleshy.

BIENNIAL A plant that ordinarily requires 2 years, or at least part of two growing seasons, with a dormant period between growth stages to complete its life cycle.

BILATERALLY SYMMETRICAL (ZYGOMORPHIC) Term describing a flower or other structure that can be divided into two similar parts by division along only one longitudinal plane.

BINOMIAL SYSTEM System of plant nomenclature in which each species is given a unique name consisting of two parts, genus name and species name.

BIOLOGICAL CONTROL Using living organisms to reduce the population of pest organisms, e.g., control of aphids by ladybird beetles.

BIOTECHNOLOGY The application of molecular biology and genetic engineering to industrial, medical, and agricultural problems; see also genetic engineering.

BLADE Usually the flattened, green, expanded portion of the leaf; also, lamina.

BLOAT An excessive accumulation of gases in the rumen of animals, commonly caused by grazing on legumes such as clover.

BOLTING Developmental process in which rosetted plants produce a flower stalk and seed and then die before the end of the season.

BRACT Small, pointed, modified leaves which subtend many flowers or inflorescences and may appear to be part of the flower.

BRAN The outer layers of a grain removed in milling, consisting of the pericarp, testa, and usually the aleurone layer.

BRANCH Lateral portion of the shoot that originates from the trunk or from another branch and gives rise to shoots, twigs, and leaves.

BREEDER'S SEED Seed directly controlled by the originating or sponsoring plant breeder.

BUDDING Type of grafting in which a vegetative bud (scion) is placed in a stock plant (stock).

BUDS Dormant and unelongated stems composed of a very short axis of meristem cells from which embryonic leaves, lateral buds, flower parts, or all three arise.

BUD SPORT Mutation which occurs in the apical meristem of a bud.

BULB A budlike structure consisting of a small stem with closely crowded fleshy or papery leaves or leaf bases.

BULBIL An aerial bulblet.

BULBLET A miniature bulb which develops from meristems in the axils of scaly bulbs.

BUNCHGRASS A grass that grows in tufts and does not spread by rhizomes or stolons; also called *tufted grass*.

BURNED LIME CaO or CaO + MgO, formed when limestone is heated to drive off CO_2, leaving the oxide; also called *quicklime*.

C_3 PLANT A plant in which the first product of CO_2 fixation is the 3-carbon compound phosphoglyceric acid. This group includes many crop plants.

C_4 PLANT A plant in which the first product of CO_2 fixation is the 4-carbon compound oxaloacetic acid. C_4 plants include many tropical grasses, corn, sugarcane, and some weed species.

C:N RATIO The relative amounts of organic materials and nitrogen present in a plant or in a soil.

CALCITIC LIME Calcium carbonate ($CaCO_3$).

CALORIE The amount of heat necessary to raise the temperature of 1 g of water 1°C. A large Calorie (1000 cal, or 1 kcal) raises the temperature of 1 kg of water 1°C.

CALVIN CYCLE The dark reactions of photosynthesis in which CO_2 is fixed and then chemically reduced.

CALYX The collective name for all the sepals in a flower.

CAM Crassulacean acid metabolism, a photosynthetic pattern followed by many succulent plants in arid environments. The CO_2 is fixed at night into an acid and then used in the daylight via the Calvin cycle.

CAMBIUM The lateral meristem or growing layer of a stem or root.

CANOPY The spatial arrangement of leaves in a community of plants.

CAPILLARITY The movement of water into soil pores caused by the forces between the water molecules and the soil particles (adhesion) and the forces between water molecules (cohesion).

CAPILLARY WATER Water held in the capillaries of the soil at soil water potentials of -0.03 to -3.1 MPa; it usually meets the water needs of plants.

CAPITULUM A globose or disk inflorescence with a short axis and sessile flowers.

CAPSULE Simple, dry, dehiscent fruit with two or more locules which split in various ways.

CARBOHYDRATE A chemical whose C:H:O ratio is 1:2:1; includes sugars, starches, and cellulose.

CAROTENOIDS Pigments which range from yellow to red; found in chromoplasts and chloroplasts.

CARYOPSIS An indehiscent dry fruit with one seed which is completely fused to the inner surface of the pericarp; typical of the grasses.

CATCH CROP An alternate crop planted after the regular crop has failed or other circumstances make its success doubtful; usually a short-season crop such as millet or buckwheat.

CATION-EXCHANGE CAPACITY The number of negatively charged sites on a soil which can react with and hold cations. It is high for clays and humus and low for sand.

CATKIN A type of spike inflorescence that has unisexual flowers with a perianth.

CELL The smallest unit of living matter capable of continued independent life and growth; a self-contained and at least partially self-sufficient unit bounded by a cell wall.

CELL CULTURE Growing dividing, unassociated cells in a liquid culture medium. Usually restricted in plant biotechnology to isolated protoplasts.

CELLULOSE A polymer of glucose molecules that forms the major structural component of cell walls; the most abundant biochemical in the world.

CENTROMERE The portion of the chromosome to which the spindle fiber is attached and by which sister chromatids are joined.

CEREAL The seeds of cereal crops. Grain is a collective term applied to cereals.

CEREAL CROP A member of the grass family grown for its edible seed, e.g., wheat, oats, barley, rye, rice, corn, grain sorghum, and millet. Buckwheat, though a dicot and therefore not a member of the grass family, is commonly included among the cereals.

CERTIFIED SEED The progeny of registered seed stock, certified with a metal seal and blue tag.

CHAFF The glumes, lemmas, paleas, and lighter plant-tissue fragments released in threshing.

CHELATE A complex organic molecule which can be combined with a cation such as Fe^{2+} but will not ionize. Chelates are used to supply micronutrients where fixation by the soil makes the unchelated ions unavailable.

CHEWING INSECTS Insects with chewing mouth parts, e.g., worms, grasshoppers, and Japanese beetles.

CHIASMA The interconnection of chromatids between two adjacent homologous chromosomes; plural *CHIASMATA; see also* crossing over.

CHILLING INJURY Damage to such horticultural products as banana, papaya, cucumber, and sweet potato from exposure to cold but above-freezing temperatures.

CHILLING REQUIREMENT A cold period required by certain plants and plant parts in order to break physiological dormancy or rest; often expressed in terms of the number of hours required at 7°C or less.

CHIMERA A plant or plant organ composed of tissues from at least two phenotypes of cells; the "new" cell types often arise from mutations.

CHLORENCHYMA Parenchyma cells that contain chloroplasts and compose the palisade and spongy mesophyll cells of leaves; principal site for photosynthesis.

CHLOROPHYLL The green pigment in plants which absorbs the radiant energy ultimately fixed in the form of reduced carbon compounds.

CHLOROPLAST The organelle of a cell in which photosynthesis occurs.

CHLOROSIS A condition in which a plant or part of a plant is light green or greenish-yellow because of poor chlorophyll development or the destruction of chlorophyll as a result of mineral deficiency or disease.

CHROMATID One of the two paired parts of a chromosome following its longitudinal replication.

CHROMATIN A dark-staining, thread-like network of genetic material in the nucleus composed of DNA, RNA, and proteins. Chromatin condenses to microscopically visible chromosomes during cell division.

CHROMOPLAST A plastid containing only carotenoid pigments (yellow, red, and orange), as distinguished from a chloroplast, which contains green pigments as well.

CHROMOSOME A nuclear structure consisting of chromatin and bearing hereditary units or genes. Chromosomes become microscopically visible during nuclear division.

CLADOPHYLL A leaflike structure which may bear flowers, fruits, and temporary leaves.

CLASS In the scientific classification system, a group of plants made up of orders.

CLAY Fine soil particles less than 0.002 mm in diameter.

CLEAN CULTIVATION Periodic soil tillage to eliminate all vegetation other than the crop being grown.

CLEISTOGAMY Self-pollination within a flower, usually the result of a closed flower.

CLIMACTERIC A developmental process undergone by certain ripening fruits that is marked by an increase in respiration and a softening of fruit tissues.

CLIMATE The long-term average weather conditions.

CLIMAX VEGETATION The ultimate, more or less stable stage of succession in which a particular combination of species comes to dominate an ecosystem.

CLONE Plants derived from a single individual and propagated entirely by vegetative means.

CODON A sequence of three nucleotides (a triplet) in a gene or mRNA that codes for a particular amino acid in a protein.

CO₂ COMPENSATION POINT The CO_2 concentration at which there is no net CO_2 flux, as photosynthesis balances respiration. In C_3 plants this is usually about 50 ppm; in C_4 plants, it is close to zero.

CO₂ FERTILIZATION The practice of adding CO_2 to the atmosphere around plants to increase their photosynthetic output and therefore yield.

COLEOPTILE A protective sheath enclosing the growing point and leaves of grass seedlings.

COLLAR A narrow band on the grass leaf where the sheath joins the blade.

COLLENCHYMA CELL Modified parenchyma cells which are characterized by thickening of the cell wall at their inner angles.

COLLOID An insoluble particle small enough to remain suspended in a liquid without agitation.

COMMERCIAL FLORICULTURE Area of horticulture which includes the commercial production and distribution of cut flowers, flowering pot plants, foliage plants, and bedding plants.

COMPANION CELL Small, slender living cell associated with the sieve cell in the phloem of angiosperms.

COMPANION CROP A crop grown with another crop; often used to secure a return from the land in the first year of a new seeding or to help the other crop get established; also known as a *nurse crop*.

COMPOUND BUD A bud containing both vegetative and floral primordia; also called *mixed bud*; cf. simple bud.

COMPLETE FERTILIZER A fertilizer containing nitrogen, phosphorus, and potassium.

COMPLETE FLOWER Flower composed of four sets of floral parts—sepals, petals, stamens, and pistils.

COMPOUND LEAF A leaf with the blade divided into several leaflets or sections.

COMPOUND OVARY An ovary with two or more locules.

COMPOUND TISSUE Tissue composed of two or more cell types.

COMPOUND UMBEL Inflorescence in which a series of simple umbels arises from the same point on the main axis.

CONCENTRATE SPRAYER A sprayer designed to deliver pesticides to a crop at normal amounts per hectare but in much lower volumes of water.

CONTACT HERBICIDE A herbicide that kills only the tissues with which it comes into contact.

CONTACT POISON An insecticide absorbed by the insect through the skin or body openings instead of being ingested.

CONTOUR TILLAGE The cultivation of land along the lines of uniform elevation, or contour lines, to reduce erosion.

CONTROLLED-ATMOSPHERE STORAGE A cold storage in which the concentration of atmospheric gases is adjusted to extend the storage life of fresh produce, usually by lowering oxygen and raising carbon dioxide.

CONVENTIONAL TILLAGE A tillage system that involves plowing. The number of operations varies with crop and area.

COOL-SEASON PLANT A plant in which peak growth occurs during the spring or fall; it does not grow well under hot conditions, cf. warm-season plant.

CORK Plant part composed of cells with suberized walls formed by the cork cambium; also called *phellem*.

CORK CAMBIUM A cylindrical layer of cells of the cortex or phloem which develops or reinitiates the capacity to divide; also called phellogen; produces cork cells.

CORM Short, fleshy, underground stem with few nodes and short internodes; a means of vegetatively propagating some plants.

COROLLA Collectively the petals which enclose and protect the pistil and stamen.

CORTEX The primary parenchymatous tissue lying between the epidermis and the vascular tissue of stems and roots.

CORYMB Raceme inflorescence in which the pedicels of the lower flowers are longer than the pedicels of the upper flowers, resulting in a flattened top.

COTYLEDON Embryonic leaves which serve as food-storing organs or develop into photosynthetic structures as the seed germinates.

COVER CROP A crop, e.g., rye, grown to reduce soil erosion, conserve nutrients, and provide organic matter. Cover crops are grown during the season when a cash crop is not being grown or between the rows of crops, e.g., peaches.

CRITICAL LEVEL A concentration of a nutrient element below which deficiency symptoms are likely or a response to additions of the nutrient can be expected.

CROP A plant which is harvested for use by people or livestock.

CROPPING SYSTEM The cropping patterns used on a farm and their interaction with farm resources, other farm enterprises, and available technology.

CROSSING-OVER An exchange of chromosomal material (genes) between nonsister chromatids which transfers genes between homologous pairs; usually occurs during prophase of meiosis. *See also* chiasma.

CROSS-POLLINATION The process by which pollen is transferred from an anther of one flower to the stigma of a different plant or cultivar.

CULM The upright stem of a grass plant.

CULTIVAR A cultivated variety that has originated and persisted under cultivation, not necessarily referable to a botanical species, and of botanical or horticultural importance, requiring a name.

CUTICLE The waxy covering on leaves or fruit, which protects the tissue against excess moisture loss.

CUTIN Any of a group of lipid substances deposited both on the inside and outside of epidermal cell walls; it minimizes water loss.

CUTTING Detached vegetative plant part which under favorable conditions develops into a complete plant with characteristics identical to those of the parent.

CYME A broad, more or less flat-topped determinate inflorescence in which the central flowers bloom first.

CYTOKINESIS Cell division; distinguished from mitosis and meiosis, which involve nuclear division.

CYTOPLASM The viscous material in which organelles float in the cell.

DARK REACTIONS The series of photosynthetic reactions in which CO_2 is actually fixed. The energy to drive the dark reactions comes from the light reactions. See Calvin cycle.

DARK RESPIRATION The complex series of reactions of mitochondria in which carbohydrates, especially glucose, are broken down to release energy, much of which is conserved in high-energy compounds like ATP. Dark respiration occurs during both dark and light periods; cf. photorespiration.

DAY-NEUTRAL PLANT A plant which may flower under any day length.

DECIDUOUS The characteristic of perennials whose leaves are shed, usually at the end of the growing season.

DEHISCENT Descriptive of dry fruit in which the carpel splits along definite seams at maturity.

DENITRIFICATION The conversion of nitrate nitrogen into gaseous forms, leading to nitrogen losses from soils.

DEOXYRIBONUCLEIC ACID (DNA) The biochemical responsible for genetic coding; it consists of the sugar deoxyribose, together with phosphate, and adenine, guanine, cytosine, or thymine.

DERMAL SYSTEM The outermost layer of cells, the epidermis, of floral parts, leaves, fruits, and seeds, and of stems and roots until they begin secondary growth.

DETERMINATE GROWTH A pattern of development in which the apical meristem differentiates into flowers, terminating the production of additional leaves and stems.

DEW POINT The temperature at which a given mixture of air and water vapor will reach 100 percent relative humidity or at which condensation will begin.

DICHASIUM Inflorescence with a peduncle having a terminal flower and a pair of lateral branches below.

DICHOTOMOUS KEY An analytical device requiring a choice between two (or more) contradictory propositions to be made in each step.

DICOTYLEDON Class of plants whose embryos have two cotyledons.

DICTYOSOME Cell organelle that is a group of flat, disk-shaped sacs formed by cytoplasmic membranes and serving as a source of plasma membrane materials and collection centers for complex carbohydrates. Also called *Golgi body*.

DIFFERENTIATION The changes a cell or tissue undergoes during growth that result in a specialized form and function.

DIOECIOUS Type of sex expression in which plants produce male and female organs on separate plants.

DIPLOID Having two complete haploid sets of chromosomes; considered typical or normal for the species.

DISBUDDING The removal of vegetative or flower buds.

DISEASE An abnormality caused by an infectious pathogen.

DIURNAL Daily, pertaining especially to actions completed within 24 h.

DIVISION Taxonomic group of the plant kingdom, made up of classes; the Latin names of divisions end in -*phyta*.

DNA See deoxyribonucleic acid.

DOLOMITIC LIME Calcium carbonate ($CaCO_3$) with sizable quantities of dolomite [$CaMg(CO_3)_2$], recommended where magnesium tends to be deficient.

DOMESTICATED PLANT A species, e.g., corn and soybean, which has been bred and selected to the extent that it is not able to survive without human cultivation.

DOMINANT A character that is manifested in the hybrid to the apparent exclusion of the contrasting character from the other (recessive) parent.

DORMANCY A state of suspended growth or the lack of outwardly visible activity caused by environmental or internal factors.

DOUBLE CROSS Combination of two single cross hybrids in hybridization between four inbred lines.

DOUBLE FERTILIZATION Union of the two male gametes with the female gamete and the polar nuclei.

DOUBLE-WORKING Type of graft in which the graft combination contains an interstock or intermediate stem piece grafted between the scion and stock.

DRIP IRRIGATION See trickle irrigation.

DRUPE Type of fruit with a thin exocarp, a thick and fleshy mesocarp, and a hard and stony endocarp, e.g., peach and olive.

DRY FRUIT Fruit in which the pericarp is often hard and brittle at maturity.

DRYLAND FARMING Production of crops that require some tillage in subhumid or semiarid regions without irrigation. The system usually involves fallow periods between crops during which water from precipitation is absorbed and retained.

DUSTER A pesticide applicator that applies the pesticide as a dry powder usually mixed in a diluent such as talc.

DWARFING The reduction or inhibition of height in certain plants caused by genetic, chemical, or other physiological factors.

ECOSYSTEM The totality of physical, biological, climatic, etc., factors acting together to create a complex community of organisms.

ECOTYPE A cultivar or strain adapted to a particular environment.

EGG The cell in the embryo sac that forms the embryo after fertilization by the sperm.

ELECTROMAGNETIC SPECTRUM The range of radiant energy from wavelengths of less than 0.001 nm to greater than 100,000 nm.

EMBRYO SAC The mature female gametophyte of angiosperms found in the ovule; contains the egg and seven other nuclei or cells.

EMITTER The water-delivery mechanism or outlet in a trickle-irrigation system.

EMULSIFIABLE CONCENTRATE A liquid formulation which mixes with water to form an emulsion but does not dissolve to form a solution.

ENDERGONIC The type of chemical reaction or physical process which proceeds only as energy is applied to it; cf. exergonic.

ENDOCARP The inner layer of the fruit pericarp; cf. exocarp and mesocarp.

ENDODERMIS Single layer of cells between the cortex and pericycle in the root.

ENDOGENOUS A naturally produced substance; one produced within the organism on which it is acting; cf. exogenous.

ENDOPLASMIC RETICULUM Structure extending throughout cytoplasm and functioning in the transport of cell products, as a surface for protein synthesis by the ribosomes, in the separation of enzymes and enzyme reactions, in support, and in moving cell-membrane components into position, as in cell division.

ENDOSPERM The starchy or nutrient-rich portion of a seed that often surrounds the embryo and stores reserve food.

ENSILAGE *See* silage.

ENSILE To make and store forage as silage.

ENTOMOLOGY The study of insects.

ENZYME A protein molecule which acts as a catalyst for specific biochemical reactions.

EPICOTYL The part of the axis of an embryo above the region of attachment of the cotyledons.

EPIDERMIS The outer layer of cells (usually covered with cutin) on all parts of a young plant and on some parts of older plants, e.g., leaves and fruits.

EPIGEAL GERMINATION The pattern of seedling emergence in which the epicotyl remains inactive at first and the hypocotyl grows, pushing up the cotyledon(s); cf. hypogeal.

EPIGYNOUS Descriptive of a flower in which the perianth and stamens are attached above the ovary; cf. hypogynous and perigynous.

EROSION Wearing away of the land surface by running water, wind, ice, or other geological agents.

ESSENTIAL AMINO ACID An amino acid that monogastric animals must obtain from their food because they cannot synthesize it.

ETHYLENE A gas (C_2H_4) which is a phytohormone involved in a number of physiological processes, e.g., stem elongation, fruit ripening, and stomatal closure.

EVAPOTRANSPIRATION The loss of soil moisture to the atmosphere by plant transpiration and evaporation from the soil surface.

EVAPOTRANSPIRATION POTENTIAL The water loss per unit time of evaporation and transpiration combined assuming unlimited water.

EXERGONIC The type of chemical reaction or physical process which releases energy (often as heat) as it proceeds; cf. endergonic.

EXOCARP The outer, usually skinlike region of the fruit pericarp; cf. endocarp and mesocarp.

EXOGENOUS External; descriptive of a substance applied to a plant to alter its growth or other processes; cf. endogenous.

EXTERNAL DORMANCY The inability to grow caused by unfavorable external conditions, e.g., moisture, temperature, or oxygen; quiescence.

FALLOW State of land left without a crop or weed growth for extended periods often to accumulate moisture.

FAMILY A group of closely related genera or (rarely) a single genus.

FAN-AND-PAD COOLING SYSTEM A cooling device used in greenhouses. Air is pulled through wet pads by means of fans. As water evaporates, large quantities of heat are absorbed (heat of vaporization).

FAR-RED LIGHT Radiant energy at a predominant wavelength of 730 nm; important in phytochrome-mediated processes.

FERMENTATION The anaerobic process by which sugars are converted into ethyl alcohol by the action of specific yeasts.

FERTILIZATION (1) The practice of adding nutrients to soil or plants for use by plants. (2) The union of egg and sperm; syngamy.

FIBER CELL Sclerenchyma cells which are slender and elongated, with tapering ends which overlap and are often fused with others.

FIBER CROP Crops grown for their fiber content, e.g., cotton, flax, hemp, ramie, sissal, and jute.

FIBROUS ROOT Root system in which primary and lateral roots develop more or less equally and have a limited quantity of cortex; cf. taproot.

FIELD CAPACITY The amount of water a soil can hold against gravity. At field capacity, the soil water potential is -0.03 MPa, the upper limit of available water. Field capacity is expressed as a percentage of the dry weight of a soil.

FILAMENT The part of the stamen which holds the anther in a position favorable for pollen dispersal.

FIRE BLIGHT A major disease of pear and apple caused by the bacterium *Erwinia amylovora*.

FLESHY FRUIT A fruit whose pericarp is soft and fleshy at maturity; includes the berry, pepo, hesperidium, drupe, and pome.

FLESHY ROOT A root that accumulates and stores a rich supply of reserve food for the plant. *See also* taproot and fibrous root.

FLORICULTURE The cultivation, management, marketing, and arranging of flowers and foliage plants.

FLORIGEN A flowering hormone, sought for decades but not yet isolated.

FLOWER A shoot of determinate growth with modified leaves that is supported by a short stem; the structure involved in the sexual reproductive processes of angiosperms.

FOLIAR FERTILIZATION The practice of applying plant nutrients in solutions to the foliage of crops.

FOLLICLE Simple, dry, dehiscent fruit having one locule which splits along one suture.

FOOTCANDLE See lux.

FORAGE Grasses, legumes, and other crops cultivated and used for feeding animals as hay, pasture, fodder, or silage.

FORB A nongrass herb.

FORCING Producing a marketable pot plant or cut flower out of season by manipulation of environmental factors.

FORM A member of a population that differs from other members to a degree too small for it to be called a *cultivar*.

FOUNDATION SEED Seed stock handled to maintain specific genetic identity and purity as closely as possible under supervised or approved production methods.

FREE WATER Water moving through the soil by the force of gravity.

FREEZE The condition that exists when the air temperature remains below 0°C over a widespread area long enough to characterize the weather.

FROST-FREE SEASON The interval between the last occurrence of freezing temperatures in the spring and the first one in the fall.

FROST HEAVING Physical uplifting of the soil surface, plant, or structure caused by daily thawing and freezing. It can be exceptionally harmful to stands of taprooted plants such as alfalfa.

FROST POCKET A depression in the terrain into which cold air drains but cannot escape, causing the area to be subject to freeze injury.

FRUIT An expanded and ripened ovary with attached and subtending reproductive structures.

FRUIT-SET The physiological process in which a fertilized ovule becomes "committed" to further development (instead of abscission).

FUMIGANT An organic compound with high vapor pressures, i.e., a gas at temperatures above 5°C, used for insect and disease control in confined areas.

FUNDAMENTAL TISSUE SYSTEM All tissues of the plant encased by the epidermis other than those of the vascular system.

GAMETE Male or female sex cell; egg or sperm.

GAMETOPHYTE In alternation of generations, the sexual phase producing gametes. Gamotophytic nuclei are haploid.

GENE A genetic unit of inheritance carried on the chromosomes of each cell which determines the heritable characteristics of the cell.

GENERATIVE NUCLEUS The nucleus of pollen grains which by mitiotic division forms two sperm nuclei.

GENETIC BASE The aggregate of genetic variability available within a species or cultivar; may be broad or narrow.

GENETIC CODE The set of 64 different possible codons (or nucleic acid triplets) and their corresponding amino acids; determines which amino acids will be added during protein synthesis.

GENETIC ENGINEERING The nonsexual manipulation of genes and genotypes; *see also* biotechnology.

GENOME One complete set of genes, containing all the genetic information to produce an individual.

GENOTYPE The genetic makeup of an individual determined by the assemblage of genes it possesses.

GEOTROPISM The directional growth of plants in response to gravity.

GERMINATION The initiation of active growth by the embryo of the seed, resulting in the rupture of seed coverings and the emergence of a new seedling plant capable of independent existence.

GERMPLASM The (sometimes multiple) sets of genes that constitute an individual or a cultivar; a source of genetic material for breeding work.

GIBBERELLIN (GIBBERELLIC ACID) A phytohormone involved in such physiological processes as germination, stem elongation, and sex determination.

GLUME Reduced leaves found in pairs at the base of a spikelet.

GLYCOLYSIS The initial phases of enzymatic breakdown of a carbohydrate, e.g., glucose and glycogen, by way of phosphate derivatives.

GOLGI BODY See dictyosomes.

GRAFTING Joining two separate structures, e.g., a root and stem or two stems, so that by tissue regeneration they form a union and grow as one plant.

GRAIN A grass fruit, especially of the larger-seeded species; *see also* cereal.

GRASSLAND Any plant community in which grasses or legumes or both constitute the dominant vegetation.

GRAVITATIONAL WATER Water in excess of capillary water and thus held at soil water potentials of more than −0.03 MPa; it usually percolates through a soil within 24 to 48 h after a rain or irrigation.

GREENHOUSE EFFECT The effect of the earth's atmosphere on incoming and outgoing radiation. Solar radiation, predominantly of short wavelengths, readily passes through the atmosphere. Terrestrial radiation is of much longer wavelengths and is trapped or reflected by the atmosphere. The atmosphere acts like the glass in a greenhouse in its selective transmission of radiant energy.

GREEN MANURE A crop that is grown for soil protection, biological nitrogen reduction, or organic matter and plowed or disked into the soil.

GROUND COVER A crop planted to provide a covering over the soil, often ornamental.

GROWING SEASON The period from the last spring freeze to the first freeze in the fall. In the United States this ranges from about 100 to 365 days.

GROWTH An increase in volume, dry weight, or both.

GROWTH HABIT The basic pattern of structural development or symmetry attained by a plant, e.g., shrubby, treelike, or viny.

GUARD CELLS Cells which bound stomata and by their turgor pressure determine whether a stoma is open or closed.

GUTTATION The exudation of water through leaves via structures called *hydathodes*; it occurs largely at night because of root pressure when little or no transpiration is occurring.

GYMNOSPERM A plant that bears naked seeds without an ovary.

GYNOECIOUS Type of sex expression in which female flowers only are produced.

GYNOMONOECIOUS Type of sex expression in which plants have perfect as well as imperfect pistillate flowers on the same plant.

HAPLOID Having one half of the complete set of chromosomes typical for the species.

HARDENING The result of a great many changes which occur in a plant as it develops resistance to adverse conditions, especially cold.

HARDENING OFF The treatment of tender plants to enable them to survive a more adverse environment, e.g., withholding nutrients, lowering temperatures, allowing temporary wilting, and other methods to slow growth rate.

HARDPAN An impervious layer in a soil which restricts root penetration as well as movement of air and water; also called *pan*.

HARDWOOD CUTTING Cutting made from woody deciduous species and narrow-leaved evergreen species, e.g., grape and hemlock.

HAY The harvested forage of the finer-stemmed crops cured by drying.

HAYLAGE High-moisture hay that has been ensiled.

HEADING BACK Type of pruning cut in which the terminal portion of the shoot is removed but the basal portion is not.

HEAT OF FUSION The amount of heat required to change 1 g of a substance at its melting point from the solid to the liquid state or vice versa; for water it is 80 cal.

HEAT OF VAPORIZATION The amount of heat required to change 1 g of a substance at its boiling point from the liquid to the vapor state or vice versa; for water it is 540 cal.

HERBACEOUS Not woody; a plant or part that contains largely parenchymatous tissue.

HERBACEOUS CUTTING Cutting made from succulent herbaceous plants, e.g., chrysanthemum, coleus, and geranium.

HERBACEOUS PERENNIALS Plants with soft, succulent stems whose tops are killed back by frost in many temperate and colder climates but whose roots and crowns remain alive and send out top growth when favorable growing conditions return.

HERBARIUM A collection of plant specimens that have been classified taxonomically, pressed, dried, and mounted on sheets of herbarium paper.

HERBICIDE A material which will kill plants; some kill essentially all plants but others are selective.

HESPERIDIUM A fruit with a leathery rind, e.g., oranges and other citrus fruits.

HETEROSIS Hybrid vigor such that an F_1 hybrid falls outside the range of the parents with respect to some character or characters.

HETEROTROPH An organism which obtains energy by degradation of complex, energy-rich organic matter; cf. autotroph.

HETEROZYGOUS Having two different alleles of a gene pair present in the same organism; having many heterozygous pairs within one's genome.

HIGH-PRESSURE SPRAYER A sprayer with a high-pressure pump to force the spray through nozzles for atomization and delivery to the plant.

HOMOLOGOUS CHROMOSOME A chromosome that is a member of a distinct morphological pair.

HOMOZYGOUS Having identical alleles of a gene pair, having essentially all homozygous pairs in one's genome.

HORTICULTURE The intensive cultivation of plants, from Latin *hortus*, "garden."

HUMUS The product of organic matter degradation by microorganisms and chemical reactions. It is dark-colored and amorphous and relatively resistant to further rapid degradation.

HYBRID In its simplest form, a first-generation cross between two genetically diverse parents.

HYBRID VIGOR See heterosis.

HYDATHODE See guttation.

HYDRATED LIME Burned lime which has been reacted with water to form $Ca(OH)_2$ or $Ca(OH)_2$ + $Mg(OH)_2$.

HYDROCOOLING A system for cooling fresh produce by flooding the product with large volumes of cold water to remove field heat.

HYDROPHYTE A plant that requires relatively large amounts of water for normal growth; cf. mesophyte and xerophyte.

HYGROSCOPIC MATERIALS Substances which attract water, e.g., salt.

HYGROSCOPIC WATER Soil moisture which exists as a very thin film around soil particles and is unavailable to plants. At the upper limit of hygroscopic water, soil water potential is −3.1 MPa.

HYGROTHERMOGRAPH A device which continuously records both temperature and relative humidity.

HYPOCOTYL The stem of an embryo below the cotyledons.

HYPOGEAL GERMINATION The pattern of seedling emergence in which the hypocotyl remains inactive and the epicotyl grows, pushing only the plumular leaves upward and leaving the cotyledons underground; cf. epigeal germination.

HYPOGYNOUS Classification of a flower in which the sepals, petals, and stamens are attached to the receptacle below the ovary; cf. epigynous and perigynous.

IMPERFECT FLOWER Flower lacking either stamens or pistils.

INBRED A genetic line of normally cross-pollinated species which has been self-fertilized for a number of generations to produce a population of similar and highly homozygous individuals; a pure line of a normally cross-pollinated species.

INBREEDING DEPRESSION Decreased vigor resulting from self-pollination of normally open-pollinated plants or species; see also heterosis.

INCOMPLETE FLOWER Flower lacking one or more of the four sets of floral parts.

INDEHISCENT Type of dry fruit in which the fruit wall does not split at any definite point or seam at maturity.

INDETERMINATE GROWTH A pattern of development in which the apical meristem remains vegetative, i.e., does not produce flowers, so that new leaves and stems continue to be produced while flowers are also forming; cf. determinate growth.

INFLORESCENCE The arrangement of the flowers on the floral axis, a flower cluster.

INSOLATION The radiation received from the sun.

INTEGRATED PEST MANAGEMENT The control of one or more pests by a broad spectrum of techniques ranging from biological means to pesticides. The goal is to keep damage below economic levels without eliminating the pest completely.

INTEGUMENT The wall of the ovule surrounding the embryo sac.

INTERCROPPING Planting two or more species together.

INTERNODE The portion of the stem between two consecutive nodes.

INTERPHASE The stage during the cell cycle when it is not undergoing microscopically visible division. Predivision events occur during interphase.

INTERSTOCK Intermediate stem piece grafted between the scion and stock; cf. double-working.

INTRODUCTION A plant brought into a geographical area. Most of our crops were introduced into this country from other parts of the world.

JUVENILE A plant that has not yet reached reproductive maturity.

LAMELLAE Membranous structures in the chloroplast. See also middle lamella.

LAMINA The expanded leaf blade.

LANDSCAPE DESIGN The profession concerned with the planning and planting of outdoor space to secure the most desirable relationship between land forms, architecture, and plants to satisfy aesthetic and functional needs.

LATH HOUSE A frame supporting strips of wood spaced to provide about 50 percent shade; snow fence is a common material.

LATICIFER CELL Parenchyma cells specialized to synthesize and store latex.

LAYERING A vegetative method of propagating new individuals by producing adventitious roots before the new plant is severed from the parent.

LEACHING The downward movement of nutrients or salts through the soil profile in soil water; it accounts for nutrient losses but can also be beneficial in ridding a soil of excess salts; also called eluvation.

LEAF-BUD CUTTING Cuttings consisting of a leaf blade, petiole, and a short piece of the stem, with the attached axillary bud placed with the bud end in the medium and covered enough to support the leaf.

LEAF CUTTING Cuttings consisting of leaves with or without the petioles.

LEAFLET One of the units of a compound leaf.

LEAF MARGIN The edge of the leaf, which may be entire, serrate, dentate, lobed, etc.

LEAF Vegetative lateral outgrowth of stems which typically develop special structural adaptations for photosynthesis.

LEGUME A simple, dry, dehiscent fruit with one locule which splits along two sutures; a member of the bean family.

LEMMA The outer bract formed at the base of a grass floret.

LEUCOPLAST A colorless plastid.

LIGHT-COMPENSATION POINT A level of irradiance at which there is no net flux of CO_2 from a leaf and photosynthesis equals respiration.

LIGHT METER An instrument to measure visible light, usually in units of lumens or lux.

LIGHT REACTIONS The series of photosynthetic reactions, in which light energy is converted into chemical energy; see also z scheme.

LIGHT SATURATION A level of irradiance above which there is no further increase in net photosynthesis.

LIGNIN A complex organic polymer which provides rigidity, toughness, and strength to cell walls.

LIGULE A membranous appendage or fringe of hairs on the inside of grass leaves where the sheath joins the blade.

LIME Ground limestone consisting of $CaCO_3$ with varying amounts of $CaMg(CO_3)_2$ used to raise soil pH.

LINE Plant cultivar of uniform appearance which has economic or practical value and can be reproduced uniformly by seed.

LINKAGE Association of characters in inheritance due to the location of genes on the same chromosome.

LIPID A biochemical that is soluble in polar solvents, e.g., fats, oils, phospholipids, and chlorophyll.

LOCULE A cavity of the ovary.

LOCUS The position occupied by a given gene on a chromosome; any one of the variant forms of a gene may be present on a particular locus.

LODGING Leaning or falling over of plants.

LONG-DAY PLANT A plant which requires a day longer than its critical day length (more exactly, a night shorter than its critical dark period) in order to flower; also called *short-night plant*.

LUMEN See lux.

LUX (lx) A unit of measure for visible light. Full midday sunshine in summer approximates 108 klx; 10.7 lx = 1 lumen (lm), formerly called a footcandle.

LUXURY CONSUMPTION The uptake of a nutrient in quantities greater than required for optimum growth and productivity; common with potassium.

LYSOSOME Single-membrane organelles in the cytoplasm which contain hydrolytic enzymes capable of digesting other cellular particles.

MACRONUTRIENT Essential nutrient needed in relatively large amounts, e.g., nitrogen and potassium.

MARKET GARDENING Growing an assortment of vegetables for local or roadside markets.

MEADOW An area covered with perennial fine-stemmed grasses and often used to produce hay; see sward.

MEGAGAMETOPHYTE The female gamotophyte; the embryo sac.

MEGASPORE MOTHER CELL A diploid cell which by meiosis produces four megaspores, one of which gives rise to the embryo sac.

MEIOSIS A type of cell division which produces the gametophytic phase in both the male and the female reproductive parts of the flower.

MENDELIAN INHERITANCE Pattern of inheritance initially described by Mendel in which pairs of alleles are segregated and recombined in independent patterns; also called *Mendelian genetics.*

MERISTEM A region of a plant in which cells are not fully differentiated and are capable of repeated mitotic divisions.

MESOCARP The center portion of the fruit pericarp; cf. endocarp and exocarp.

MESOPHYLL A tissue consisting of largely parenchyma cells between the upper and lower epidermal layers of the leaf; the major photosynthetic tissue in most plants.

MESOPHYTE A plant that requires a moderate amount of water for normal growth; includes most crop plants.

METABOLISM The biochemical processes an organism must accomplish in order to live.

METAPHASE That stage of nuclear division in which the chromosomes are arranged at the equatorial plane of the spindle.

MICROENVIRONMENT The physical conditions that exist at a particular site.

MICROGAMETOPHYTE The male gametophyte; the pollen grain.

MICRONUTRIENT An essential nutrient needed in small amounts, e.g., boron, molybdenum; also called a *trace* or *minor element.*

MICROSPORANGIUM A chamber of the anther in which pollen is formed.

MICROSPORE MOTHER CELL A diploid cell which by meiosis produces four microspores, each of which becomes a pollen grain.

MICROTUBULE Cell structures located in the cytoplasmic matrix of nondividing cells and in the spindle fibers of dividing cells which may be involved in the growth of the cell wall, cell-plate development, and mitosis.

MIDDLE LAMELLA A viscous (jellylike) substance, consisting of calcium and magnesium pectates, binding the primary cell wall of one cell to that of adjacent cells.

MINIMUM TILLAGE A soil-management system in which the crop (often corn) is seeded directly without plowing or disking. *See also* no tillage.

MITOCHONDRION Organelle serving as the cell's powerhouse; it is where organic molecules are oxidized, releasing energy, and where that energy is incorporated into molecules of ATP, the main chemical-energy source for all cells.

MITOSIS Nuclear division, involving replication of chromosomes and their separation into two equal groups to form two daughter nuclei.

MIXED BUD See compound bud.

MONOCOTYLEDON Class of plants with embryos having one cotyledon.

MONOCULTURE Repeated growing of a single crop on the same land.

MONOECIOUS Sex expression in which male and female structures are produced on the same plant.

MONOMER A basic subunit or building block of a polymer; amino acids are the monomers of protein polymers.

MOSAIC A disease caused by virus infection; symptoms include mottling of leaves and flowers.

MOUND OR STOOL LAYERING A layering procedure in which soil is mounded around the base of new shoots in the spring, excluding light and enhancing root formation.

mRNA Messenger ribonucleic acid; the nucleic acid on which the sequence of amino acids in a protein is coded.

MULCH A material applied to the surface of a soil for conservation of moisture, stabilization of soil temperature, and suppression of weeds.

MULTIPLE CROPPING Growing two or more crops during one cropping year.

MUTAGEN A chemical or other agent that increases the rate of mutation.

MUTATION A spontaneous change in the genetic makeup of the cell.

MYCORRHIZA A fungus living in a mutualistic relationship with the roots of a vascular plant.

NATURAL SELECTION Differential reproduction in nature, leading to an increase in the frequency of some genes or gene combinations and to a decrease in the frequency of others. This process can also be artificially imposed to produce a desired characteristic.

NATURAL SYSTEM Classification system which attempts to show relationships between plants through the use of selected morphological structures.

NEMATODE Microscopic animal, mostly worm-shaped, which can be parasitic on plants as well as animals. Nematode damage to many crops can be severe.

NET PHOTOSYNTHESIS Gross photosynthesis minus respiration. Net photosynthesis is determined by measuring net CO_2 uptake.

NICHE (1) A plant's particular location (with its microenvironment); (2) a plant's physiological adaptation to a particular microenvironment.

NITRIFICATION The conversion of ammonium ions to nitrite and then to nitrate ions: $NH_4 \rightarrow NO_2^- \rightarrow NO_3^-$.

NITROGEN FIXATION A process carried out by the symbiotic relationship between higher plants, particularly legumes, and certain bacteria by which atmospheric N_2 is fixed into a form usable by plants. Certain free-living bacteria and blue-green algae can also fix nitrogen.

NODE The often enlarged portion of a stem from which buds and leaves arise.

NODULE A tubercle formed on legume roots by association of the plant with symbiotic nitrogen-fixing bacteria of the genus *Rhizobium.*

NO TILLAGE Method of planting crops that involves no seedbed preparation other than opening small slits in the soil so that seed can be placed at the intended depth. There is generally no cultivation during crop production, but chemicals are often used for vegetation control; also called *zero tillage.*

NUCLEIC ACID The biochemicals DNA and RNA responsible for genetic transmission from cell to cell and generation to generation.

NUCLEOLI Spherical bodies in the nucleus, composed of ribosomal RNA and proteins.

NUCLEOTIDE The monomer of a nucleic acid; of two types, purines and pyrimidines.

NURSE CROP See companion crop.

NURSERY An enterprise that produces or distributes ornamental plants.

NUT An indehiscent dry fruit similar to the achene except that the pericarp is hard throughout.

NUTRIENT A chemical element taken into a plant that is essential for growth, development, or reproduction.

OFFSHOOT Short, horizontal stems which develop from the crown of stems.

OIL CROP A crop grown for its oil content, e.g., soybean, peanut, sunflower, safflower, sesame, castorbean, mustard, rape, cottonseed, flax, jojoba, and corn.

OPPOSITE LEAVES Leaves arising from opposite sides of the same node.

ORDER A group of closely related families with some common traits but some marked differences. The Latin names of orders end in *ales*.

ORGANIC MATTER Carbon-containing materials of either plant or animal origin; exists in all stages of decomposition in soil.

OVARY The usually expanded portion of the pistil that contains one or more ovules; matures into a fruit.

OVULE The embryo-sac-bearing structure(s) in the ovary; matures into a seed.

PALATABILITY The quality of being agreeable to the taste; used for feed crops.

PALEA The inner bract formed at the base of a grass floret; often present on threshed caryopses.

PALISADE PARENCHYMA Elongated, aligned cells that form the upper part of the mesophyll in many leaves.

PALMATE LEAF A compound leaf with all the leaflets arising from a common point at the end of the petiole.

PALMATE VENATION The pattern of vein development in leaves with netted venation having several large veins radiating into the blade from the petiole at the point where petiole and blade join.

PAN Layers in soils that are compacted or have high clay content which restrict root growth and water movement; also called *hardpan*.

PANICLE Type of inflorescence with either a cluster of racemes or corymbs; distinctly branched.

PARALLEL VENATION The pattern of vein development in leaves with large veins that are essentially parallel and not obviously interconnected by lateral veins.

PARENCHYMA Thin-walled cells having a protoplast. Their functions include food storage, photosynthesis, wound healing, and formation of adventitious structures.

PARTHENOCARPY Fruit development without fertilization.

PARTHENOGENESIS Production of offspring by females in the absence of males, as can occur with aphids.

PASTURE An area of untilled ground covered with grass or some other forage and used for grazing.

PEAT-LITE MIX A growth medium of peat and either vermiculite or perlite, developed at Cornell University.

PEDUNCLE The short stem of the flower cluster.

PEPO Berry that has a hard rind around the fruit, e.g., watermelon.

PERCOLATION The movement of water downward through the soil under the influence of gravity.

PERENNIAL A plant which does not die after flowering but lives from year to year.

PERFECT FLOWER A flower that has both pistil (or pistils) and stamens but may lack sepals, petals, or both.

PERIANTH The floral parts surrounding the anthers, usually consisting of sepals and petals.

PERICARP The fruit wall, consisting of three distinct layers: the exocarp, the mesocarp, and the endocarp.

PERICYCLE Thin layer of parenchyma cells that separates the endodermis from the vascular components in roots.

PERIDERM A tissue consisting of cork, cork cambium, and phelloderm; *see* bark.

PERIGYNOUS Descriptive of a flower in which the receptacle is extended to form a cuplike structure around a portion of ovary; cf. epigynous and hypogynous.

PERMANENT SOD A soil-management system in which a sod is periodically mowed but no tillage is carried out; benefits include prevention of soil erosion, maintenance of good organic-matter levels, and improved soil structure.

PERMANENT WILTING POINT The lower limit of available water; like field

capacity, expressed as a percentage of the dry weight of the soil; cf. wilting coefficient.

PETAL Usually whorled floral structures which collectively make the corolla. They protect the inner reproductive structures and often attract insects by their color or nectar and thus facilitate pollination.

PETIOLE The leaf stalk.

PGR Plant-growth regulator.

pH A measure of acidity or alkalinity, expressed as the negative logarithm of the hydrogen-ion concentration. A pH of 7 is neutral; less than 7 is acidic; more than 7 is alkaline.

PHELLEM See cork.

PHENOTYPE The visible appearance of a plant or a particular trait, governed by its genotype as modified by the environment.

PHLOEM A compound vascular tissue in plants composed of sieve tubes, companion cells, fibers, and parenchyma cells.

PHLOEM FIBER Nonliving, thick-walled elongated cells which are in groups within the phloem.

PHOTOMORPHOGENESIS A nondirectional developmental response to a nonperiodic, nondirectional light stimulus. As a seedling plant emerges from the soil and is exposed to light, its leaves expand, stem growth slows, and the stem thickens.

PHOSPHOLIPID A compound made up of fatty acids, glycerol, and phosphate; an important constituent in membranes.

PHOTOPERIOD A cycle of light and dark intervals, usually with a period of 24 h.

PHOTOPERIODISM The developmental responses of plants to the relative lengths of the light and dark periods.

PHOTORESPIRATION Release of CO_2 and consumption of O_2 which occur only in the light. C_3 plants have high rates of photorespiration, whereas C_4 plants have little or none. Also called *light respiration*.

PHOTOSYNTHETICALLY ACTIVE RADIATION (PAR) Radiant energy between 400 and 700 nm to which the photosynthetic apparatus responds.

PHOTOSYSTEM A collection of pigments and other compounds which captures light energy and transfers it (as electrons) in photosynthesis.

PHOTOTROPISM The orientation of plants toward or away from more intense light or illumination from one side.

PHYLLOTAXY The arrangement or angular displacement of succeeding leaves on the stem.

PHYLOGENETIC SYSTEM Taxonomic system which classifies plants according to their evolutionary pedigree, reflecting genetic relationships between plants and establishing their presumed kinship.

PHYTOCHROME A photoreceptive pigment which receives radiant stimuli leading to many photomorphogenic responses.

PHYTOHORMONE A natural substance produced in small amounts which modulates growth or other plant developmental processes. *See also* plant-growth regulator.

PICK YOUR OWN A system of direct marketing in which the customer harvests the products; well adapted to strawberries, raspberries, some tree fruits, and many vegetables.

PINCHING Breaking off the terminal growing point, allowing the axillary buds to start growing.

PINNATE LEAF A compound leaf with the leaflets arranged along both sides of the midrib.

PINNATE VENATION The pattern of vein development in leaves with netted venation where secondary veins extend laterally from a single major midrib.

PISTIL The female reproductive organ, consisting of the stigma, style, and ovary.

PISTILLATE FLOWER Flowers in which only the pistils are present; there are no stamens.

PITH RAY Areas of nonvascular cells between vascular bundles.

PLACENTA The tissue of the ovary to which ovules are attached.

PLANT-GROWTH REGULATOR A natural or synthetic substance which acts endogenously or exogenously at very low concentrations to modulate plant development.

PLANT PATHOGEN A microorganism which causes a plant disease.

PLANT TAXONOMY The science of classification, systematic nomenclature, and identification of plants.

PLASMA MEMBRANE A differentially or selectively permeable and flexible membrane enveloping the protoplast; composed primarily of lipids and proteins; also called *plasmalemma*.

PLASMODESMA Pore or pit in plant cell walls through which cytoplasmic strands extend. Plural is plasmodesmata.

PLASTIC MULCH Thin clear or black polyethylene film, used as a mulch, especially on vegetables, for moisture retention, increased soil temperature, and (with black plastic) weed control.

PLASTID Cell organelle involved in food synthesis or in storage of fats, starches, proteins, and various pigments, or both.

PLUMULE The true leaves in a seed at the stem apex on an embryo, above the cotyledon.

POLLEN TUBE The tube, formed by the pollen grain, which grows through stigma, style, ovary, and micropyle to the female gametophyte.

POLLINATION The transfer of pollen from an anther to a stigma either in the same flower or between two flowers.

POLYMER A complex compound formed from the chemical bonding or polymerization of many similar subunits; cf. monomer.

POLYPLOIDY Condition in which cells have more than two sets of chromosomes.

POME Type of fruit in which portions of the fleshy parts are derived from parts of the flower other than the ovary.

POMOLOGY The branch of horticulture dealing with fruits.

PORE SPACE The voids in a soil occupied not by solid particles but by varying proportions of water and air.

POT-BOUND Having a restricted, circular pattern of root growth caused by extended container culture.

PRIMARY TISSUE Tissue developed from cells produced in the apical meristem; *see also* secondary tissue.

PRIMORDIUM An embryonic or early developmental stage of an organ such as a leaf or flower.

PROPAGATION Increase in numbers or perpetuation of a species by reproduction.

PROPHASE The phase of nuclear division after DNA replication and during which the chromosomes become microscopically visible as they shorten and thicken.

PROTEIN A polymer of amino acids; the type of biochemical which includes enzymes and various other structural or storage forms.

PROTOPLASM The living substance of a cell.

PROTOPLAST The living unit inside the cell wall, composed of protoplasm and plasma membrane.

PRUNING Removal of plant parts, e.g., buds, shoots, and roots, to maintain a desirable form by controlling the direction and amount of growth.

PUDDLED SOIL A soil whose structure has been destroyed; the pore spaces are closed up, and both aeration and water movement are poor.

PULSE Leguminous plants or their seeds; mainly plants with large seeds used for food, e.g., peanut, field pea, field bean, cowpea, soybean, lima bean, mungbean, chickpea, pigeonpea, and lentil.

PURINE One of the two basic types of nucleotides. Adenine and guanine are purines; cf. pyrimidine.

PYRIMIDINE One of the two basic types of nucleotides. Cytosine, thymine, and uracil are pyrimidines; cf. purine.

RACEME Type of inflorescence in which stalked flowers are on pedicels approximately equal in length on a single floral axis.

RADIALLY SYMMETRICAL Characterizing a flower or other organ or organism which can be divided into similar parts by division along more than one plane and in which all the parts of each group are alike in size and shape; also called *actinomorphic*.

RADIATION One of the forms in which energy is transferred; it is characterized by both wavelength and frequency and includes the range from short wavelength and high energy, e.g., gamma rays, to long wavelength and low energy, e.g., radio waves. Visible light is a small part of the total spectrum of radiation.

RADIATIONAL FREEZE A freeze associated with calm conditions, radiational cooling, and a temperature inversion.

RADICLE Lower portion of the embryonic axis; the embryonic root.

RANGE A natural or well-adapted stand of plants growing over extensive areas upon which livestock can graze.

RATOON Sprouting or regrowth from buds or young tillers; crops originating from ratoon growth.

RECEPTACLE The enlarged apex of the pedicel where the floral parts arise, e.g., the fleshy portion of a strawberry; also called a *torus*.

RECESSIVE A transmissable trait not expressed phenotypically when masked by the presence of a dominant allele for the trait.

RECOMBINATION A new combination of alleles resulting from rearrangement following crossing-over. Also descriptive of the alteration of a genotype by gene splicing techniques.

RED LIGHT Radiant energy at a predominant wavelength of 660 nm.

REGISTERED SEED Seed produced from foundation and other registered seed and used to produce certified seed; guaranteed to meet established standards of purity and quality.

RELATIVE HUMIDITY The amount of water vapor present in the air expressed as a percentage of the maximum water vapor the air can hold at the same temperature and pressure.

REPLICATION The duplication of genetic material, both DNA and chromosomes.

RESPIRATION The oxidation of compounds to release energy and CO_2; see also dark respiration and photorespiration.

REST A state of suspended growth or outwardly visible activity due to internal physiological factors. Rest is broken by extended exposure to temperatures of 7°C or less (chilling requirement); also called *physiological dormancy.*

RHIZOBIUM A bacterium which can infect legumes and establish a symbiotic relationship through which it fixes atmospheric nitrogen.

RHIZOME A horizontal stem growing partly or entirely underground; often

thickened, it can serve as a storage organ.

RIBONUCLEIC ACID (RNA) A nucleic acid consisting of the sugar ribose, together with phosphate and adenine, guanine, cytosine, or uracil.

RIBOSOME A small, dense, globular particle floating in the cytoplasm or associated with the membranes of the endoplasmic reticulum; the site of synthesis of proteins.

RNA See ribonucleic acid.

ROGUE (1) An off-type plant. (2) To remove off-type plants by pulling or cutting, to eliminate inferior individuals.

ROOT Vegetative plant part which anchors the plant, absorbs water and minerals in solution, and often stores food.

ROOT CUTTING Cutting 5 to 15 cm long made from root sections in the fall or winter.

ROOT HAIR An extension of the epidermal cells on young roots immediately behind the root tip. Root hairs are important in uptake of both water and nutrients.

ROOT PRESSURE A force generated in the xylem of roots and stems partially accounting for the rise of water in plants.

ROOT ZONE Region of the soil being exploited by plant roots.

ROTATION The repetitive cultivation of an ordered succession of crops or crops and fallow on the same land. One cycle may take several years to complete.

rRNA Ribosomal RNA, the type constituting the structural and catalytic fractions of ribosomes.

RUMEN The first compartment of the multichambered stomach of a ruminant, or cud-chewing, animal.

RUMINANT Of or relating to a suborder of mammals having a complex multichambered stomach.

RUNNER A slender stolon with elongated internodes; it roots at nodes which touch the ground.

SAMARA An indehiscent dry fruit with one or two seeds, in which the pericarp bears a flattened winglike outgrowth.

SAND Coarse soil particles 0.05 to 2 mm in diameter.

SCALE Modified leaves that often serve as protective structures.

SCARIFICATION Chemical or physical treatment of seeds to break or weaken the seed coat sufficiently for germination to occur. The unscarified seed coat may prevent penetration of water or oxygen or emergence of the embryo.

SCION The upper part of the union of a graft; also called *cion; see also* stock.

SCLERENCHYMA Cell(s) with a thick wall, usually lignified, and without protoplast at maturity.

SECONDARY TISSUE Tissue developed from cells produced in the cambium or other nonapical meristems; see also primary tissue.

SEED Plant embryo with associated stored food encased in a protective seed coat.

SELECTION A natural or artificial process whereby certain genetic traits come to predominate in a population.

SELECTIVE HERBICIDE A herbicide which kills only certain groups of plants; e.g., 2,4-D kills broadleaf plants but not grasses.

SELF-FERTILE Characteristic of a plant able to produce viable seed when pollinated by itself or another plant of the same cultivar.

SELF-POLLINATION The process by which pollen is transferred from an anther to a stigma of the same flower or another flower of the same plant or cultivar.

SELF-STERILITY The inability of a plant to set viable seed or fruit with pollen from the same cultivar.

SEMIARID REGION A geographical region receiving 25 to 50 cm of precipitation annually. Production of most crops requires water conservation or irrigation.

SEMIHARDWOOD CUTTING Cutting made from woody, broad-leaved evergreen species, e.g., ligustrum and holly.

SEMIHUMID REGION A geographical region receiving 75 to 100 cm of precipitation annually. Crop production of most species is not usually limited by natural precipitation.

SEMIPERMEABLE MEMBRANE A membrane allowing free movement of water but restricting passage of solutes.

SENESCENCE The stage of development during which deterioration occurs, leading to the death of an organism or organ. Sometimes defined from specific criteria such as a decline in chlorophyll or dry weight.

SEPAL Structure which usually forms the outermost whorl of the flower; collectively, the calyx.

SEQUENTIAL CROPPING A pattern of multiple cropping in which one crop follows another on the same land in the same year.

SERRATE Notched or toothed, used for margins of the leaf.

SESSILE Without a petiole (as in some leaves) or without a pedicel (as in some flowers and fruits).

SET A plant, e.g., tomato and tobacco, which is first sown in flats or beds and later transplanted in the field; a small bulb (onion, garlic) for planting.

SEXUAL REPRODUCTION The reproduction of plants through a sexual process involving meiosis.

SHADE CLOTH A fabric woven from Saran fibers or other material to provide shade levels ranging from less than 20 percent to more than 90 percent.

SHATTER Falling off and consequent loss of potentially harvestable foliage or fruit; also called *shattering*.

SHEATH The basal portion of a grass leaf that surrounds the stem; it may be split, overlapping, or closed.

SHOOT Stem, 1 year old or less, that has leaves. Also used to describe the entire aboveground plant structure, leaves and stems.

SHORT-DAY PLANT A plant which requires a day shorter than its critical day length or, more exactly, a night longer than its critical dark period in order to flower; also called *long-night plant*.

SIEVE CELL Elongated, slender cells with thick, nonlignified walls; the main component of the phloem.

SILAGE Forage, usually corn or sorghum, preserved in a moist condition and prepared by partial fermentation in an anaerobic environment in silos; also called *ensilage*.

SILIQUE Simple, dry, dehiscent fruit with two fused locules which separate at maturity, leaving a persistent partition between them.

SILT Soil particles 0.002 to 0.005 mm in diameter.

SIMPLE BUD Bud containing either leaf or flower primordia but not both; cf. compound bud.

SIMPLE LAYERING A method of asexual propagation in which the stem behind the end of the branch is covered with soil to promote rooting and the tip remains above ground.

SIMPLE LEAF Leaf in which the blade consists of one undivided unit.

SIMPLE OVARY Ovary having only one locule.

SIMPLE TISSUE Tissue composed of one cell type.

SINGLE CROPPING See sole cropping.

SINGLE CROSS Hybridization between two inbred lines.

SINGLE FRUIT Fleshy or dry fruit which forms a single ripened ovary.

SI UNIT Unit of measure in the Système Internationale, a complete and coherent modification of the metric system; see the Appendix.

SLASH AND BURN A pattern of agriculture in which existing vegetation is cut, stacked, and burned to provide space and nutrients for cropping; also called *swidden cultivation*.

SLOW-RELEASE FERTILIZER A fertilizer made by coating the particles with a wax or other insoluble or slowly soluble material to provide a predictable, slow release of the encapsulated materials.

SMALL GRAIN Any of the four cool-season grass crops wheat, barley, rye, or oats.

SOD (1) The top few inches of soil which is permeated and held together by grass roots. (2) The aggregated soil and vegetative matter sometimes harvested for replanting elsewhere.

SOD-FORMING GRASS A grass which usually spreads by stolons and rhizomes forming a densely matted sod; contrasted with a bunchgrass.

SOFTWOOD CUTTING Cutting taken from soft, succulent new spring growth of deciduous or evergreen species of woody plants.

SOIL The outer, weathered layer of the earth's crust which can potentially support plant life; it is made up of

inorganic particles, organic matter, organisms, water, and air.

SOIL HORIZON Layers of soil which constitute the soil profile from its surface to bedrock.

SOIL MANAGEMENT The practices used in conserving or employing a soil; may include various types of tillage and production systems.

SOIL PASTEURIZATION The heat or chemical treatment of soil to destroy all harmful organisms but not necessarily all soil organisms.

SOIL STRUCTURE The aggregation of individual soil particles into larger units—crumbs, peds, granules, or aggregates.

SOIL TEXTURE The distribution of particle sizes of soil in terms of sand, silt, and clay.

SOLE CROPPING One crop variety grown alone in pure stands at normal density; also called *monoculture* or *single cropping*.

SOMOCLONAL VARIATION The production of new genotypes (and phenotypes) following propagation of cell cultures into whole plants.

SOMATIC Descriptive of cells or tissues that do not function in sexual reproduction, as distinct from gametic cells.

SPECIES A group of organisms bearing great similarity capable of interbreeding to produce similar and fertile offspring.

SPECIFIC LEAF WEIGHT Dry weight per unit leaf area.

SPERM A male gamete; the cell or nucleus which fuses with an egg (or two polar nuclei) in fertilization.

SPERMATOPHYTA Division of the plant kingdom which includes plants reproduced by seeds.

SPIKE Inflorescence of an indeterminate raceme of sessile flowers attached to the floral axis with the oldest flowers at the base.

SPIKELET A basic morphological unit of grass inflorescences; may contain one or more florets.

SPINDLE An aggregation of microtubules (spindle fibers) essential for positioning and distributing the chromosomes at nuclear division.

SPINE A sharp-pointed woody structure, usually modified from a leaf or part of a leaf; see also thorn.

SPORE Unicellular or few-celled structure of many types and forms, usually involved in asexual reproduction.

SPOROPHYTE In alternation of generations, the asexual phase. Nuclei of the sporophytic phase are typically diploid.

SPRAYER Applicator in which the pesticide is mixed in water and distributed on the plant.

SPRINKLER IRRIGATION Overhead application of water to a crop by any of a wide range of systems; also called aerial irrigation.

SPUR A stem with short internodes, usually from older wood; normally bears leaves or both leaves and fruit.

SPUR-TYPE TREE A fruit tree with a compact growth habit caused by shorter internodes and more spurs than a standard tree of the same cultivar.

STAMEN Part of the flower consisting of the anther, in which pollen is produced, and a slender filament which holds the anther in a position favorable for pollen dispersal; male reproductive organ.

STAMINATE FLOWER Flower in which only the stamens are present; there are no pistils.

STARCH A polymer of glucose which may be relatively unbranched (amylose) or branched (amylopectin).

STELE The vascular tissues and associated fundamental tissues of the root and stem.

STEM An aboveground (usually) axis of a plant which develops from the epicotyl of the embryo or from a bud of an already existing stem or root.

STEM CUTTING Segment of shoots containing lateral or terminal buds.

STIGMA The pollen-receptive site of the pistil.

STIPULE Leaflike appendage often found on either side of the base of the petiole.

STOCK The lower part of a graft; also called rootstock.

STOLON An aboveground prostrate stem; may form roots at the nodes that come into contact with the ground; cf. runner.

STOMA A microscopic opening in leaves through which gas exchange for photosynthesis, respiration, and transpiration occurs; also called stomate. Stomata are bounded by guard cells, whose turgor pressure opens or closes the opening.

STOMACH POISON An insecticide which must be ingested to be effective; used against chewing insects.

STOMATE See stoma.

STRATIFICATION Storing seeds at 2 to 4°C under moist conditions to break physiological dormancy or rest.

STROMA The viscous portion of the chloroplast and mitochondria where transformations of carbon-containing compounds occur.

STYLE The slender part of a pistil between the stigma and the ovary, through which the pollen tube grows.

SUBERIN A thin, varnishlike layer which helps seal moisture inside the tissue and keep rot-producing organisms out.

SUBSTRATE The compound(s) upon which an enzyme acts to produce new product(s).

SUBSURFACE IRRIGATION Application of irrigation water to a crop by artificially elevating the water table. Water moves upward into the root zone by capillarity.

SUBTROPICAL FRUIT A fruit plant intermediate between tropical and temperate species, e.g., citrus fruits. Some subtropicals are evergreen, others deciduous. Minimum temperature which can be survived is usually about −5°C.

SUCCULENT A plant that stores water, more common in arid and semiarid regions, e.g., cactus.

SUCKING INSECT Insect with sucking mouth parts which sucks sap from plants; e.g., aphids, scale insects, and leafhoppers.

SUMMER ANNUAL A plant which germinates, flowers, produces seed, and dies in one growing season.

SURFACE IRRIGATION The application of water directly to the soil surface, e.g., flooding the entire area or flooding furrows.

SWARD A grassy surface of a lawn, pasture, or playing field, not necessarily a pure stand.

SWIDDEN See slash and burn.

SYSTEMIC Chemical (fertilizer or pesticide) readily translocated throughout a plant; it may be applied to the soil or sprayed on the aboveground parts.

TAPROOT Primary root that persists and maintains its dominance; often swollen and provides for storage.

TAXON A category or taxonomic group; plural, taxa.

TAXONOMY See plant taxonomy.

TELOPHASE The stage of nuclear division during which the nuclear membrane reforms and the chromosomes gradually become less evident.

TEMPERATE FRUIT A fruit plant which requires a cool period and is deciduous; e.g., apple, pear, and peach.

TEMPERATURE INVERSION A condition at night in the lower atmosphere in which there is warm air above the cooler surface air.

TENDRIL Slender, twining leaf modification used for support, as in the grape.

TERMINAL BUD Large and vigorous bud at the tips of stems, responsible for terminal growth.

TESTA The seed coat.

TETRAD The four microspores collectively.

THERMISTOR An electrical temperature sensor based on the changing resistance of certain materials as their temperature changes.

THERMOCOUPLE An electric temperature sensor obtained by joining two different metals (e.g., copper and constantin).

THINNING Type of pruning cut in which entire shoots are removed; an extension of, or complement to, heading back.

THORN A short, sharp-pointed branch; see also spine.

THYRSE A compact, condensed, determinate cyme or panicle inflorescence.

TILLAGE Mechanical soil-stirring action to benefit crops by supplying a suitable soil environment for seed germination, root growth, and weed control.

TILLER   A branch which arises at soil level; usually restricted to the grasses.

TILTH   The physical condition of the soil with respect to its fitness for planting or growing a crop.

TIP LAYERING   Layering in which rooting takes place near the tip of the current season's shoot, which falls to the ground naturally.

TISSUE   A group of cells with similar origin and function.

TISSUE CULTURE   Growing masses of unorganized cells (callus) on agar or in liquid suspension. Useful for the rapid asexual multiplication of plants.

TONOPLAST   Membrane surrounding the vacuole.

TOPSOIL   The upper layer of soil rich in organic matter; it forms naturally but is easily lost by disturbance.

TOPWORKING   Grafting procedure by which branches of trees are changed to a more desirable cultivar.

TOTIPOTENCY   The ability of a cell or cells to generate or regenerate a whole organism.

TRACHEID   The major cell type in the xylem of gymnosperms and other non-flowering vascular plants. A long, slender, hollow cell with end walls.

TRANSLOCATED HERBICIDE   A herbicide which is absorbed and translocated throughout the plant.

TRANSPIRATION   Release of water vapor by plants. Transpiration occurs through the cuticle, stomata, and lenticels, but most is through the open stomata.

TRANSPIRATION RATIO   The ratio of units of water taken up per unit of dry matter produced by a plant; also called the *water requirement*.

TRANSPIRATIONAL PULL   Tension within a plant generated by transpiration and exerting a pulling force; a major factor in the rise of water in plants.

TRIBE   A subdivision of a subfamily.

TRICKLE IRRIGATION   The application of small quantities of water, usually on a daily basis, directly to the root zone through various types of delivery systems.

tRNA   Transfer RNA; the form of RNA that brings specific amino acids to the ribosome for protein synthesis.

TROPICAL FRUIT   A fruit plant which is evergreen and cannot withstand freezing temperatures, e.g., banana, pineapple, and mango.

TROPISM   A growth response that orients a plant or organ toward or away from a directional stimulus or gradient.

TRUCK CROP   Vegetable crops grown for wholesale markets and shipping.

TRUE-BREEDING   Characteristic of a line of plants that is homozygous for each trait and therefore produces offspring identical to itself and each other.

TUBE NUCLEUS   The nucleus of a pollen grain that directs the growth and development of the pollen tube.

TUBER   Enlarged underground stem serving as a storage organ.

TUFTED GRASS   See bunchgrass.

TURF   The grassy surface and matted upper strata of earth filled with grass roots.

TURGID   Condition in which a cell or plant is fully expanded by hydrostatic pressure exerted on the cell wall by the protoplast.

TWIG   Stem 1 year old or less (usually applied to stem without leaves).

UMBEL   Type of inflorescence in which the pedicels arise from a common point and are about equal in length.

VACUOLE   Part of the cell, filled with cell sap composed of a water solution of inorganic salts, various organic solutes, and undissolved crystals.

VACUUM COOLING   A cooling system for fresh leafy vegetables, e.g., lettuce. The wet product is put into a vacuum chamber, and the atmospheric pressure is lowered to about 0.016 MPa. As water evaporates, the heat of vaporization quickly removes heat from the product.

VAPOR-PRESSURE DEFICIT   The difference between the actual water vapor pressure and the vapor pressure needed to saturate the air at the same temperature.

VARIETY   A subdivision of a species which differs from other plants of the species in one or more recognizable characters; *see also* cultivar.

VASCULAR BUNDLE   A strand of conducting tissue consisting of xylem, phloem, and often cambium.

VASCULAR CAMBIUM   The lateral meristem that gives rise to secondary phloem and xylem.

VASCULAR SYSTEM   System composed of the xylem and phloem which conducts water, mineral solutes, and foods in solution in addition to providing support and strength.

VASCULAR TISSUE   Tissue involved with internal transport; xylem and phloem.

VEGETABLE   A nonbotanical term used to describe the edible portion of any herbaceous garden plant.

VEGETATIVE   Not flowering; the stage of development when a plant is producing only roots, stems, and leaves, the vegetative organs.

VERNALIZATION   The induction of certain plants to flower following a cold period.

VESSEL   Long tube in the xylem composed of vessel elements through which water and minerals move.

VIRUS   An infectious, submicroscopic particle containing both protein and RNA.

VISIBLE WAVELENGTHS (LIGHT)   The part of the electromagnetic spectrum with wavelengths between approximately 380 and 760 nm, to which the human eye responds.

WARM-SEASON PLANT   A plant in which peak growth occurs during the warmest months of the year. It does not perform well under cool conditions; cf. cool-season plant.

WATER TABLE   The upper edge of free water in the soil. When a hole is dug, water will fill the hole to the level of the water table.

WATER POTENTIAL   The tendency of water molecules to diffuse, evaporate, or be absorbed from one area or entity to another; usually expressed in terms of pressure.

WEATHER   The short-term atmospheric conditions, including temperature, relative humidity, wind, sky conditions, precipitation, and atmospheric pressure; *see also* climate.

WEED   A plant growing where it is not wanted.

WETTABLE POWDER A powder which mixes with water to form a suspension but does not dissolve, so that continuous agitation is required.

WHORLED ARRANGEMENT Three or more leaves at a node; cf. alternate and opposite.

WILTING COEFFICIENT A measure of the amount of water present in a soil when a plant cannot obtain enough water to regain turgidity in a saturated atmosphere. Soil water potential is about −1.5 MPa; cf. permanent wilting point.

WIND MACHINE A large fan, usually permanently mounted on a tower, which mixes air in a temperature inversion to prevent freeze damage by raising the air temperature at the crop level.

WINTER ANNUAL A plant which germinates in the fall, then flowers and dies the following spring or summer.

WINTER DESICCATION Injury to plants, particularly evergreens, by loss of moisture from the aboveground portions which cannot be replaced because of frozen or very cold soil. Symptoms are browning and death of leaves and small branches.

XEROPHYTE A plant that normally requires relatively small amounts of water for growth; cf. hydrophyte and mesophyte.

XYLEM Compound vascular tissue of plants composed of tracheids, vessels, fibers, and parenchyma cells.

XYLEM FIBER Elongated strengthening cell with thickened walls.

YELLOWS A group of diseases caused by virus or viruslike infections. Symptoms include yellowing, curling, and stunting of plants.

Z SCHEME The flow of electrons that results in light energy being converted into chemical energy in the chloroplast membrane; *see also* light reactions.

ZYGOMORPHIC See bilaterally symmetrical.

ZYGOTE A fertilized egg; a cell resulting from the fusion of gametes.

INDEX

INDEX

Winter annual, 69, 361–362
Winter desiccation, 177
Winter injury, 176–178, 184–185, 492
Witches'-broom, 68
World population trends, 8

Xanthophylls, 37
Xenia in sweet corn, 311

Xerophytes, 186
Xylem, 41–51, 192, 498

Yams (*Dioscorea* sp.), 408

z Scheme (light reactions), 86–88, 93, 97

Zea mays (corn), 24–25, 67–69, 77, 287–288, 378, 392–394, 413–414, 442
Zinc, 236–237
Zizania aquatica (wild rice), 390
Zoysia (zoysiagrasses), 470
Zygocactus truncatus (Christmas cactus), 481